ANATOMY & PHYSIOLOGY

Laboratory Manual

The Unity of Form and Function

Tenth Edition

Eric Wise

Santa Barbara City College

McGraw Hill

LABORATORY MANUAL FOR SALADIN: ANATOMY & PHYSIOLOGY: THE UNITY OF FORM AND FUNCTION, TENTH EDITION

Published by McGraw Hill LLC, 1325 Avenue of the Americas, New York, NY 10019. Copyright ©2024 by McGraw Hill LLC. All rights reserved. Printed in the United States of America. Previous editions ©2021, 2018, and 2015. No part of this publication may be reproduced or distributed in any form or by any means, or stored in a database or retrieval system, without the prior written consent of McGraw Hill LLC, including, but not limited to, in any network or other electronic storage or transmission, or broadcast for distance learning.

Some ancillaries, including electronic and print components, may not be available to customers outside the United States.

This book is printed on acid-free paper.

1 2 3 4 5 6 7 8 9 LMN 28 27 26 25 24 23

ISBN 978-1-266-04253-9 (bound edition)
MHID 1-266-04253-9 (bound edition)

Portfolio Manager: *Matt Garcia*
Product Developer: *Melisa Seegmiller*
Marketing Manager: *Valerie Kramer*
Content Project Manager: *Jeni McAtee*
Manufacturing Project Manager: *Sandy Ludovissy*
Content Licensing Specialist: *Lori Hancock*
Designer: *David Hash*
Cover Image: *Ysbrand Cosijn/Shutterstock*
Compositor: *MPS Limited*

mheducation.com/highered

CONTENTS

Anatomy and physiology can be the jewel in our students' education or the bane of their college career. As an instructor of anatomy and physiology for many years, I decided to write a lab manual that was student-friendly and with a singular focus on the lab portion of the course. This lab manual was written for the undergraduate student of anatomy and physiology, and it consists of 41 exercises designed to help students learn basic human anatomy and the practical lab applications in physiology.

The diversity of interests in today's anatomy and physiology students is due, in part, to the number of majors that either require or recommend the subject. This lab manual provides a framework for understanding anatomy and physiology for students interested in nursing, radiology, physical or occupational therapy, physical education, dental hygiene, or other allied health majors.

This manual was written to be used with Saladin, *Anatomy & Physiology*, tenth edition. The illustrations are labeled; therefore, students do not need to bring their lecture text to the lab. The lab manual accompanies the lecture text and lecture portion of the course and can be used in either a one-term or full-year course. The illustrations are outstanding, and the balanced combination of line art and photographs provides effective coverage of material. The amount of lecture material in the manual is limited, so there is little material included that is not part of the lab experience.

Practical lab experience is an invaluable opportunity to reinforce lecture concepts, enrich students' understanding of anatomy and physiology, and allow them to explore new dimensions in the subject area. The educational benefit of reinforcing lecture material with hands-on experiments and acquiring knowledge with a learn-by-doing philosophy makes the anatomy and physiology lab a very special educational environment. Many of us use lab experiences to present conceptually difficult material in physiology, and to provide students with different learning styles another avenue for learning.

The 41 exercises in this lab manual provide a comprehensive overview of the human body. Each exercise presents the core elements of the subject matter. You can tailor this manual to match your vision of the course, or use it in its entirety. There are significant differences in the laboratories found around the country, and the advances in physiology equipment, especially computer modules, are numerous and continually evolving. The materials section in each lab is designed for a lab of 24 students and includes the amounts and types of reagents to be used. The labs generally take between 2 and 3 hours to complete.

This lab manual was written for three types of anatomy and physiology courses. For those courses that use the cat as the primary dissection animal, cat dissections or mammalian organ dissections follow the material on humans. For those courses that use

models or charts, numerous cadaver photographs are included so that students can see the representative structures as they exist in the cadaver material. Finally, for those courses that use cadavers, this lab manual can be used by studying the human material and omitting the cat dissection sections.

Key Features

1. **Dynamic art program.** All the illustrations have been rendered by a state-of-the-art digital illustration company. They are extremely accurate, use bold and appealing colors, and offer a unique, three-dimensional look.
2. **Labels.** Illustrations are labeled for students to learn the names and terminology by looking at real-life examples or models and by referring to the illustrations in the manual.
3. **Instructional photographs.** Numerous full-color photographs, including detailed histological light micrographs and cadaver and cat dissections, show detailed structures.
4. **Focus on the laboratory.** This manual focuses primarily on the material necessary for the laboratory and does not repeat the material presented in the lecture text, with the expectation that students can look up material in the lecture text when necessary.
5. **Integrated use of the cat for dissection specimen.** The cat is used as the dissection animal; however, it is integrated with material on human anatomy, so that animals do not have to be relied upon as dissection specimens unless so desired.
6. **PhILS Virtual labs.** An icon appears in some of the exercises after the materials section. This icon represents Ph.I.L.S. (Physiology Interactive Lab Simulations) and signals the reader that a supplemental laboratory exercise can be found in Ph.I.L.S. 4.0. This online resource, available through McGraw Hill Education, offers 42 lab simulations that may be used to supplement or substitute for wet labs. Users may adjust variables, view outcomes, make predictions, draw conclusions, and print lab reports.
7. **Safety.** Safety guidelines appear on page xiii. The international symbol for caution (CAUTION) is used throughout the manual to identify material that the reader should pay close and special attention to when preparing for or performing the laboratory exercise.
8. **Activity icons.** This key symbol (A) is a marker for active student participation in the lab.
9. **Clean up.** At the end of many laboratory exercises an icon for clean up reminds the student to clean up the laboratory. Special instructions are given where appropriate.

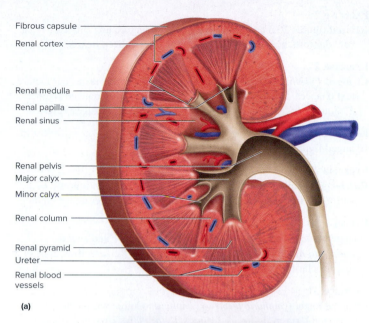

Fibrous capsule
Renal cortex
Renal medulla
Renal papilla
Renal sinus
Renal pelvis
Major calyx
Minor calyx
Renal column
Renal pyramid
Ureter
Renal blood vessels

(a)

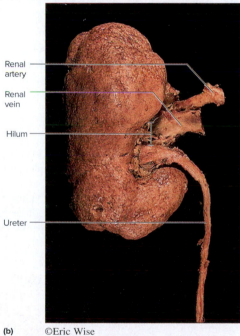

Renal artery
Renal vein
Hilum
Ureter

(b)   ©Eric Wise

10. **Data collection.** Collection of data is embedded within each exercise instead of in a separate table at the back of the manual.
11. **Summary data.** These sections can be found at the end of appropriate exercises for ease of handing in material.
12. **User-friendly format.** Each exercise begins on a right-hand page, and the pages are perforated to allow students to more easily remove the exercises to turn them in and later store them.
13. **Study hints.** Study hints are located in selected chapters where additional information is provided to help students comprehend and retain the more difficult material presented.
14. **Key terms.** Current anatomical terminology is used through-out the laboratory manual. Key terms are boldfaced.

What's New in the Tenth Edition

The exercises in the tenth edition of the Lab Manual to Accompany Saladin's *Anatomy and Physiology, The Unity of Form and Function* have all been updated by the key changes described below.

- The artwork has been updated to correlate to the lecture text.
- Artwork has been updated to have more inclusiveness to the manual.
- Each exercise has been updated for clarity, as well as for content and terminology to align more closely to the lecture text.

Changes to individual exercises are described below.

Exercise 1
Changed gender-oriented language. Minor changes in text for clarity. Changed pH scale image (figure 1.2) for better readability.

Exercise 2
Updated figure 2.1 to agree with lecture text for more ethnic diversity. Changed *median* or *midsagittal* to *sagittal*. Added *inguinal* as synonym to *iliac*.

Exercise 3
Changed *stage* to *mechanical stage*. Minor adjustments to text for clarity.

Exercise 4
Modified figure 4.11 for better visibility. Added a review question for experimental procedure.

Exercise 5
Minor changes in text. Added the term *myocytes* as a synonym for *muscle cell*. Deleted *neurosoma* and replaced it with *cell body* to agree with lecture text.

Exercise 6
Arranged text to provide more space for student drawings. Added more detail to text for better comprehension. Described mammary glands and ceruminous glands as more localized glands to distinguish their placement from sweat and sebaceous glands. Rewrote the growth phases of hair to correlate to lecture text. Revised illustration of hair in follicle.

Exercise 7
Modified color on figure 7.1 for better visualization. Changed *pelvic girdle* to *hip bone* for agreement with lecture text. Changed *epiphysial* back to *epiphyseal* to agree with changes in lecture text.

Exercise 8
Modified figures 8.2, 8.5, 8.14, and 8.19 to correlate to lecture text. Changed *hypophysial* back to *hypophyseal* as in previous editions to agree with lecture text. Made small changes in text for clarity.

Exercise 9
Modified color on figure 9.1 for better visualization. Changed *epiphysial* back to *epiphyseal* as per changes in lecture text. Clarified which side of the body clavicles in figure 9.3 came from for clarity. Modified hand and foot art to correlate with lecture text.

Exercise 10
Changed *epiphysial* back to *epiphyseal*. Made arrows in many figures more defined for easier visibility.

Exercise 11
Figures 11.10 and 11.11 were modified for easier visibility of leader lines.

Exercise 12
Adjusted leader lines in figure 12.1 for more accuracy. Changed ethnicity of individual in figure 12.4.

Exercise 13
Figure 13.1 indicated in legend that trapezius was removed. Similarly clarified legend in figure 13.3. Made changes in text throughout for clarity. Modified figure 13.5 (a) to correlate with lecture text.

Exercise 14
Changed ethnicity of models in this exercise. Made modifications in text for better clarity.

Exercise 15
Minor changes in text for more accurate descriptions.

Exercise 16
Added *enteric plexus* as additional nervous division. Changed *neurosoma* to *cell body*. Changed *glial cells* to *neuroglia* as in lecture text. Changed ethnicity of individuals throughout the exercise. Made color modifications to plexuses and nerves as in lecture text.

Exercise 17
Deleted figure 17.4b because it was redundant. Made changes to text for better comprehension. Indicated that the subarachnoid space hold CSF. Changed colors in brain stem to agree with lecture text. Made changes to figure 17.14 for clarity.

Exercise 18
Changed *interneuron* to *integrating center*. Made changes to text for better clarity. Modified figure 18.3 to agree with text.

Exercise 19
Updated gender descriptions in experimental procedures.

Exercise 20
Modified figure 20.1 to clarify the arrow. In projection pathways section, changed *fibers* to *neurons*.

Exercise 21
Indicated that the pupil is an opening in the first paragraph of the procedure section. Indicated that the conjunctiva extends across the outer surface of the exposed eye. Described eye color to reflect changes in lecture text. Made changes in the text for clarity.

Exercise 22
Made changes to figures 22.1, 22.4, 22.7, and 22.8 to agree with changes in lecture text. Changed *his* and *hers* to *your lab partner* or *their*. Changed *earlobe* to *lobule* to agree with lecture text.

Exercise 23
Added the true translation of the *pineal gland* as resembling a pine nut from the erroneous perception that it resembled a pine cone. Changed *hypophysial* back to *hypophyseal,* and changed *hypothalamohypophysial* to *hypothalamohypophyseal* as per changes in lecture text. Modified figures 23.2 and 23.10 to agree with lecture text. Referenced the reproductive functions of the ovary to Exercise 41. Modified text for clarity.

Exercise 24
Made minor changes to text such as changing *lymphatic* to *lymphoid*.

Exercise 25
Changed the blood type percentages (Table 25.2) to reflect updates in lecture text.

Exercise 26
Revised figures 26.5 and 26.10 to agree with lecture text. Indicated that valvular tests may not work well on preserved sheep hearts.

Exercise 27
Changed *Purkinji* to synonym of *subendocardial branches* as per lecture text changes. Made similar changes to figure 27.1. Changed figures 27.5 and 27.6 to make more visible. Made gender changes in text.

Exercise 28
Changed figure 28.1 as per lecture text.

Exercise 29
Rearranged graphics to put them closer to text. Changed figures 29.7a and 29.11 for better clarity. Rewrote material for better understanding.

Exercise 30
Changed *ulnar* and *radial recurrent arteries* to *ulnar* and *radial collateral arteries* as per lecture text changes. Reworded sections of text for clarity.

Exercise 31
Changed many *lymphatic terms* to *lymphoid* to agree with lecture text. Changed term in figure 31.2 (*intercellular cleft*). Changed ethnicity of figure 31.4.

Exercise 32
Changed hypertension values to agree with changes in lecture text. Changed *women* to *females* and *men* to *males*. Made minor changes to text.

Exercise 33
Made minor changes to text and graphics for clarity.

Exercise 34
Made changes to preparation and procedures due to COVID-19. Added more detailed caution material to make people aware of proper procedures. Added a virtual section of lab that avoids direct respiratory exercises. Indicated that all respiratory equipment should be sterile and directed the sterilization procedures per CDC guidelines. Modified CPR directions and cautions per updates at the time of this writing.

Exercise 35
Made changes to preparation and procedures due to COVID-19. Added more detailed caution material to make people aware of proper procedures. Rewrote sections of text for clarity.

Exercise 36
Made changes to figures 36.2a and 36.9a to agree with main lecture text.

Exercise 37
Made small changes in text to correlate to main lecture text.

Exercise 38
Made changes to figures 38.8 and 38.10 to agree with main lecture text.

Exercise 39
No changes

Exercise 40
Changed figures 40.1, 40.2, and text art in review section to agree with main lecture text.

Exercise 41
Changed figures 41.1, 41.12, and text art in review section to correlate to main lecture text.

Create

Your Book, Your Way

McGraw Hill's Content Collections Powered by Create® is a self-service website that enables instructors to create custom course materials—print and eBooks—by drawing upon McGraw Hill's comprehensive, cross-disciplinary content. Choose what you want from our high-quality textbooks, articles, and cases. Combine it with your own content quickly and easily, and tap into other rights-secured, third-party content such as readings, cases, and articles.

Content can be arranged in a way that makes the most sense for your course, and you can include the course name and information as well. Choose the best format for your course: color print, black-and-white print, or eBook. The eBook can be included in your Connect course and is available on the free ReadAnywhere® app for smartphone or tablet access as well. When you are finished customizing, you will receive a free digital copy to review in just minutes! Visit McGraw Hill Create®—www.mcgrawhillcreate.com—today and begin building!

MCGRAW HILL Teaching and Learning Tools

Anatomy & Physiology Revealed® 4.0

Anatomy & Physiology Revealed® 4.0: An Interactive Cadaver Dissection Experience

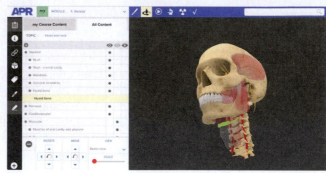

APR/McGraw Hill

Available in Connect and online at aprevealed.com, Anatomy & Physiology Revealed (APR) is an interactive human cadaver dissection tool built to enhance both lecture and lab. APR contains all the systems covered in A&P and Human Anatomy courses, including Body Orientation, Cells and Chemistry, and Tissues. Detailed cadaver photographs blended with a state-of-the-art layering technique provide a uniquely interactive dissection experience.

With a new streamlined, user- and mobile-friendly interface, increased accessibility, updated animations and 3D interactive models, APR was built to increase the success of your A&P laboratory course.

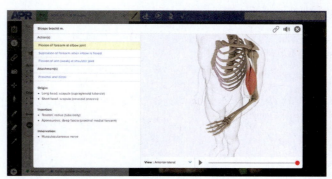

APR/McGraw Hill

- Dissection: Peel away layers of the human body to reveal structures. Structures can be pinned and labeled, just as in a real dissection lab. Each labeled structure is accompanied by detailed information, audio pronunciation, and alternate views. Dissection images can be captured and saved. A direct link tool also can be used to bring students to the exact view you specify.
- Animations: Modern, updated animations demonstrate muscle action, show detailed attachments, clarify anatomical relationships, and explain difficult concepts.
- Histology: Labeled light micrographs presented with each body system allow students to study the cellular detail of tissues.
- Imaging: Labeled x-ray, magnetic resonance imaging (MRI), and computed tomography (CT) images familiarize students with the appearance of key anatomical structures as seen through different medical imaging techniques.
- Self-Quizzing: Challenging exercises allow students to test their ability to identify anatomical structures in a timed practical exam format or with traditional multiple choice questions. A results page provides an analysis of test scores and links back to all incorrectly identified structures for review.
- Rotatable 3D Models: Interactive, rotatable 3D models enhance the learning experience and allow students to see the spatial relationship of structures in the human body. Side-by-side corresponding cadaver images provide perspective.
- My Course Content: Instructors may customize APR 4.0 to their course by selecting the specific structures they require in their course. Once the structure list is generated, APR highlights these selected structures for students.

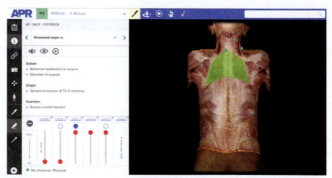

APR/McGraw Hill

Connect Virtual Labs is a fully online lab solution that can be used in conjunction with the lab manual as a preparation, supplement, or make-up lab. These simulations help students learn the practical and conceptual skills needed, then check for understanding and provide feedback. With adaptive pre-lab and post-lab assessment available, instructors can customize each assignment.

⚗️ PhILS

Physiology Interactive Lab Simulations© 4.0 (Ph.I.L.S.) offers 42 lab simulations that may be used to supplement or substitute for wet labs. Users may adjust variables, view outcomes, make predictions, draw conclusions, and print lab reports. Ph.I.L.S is now more readily available than ever with a new user- and mobile-friendly interface, along with increased accessibility features.

McGraw Hill Connect® empowers students to learn and succeed in the Anatomy and Physiology course with user-friendly digital solutions.

Anatomy & Physiology Revealed (APR) is an interactive cadaver dissection tool to enhance lecture and lab. Featuring real cadaver photography, animations, interactive 3D models, histology, imaging and more. Now assignable in Connect! The result? Students are prepared for lab, engaged in the material, and utilize critical thinking.

Anatomy & Physiology Revealed 4.0

Virtual Labs

Virtual Labs helps connect the dots between lab and lecture, boosts student confidence and knowledge, and improves student success rates. Interactive animations and simulations also encourage students to explore key physiological processes and difficult concepts. The result? Students are engaged, prepared, and utilize critical thinking skills.

Practice ATLAS

Practice Atlas for A&P is integrated into APR, pairing images of common anatomical models with stunning cadaver photography, which allows students to practice naming structures on both models and human bodies. The result? Students are better prepared, engaged, and move beyond basic memorization.

A&P Prep

A&P Prep helps students thrive in college-level A&P by helping solidify knowledge in the key areas of cell biology, chemistry, study skills, and math. The result? Students are better prepared for the A&P course.

PhILS

PhILS (Physiology Interactive Lab Simulations) is the perfect way to reinforce key physiology concepts with powerful lab experiments. The result? Students gain critical thinking skills and are better prepared for lab.

SmartBook®

SmartBook 2.0 provides personalized learning to individual student needs, continually adapting to pinpoint knowledge gaps and focus learning on concepts requiring additional study. The result? Students are highly engaged in the content and better prepared for lecture.

This laboratory manual was written to help you gain experience in the lab as you learn human anatomy and physiology. The 41 exercises explore and explain the structure and function of the human body. You will be asked to study the structure of the body using the materials available in your lab, which may consist of models, charts, mammal study specimens such as cats, preserved or fresh internal organs of sheep or cows, and possibly cadaver specimens. You may also examine microscopic sections from various organs of the body. You should familiarize yourself with the microscopes in your lab very early, so that you can take the best advantage of the information they can provide.

The physiology portion of the course involves experiments that you will perform on yourself or your lab partner. They may also involve the mixing of various chemicals and the study of the functions of live specimens. Because the use of animals in experiments is of concern to many students, a significant attempt has been made to reduce (but, unfortunately, not eliminate) the number of live experiments in this manual. Until there is a sound replacement for live animals, their use will continue to be part of the college physiology lab. Your instructor may have alternatives to live animal experimentation exercises. It is important to get the most out of what live specimen experimentation there is. Coming to the lab unprepared and then sacrificing a lab animal while gaining little or no information is an unacceptable waste of life. Use the animals with care. Needless use or inhumane treatment of lab animals is not acceptable or tolerated.

As a student of anatomy and physiology you will be exposed to new and detailed information. The time it takes to learn the information will involve *more* than just time spent in the lab. You should maximize your time in the lab by reading the assigned lab exercises before you come to class. You will be doing complex experiments, and if you are not familiar with the procedure, equipment, and time involved you could ruin the experiment for yourself and/or your lab group. The exercises in physiology are written so you can fill in the data as you proceed with the experiment. At the end of each exercise are review sheets that your instructor may wish to collect to evaluate the data and conclusions of your experiments. The illustrations are labeled except on the review pages. All review materials can be used as study guides for lab exams or they may be handed in to the instructor.

The anatomy exercises are written for cat and human study, though these exercises can be used with or without cats or cadavers. Get *involved* in your lab experience. Don't let your lab partner do all the dissections or all the experimentation; likewise, don't insist on doing everything yourself. Share the responsibility and you will learn more.

Safety

Safety guidelines appear on page xiii for reference. The international symbol for caution (CAUTION) is used throughout the manual to identify material that you should pay close and special attention to when preparing for or performing laboratory exercises.

Activity Icons

This symbol (A) indicates the need for your active participation in the lab exercise.

Clean Up

Special instructions are provided for clean up at the end of appropriate laboratory exercises and are identified by this unique icon (🖐).

How to Study for This Course

Some people learn best by concentrating on the visual, some by repeating what they have learned, and others by writing what they know over and over again. In this course, you will have to adapt your learning style to different study methods. You may use one study method to learn the muscles of the body and a completely different method for understanding the function of the nervous system. Some students need only a few hours per week to succeed in this course, while others seem to require more time.

You need to go to class. Go to lab on time. The beginning of the lab is when most instructors go over the material and point out what material to omit, what to change, and how to proceed. If you do not attend lab, you do not get the necessary information.

Read the material ahead of time. The subject matter is visual, and you will find an abundance of illustrations in this manual. Record on your calendar all the lab quizzes and exams listed on the syllabus provided by your instructor. Budget your time so you study accordingly.

Work hard! There is absolutely no substitute for hard work to achieve success in a class. Some people do math more easily than others, some people remember things more easily, and some people express themselves better. Most students succeed because they work hard at learning the material. It is a rare student who gets a bad grade because of a lack of intelligence. Working at your studies will get you much farther than worrying about your studies.

Be *actively* involved with the material and you will learn it better. Outline the material after you study it for a while. Read your notes, go over the material in your mind, and then make the information your own. There are several ways that you can get actively involved.

Draw and doodle a lot. Anatomy is a visual science, and drawing helps. You do not have to be a great illustrator. Visualize the material in the same way you would draw a map to your house for a friend. You do not draw every bush and tree but, rather, create a *schematic* illustration that your friend could use to get the *pertinent* information. As you know, there are differences in maps. Some people need more practice than others, but anyone can do it. The head can be a circle, which can be divided into pieces representing the bones of the skull. Draw and label the illustration after you have studied the material and without the use of your text! Check yourself against the text to see if you really know the material. Correct the illustration with a colored pen, so that you highlight the areas you need work on. Go back and do it again until you get it perfect. This does take some time, but not as much as you might think.

Write an outline of the material. Take the mass of information to be learned and go from the general to the specific. Let's use the skeletal system as an example. You may wish to use these categories:

1. Bone composition and general structure
2. Bone formation
3. Parts of the skeleton
   a. Appendicular skeleton
      (1) Pectoral girdle
      (2) Upper limb
      (3) Pelvic girdle
      (4) Lower limb
   b. Axial skeleton
      (1) Skull
      (2) Hyoid
      (3) Ribs
      (4) Vertebral column
      (5) Sternum

An outline helps you organize the material in your mind and lets you sort the information into areas of focus. If you do not have an organizational system, then this course is a jumble of terms with no interrelationships. The outline can get more detailed as you progress, so you eventually know that the specific nasal bone is one of the facial bones and the facial bones are skull bones, which are part of the axial skeleton, which belongs to the skeletal system.

Test yourself before the exam or quiz. If you have practiced answering questions about the material you have studied, then you should do better on the real exam. As you go over the material, jot down possible questions to be answered later, after you study. If you compile a list of questions as you review your notes, then you can answer them later to see if you have learned the material well. You can also enlist the help of friends, study partners, or family (if they are willing to do this for you). You can also study alone. Some people make flash cards for the anatomy portion of the course. It is a good idea to do this for the muscle section, but you may be able to get most of the

information down by using the preceding technique. Flash cards take time to fill out, so use them carefully.

Use memory devices for complex material. A mnemonic device is a memory phrase that has some relationship to the study material. For example, there are two bones in the wrist right next to one another, the trapezium and the trapezoid. The mnemonic device used by one student was that trapez*ium* rhymes with th*umb* and it is the one under the thumb.

Use your study group as a support group. A good study group is very effective in helping you do your best in class. Study with people who will push you to do your best. If you get discouraged, your study partners can be invaluable support people. A good group can help you improve your test scores, develop study hints, encourage you to do your best, and let you know that you are not the *only* one living, eating, and breathing anatomy and physiology.

Just as a good study group can really help, a bad group can drag you down farther than you might go on your own. If you are in a group that constantly complains about the instructor, that the class is too hard, that there is too much work, that the tests are not fair, that you don't really need to know this much anatomy and physiology for your field of study, and that this isn't medical school, then get yourself out of that group and into one excited by the information. Don't listen to people who complain constantly and make up excuses instead of studying. There is a tendency to start believing the complaints, and that begins a cycle of failure. Get out of a bad situation early and get with a group that will move forward.

Do well in the class and you will feel good about the experience. If you set up a study time with a group of people and they spend most of the time talking about parties, sports, or personal problems, then you aren't studying. There is nothing wrong with talking about parties, sports, or helping someone with personal problems, but you need to address the task at hand, learning anatomy and physiology. Don't feel bad if you must get out of your study group. It is *your* education, and if your partners don't want to study, then they don't really care about your academic well-being. A good study partner is one who pays attention in class, who is prepared ahead of the study time session, and who can explain information that you may have gotten wrong in your notes. You may want to get the phone number of two or three such classmates.

Test-Taking

Finally, you need to take quizzes and exams in a successful manner. By doing practice tests, you can develop confidence. Do well early in the semester. Study extra hard early (there is no such thing as overstudying!). If you fail the first test or quiz, then you must work yourself out of an emotional ditch. Study early and consistently, and then spend the evening before the exam going over the material in a general way and solving those last few problems. Some people do succeed under pressure and cram before exams; however, the information is stored in short-term memory and does not serve them well in their major field! If you study on a routine basis, then you can get up on the morning of a test, have a good breakfast, listen to some encouraging music, maybe review a bit, and be ready for the exam.

Your instructor is there to help you learn anatomy and physiology, and this laboratory manual was written with you in mind. Relate as much of the material as you can to your body and keep an optimistic attitude.

Please feel free to write me or e-mail me with your comments, suggestions, and criticisms. I value your input and hope that your comments will lead to an even better next edition of this laboratory manual.

Eric Wise
Santa Barbara City College
721 Cliff Drive
Santa Barbara, CA 93109
wise@sbcc.edu

Acknowledgments

Many people have been involved in the production of this lab manual, and I would like to thank the editorial and marketing teams at McGraw Hill, including Matt Garcia, Valerie Kramer, and Melisa Seegmiller. Thanks also go to the production team—Jeni McAtee, Lori Hancock, and David Hash—for their input and encouragement.

I would like to dedicate this book to my children Sarah, Caitlin, and Zen, whom I love so very much.

LABORATORY
SAFETY GUIDELINES

The following is a partial list of safety guidelines for you to follow in the anatomy and physiology lab. More complete descriptions of safety procedures are found throughout the manual.

1. Read all of the lab material prior to coming to class. This is a safety issue. Failure to read or understand the lab can result in hazards. Unauthorized experiments are not allowed in the lab.

2. Locate the first-aid kit, eyewash station, shower station, fire blanket, fire extinguisher, and other safety areas in the lab prior to beginning the first lab. Be familiar with how to use the equipment in the event of an emergency.

3. Clean up spills. Inform your instructor of any spill in the lab. Be careful if the material is toxic or caustic. If you are not sure if the material is hazardous, ask your instructor for the proper procedure for the cleanup.

4. Assume all bodily fluids in the lab are infectious. Follow precautions when handling bodily fluids, such as wearing protective gloves, lab coats, and protective eye wear. Never use any instrument twice that comes into contact with bodily fluid! Once the instrument is used, dispose of it in either a biohazard bag or in a container of 10% bleach or other disinfecting solution. Clean all lab surfaces with a bleach or disinfecting solution at the end of a lab involving bodily fluids, even if you think no fluid has come in contact with the table surface.

5. Keep the lab clean and free of clutter. Place all backpacks, purses, and umbrellas in safe areas and not on the lab tables.

6. Do not eat, smoke, or chew gum in the lab. Many reagents in the lab are toxic, so do not drink them. Never pipette anything by mouth. Use a pipette bulb or pipette pump when pipetting.

7. Keep your hair secured so it does not catch fire or dip into beakers containing solutions. Never heat volatile material over an open flame. An explosion might occur.

8. Do not wear contact lenses in the lab. Notify your instructor if you wear contact lenses.

9. Do not throw sharp material such as glass or cutting blades in the normal trash containers in the lab. They are to be disposed of in an appropriate container such as a "sharps" container. Report any glassware breakage to your instructor, and dispose of it in the appropriate container.

10. Never point a test tube that is heating over a Bunsen burner in the direction of someone else. Never walk away from anything that is being heated. Pay attention to material on hot plates and remove material with appropriate mitts or tongs. Heat material only in appropriate heat-resistant containers. Turn off and unplug hot plates immediately after use.

11. Dissect with the blade cutting away from you and your lab partners. If you do cut yourself, make sure you wash the wound well with soap and water and notify your instructor.

12. If you have an allergic reaction to the preserving fluid (usually restricted breathing, a flushed feeling, or a skin rash), notify your instructor immediately. Notify your instructor if you are pregnant or have any medical condition.

13. Do not apply cosmetics in lab.

14. Wear closed-toed shoes in lab, not sandals.

15. Wash your hands thoroughly after lab, especially before eating or going to the restroom.

Notes

Introduction to Lab Science, Measurement, and Chemistry

INTRODUCTION

Science is the study of physical phenomena and follows specific guidelines that make it unique among all disciplines. The human body's structure and function fall within the realm of scientific investigation; for example, **human anatomy** involves the understanding of the structure of the human body, while **human physiology** is the study of the function of the body. There are many ways people can understand anatomy and physiology. One of these involves use of the **scientific method.**

The scientific method often begins with a question, for example, how does the body digest food? The next part in this method frequently involves the development of a **hypothesis,** which is a testable proposal that seeks to explain a scientific question. Testing done to prove or disprove the hypothesis is usually carried out in the form of an **experiment.** A more lengthy discussion of the scientific method is provided in the Saladin text in chapter 1, "Major Themes of Anatomy and Physiology," and chapter 2, "The Chemistry of Life."

Much of science involves measurement and collection of **data.** Experimental data are the pieces of information, or "facts," obtained and later examined to support or reject the proposed hypothesis. In this exercise, you collect data and examine how to graph data so the information becomes more comprehensible.

OBJECTIVES

At the end of this exercise, you should be able to

1. describe the advantages of the metric system over the U.S. customary system;
2. define *independent variable* and *dependent variable*;
3. list four base units of the metric system;
4. convert fractions into decimal equivalents;
5. collect and graph data taken in class; and
6. discuss pH, acid, base, ionic bond, and covalent bond.

MATERIALS

Acid/Base

Five 10 mL test tubes and test tube racks

Safety goggles and protective gloves

10 mL graduated cylinder

Permanent marker

Distilled water in dropper bottle

0.1 M HCl in dropper bottle

0.1 M NaOH in dropper bottle

Baking soda (sodium bicarbonate)

Sodium chloride (table salt)

Wide-ranging pH paper (pH 1–14)

Parafilm®

Small metal spatula

Balance and weigh paper

Ionic and Covalent Bonded Molecules

18-gauge wire

Alligator clips

9-volt battery

6-volt flashlight bulb

Miniature screw lamp receptacle (Carolina #756481 or Sargent Welch #CP 33008-00)

Two 50 mL beakers

15% sucrose solution in dropper bottle

15% sodium chloride solution in dropper bottle

Hydrogen Bonds

Graduated cylinder

Two 50 mL beakers

Small bottle of distilled water

Small bottle of ethanol (70% or greater)

Hot plate (do not use open flame)

Heart Rate and Exercise

Clock or watch with accuracy in seconds

Calculator

TABLE 1.1	Metric System and Equivalents	
Quantity	**Base Unit**	**U.S. Equivalent**
Length	Meter (m)	1.09 yards (39.4 inches)
Volume	Liter (L)	1.06 quarts
Mass	Gram (g)	.036 ounce ($\frac{1}{454}$ of a pound)
Time	Second (s)	Second

PROCEDURE

Measurements in Science

Members of the scientific community and people of many nations of the world use the **metric system** to record quantities such as length, volume, mass (weight), and time. This is because the metric system is based on units of 10, and conversion to higher or lower values is relatively easy when compared to using the U.S. customary system. For example, assume you are working on a bicycle and are using a $\frac{1}{2}$-inch wrench. If you need a larger wrench you move to a $\frac{9}{16}$-inch, then a $\frac{5}{8}$-inch, then an $\frac{11}{16}$-inch, or perhaps as large as a $\frac{3}{4}$-inch wrench. This requires a bit of computation as you move from one size to the next. On the other hand, if you are using the metric system and a 12-millimeter (mm) wrench is too small, you progressively move to a 13 mm, 14 mm, or 15 mm wrench.

The same idea can be applied to volume, temperature, or weight. In the case of volume, there are 8 ounces per cup, 128 ounces per gallon. The calculation for the number of ounces in 7 gallons is a little cumbersome (7 gallons × 128 ounces). In the metric system, there are 1,000 milliliters in 1 liter, so there are 7,000 milliliters in 7 liters. The conversions are much easier.

Medical dosages are given frequently in milliliters or cubic centimeters (cc). One milliliter occupies 1 cubic centimeter, so these values are interchangeable. Examine table 1.1 and compare the quantity, base unit, and U.S. equivalent. Additional conversions can be found in appendix A.

If the quantity measured is much larger or smaller than the base unit, then the base unit can be expressed in multiples of 10.

For example, if you had 1,000 grams, you would have a **kilo**gram. If you had one-thousandth of a gram ($\frac{1}{1,000}$ gram), you would have a **milli**gram. Examine table 1.2 as you answer the following questions. For those questions highlighted with a question mark, fill in the Chapter Summary Data section at the end of the exercise with your results if your instructor wants you to hand in your results.

? What is $\frac{1}{100}$ gram? _____ (1)

? What is 1,000 seconds? _____ (2)

? What is 10 meters? _____ (3)

? What is $\frac{1}{1,000,000,000}$ liter? _____ (4)

As you can see from table 1.2, some measurements in science are very small. For example, the amounts of hormones circulating in the blood are minute. To provide a shortened notation of very large or small numbers, we use scientific notation. A number such as 60,000 is written as 6×10^4. You move the decimal point four places to the left, and thus the superscript above the 10 is a 4. For very small numbers, the superscript is written as a negative number. The number 0.00006 is written as 6×10^{-5}, because you move the decimal point five places to the right.

TABLE 1.2	Decimals of the Metric System		
Name	**Description**	**Multiple/Fraction**	**Scientific Notation**
Kilo	One thousand times greater	1,000	1.0×10^3
Deca	Ten times greater	10	1.0×10^1
Deci	One-tenth as much	$\frac{1}{10}$ 0.1	1.0×10^{-1}
Centi	One-hundredth as much	$\frac{1}{100}$ .01	1.0×10^{-2}
Milli	One-thousandth as much	$\frac{1}{1,000}$ .001	1.0×10^{-3}
Micro	One-millionth as much	$\frac{1}{1,000,000}$ .000001	1.0×10^{-6}
Nano	One-billionth as much	$\frac{1}{1,000,000,000}$ .000000001	1.0×10^{-9}

(A) Activity

Convert the following numbers into scientific notation:

❓ 4,300,000 _____ (5)

❓ 0.000034 _____ (6)

❓ 2,200 _____ (7)

❓ 0.0019 _____ (8)

Data Collection

Before you begin this section, read *Lab Reports* in appendix C. Scientists experiment by altering one **variable** and seeing what effect occurs. For example, the change in the weight of a person can be altered by diet or exercise. In an experiment, only one variable (diet or exercise, but not both) should be changed to see how that change affects the weight of the subject. The variable that scientists change in an experiment is called the **independent variable.** The exercise regimen of a person is the independent variable. The result caused by the change in the independent variable is known as the **dependent variable.** The change in weight as a result of exercise is the dependent variable.

By graphing data, you can more easily see potential correlations in a sample size. One problem in sampling is that you need to have a large enough number to have a representative sample of a group. Let's suppose that several members of the basketball team are enrolled in your lab section. This might have a rather unusual effect on your graph (fig. 1.1). On the other hand, if you are able to sample your entire school, the effects of the size of the basketball players in the sample would be minimized (fig. 1.2).

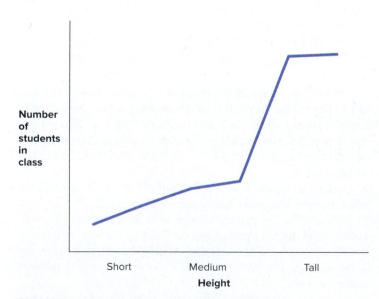

FIGURE 1.1 Distribution of Students in a Class.

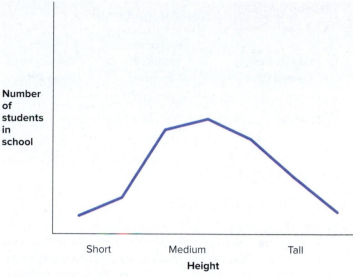

FIGURE 1.2 Distribution of Students in Entire School.

(A) Activity

In this part of the exercise, you will look for a correlation between heart rate and exercise. Although you may be familiar with this response, make a hypothesis about the relationship between heart rate and exercise.

❓ Record your hypothesis in the following space.

_____

_____ (9)

While you are sitting quietly in lab, have your lab partner measure your heart rate by placing their fingers on the thumb side of your wrist. The pulse should be counted for a full minute. You should not have done any strenuous exercise for at least 10 minutes prior to taking your pulse.

❓ Your heart rate, bpm: _____ (10)

Write your results on the board or a designated piece of paper in lab. Record the distribution of the entire class in terms of resting heart rate in the left hand column of table 1.3.

Range and Mean in Measurements The extremes of measurement, which represent the highest and lowest values, represent the **range** of the values.

(A) Activity

After table 1.3 is completed by all members of the class, use these data obtained by members of your lab class and record

TABLE 1.3	Heart Rate (BPM)					
	Subject Resting Heart Rate	Number of Seconds of Exercise				Percent Change in Heart Rate
		30	60	90	120	
0						
1						
2						
3						
4						
5						
6						
7						
8						
9						
10						
11						
12						
13						
14						
15						
16						
17						
18						
19						
20						
21						
22						
23						
24						

the range of the resting heart rate in beats per minute (bpm) for the entire class.

❓ Highest value: bpm _____ (11)

❓ Lowest value: bpm _____ (12)

The **mean** heart rate of the class is the average rate for the group. Using the data obtained for the heart rate in table 1.3, add all of the values together (the sum of the heart rates) and divide that number by the number of participating individuals.

This is the mean heart rate. Record the number in beats per minute.

❓ Mean heart rate, bpm: _____ (13)

_____

Caution!   If you have a heart condition or any other reason for which you should not do exercise, please let your instructor know.

_____

Your instructor should divide the class into four groups for the next part of the exercise. These groups should be balanced so there is an even distribution of students who are athletic and students who do not exercise in each group.

A   Activity

Have the four groups do the following activity.

Group one does jumping jacks or other similar exercise for 30 seconds.

Group two does same exercise as group one for 60 seconds.

Group three does same exercise as group one for 90 seconds.

Group four does same exercise as group one for 120 seconds.

Immediately after finishing your prescribed exercise, record your heart rate for 1 minute.

❓ Duration of time of exercise: _____ (14)

❓ Heart rate after exercise: _____ (15)

Determine the percent increase or decrease in your heart rate after exercise by dividing the rate after exercise by the resting rate and multiplying it by 100, as illustrated in the following equation.

$$\frac{\text{Heart rate after exercise}}{\text{Resting heart rate}} \times 100 = \frac{\text{Percent change}}{\text{in heart rate}}$$

Write your results on the board or a designated piece of paper in lab. Record the distribution of all members of the entire class in terms of percent change in heart rate after exercise by filling in the rest of table 1.3. Make sure that you enter the data in the column that correlates to the amount of exercise done (30, 60, 90, or 120 seconds of exercise).

Take the results from each group (1–4) and determine the mean percent change in heart rate for each group. This is done by adding all of the percent change in heart rate values for individuals who did the exercise and dividing by the total number of individuals who did that exercise. Make a bar graph by drawing a line representing the mean in table 1.4 and shading in the area under the line.

TABLE 1.4	Mean Percent Change in Heart Rate		
% Change			
200			
190			
180			
170			
160			
150			
140			
130			
120			
110			
100			
	30	60	90	120

Number of Seconds of Exercise

Chemistry

In order to fully understand the functions of the body, a fundamental knowledge of chemistry is essential. For those of you who have studied chemistry, the following pages are a simplified review. For those of you who have never had chemistry, a beginning chemistry book or online resources may be invaluable to you for the rest of the course. In the following section you will experiment with acids, bases, and chemical bonds.

Acid/Base Relationships Human cells exist within a narrow range of temperature, salinity, and pH. Significant changes in any of these environmental conditions can lead to the death of the cell. In terms of pH, solutions can be **acidic, neutral,** or **basic (alkaline).**

Pure water has a pH of 7 and is neutral. It is neither acidic nor basic (fig. 1.3). Pure water dissociates evenly into H^+ (hydrogen ions) and OH^- (hydroxide ions). Any solution with the same concentration of H^+ and OH^- is a neutral solution. Salt or sugar added to pure water form solutions that are also neutral. Solutions become more acidic as the hydrogen ion concentration increases. This lowers the pH. Adding more hydrogen ions to the solution, such as when the stomach wall adds hydrochloric acid (HCl) to the stomach cavity, causes the pH to drop to about 2. If you increase the alkalinity of a solution (such as adding ammonia to water), then the pH increases. Household ammonia has a pH of about 11.

Solutions such as hydrochloric acid and carbonic acid are common. Basic solutions, such as bicarbonate and ammonia, also play important roles in the body. The acidic or basic condition of a solution is measured in **pH** units. The term pH refers to the hydrogen ion concentration. In simple terms, one whole pH unit varies by a factor of 10 compared to the next whole pH value. Thus, a solution of pH 5 is 10 times more acidic than a solution of pH 6. **Buffers** are materials that resist changes in pH. Cells often secrete buffers so that when conditions around them change the pH is "buffered," and great fluctuations in the acid or base conditions do not change significantly. Introductory chemistry books

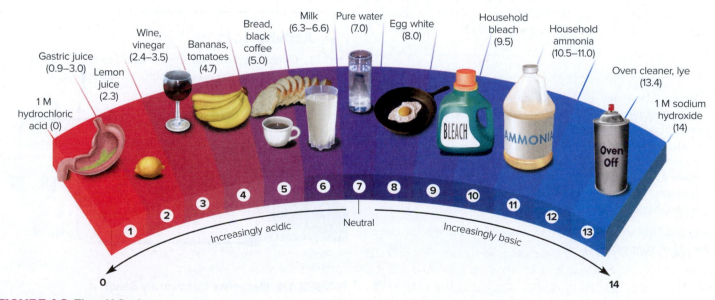

FIGURE 1.3 The pH Scale.

will give you a more detailed description of pH, and there are many online tutorials you can find if you search using the words *acid base*.

Ⓐ **Activity**

For this experiment, select five test tubes and label them 1–5 with a permanent marker. Place them in a test tube rack. Use pH paper to measure pH.

1. Use a graduated cylinder and put 6 mL of distilled water into each of the five test tubes.
2. Measure the pH of test tube 1 (distilled water). Record this pH value: _____
3. To test tube 2, add 10 drops of HCl solution. Place a small square of Parafilm® over the top and mix the contents by inverting the test tube. Do not shake the tube. Measure the pH and record the value: _____ Is this solution acidic, basic, or neutral (circle one)?
4. To test tube 3, add 0.1 g sodium bicarbonate. Place a small square of Parafilm® over the top and mix the contents. Record the pH: _____
5. To test tube 3, add 10 drops of HCl solution. Place a small square of Parafilm® over the top and mix the contents as in step 3. Record the pH of the solution: _____ Does the pH drop to the same level as test tube 2? _____
6. Add 10 more drops of HCl solution to test tube 3 and record the pH value: _____ How might you describe the action of sodium bicarbonate with respect to HCl?

_____

_____

7. To test tube 4, add 10 drops of sodium hydroxide solution. Place a small square of Parafilm® over the top and mix the contents. Record the pH value: _____ Is this solution acidic, basic, or neither (circle one)?
8. To test tube 5, add 0.2 g of sodium chloride (table salt). Place a small square of Parafilm® over the top and mix the contents. Record the pH value: _____ Is this solution acidic, basic, or neutral (circle one)?

Tissue fluid has a pH between 7.35 and 7.45. What was the range in the pH of your experiment?

❓ _____ (16)

Do your results fall within the range of normal tissue fluid pH levels?

❓ _____ (17)

Chemical Bonds   Most of the material around us is held together by chemical bonds. These bonds can be strong, intermediate, or weak. There are several types of bonds, but the three we will examine are covalent bonds, ionic bonds, and hydrogen bonds. **Covalent bonds** are those molecules where bonded atoms share electrons. These are strong bonds. The nucleus of the atom is positively charged due to the presence of protons. The positive charges of two separate atoms repel one another. When an electron is shared between these atoms, the electrons are attracted to both of the positive nuclei, and this holds them together.

In **ionic bonds**, electrons from one atom are transferred to another atom, and the result is an atom with a positive charge and an atom with a negative charge. These oppositely charged particles, called **ions** or **electrolytes**, attract one another and form ionic bonds. The bond is due to the electrostatic attraction between the two ions.

You can examine the differences between covalent and ionic bonds with a simple experiment. Ionic bonds are pulled apart (dissociate) by the action of water. Electrolytes (ions) in solution conduct electricity. Molecules that are held together with covalent bonds do not conduct electricity so easily. This can be demonstrated with the use of a 6-volt flashlight bulb, appropriate cables and switches, and a DC generator, or a 9-volt battery. Examine fig. 1.4 for the proper setup.

Ⓐ **Activity**

1. Connect a 9-volt battery terminal, using one alligator clip and wire, to the 6-volt flashlight bulb apparatus.

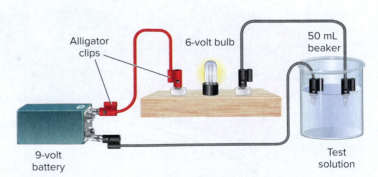

FIGURE 1.4 Electrolyte Conductivity Setup.

2. Connect the other terminal to an insulated wire that is placed in a glass beaker with 25 mL of 15% sucrose solution. The wire should have about 2 cm of insulation removed.
3. Connect the vacant terminal of the flashlight bulb setup to an insulated wire.
4. Do not touch the ends of the wires together, but insert the ends of the insulated wire into the water about 4 cm apart.
5. Record the results (light on or light off).

_____

6. Repeat the experiment, but insert the wires into a beaker with 25 mL of 15% sodium chloride solution.
7. Record the results (light on or light off).

_____

? Which one of the solutions has solute particles composed of ions? _____ (18)

? Which solution has solute particles with covalent bonds? _____ (19)

Hydrogen bonds arise from molecules, such as water, in which the electrons from the hydrogen atoms are strongly attracted to another atom (in water it is oxygen). The hydrogen atoms become more positively charged because of this, and oxygen is more negatively charged. When two or more of these molecules come within close proximity to one another, the positively charged hydrogen end of the molecule forms a weak bond with the negatively charged end of another molecule, as seen in fig. 1.5. This bond is known as a hydrogen bond.

 Activity

We can demonstrate the impact of hydrogen bonds by comparing the evaporation rates of water to those of ethanol (alcohol). Water is an example of a molecule with significant hydrogen bonding. Ethanol does have some hydrogen bonding, but it is not as strong as that in water.

1. Pour 5 mL of water into a 50 mL beaker labeled "water."
2. Pour 5 mL of ethanol into a 50 mL beaker labeled "alcohol."
3. Make sure that both liquids are at room temperature before you begin the experiment.
4. Warm the two beakers on a hot plate (NOT an open flame).

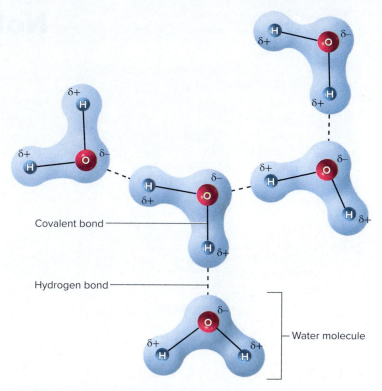

FIGURE 1.5 **Hydrogen Bonding.** Positive ends of hydrogen atoms are attracted to negative ends of oxygen atoms.

NOTE: You can use 70% lab alcohol, but the remaining 30% is water; therefore, the results will not be as dramatic. Which fluid evaporates more slowly, alcohol or water (circle one)?

Slow evaporation is due to the greater number of hydrogen bonds "holding" the molecules together.

Note

CAUTION

Despite the fact that pure water, or water with covalent molecules, does not conduct electricity as well as electrolytes, it still can conduct electricity. If the voltage is high enough, it can kill you. Which is why you do not swim outside during lightning storms or place electrical appliances on the edge of a bathtub when you are taking a bath.

Notes

REVIEW SECTION

Introduction to Lab Science, Measurement, and Chemistry

Name _____ *Date* _____

Lab Section _____ *Time* _____

❓ Chapter Summary Data

Use this section to record your results from questions within the exercise.

1. _____ 10. _____

2. _____ 11. _____

3. _____ 12. _____

4. _____ 13. _____

5. _____ 14. _____

6. _____ 15. _____

7. _____ 16. _____

8. _____ 17. _____

9. _____ 18. _____

   _____ 19. _____

Review Questions

1. The scientific discipline that studies the function of the human body is known as _____.

2. In terms of base units,

   a. what is the base unit of length in the metric system? _____

   b. what is the base unit of volume in the metric system? _____

3. How many cubic centimeters are there in 200 milliliters? _____

4. Assume a pill has a dosage of 350 mg of medication. How much medication is this in grams? _____

5. How would you write 0.000345 liter in scientific notation? _____

6. How many milligrams are there in 4.5 kilograms? _____

7. How much of a meter is 250 millimeters? _____

8. If given a length of 1/10,000 of a meter,

    a. convert this number into a decimal _____

    b. convert it into scientific notation _____

9. Use a word to describe

    a. one-thousandth of a second _____

    b. one thousand liters _____

    c. one-hundredth of a meter _____

10. Did you see a trend in your results with heart rate and exercise? If so, what do you predict for additional exercise?

    _____

    Would the trend continue indefinitely? Why or why not? _____

    _____

11. In terms of heart rate and exercise, which one is the dependent variable and which one is the independent variable?

    _____

12. According to your bar chart (table 1.4), what would be the mean heart rate at 150 seconds of exercise? _____

    _____

13. Define the term *buffer*. _____

14. a. In the pH experiment (step 3), when you added 10 drops of HCl, the pH changed by how much? _____

    b. How did the addition of buffer (steps 4 and 5) prior to adding 10 drops of HCl alter the change in pH? _____

15. What is the neutral pH? _____

16. Is a pH of 8 more basic or less basic than a pH of 6? _____

17. As a solution becomes more acidic, what happens to the concentration of hydrogen ions?

    _____

18. The term *electrolyte* is derived from the words *electro* ("electricity") and *lyte* ("to separate"). How does this term correlate with your experiment?

    _____

19. In which bond are electrons significantly shared between atoms? _____

20. Which bond—covalent, ionic, or hydrogen—is a weak bond? _____

Organs, Systems, and Organization of the Body

INTRODUCTION

The study of human anatomy and physiology requires a thorough understanding of the organs and organ systems of the human body and how they interact with one another. The first step in understanding an organ is to locate it within the body and see how it looks from different perspectives, or when it is sectioned. What you learn in this exercise will be used as a foundation for the material in the rest of the lab manual, so a thorough knowledge of the information in this exercise is vital for understanding material in future studies. In this exercise you examine the major organ systems of the body, learn directional terms, identify body cavities, and describe the major regions of the body. These topics are covered in your lecture text in Atlas A, "General Orientation to the Human Anatomy."

OBJECTIVES

At the end of this exercise, you should be able to

1. list the 11 organ systems;
2. place major organs, such as the heart, lungs, and stomach, in the proper organ system;
3. identify each body cavity and the major organ or organs it contains;
4. give directional terms equivalent to up, down, front, back, toward the midline or edge, and toward the core or surface of the body;
5. determine from an illustration whether a section is in the frontal, transverse, or sagittal plane;
6. list the major regions of the body and provide a common name, if known;
7. identify the quadrants and nine abdominal regions.

MATERIALS

Models of human torso

Charts of human torso

PROCEDURE

Organ Systems

Anatomy can be studied in many ways. **Regional anatomy** is the study of particular areas of the body, such as the head or leg. Most undergraduate college courses in anatomy and physiology (and the format of this lab manual) describe **systemic anatomy,** the study of **organ systems,** such as the skeletal system and nervous system. Although organ systems are studied separately in this organization, it is important to realize the intimate connections between the systems. If the heart fails to pump blood as part of the circulatory system, then the lungs do not receive blood for oxygenation and the intestines do not transfer nutrients to the blood as fuel. The brain and other organs are no longer capable of functioning, and the result is death. From a clinical standpoint, the failure of one system has impacts on many other organ systems. There are 11 organ systems that have functions unique to them. A brief overview of these systems is presented in the following paragraphs.

Reproductive The gonads (testes and ovaries) contain the sex-producing cells of the body, and the accessory organs, such as the uterus, vagina, penis, and seminal vesicles, play a part in the transport of the sex cells and the development of the fetus.

Urinary The kidneys are filters in the body, and the urinary bladder is a storage organ. The ureters connect the kidneys to the bladder, and the urethra is the exit tube from the body. The urinary system plays an important role in ridding the body of nitrogenous wastes, adjusting the chemical balance of body fluids, and maintaining blood volume.

Nervous The brain, spinal cord, and nerves make up the nervous system. The nervous system coordinates body regions, interprets environmental cues, and integrates information.

Muscular Individual muscles are the organs of this system. Muscles move and strengthen joints and generate heat, along with other functions, such as abdominal compression.

Respiratory The nose, larynx, trachea, and lungs are part of the respiratory system. The lungs exchange gases (oxygen and carbon dioxide) between the blood and the air.

Skeletal Each bone is considered an organ, with blood vessels and nerves found in each bone. The skeletal system supports the body, protects delicate organs, and produces blood.

Lymphatic The lymph nodes, spleen, thymus, and tonsils are part of the lymphatic system. A major function of the lymphatic system is to protect the body from foreign particles, such as bacteria, viruses, and fungi. Cells from the lymphatic system make up the **immune system,** which is not an organ system but a functional system consisting of an assemblage of cells that defend the body.

Integumentary The skin is the largest organ of the body and makes up most of this system. The integumentary system also contains associated structures, such as hair follicles, hair, nails, and the glands of the skin. The skin protects the body against microorganisms, keeps it from drying out, and produces vitamin D.

Digestive The mouth, esophagus, stomach, liver, and intestines are parts of the digestive system, which provides nutrients and water to the body and removes waste.

Endocrine This system is composed of organs that produce hormones. Hormones are vital in regulating growth and development and maintaining a constant internal body condition. Organs such as the thyroid gland and the adrenal glands are endocrine glands (glands that secrete hormones without the use of ducts). Organs such as the pancreas and the gonads have dual functions; one is endocrine and the other is exocrine (secretion of material through ducts).

Circulatory The heart, blood, and blood vessels make up this system. The heart is the pump of the system, and the blood vessels are the delivery and return portion of the system. The circulatory system is primarily involved in transporting oxygen, carbon dioxide, and other materials throughout the body.

A quick way to recall all 11 systems is to remember this phrase: "Run Mrs. Lidec." Each letter of the phrase represents the first letter of the name of one of the organ systems.

Ⓐ Activity

Examine models or charts in lab and locate the organs of the various organ systems. Fill in fig. 2.1 with the appropriate names of the organ systems as well as the Chapter Summary Data section.

Anatomical Position

In clinical settings, it is vital to have a proper orientation when dealing with patients. If two physicians are operating on a patient and one tells the other to make an incision to the left, the physician making the cut does not have to ask, "My left or your left?" because the cut is always to the *patient's* left side. When referring to the human body you will orient the body in **anatomical position.** In this position, the body is upright, facing forward, arms and legs straight, palms facing forward, feet flat on the ground, and eyes open (fig. 2.2a).

Directional Terms

With the body in anatomical position there are specific terms used to describe the location of one part with respect to another. These terms are somewhat different for quadrupeds (four-footed animals) and for humans, so both examples are given. Table 2.1 lists the directional terms used for humans.

Locate the terms in fig. 2.2. Quadrupeds do not have a superior/inferior designation. Dorsal is the back and ventral is on the belly side, while anterior (or cephalic) is the front, or head end, of the animal and posterior (or caudal) is the rear, or tail end, of the animal. There are other terms for location that have unique meanings. For the digestive system, **proximal** refers to regions closer to the mouth, while **distal** is in reference to regions closer to the anus. **Parietal** is in reference to the body wall when compared to **visceral,** which refers to areas closer to the internal organs (fig. 2.3). The heart, for example, has a visceral layer closer to the heart proper called the **visceral pericardium,** while it also has a parietal layer farther from the heart called the **parietal pericardium.** Likewise, the lungs have a **visceral pleura** and a **parietal pleura. Ipsilateral** refers to being on the same side of the body, and **contralateral** refers to being on the other half (left side/right side). The right hand and right arm are ipsilateral, while the ears are contralateral. Directional terms do not change if the body position changes. If you stand on your head, it is still superior to your feet because you reference the body as if it were in anatomical position. However, the directional terms can be relative. For the limbs, the elbow is proximal to the hand, but the elbow is distal to the shoulder.

Planes of Sectioning

When viewing a picture of an organ that has been cut, it is important to understand how the cut was made. Just as an apple looks different when cut crosswise as opposed to lengthwise, so do some organs.

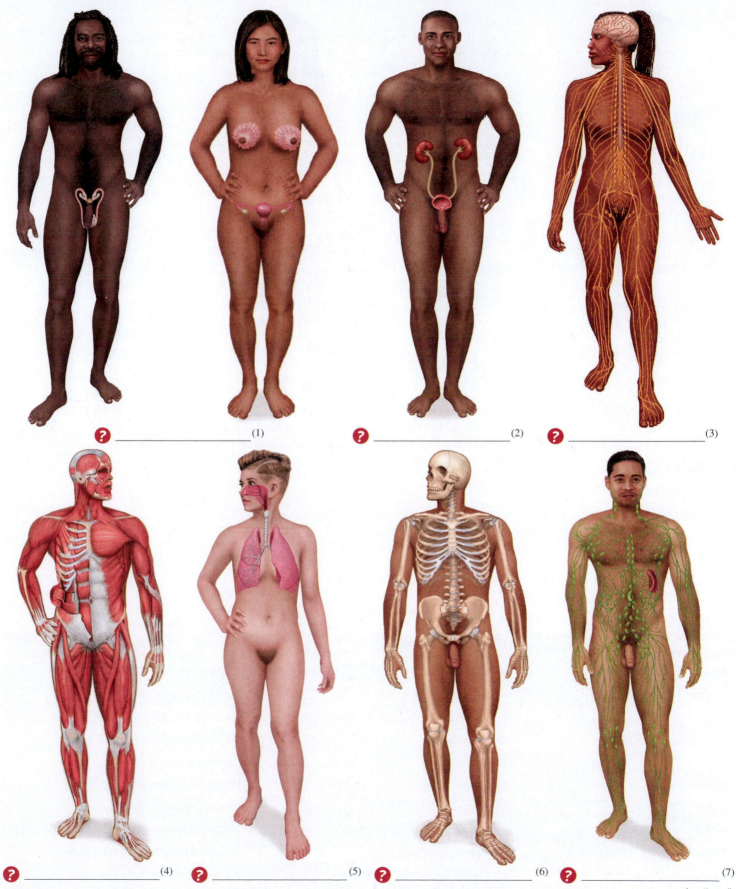

? _____ (1)   ? _____ (2)   ? _____ (3)

? _____ (4)   ? _____ (5)   ? _____ (6)   ? _____ (7)

FIGURE 2.1 **Organ Systems of the Human Body.** Fill in the terms for the organ systems here and in the Chapter Summary Data section at the end of the exercise.

(continued)

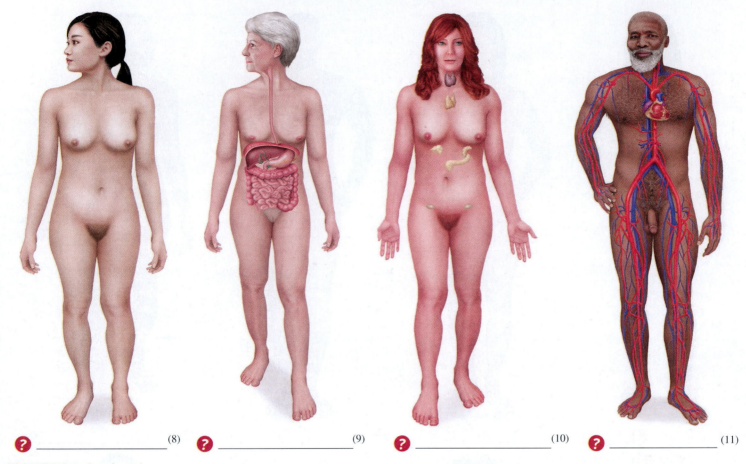

❓ _____ (8)   ❓ _____ (9)   ❓ _____ (10)   ❓ _____ (11)

FIGURE 2.1 *Continued*

TABLE 2.1	Directional Terms Used for Humans	
Term	**Meaning**	**Example**
Superior	Above	The nose is superior to the chin.
Inferior	Below	The stomach is inferior to the head.
Medial	Toward the midline	The sternum is medial to the shoulders.
Lateral	Toward the side	The ears are lateral to the nose.
Superficial	Toward the surface	The skin is superficial to the heart.
Deep	Toward the core	The lungs are deep to the ribs.
Anterior (or ventral)	To the front	The toes are anterior/ventral to the heel.
Posterior (or dorsal)	To the back	The spine is posterior/dorsal to the sternum.
Proximal	For extremities, meaning near the trunk	The elbow is proximal to the wrist.
Distal	For extremities, meaning away from the trunk	The toes are distal to the knee.

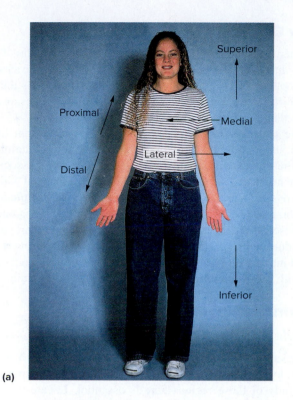

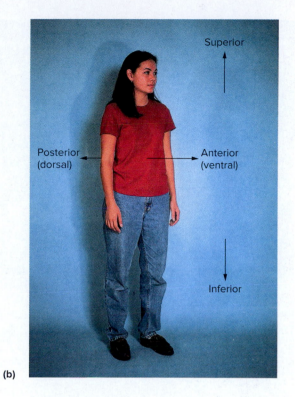

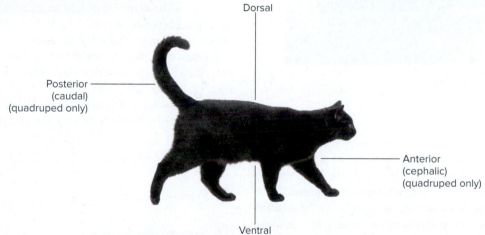

FIGURE 2.2 **Anatomical Position and Directional Terms.** (a) For humans, anterior view; (b) for humans, 3/4 view; (c) for quadrupeds, lateral view.

(a, b) ©Eric Wise; *(c)* Life on white/Alamy Stock Photo

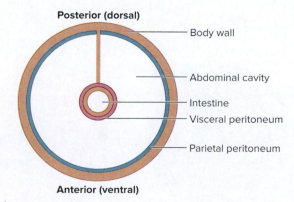

FIGURE 2.3 **Idealized Cross Section of the Body with Relationships of Visceral and Parietal Terms.**

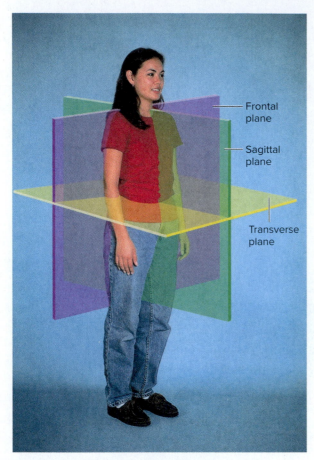

FIGURE 2.4 Sectioning Planes.
©Eric Wise

Examine fig. 2.4 for the following sectioning planes. A cut that divides the body or organ into superior and inferior parts is in the **transverse (horizontal) plane.** A cut that divides the body into anterior and posterior portions is in the **frontal (coronal) plane.** A cut that divides the body into left and right portions is in the **sagittal plane.** A cut that divides the body or organ equally into left and right halves is in the **median (midsagittal) plane,** while one that divides the body into unequal left and right parts is in the **parasagittal plane.** Examine fig. 2.5 for examples of sectioning planes.

Body Cavities

The outer regions of the body enclose several body cavities. These cavities have linings that surround internal organs. The **cranial cavity** houses the brain, and the **vertebral canal** encloses the spinal cord (fig. 2.6). These cavities are lined with meninges and are discussed in the exercise on the nervous system.

The **thoracic cavity** is superior to the diaphragm housing the lungs and mediastinum. The mediastinum contains the heart, the **pericardial membranes,** the large vessels associated with the heart, and the trachea and esophagus. Associated with the lungs are the pleura and pleural cavities. The **visceral pleura** is a membrane close to the lung, and the **parietal pleura** is outside of the visceral pleura and separated from it by the **pleural cavity.** The **abdominopelvic cavity** is inferior to the diaphragm and is subdivided into the **abdominal** and **pelvic cavities.** These cavities are lined with **serous membranes** that secrete a lubricating fluid.

Regions of the Body

Overview

Ⓐ Examine fig. 2.7 for specific areas of the body. You will refer to these areas throughout this lab manual, so a complete study of these regions here is essential. Locate these regions on a torso model in lab. In anatomical usage, the **arm** is the region between the shoulder and the elbow, and the **leg** is the region between the knee and the ankle.

Examine a muscle model in lab, and name the region where the following muscles are found. You can find these muscles in figs. 12.1, 12.5, 13.1, 13.3, 13.6, 14.1, 14.4, and 14.6 in the exercises on muscles in this lab manual. Fill these in at the end of this exercise in the Chapter Summary Data section as well.

❓ Pectoralis major _____ (12)

❓ Trapezius _____ (13)

❓ External oblique (front) _____ (14)

❓ Rectus femoris _____ (15)

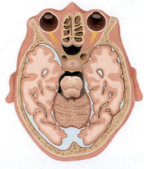
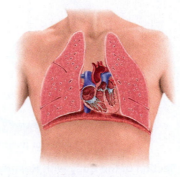
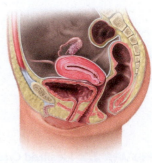

FIGURE 2.5 Planes of Sectioning.

(a) Transverse section   (b) Frontal section   (c) Sagittal section

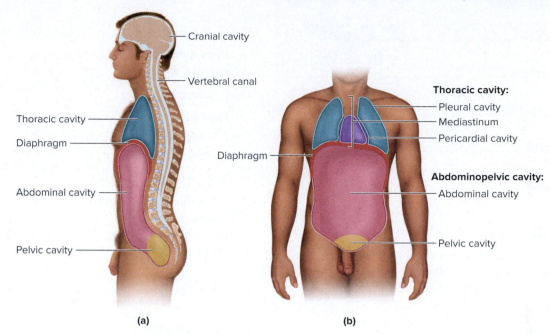

FIGURE 2.6 Body Cavities. (a) Left lateral view; (b) anterior view.

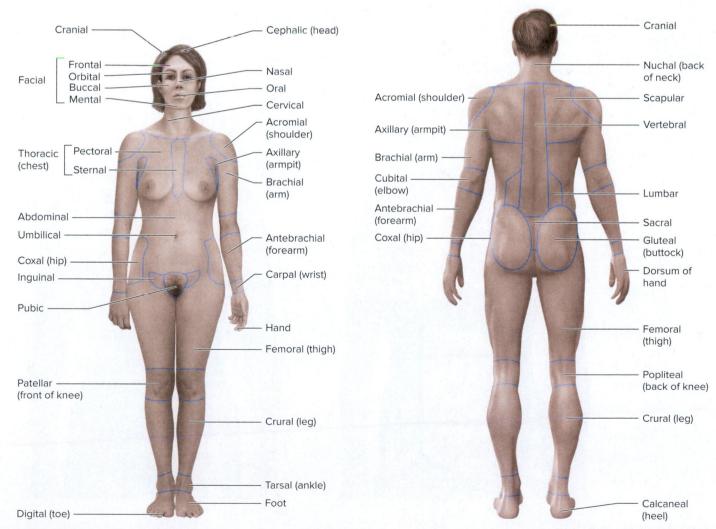

FIGURE 2.7 Regions of the Body. The anatomical regions of the body are indicated with the common name (where appropriate) in parentheses.

? Gastrocnemius _____ (16)

? Gluteus maximus _____ (17)

? Flexor carpi radialis _____ (18)

? Triceps brachii _____ (19)

? Latissimus dorsi _____ (20)

Abdominal Regions

The abdomen can be divided into either four quadrants or nine regions (fig. 2.8). Clinicians typically use four quadrants terminology, while anatomists generally use nine regions.

Nine Regions

    Right hypochondriac

    Left hypochondriac

    Epigastric

    Right lumbar (lateral abdominal)

    Left lumbar (lateral abdominal)

    Umbilical

    Hypogastric

    Right iliac (inguinal)

    Left iliac (inguinal)

Four Quadrants

    Right upper quadrant (RUQ)

    Left upper quadrant (LUQ)

    Right lower quadrant (RLQ)

    Left lower quadrant (LLQ)

(A) Examine a torso model in lab. For each of the following organs, write what abdominal region(s) contain(s) that organ. Fill in the following spaces here and in the Chapter Summary Data section at the end of the exercise.

? Liver _____ (21)

? Urinary bladder _____ (22)

? Small intestine _____ (23)

? Descending colon _____ (24)

? Left kidney _____ (25)

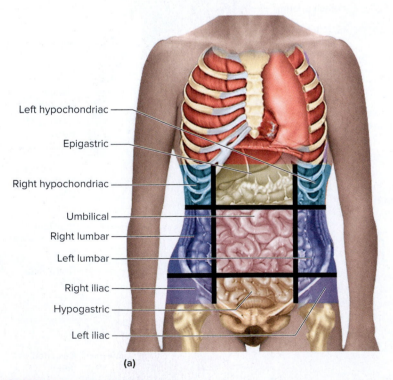
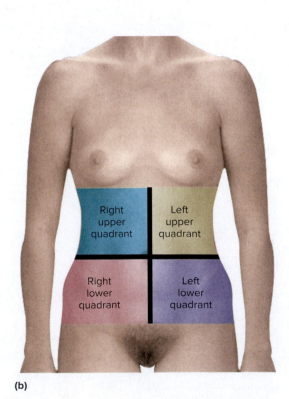

Left hypochondriac

Epigastric

Right hypochondriac

Umbilical

Right lumbar

Left lumbar

Right iliac

Hypogastric

Left iliac

Right upper quadrant

Left upper quadrant

Right lower quadrant

Left lower quadrant

(a)

(b)

FIGURE 2.8 Abdominal Regions. (a) Nine regions; (b) four quadrants.

REVIEW SECTION

Organs, Systems, and Organization of the Body

Name _____ *Date* _____

Lab Section _____ *Time* _____

❓ Chapter Summary Data

Use this section to record your results from questions within the exercise.

1. _____   14. _____

2. _____   15. _____

3. _____   16. _____

4. _____   17. _____

5. _____   18. _____

6. _____   19. _____

7. _____   20. _____

8. _____   21. _____

9. _____   22. _____

10. _____   23. _____

11. _____   24. _____

12. _____   25. _____

13. _____

Review Questions

1. Organs are grouped into functionally related associations known as _____.

2. The kidneys belong to the _____ system.

3. The liver belongs to the _____ system.

4. In anatomical position, the palms of the hand are facing _____.

5. In anatomical terms, referring to front and back, the pectoral region is _____ to the scapular region.

6. In terms of nearness to the trunk, the antebrachium is _____ to the carpal region.

7. In terms of front to back, the shoulder blades are _____ to the nipples.

8. What type of section plane is illustrated in fig. 2.6a? _____

9. What type of section plane is illustrated in fig. 2.6b? _____

10. The stomach is found in what specific cavity? _____

11. Name the body cavities in the illustration to the right.

    a. _____

    b. _____

    c. _____

    d. _____

    e. _____

12. The term *arm* in anatomy refers to the region _____

    a. between the hand and elbow.

    b. between the elbow and shoulder.

    c. between the hand and shoulder.

13. The heart is found in what specific cavity? _____

14. The brain is located in what specific cavity? _____

15. If you were to sit on a horse's back, you would be on the _____ aspect of the horse.

    a. anterior   b. ventral   c. posterior   d. dorsal

16. Pain in the appendix would be felt in what quadrant of the abdomen? _____

17. What is the difference between the abdomen and the abdominal cavity? _____

18. The region of the abdomen directly under the right side of the rib cage is the _____ region.

19. If a hairline fracture occurred in the proximal humerus (arm bone), would the injury be closer to the shoulder or to the elbow? _____ Why?

    _____

    _____

20. In a clinical report there is a note of a laceration (cut) on the posterior crural region. Where does this occur in layperson's terms? _____

    _____

21. Name the regions of the body in the illustration to the right.

a. _____

b. _____

c. _____

d. _____

e. _____

f. _____

g. _____

h. _____

i. _____

j. _____

k. _____

l. _____

m. _____

n. _____

Microscopy

INTRODUCTION

Originally, the study of anatomy and physiology was based on macroscopic, or gross, observation. This study was limited by the **resolution** of the human eye, which is the amount of detail seen in a visual image. Resolution is not the same as magnification. An image can be magnified but still be blurry (have low resolution). If you magnify print media, such as a magazine picture, with a hand lens or microscope, you will see that the image is not only larger but made of many dots (pixels in digital media). You have increased both the magnification and the resolution of the image. With the invention and use of the compound light microscope (microscopes increase the resolution), much greater detail was seen and thus began the study of cells and tissues. **Light microscopy** involves the use of visible light and glass lenses to magnify and observe a specimen. These topics are covered in the Saladin text in chapter 1, "Major Themes of Anatomy and Physiology," and chapter 3, "Cellular Form and Function." This exercise involves the use of the compound light microscope, how to examine prepared slides under the microscope, and how to make slides of fresh material for study.

OBJECTIVES

At the end of this exercise, you should be able to

1. explain the rules for proper microscope use;
2. name the parts of the microscope and their functions;
3. place a slide on the microscope and observe the material, in focus, under all magnifications of the microscope;
4. calculate the total magnification of a microscope based on the specific lenses used;
5. prepare a wet mount for observation.

MATERIALS

Compound light microscope

Prepared slide with the letter *e* (or printed material and scalpel)

Transparent ruler or slide with grid etched on it (grid slide)

Glass microscope slides and coverslips

Lens paper and lens cleaner

Kimwipes or other cleaning paper

Small dropper bottle of water

1% methylene blue solution

Clean, food-grade toothpicks

Histological slides of kidney, stomach, or liver

Slide with silk threads

PROCEDURE

Care of the Microscope

Microscopes are very expensive pieces of equipment, and you should always take great care handling them. There are a few rules concerning the care of a microscope.

1. When carrying the microscope, hold it securely with two hands—one hand under the base and one hand on the arm.
2. Keep the microscope upright at all times.
3. Keep lenses clean with lens cleaner and softened lens paper. Do *not* use paper towels or your clothing.
4. Use only the fine-focus knob when using the high-power objective lens.

5. Remove slides from the microscope before putting it away.
6. Secure the cord with a rubber band or wrap the cord carefully around the microscope.
7. Store the microscope with the *low-power* (scanning) objective lens in place.
8. Put the microscope away in its proper location.

Using the Microscope

 Activity

1. Examine fig. 3.1 and familiarize yourself with the parts of the microscope.
2. Remove the microscope from the storage area. When you carry the microscope, place one hand under the base and the other hand on the arm of the microscope. Never tilt the microscope from an upright position, as lenses or filters may fall and break.
3. Take the microscope to your desk. Unwrap the electrical cord from around the microscope and compare the microscope that you have in lab to the one illustrated in fig. 3.1. There may be differences between your microscope and the one in fig. 3.1, but you should be able to locate the parts listed in the checklist. Use the checklist to make sure you find all of the listed parts. Place a check mark next to the appropriate space when you locate each part of the microscope.

Microscope Parts Checklist		
_____ Base	_____ Arm	_____ Objective lens
_____ Condenser	_____ Iris diaphragm lever	_____ Body tube
_____ Nosepiece	_____ Coarse-focus knob	_____ Fine-focus knob
_____ Mechanical stage	_____ Ocular (eyepiece) lens	_____ Light source

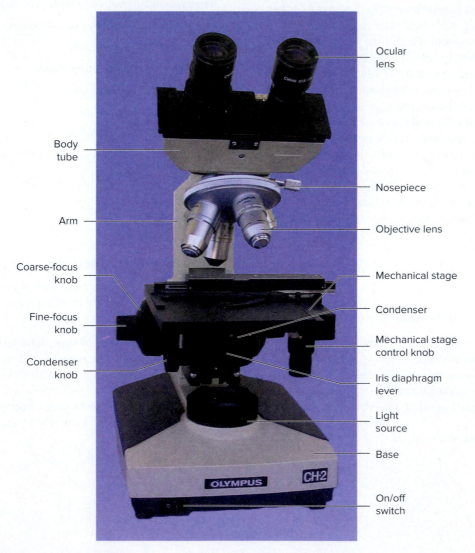

FIGURE 3.1 Compound Light Microscope.
©Eric Wise

4. With the microscope in front of you, plug it in, making sure the ocular lens or lenses are facing you. Make sure the cord is not hanging over the counter or in the aisle where someone might trip on the cord or pull the microscope off the counter. If the microscope has an illuminator (light source) dial, make sure the setting is on the lowest level.

5. The low-power objective lens (typically 4 magnifications or 4×) should be facing down. This lens is often the shortest one and, in many microscopes, has a red line around the barrel of the lens. If it is not in place, turn the nosepiece until the low-power objective clicks into place. Always use the low-power objective lens when you first look at a microscope slide.

6. Use the knob on the microscope to raise the condenser lens so it is at its highest level.

7. Take a piece of printed material and, using a scalpel, cut out a small section with some letters from the paper. Place the piece of paper on a glass microscope slide and add a drop of water to the piece of paper on the slide.

8. Place a thin coverslip on the slide by touching one edge of the coverslip to the water and lowering it slowly over the printed material as seen in fig. 3.2. If you drop the coverslip on top of the slide, you will probably trap several air bubbles, which may obscure some of your specimen.

9. Locate the light switch on your microscope and turn it on.

10. Place the slide on the microscope stage with the coverslip on the top and the letter centered in the circle on the stage of the microscope. Make sure the printing is upright so you can read it from where you are sitting. Most microscopes have a slide clip that holds the microscope slide in place. Use the stage control knobs to move the specimen so that light is coming through the specimen. This may require that you look at the specimen from the side to see if it is centered in the middle of the microscope stage.

11. Examine the specimen under low power. The **field of view** is the circle that you see as you look into the microscope. Use the coarse-focus knob to bring the specimen into focus. This is done by raising the mechanical stage of

the microscope completely and then slowly lowering the stage as you look into the ocular lenses. On some microscopes it is the objective lens that is raised or lowered with the coarse-focus knob. Focusing the microscope requires a little bit of patience. When first looking at microscope slides, always use the low-power objective lens. The coverslip should be close to the objective lens. Look through the ocular lens and rotate the coarse-focus knob slowly so the objective lens and the slide begin to move away from each other. This should bring the object into focus.

12. Adjust the ocular lenses. Binocular microscopes usually have an adjustable left ocular lens. Focus the coarse-focus knob so that the right eye is in focus. If this is difficult to do with both eyes open, you can place a piece of paper in front of your left eye while you focus your right eye. Once the right eye is in focus, use the knurled ring on the ocular lens (NOT the coarse-focus knob) and adjust it so the left eye is in focus. You can block the right ocular lens with a piece of paper if you need to while adjusting the left ocular. It is important that both eyes are adjusted correctly so you do not have eyestrain, which can lead to headaches.

13. Adjust the light. Too much or too little light makes a specimen difficult to see. You can change the light level by either adjusting the light from the rheostat knob on the base or adjusting the iris diaphragm lever in the front of the microscope. The iris diaphragm changes how much of the lens diameter is available for light to pass through. As the diaphragm is closed down, there is an increase in the **depth of field** (which refers to how much of the thickness of the specimen is in focus) as well as a reduction in the overall amount of light passing through the specimen. The condenser lens gives even light intensity to the specimen. For most observations, put the condenser lens close to the specimen.

14. Draw what the specimen looks like in the space provided.

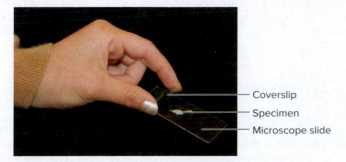

FIGURE 3.2 **Preparation of a Wet Mount.**
©Eric Wise

Does the printed material appear right side up, or is the image inverted? _____

Is the material oriented correctly, or is the image flipped horizontally? _____

How much of the material occupies the field of view? _____

15. After you get the specimen in focus under low power, you should examine the slide under the next higher power. This is done by centering the image that you observe in the field of view and then switching the objective lens to the next higher power. The image should be close to being in focus if your microscope is *parfocal*. Do not adjust the height of the mechanical stage as you change the lenses. The next higher-power objective lens should clear the slide. Once you rotate the lens, adjust the focus by using the *fine-focus knob only*. You will probably need to adjust the light by moving either the iris diaphragm lever at the front of the microscope or the rheostat knob. As magnification increases, does the field of view increase or decrease in size?

16. You can look at the specimen under greater magnification via the same procedure by turning the nosepiece to the high-power dry lens (usually around 40× on most microscopes). If you cannot focus on the high power or have lost the image you were looking for, return to low power, make sure the specimen is centered in the field of view, and try the process again. If you still cannot find the object under high power, ask your instructor to help you.

If you are having a difficult time seeing anything or seeing things in focus, this could be for several reasons. Use the following troubleshooting list to help you.

Problem	Solution
Nothing is visible in the lens.	Plug in the microscope. Turn the power supply on.
	Rotate the nosepiece until you hear it click into place.
	The bulb is burned out; replace bulb.
You see a dark crescent.	The objective lens is not in the proper position; click the lens into place.
All you see is a light circle.	The microscope is out of focus; adjust the coarse-focus knob.
	The light is up too high; turn down the light.
	The iris diaphragm is open too much; close it down.

Focusing Tips

On some faintly stained specimens, the material may be difficult to see on low power. One trick is to locate the edge of the cover-slip and turn the coarse-focus knob up and down until the edge is in sharp focus. This lets you know you are in the approximate focal plane for examining the material on the slide. Move the slide to where the specimen should be, adjust the light, and re-examine it.

Magnification and Field of View

You can estimate the size of the object you are observing if you know the diameter of the field of view. The field of view can be measured directly when under low magnification by using a clear ruler or a glass slide with a grid on it. If higher magnifications are used, rulers won't work, and you have to calculate the field of view. Remember, a millimeter is $1/1{,}000$ of a meter and a micrometer is $1/1{,}000{,}000$ of a meter. If you need to review these units, refer to Exercise 1. There is a relationship between the diameter of the field of view and the magnification used.

(A) Activity

You can first calculate the total magnification using the following procedure. Look at the barrels of your microscope and determine the magnification of each lens.

Eyepiece (ocular) magnification: _____

Low-power objective magnification: _____

Total magnification (= ocular magnification × objective magnification): _____

Place a transparent section of ruler or a glass slide with a grid on the stage of the microscope. The space between each dark line that runs vertically on a ruler is 1 millimeter (mm). Count the number of millimeters at the broadest part of the field of view and enter this number as the diameter of the field of view in the following space.

Diameter of the field of view (mm): _____

You can calculate the length of an object by determining how much of the diameter of the field of view it occupies. Let's say that the diameter of the field of view is 10 mm. If an object takes up one-half of the field of view, you can estimate its size at 5 mm. If the object took up only one-third of the field of view, how large would it be? Record your answer in the following space.

Object size (mm): _____

As the magnification increases, the field of view decreases proportionally. Thus, if the diameter of the field of view is 10 mm at one magnification and you double the magnification by changing

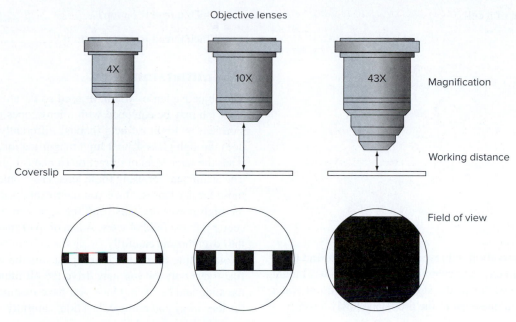

FIGURE 3.3 Increasing Magnification and Decreasing Field of View.

lenses, the field of view is reduced to a diameter of 5 mm. If you switch to a new lens and increase the magnification by 10 times, the field of view is reduced to one-tenth of the original field of view. Look at fig. 3.3 for a representation of this proportional change in the field of view.

Keep the clear ruler or a glass slide with a grid under the microscope and increase the magnification to the next higher power by moving the next larger objective lens into place. Record the total magnification of your microscope with this objective lens.

Total magnification: _____

Examine the ruler or grid under the microscope and record the diameter of the field of view in millimeters.

Diameter of the field of view in millimeters: _____

Is the decrease in the field of view related to the magnification?
_____

Now calculate the total magnification of the microscope using the high-power objective lens.

Magnification with high-power objective lens: _____

You will not be able to measure the field of view accurately under high power with a ruler; however, you should be able to calculate the diameter of the field of view. For example, if the diameter of the field of view is 2.5 mm at 40 power (40×), then it would be 0.25 mm at 400× (10 times more magnified yet one-tenth the field of view). Calculate the diameter of the field of view under high power.

Diameter of the field of view under high power: _____

Preparation of a Wet Mount

You can make relatively quick and easy observations under the microscope as long as the material is thin enough and small enough. One technique for cell observation is to examine cells from the inside of the oral cavity. Ask your instructor for permission first.

Ⓐ Activity

1. Place a drop of methylene blue (a stain) on the slide. Methylene blue is a stain that makes the nucleus of the cell more apparent.
2. With a toothpick, *gently* scrape the inside of your cheek.
3. Smear the material from the toothpick onto the slide.
4. Place a coverslip on the slide by holding the coverslip at a 45-degree angle and slowly lowering it on the slide to avoid trapping air bubbles. Examine the specimen under the microscope. The small blue structures inside the cells are the nuclei.
5. Draw a single cell in the space provided and label the plasma membrane, nucleus, and cytoplasm.

Your illustration of a cell:

For another observation, remove a hair from your head (preferably one with a split end) and examine it by making a wet mount. Place the hair in the center of the slide and add a drop of water. Place the coverslip on one edge of the drop and slowly lower it.

Observation of a Prepared Slide

Ⓐ **Activity**

Examine a prepared slide of tissue provided by your instructor. Examine the entire sample using the low-power objective lens. You should scan the entire sample, looking for areas you want to observe more closely. Move to the next higher power and adjust the focus using the fine-focus knob. Finally, examine the material with the high-power dry objective lens and draw what you see in the space provided.

Draw an illustration of material from a prepared slide:

Name of the sample drawn: _____

Observation of Silk Threads

Ⓐ **Activity**

Examine a slide that has crossed silk threads. Focus the slide on low power. How many of the threads can you see in focus without adjusting the focus knob further?

Move to medium and high power. What happens to the depth of field with increasing magnification? _____

_____

Which thread is on top? _____

Which thread is on the bottom? _____

Oil Immersion Lens

The objective lenses you have used so far are called "dry" lenses. Your lab may be equipped with microscopes that have oil immersion lenses. Light refracts (bends) differently through air than it does through glass. Under high magnification, oil is used because it has the same level of refraction (it bends light the same) as glass. The techniques for using these lenses are somewhat different from those for dry lenses. Once you have examined the specimen using the high-power dry lens, find the spot you want to examine and center it in the field of view. Add a drop of immersion oil on top of the coverslip and carefully swing the oil immersion lens into place. Avoid getting oil on the other lenses of the microscope. Use the fine focus only, or you may drive the oil immersion lens through the slide and break it. Once you have examined the slide, swing the lens away and remove the slide, carefully wiping away the immersion oil with a clean piece of lens paper (do not use your shirt or a paper towel; these can scratch the lens). Use only lens paper to clean the oil from the oil immersion lens. Use another paper to remove any remaining oil.

Cleaning the Microscope

Smudges on the images you view through the microscope may be due to several things. There may be makeup or dirt on the ocular lens (or lenses). There may be dirt, oil, salt, stains, or other material on the objective lenses. To clean a lens, place a small amount of lens-cleaning fluid on a clean sheet of lens paper. Make one circular pass on the lens and throw away the paper. If you continue to clean the lens with the same lens paper, you can grind dirt or dust into the lens. Use a fresh piece of lens paper and repeat the procedure if further cleaning is needed.

You may want to clean the microscope slide before you examine it. Use a cleaning paper, such as a Kimwipe, to clean oil or dust from the slide.

Finally, dust may have collected inside the microscope over the years, or the lenses may be scratched. There is nothing you can do about this, though you may want to bring it to your instructor's attention.

Putting the Microscope Away

When you are finished using the microscope, make sure to do the following:

1. Remove the slide from the stage.
2. Rotate the nosepiece so the low-power (4×) lens is down.
3. Make sure the slide clip is centered so the bar is not sticking out from the side of the microscope.
4. Wrap the cord loosely around the microscope or secure the cord with a rubber band.
5. Hold the microscope with two hands, one on the arm and one on the base.
6. Put the microscope away in the correct place.

REVIEW SECTION

Microscopy

Name _____ *Date* _____

Lab Section _____ *Time* _____

Review Questions

1. If the ocular lens is 10×, what would be the total magnification for the following objective lenses?

   a. 5× _____   b. 17× _____   c. 35× _____

2. What is the function of the iris diaphragm of the microscope? _____
   _____

3. If the diameter of the field of view is 5.6 mm at 40×, what would the diameter be at 80×? _____

4. What is the name of the thin glass plate placed on top of a specimen? _____

5. The microscope that you use in lab is a(n) _____

   a. compound light microscope.   b. dissecting microscope.   c. electron microscope.

6. What is the name of the circle you see when you look through the ocular lens of the microscope? _____
   _____

7. When you switch from a low-power objective lens (for example, 4×) to a higher-power objective lens (for example, 10×), what
   happens to the working distance between the lens and the coverslip? _____
   _____
   _____

8. What happens to the field of view when you change from a low-power objective lens to a higher-power objective lens? Does it
   increase or decrease? _____
   _____

9. When should you use the scanning lens on the microscope? _____
   _____

10. How should you clean the lenses of a microscope? _____
    _____

11. What is the proper way to carry the microscope in lab? _____
    _____

12. Examine the following "field of view" and determine what the size of the object is.

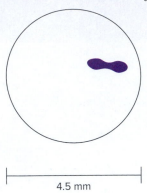

4.5 mm

13. Label the parts of the microscope illustrated, using the numbers for the terms provided.

1. Arm
2. Base
3. Body tube
4. Coarse-focus knob
5. Iris diaphragm lever

6. Light source
7. Objective lens
8. Ocular lens
9. Mechanical stage

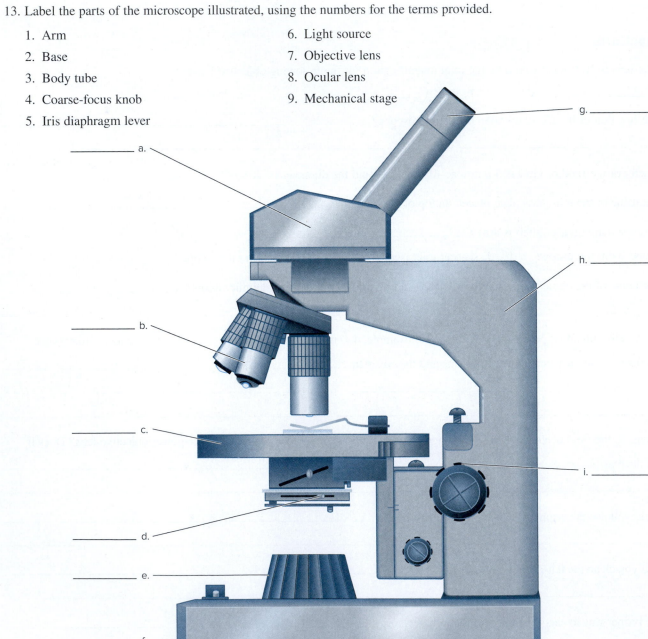

Cell Structure and Function

INTRODUCTION

The cell is the structural and functional unit of the human body and is the fundamental unit of living organisms. Many diseases can be traced to some type of cellular change. Cells grow, divide, acquire nutrients, release wastes, and respond to local stimuli. Cells perform many functions, some unique to the particular organ where they are found. **Cytology** is the scientific study of cells.

The membrane of the cell is the dynamic interface between the internal environment of the cell and the external environment. In humans, most cells are bathed in a liquid medium called extracellular fluid (ECF), which provides nutrients, oxygen, hormones, water, and other materials to the cell. From the interior of the cell, the cell releases ammonia, carbon dioxide, and other metabolic products into this liquid. The plasma membrane in large part regulates the exchange of materials between the cell's interior and the environment surrounding it. The exchange of materials between the cell and the ECF maintains the homeostatic balance the cell must have to survive. This is done passively in some cases, and in others, ATP is used to actively transport material. Even small changes in the concentration of certain materials in the cell might lead to cellular death, so the constant adjustment of water, ions, and other metabolic products is extremely important. The membrane also has embedded enzymes that play a significant role in cellular function.

In this exercise, you examine the structure of animal cells, learn how cells of the body divide to make new cells, look at the physical processes that influence plasma membrane dynamics, and study the nature of membrane transport. These topics are discussed in chapter 3, "Cellular Form and Function," and chapter 4, "Genes and Cellular Function," in the Saladin text.

OBJECTIVES

At the end of this exercise, you should be able to

1. describe the importance of cells to the body;
2. describe the plasma membrane in terms of location, composition, and function;
3. describe the structure and function of the organelles;
4. draw a representation of each of the organelles;
5. describe the three main events of the cell cycle;
6. name the four phases of mitosis and what occurs during each phase;
7. describe the processes by which substances move across membranes;
8. describe the differences between *hypertonic, hypotonic,* and *isotonic;*
9. explain *diffusion, osmosis,* and *filtration;*
10. describe the movement of water across a selectively permeable membrane;
11. compare and contrast diffusion and osmosis.

MATERIALS

Models or charts of animal cells

Electron micrographs of cells or textbook with electron micrographs

8.5- by 11-inch blank notebook paper

Prepared slides of whitefish blastula

Microscopes

Modeling clay (Plasticine™)—two colors

Marbles

Brownian Motion

India ink in dropper bottles

Dropper bottle of water

Microscopes

Microscope slides

Coverslips

Hot plate

Diffusion Demonstration

Potassium permanganate crystals

100 mL beaker (one per table)

Water

Small forceps or spatula

Diffusion

Agar plates (three per table)

0.01 Molar (M) potassium permanganate solution in dropper bottles

0.01 M methylene blue solution in dropper bottles

Dishpan filled with crushed ice

Plastic drinking straws

Millimeter ruler

Fine probe, spatula, or small forceps

Warming tray (35–40°C) or incubator

Osmosis Demonstration

String

Dialysis tubing

Glass tube

Dark corn syrup or concentrated, colored sucrose solution (20%)

1% starch solution

Ring stand and clamp

250 mL beaker

Distilled water

Permanent marker

Osmosis Experiment

Four strips of 20-cm-long dialysis tubing (one set of four per table)

Four 200 mL beakers

String

Scissors

Four solutions (2 L each) of 0%, 5%, 15%, and 30% sucrose (color each one with a different color of food coloring)

One liter of 15% sucrose solution

Balances

Towels

Pipettes (10 mL)

Pipette pumps

Osmosis and Living Cells

Clean glass microscope slides

Coverslips

5 mL of mammal blood (check with local veterinarian's office)

Distilled water in dropper bottle (one per table)

0.9% saline solution in dropper bottle (one per table)

5% sodium chloride solution in dropper bottles (one per table)

Protective gloves

Virtual Lab

PhILS software and computer

Filtration

Filter paper

Funnel

Ring stand with ring clamp

10 mL graduated cylinder

500 mL beaker

Iodine solution in dropper bottles (one per table)

Filtration solution (500 mL of 1% starch, 1% charcoal, and water with blue food coloring), consists of 5 g of starch and charcoal in water colored blue in one bottle or flask

Stopwatch or clock with second hand

PROCEDURE

Overview of the Cell

There are many types of cells in the body. Some are long and thin, some are spherical, and some are flat. You will examine a representative cell as an example, but realize that there is tremendous diversity in cell shapes and functions.

Cells consist of two main parts: the **plasma** (or **cell**) **membrane** and the **cytoplasm.** The plasma membrane is the outer boundary of the cell, and although it cannot be seen using the light microscope, its location is determined by the difference in color between the cytoplasm and the surrounding liquid on the microscope slide. The cytoplasm is the portion of the cell in which water, dissolved materials, and cellular **organelles** are found. The fluid in which the organelles are suspended is called the **cytosol.** Small filaments and tubules make up the **cytoskeleton** (also considered part of the cytoplasm). The nucleus directs the cell's activities and stores its genetic information. It is visible with the light microscope and frequently appears as a spherical or an oblong structure.

A Activity

Figure 4.1 shows an overview of the cell. Locate the cellular structures on the models or charts in the lab. Take a single sheet of notebook paper and make a sketch of a representative cell with all of the organelles. Use three quarters of the sheet of paper for your drawing.

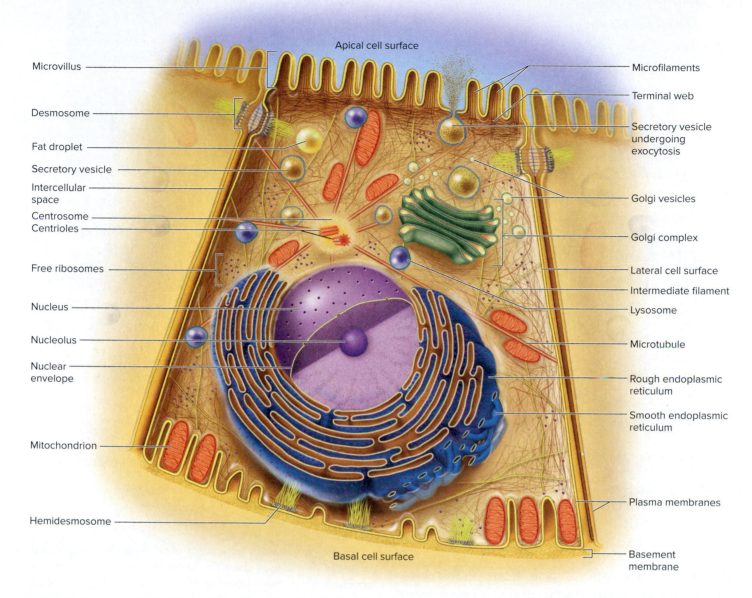

Apical cell surface

Microvillus

Desmosome

Fat droplet

Secretory vesicle

Intercellular space

Centrosome
Centrioles

Free ribosomes

Nucleus

Nucleolus

Nuclear envelope

Mitochondrion

Hemidesmosome

Basal cell surface

Microfilaments

Terminal web

Secretory vesicle undergoing exocytosis

Golgi vesicles

Golgi complex

Lateral cell surface

Intermediate filament

Lysosome

Microtubule

Rough endoplasmic reticulum

Smooth endoplasmic reticulum

Plasma membranes

Basement membrane

FIGURE 4.1 Overview of a Representative Cell.

Plasma Membrane

The plasma membrane is the "gatekeeper" of the cell. It is selective in what it allows into the cell. The plasma membrane is composed of a **phospholipid bilayer,** proteins, cholesterol, and other molecules. Phospholipids consist of a hydrophilic phosphate group attached to hydrophobic lipid groups (fig. 4.2). Interspersed among the phospholipid molecules are **cholesterol molecules,** which provide stability to the membrane or, in greater concentrations, make the membrane more fluid. Two major types of proteins are also found in the membrane. **Peripheral proteins** are found on the inner or outer surface of the membrane, whereas those passing into the membrane are known as **transmembrane proteins.** Transmembrane proteins may have carbohydrates or

other molecules associated with them and frequently serve as cell markers. Some proteins function as channels by which specific materials can pass through the membrane. Proteins may anchor one cell to another, provide a place for metabolic reactions to take place, act as cell markers that identify a particular cell, or act as membrane receptors or channels. The plasma membrane is important in establishing electrochemical charge differentials, which allows for signals to be passed along the membrane. It is also a selectively permeable membrane that provides an entrance to or exit from the cell for some materials while excluding other material from entering or exiting the cell's interior. Use the remaining one quarter of the sheet of paper of your cell drawing to draw the plasma membrane, associated proteins, and cholesterol molecules.

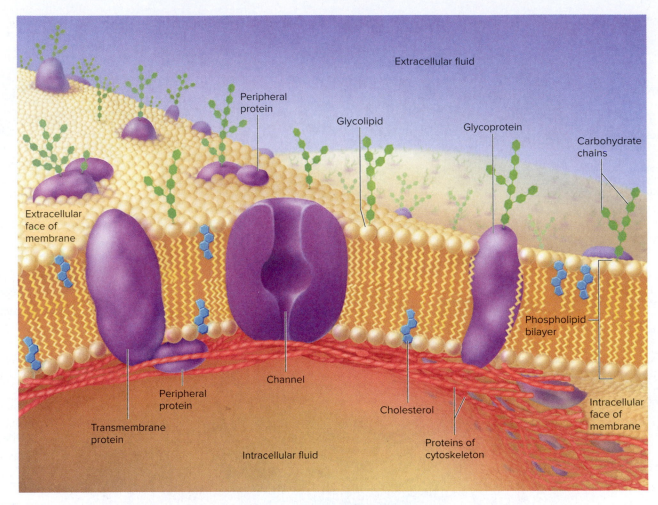

Extracellular fluid

Peripheral protein

Glycolipid

Glycoprotein

Carbohydrate chains

Extracellular face of membrane

Phospholipid bilayer

Channel

Peripheral protein

Transmembrane protein

Intracellular fluid

Cholesterol

Proteins of cytoskeleton

Intracellular face of membrane

FIGURE 4.2 Plasma Membrane.

Cytoplasm

Most of the inside of the cell is cytoplasm, which consists of the inner fluid portion of the cell, known as the cytosol; the inner framework of the cell, called the cytoskeleton; and the small, specialized units of the cell, called organelles. The cytosol is composed of water with dissolved materials, such as sugars, ions, proteins, and amino acids.

Organelles

The term *organelle* literally means "small organ." Look at illustrations of organelles in fig. 4.1. Each organelle has a particular function in the cell. There are two types of organelles—membranous and nonmembranous. Organelles represent a wonderful example of **specialization** on a microscopic scale—the function of the organelle is determined by its structure.

Mitochondria are rod-shaped organelles with a double membrane. The major function of the mitochondrion (plural *mitochondria*) is to convert the stored chemical energy in food molecules to stored chemical energy in molecules of adenosine triphosphate (ATP).

Ribosomes are the smallest of the organelles (about 25 nm in diameter) and are nonmembranous. They produce proteins.

The **endoplasmic reticulum** is an organelle composed of a network of enclosed channels. There are two types of endoplasmic reticulum: rough endoplasmic reticulum, which has attached ribosomes, and smooth endoplasmic reticulum, which does not have ribosomes on its surface. The **rough endoplasmic reticulum (RER)** is associated with the nucleus and produces proteins for transport and use outside the cell. It also makes lipids, including the phospholipids and steroids of the plasma membrane.

The **smooth endoplasmic reticulum (SER)** is a distal extension of the rough endoplasmic reticulum. It produces lipid compounds and detoxifies material.

The **Golgi** (GOAL-jee) **complex** receives material from the endoplasmic reticulum and other parts of the cytoplasm and serves as an assembly and packaging organelle. It has **cisterns,** which are flattened, membranous sacs, and it forms vesicles to transport the molecules it assembles.

The **nucleus** has two major functions: One is to house the genetic information of the cell, and the other is to direct many cellular functions. These two functions are carried out by DNA (deoxyribonucleic acid), which combines with proteins to form a material called **chromatin** in the nucleus. The nucleus is bounded by a **nuclear envelope,** which is a double membrane. The nuclear envelope contains nuclear pores, which allow the movement of materials into or out of the nucleus.

One or more structures known as the **nucleoli** (sing. *nucleolus*) are inside the nucleus. The nucleoli consist of portions of chromosomes and thus contain DNA and protein. They make rRNA, which forms ribosomes. Examine a model or chart in lab and find the nucleus, the nuclear envelope, and the nucleolus.

Vesicles are membrane-bounded sacs inside the cell that digest subcellular material, transport material out of the cell, and carry on enzymatic activities. Vesicles can transfer material from one organelle to another. They can also fuse with the plasma membrane protecting the integrity of the membrane. If a substance were to be removed from the cell by simply opening a hole in the plasma membrane, the cell would probably burst.

The vesicle fuses with the plasma membrane, ejecting the larger molecules without disrupting the plasma membrane. Two specialized vesicles in the cytoplasm are **lysosomes** and **peroxisomes.**

Lysosomes (LY-so-somz) are vesicles filled with digestive enzymes. Some engulf foreign particles and release digestive enzymes that hydrolyze the foreign particles. Damaged or old organelles are also removed by lysosomes.

Peroxisomes use the enzyme peroxidase that converts hydrogen peroxide to water and oxygen. Hydrogen peroxide is formed in cells by the metabolism of fatty acids and amino acids, as well as by the interaction of water with an unstable form of oxygen known as oxygen free radicals.

 Activity

Look at fig. 4.1 and fill in the functions of the cellular structures in table 4.1. Your text has more detailed descriptions of these structures, and you should refer to it for more information.

Other Cellular Components

Most cells have extensions on their surface, such as **microvilli, cilia,** and, in human sperm cells, **flagella. Microvilli** are small extensions of the plasma membrane of some cells that increase the

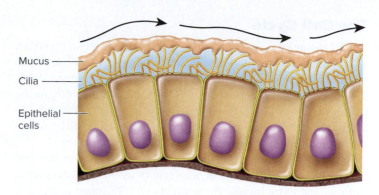

FIGURE 4.3 Cilia.

surface area of the cell. They are found in many cells, including those of the digestive and urinary systems. In some cells, such as some specialized sense organs, they serve a sensory function. Cilia (fig. 4.3) and flagella extend from the body of the cell and consist of microtubules covered by the plasma membrane. Cilia and flagella have a similar structure, but cilia are shorter than flagella and flagella have additional cytoskeletal filaments. Add cilia to your cell drawing.

Centrosomes are also unique cellular structures. They are typically found close to the nucleus and contain centrioles, which play a part in microtubule formation. They are involved in the formation of a structure known as the spindle apparatus, which is associated with cellular division.

Extracellular Matrix

So far you have examined cells and their internal structures. Humans are composed of not only cells but also nonliving material known as the **extracellular matrix (ECM).** The ECM supports cells, anchors cells, separates tissues from one another, regulates communication between cells, and assists in wound healing. The extracellular matrix consists of fibers, binding proteins, and ground substance.

TABLE 4.1	Cellular Structures and Their Functions
Structure	**Function**
Nucleus	
Mitochondrion	
Ribosome	
Rough endoplasmic reticulum	
Smooth endoplasmic reticulum	
Golgi complex	
Lysosome	
Peroxisome	

The Cell Cycle

One of the great wonders of science is the mechanism by which a single cell, the result of the fusion of egg and sperm, develops into a complex, multicellular organism, such as a human. Various estimates put the number of cells in the human body in the trillions. All of these cells came from the first cell, or **zygote.** In this part of the lab exercise we examine the mechanism by which this duplication of cells occurs.

Cells produce more cells by a process known as the cell cycle. The cell cycle can be divided into three events: **interphase, mitosis,** and **cytokinesis.** These events are illustrated in fig. 4.4. An average cell spends most of its time in interphase.

Interphase

Interphase is the time when a cell undergoes growth and duplication of DNA in preparation for the next cell division. If a cell is not actively dividing, then interphase is regarded as the time when a cell carries out normal cellular function.

Interphase has three separate phases known as the G_1 phase, S phase, and G_2 phase. In the G_1 phase (*G* stands for *gap*), cells are growing in size and producing organelles. In the S phase (*S* stands for *synthesis*), the DNA of the cell is duplicated. The double helix of the DNA molecule unzips and two new, identical DNA molecules are produced. In the final phase of interphase, the G_2 phase, the cell continues to grow and prepares for the process of mitosis. Some cells, such as muscle cells and brain cells, do not undergo further division and are said to be in the G_0 (G zero) phase. Cells in interphase have a distinct nuclear envelope, and the genetic information is dispersed in the nucleus as chromatin.

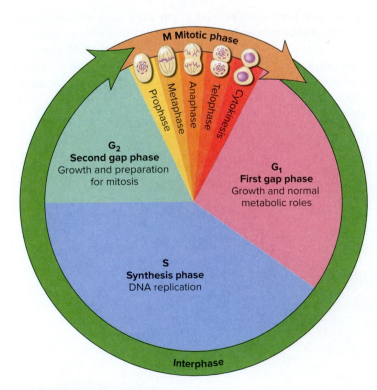

FIGURE 4.4 The Cell Cycle.

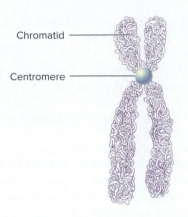

FIGURE 4.5 Structure of an Isolated Chromosome.

Mitosis

Mitosis (my-TOE-sis) is a continuous event divided into four distinct phases. Mitosis is nuclear division, and it involves the division of genetic information to produce two identical nuclei. In order for mitosis to occur, the chromatin in the nucleus of the cell must condense into compact units called **chromosomes.** A chromosome consists of two **chromatids** held at the center by a **centromere.** Examine fig. 4.5 for the structure of a chromosome. You should also note the structure of chromosomes as you study the cells undergoing mitosis. The four phases of mitosis— **prophase, metaphase, anaphase,** and **telophase**—are described next. Refer to fig. 4.6 as you read the descriptions.

Prophase

The first indication that a cell is undergoing mitosis is the condensation of chromatin into chromosomes. In addition to the thickening of the chromosomes, the nucleolus disappears and the nuclear envelope begins to disassemble. In order for the chromosomes to separate and move away from each other, the nuclear envelope, which normally forms a barrier, must not be present. During prophase, the mitotic apparatus becomes apparent.

The **mitotic apparatus** consists of two **asters,** which are points of radiating fibers at each end (pole) of the cell, **centrioles** in the middle of the asters, and **spindle fibers,** which attach to chromosomes. The spindle fibers, which are **microtubules,** attach to the chromosomes at the **centromere.**

Metaphase

In this phase the chromosomes align between the poles of the cell in a region known as the **metaphase plate.**

Anaphase

In anaphase, the chromatids separate at the centromere, and each chromatid is now known as a **daughter chromosome.** The spindle fibers attach to the region of the centromere known as the kinetochore and pull the daughter chromosomes toward opposite poles of the cell. The centromere regions of the chromosomes move first, and the arms of the chromosomes follow.

Telophase

Once the daughter chromosomes reach the poles, telophase (TEE-lo-faze) begins. The daughter chromosomes begin to unwind into chromatin, the nucleolus reappears, and the nuclear envelope begins to re-form. The mitotic apparatus disassembles, thus terminating mitosis.

Cytokinesis

The splitting of the cell's cytoplasm into two parts is known as **cytokinesis** (SY-toe-kih-NEE-sis). Although cytokinesis is a distinct process, it frequently begins during late anaphase or early telophase. In late anaphase, as the chromosomes are moving to the

Drawing of Cells in Phases of Mitosis

poles, the plasma membrane begins to constrict at a region known as the **cleavage furrow** (fig. 4.6e). This begins the process of dividing the cytoplasm, ending as the cell splits into two daughter cells. The cytoplasm and the organelles are effectively divided into two parts.

Ⓐ Activity

Examine a slide of whitefish blastula and look for the various phases of the cell cycle in those cells. Most of the cells you see are in a particular part of the cell cycle. What is this phase, and why are most of the cells in this phase?

Compare the slide with fig. 4.6. Draw representative cells in each phase of mitosis in the following space.

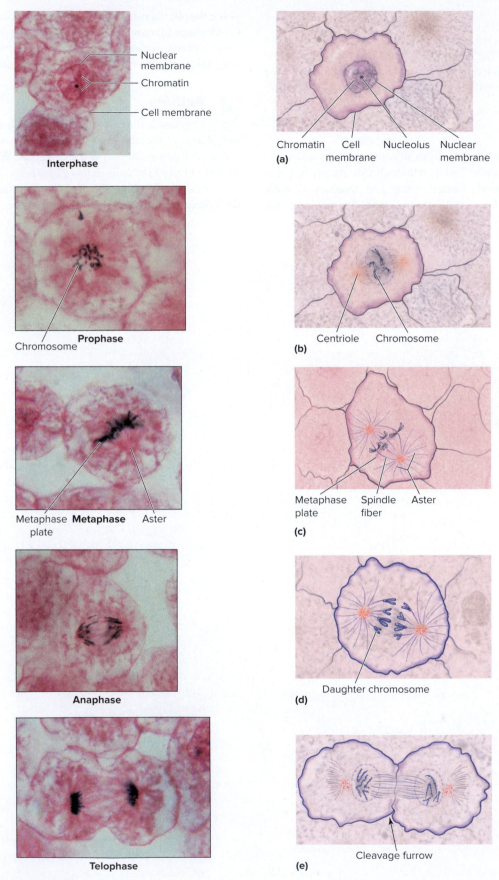

Interphase

Chromosome — **Prophase**

Metaphase **Metaphase** Aster
plate

Anaphase

Telophase

Nuclear
membrane
Chromatin
Cell membrane

Chromatin   Cell   Nucleolus   Nuclear
         membrane   membrane
(a)

Centriole   Chromosome
(b)

Metaphase   Spindle   Aster
plate   fiber
(c)

Daughter chromosome
(d)

Cleavage furrow
(e)

FIGURE 4.6 **The Cell Cycle (1,000×).** (a) Interphase, mitosis; (b) prophase; (c) metaphase; (d) anaphase; (e) telophase with cytokinesis.
(left a–e) ©Eric Wise

TABLE 4.2	Major Events of Mitosis
Prophase	Chromatin condenses to form visible chromosomes.
	Nuclear envelope disappears.
	Spindle apparatus forms.
	Nucleolus disappears.
Metaphase	Chromosomes align on the metaphase plate.
Anaphase	Chromosomes split and daughter chromosomes migrate to poles; cytokinesis often begins.
Telophase	Chromosomes reach poles; nuclear envelope re-forms.
	Chromosomes unwind to chromatin; cytokinesis divides the cytoplasm.
	Nucleolus reappears.

(A) Activity

Review the phases of mitosis in table 4.2. Take two colors of modeling clay and make chromosomes from the clay. You should have a long chromosome and a short chromosome of each color, for a total of four chromosomes. Each chromosome should have two chromatids, and the chromosomes should be joined by a marble, which represents the centromere. Draw a large circle on a sheet of paper to represent a cell. Manipulate the clay chromosomes to show how mitosis occurs. After you have done this, describe the process of mitosis on that page. List each stage and what happens in that stage.

Cellular Function

This portion of the lab is most efficiently done if the timing of the experiments overlaps. While you are waiting for the results of one experiment, begin another.

Brownian Motion

All matter has **kinetic energy** unless it is at absolute zero (–273°C). Kinetic energy is the energy of motion, and it is the driving force of the movement of atoms and molecules. Atoms and molecules are too small to be seen, even with the use of a light microscope, yet their movement can be inferred as they hit large particles that are visible under the microscope. Items such as dust or ink particles can be seen vibrating or jiggling, and we interpret this as the collective collisions of many molecules striking a larger structure and causing it to move. This is named **Brownian motion,** after the botanist Robert Brown, who first described this in the nineteenth century.

(A) Activity

You will observe Brownian motion by examining ink particles in the following activity:

1. Place a drop of India ink on a clean glass microscope slide.
2. Add a drop of water to the slide.
3. Carefully set a coverslip on the slide.

4. Examine the slide under high power and describe the movement of the ink particles in the space provided.

? Nature of the movement of ink particles: _____ (1)

5. Remove the slide from the microscope and place it on a warm surface (such as a hot plate on the low-temperature setting) for a few seconds until the slide becomes warm. If you leave it on the hot plate for too long, the water will evaporate and you will see no motion at all.
6. Quickly return the slide to the microscope and note the change in speed of the particles, compared with the initial observation.
7. Describe any difference in the space provided.

? Speed of movement or particles: _____ (2)

? How might kinetic energy play a part in the differences between the first observation and the second observation?

_____

_____ (3)

Diffusion

Kinetic energy moves particles in liquid solution or in a gas. Particles in a high concentration in one part of the solution are struck by chance collisions with other molecules, and those particles would begin to disperse. This process is known as **diffusion,** which is defined as the movement of particles from regions of high concentration to regions of low concentration. The difference between the two concentrations is known as the **concentration gradient.** Molecules move down the concentration gradient. If the particles become uniformly dispersed, then the system has reached **equilibrium.** An example of the essential nature of diffusion is the movement of oxygen into the blood vessels of the lungs. Oxygen in the air is at a higher concentration than in the blood of the lung capillaries; consequently, oxygen moves from the air to the blood.

Many liquid solutions consist of a liquid portion, the solvent, and the dissolved portion, the solutes. If the solutes are concentrated in one area, they will diffuse through the solvent. There are many solvents, but water is a vital solvent for the body because it is the most abundant solvent and frequently contains sugars, ions, and amino acids.

(A) Activity

You can demonstrate diffusion by placing a crystal of potassium permanganate in a 100 mL beaker of water. Do this as a group at each table.

1. Fill a 100 mL beaker almost to the top with tap water.
2. Place the beaker on your table and drop a small crystal of potassium permanganate into it.
3. Leave the beaker undisturbed, but note the changes that occur during 1 hour.

4. Record your observations in the place provided. While you are waiting, continue with the other experiments.

? What process drives the movement of potassium permanganate particles?

_____

_____ (4)

Many factors can affect the rate of diffusion, including changes in temperature, changes in concentration of the solute, the size or weight of the solute particles, and interactions between the solute and the solvent. In the following two experiments, you will examine the effects of the weight of the particle and of temperature on diffusion rate.

Diffusion Rates and Particle Weight

Caution! Dyes such as potassium permanganate and methylene blue stain clothes, skin, and lab notebooks.

(A) **Activity**
Unless your instructor states otherwise, do this experiment as a group of three or four students. Agar, a gel, can be used as a medium in which to measure diffusion rates of materials. Agar consists of water and dissolved algal polysaccharides. The liquid nature of agar is such that diffusion occurs in the gel at a slow rate.

1. Using a plastic drinking straw and a petri dish filled with agar, gently make two stabs into the dish opposite one another (fig. 4.7). Do not twist the straw, or you

FIGURE 4.7 Placement of the Wells in Agar.
©Eric Wise

? **TABLE 4.3**   **Diameter of Diffusion (in mm)**

Fill in the data below and in the Chapter Summary Data section at the end of the exercises.

	20 min	40 min	60 min	80 min
Potassium permanganate	____	____	____	____ (5a)
Methylene blue	____	____	____	____ (5b)

risk breaking the agar, leaving a crack into which fluid will run.
2. Remove the small plug of agar with a fine probe if needed so that a well is left in the agar.
3. Into one of the wells, place three drops of 0.01 M potassium permanganate solution (molecular weight 158).
4. Into the other well, place an equal amount of 0.01 M methylene blue solution (molecular weight 320).
5. Potassium permanganate is a purple solution; methylene blue is blue.
6. Leave the petri dish on your desk and make observations of the diffusion rate every 20 minutes for 80 minutes.
7. Record the diameter of the diffusion in table 4.3 and continue with the following experiments.

Effects of Temperature on Diffusion Rates

1. Prepare two more petri dishes in the same way, except this time take one of the petri dishes from a refrigerator and another from a warming tray (such as an electric warming tray on the lowest setting or an incubator set at 37°C).
2. Make two wells in the cold petri dish and place three drops of the respective dyes, as you did previously.
3. This time, however, place the cold petri dish in a dishpan filled with crushed ice and examine after 80 minutes. You only need to record the diameter of diffusion at the end of the 80 minutes.
4. Remove a petri dish from the warming tray or incubator, make two wells in the dish, fill it with appropriate solutions, and return it to the warming tray or incubator. Examine after 80 minutes.
5. Record your results in table 4.4.

? **TABLE 4.4**   **Diameter of Diffusion (in mm)**

Fill in the data below and in the Chapter Summary Data section at the end of the exercises.

	80 min (cold)	80 min (warm)
Potassium permanganate	_____	_____ (6a)
Methylene blue	_____	_____ (6b)

? Compare the diffusion rates of the cold and warm temperatures with the dish left at room temperature for 80 minutes. How does an increase or a decrease in temperature affect the diffusion rate?

_____

_____ (7)

? What can you say about temperature and the kinetic energy of the system?

_____

_____ (8)

Osmosis

In the preceding diffusion experiments there were no barriers to the movement of the particles. Plasma membranes are barriers to certain molecules, while they allow other molecules to pass through. This type of membrane is called a **selectively permeable membrane.** Water, some alcohols, oxygen, and carbon dioxide move easily across the plasma membrane, while larger molecules, such as proteins, or charged particles (ions) are prevented from crossing the membrane. Ions do move across the membrane proper. They typically travel through gates or channels.

The movement of water across a selectively permeable membrane from solutions of higher water concentration (water with less solutes) to lower water concentration (water with more solutes) is known as **osmosis.** Osmosis is a type of diffusion, and the process can be viewed from the perspective of the solvent or the perspective of the solute.

The Solvent Perspective of Osmosis

In diffusion, material moves from higher concentrations to lower concentrations. The same can be seen with osmosis except that you measure the movement of water instead of the movement of solutes. If you have two solutions separated by a selectively permeable membrane, a 10% sugar solution (for example, 90% water) and a 5% sugar solution (for example, 95% water), then the water will move from higher water concentration (95% water) to lower water concentration (90% water). The greater the difference between the two solutions, the greater the concentration gradient, which in this case is known as the **osmotic potential.**

The Solute Perspective of Osmosis

Even though the movement of water is measured in osmosis, it is the concentration of solute particles that influences that movement. A 10% sugar solution has more solutes than a solution of 5% sugar. The 10% sugar solution is said to be **hypertonic** to the 5% sugar solution. The 5% sugar solution is said to be **hypotonic** to the 10% sugar solution. If these two solutions are separated by a selectively permeable membrane, water flows *from* the hypotonic solution *to* the hypertonic solution. If enough water flows across the membrane and the two solutions reach the same concentration of sugar, **equilibrium** is established and the net movement of water stops. If solutions have the same concentration of solutes, they are said to be **isotonic** to one another.

Demonstration of Osmosis

Observe osmosis with an **osmometer.** The long tube (usually made of glass or plastic) is filled by using a funnel with dark corn syrup or sugar solution, that flows into a dialysis bag. Sugar does not cross dialysis membranes, but water does, so the dialysis membrane is a selectively permeable membrane. The osmometer is placed in a beaker of water and clamped to a ring stand (fig. 4.8). The level of the corn syrup or sugar solution is indicated with a mark from a permanent marker.

Examine the setup throughout the lab period. You may find that the liquid in the tube eventually stops rising. This occurs when the gravitational pressure equals the force exerted by the process of osmosis. The amount of force required to balance, or equilibrate, osmosis is the **osmotic pressure.**

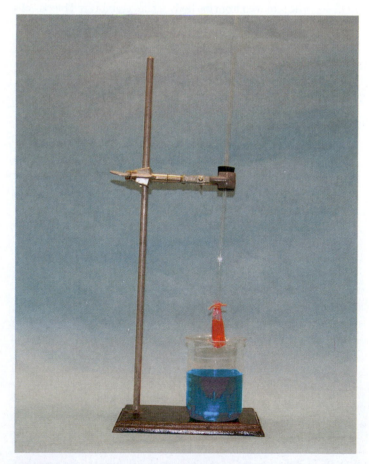

FIGURE 4.8 Osmosis Demonstration. Setup prior to osmometer being lowered into the solution.
©Eric Wise

Examine an osmometer (or prepare one, if your lab instructor indicates for you to do so) that contains a starch solution.

? Does the liquid move up the tube? Why or why not?

_____

_____ (9)

? What does this say about the osmotic activity of starch?

_____

_____ (10)

Osmosis and the Concentration Gradient

(A) Activity

The osmotic potential varies, depending on the concentration gradient between the two solutions, and this affects the rate of osmosis. In this experiment, you determine the effects of various concentrations of sucrose solutions on the rate of osmosis by studying the change in weight of bags made from dialysis tubing. This dialysis tubing is normally used to help kidney patients as it allows waste products to be removed from the body while retaining red blood cells.

1. Take four 200 mL beakers and label each one with a wax pencil or permanent marker. The beakers should be labeled "0%," "5%," "15%," and "30%."
2. Fill each beaker about 2/3 full with the appropriate sugar solution.
3. Tie one end of each of the dialysis tubes with string to make an open-ended bag. The dialysis tubes should be soaked in water for a few minutes prior to tying them.
4. Fill four dialysis bags with 10 mL of a 15% sucrose solution. You can do this by placing a pipette pump (or bulb) on the end of a 10 mL pipette, drawing liquid to the 10 mL mark, and filling the dialysis bag. Remove as much air as you can from the bag prior to filling it with sucrose solution and then tie or clamp the open end of the bags (fig. 4.9).
5. Rinse each bag with distilled water and blot the bag gently with a towel.
6. Weigh each bag to the nearest tenth of a gram and record the weights in table 4.5. This weight is the initial weight of the bag.
7. Place bag 1 in a beaker with the pure water (0% sugar solution).
8. Place bag 2 in a 5% sugar solution.
9. Place bag 3 in a 15% sugar solution.
10. Place bag 4 in a 30% sugar solution.
11. Make sure all the bags are covered with their respective sugar solutions and leave them there for 20 minutes.
12. After 20 minutes, remove the bags from the beakers and blot and weigh each bag. Remember to place each bag back in its proper solution!
13. Record the weight of the bags each 20 minutes for a total of 80 minutes in table 4.6.

Calculate the **change of weight** of each bag from the initial weight for each of the time periods. For example, let's assume a

(a)

(b)

FIGURE 4.9 Dialysis Bags. (a) Filling; (b) tying.
©Eric Wise

? TABLE 4.5 | Initial Weight of Dialysis Bags

Fill in the data below and in the Chapter Summary Data section at the end of the exercises.

Bag 1 _____ grams (11a)

Bag 2 _____ grams (11b)

Bag 3 _____ grams (11c)

Bag 4 _____ grams (11d)

? TABLE 4.6 | Weight of Bags (in g) for Each Time Period

Fill in the data below and in the Chapter Summary Data section at the end of the exercises.

	Time			
Bag	20 min	40 min	60 min	80 min
1	_____ (12a)	_____ (12e)	_____ (12i)	_____ (12m)
2	_____ (12b)	_____ (12f)	_____ (12j)	_____ (12n)
3	_____ (12c)	_____ (12g)	_____ (12k)	_____ (12o)
4	_____ (12d)	_____ (12h)	_____ (12l)	_____ (12p)

bag weighed 20.5 grams as the initial weight and the recorded weights are as follows:

20 min	40 min	60 min	80 min
23.5 g	24.2 g	25.0 g	25.6 g

The change in weight is as follows:

3.0 g	3.7 g	4.5 g	5.1 g

Graph the change in weight for each bag of your experiment in chart 1 and connect the points with a line. Indicate which line represents which solution.

Change in Weight with Time | Chart 1

Which of the bags (if any) gained weight? _____ (13)

Which of the bags (if any) lost weight? _____ (14)

Determine the osmotic relationship (hypertonic, hypotonic, isotonic) of the beaker to the solution in the bag.

? Bag 1: The beaker solution is _____ to the bag solution. (15)

? Bag 2: The beaker solution is _____ to the bag solution. (16)

? Bag 3: The beaker solution is _____ to the bag solution. (17)

? Bag 4: The beaker solution is _____ to the bag solution. (18)

? Does the change in weight correlate to what you have learned about osmosis? If so, how does it correlate?

_____

_____

_____ (19)

Osmosis and Living Cells

It is important that cells are bathed in isotonic solutions. This experiment demonstrates that importance by placing blood cells in three different solutions and observing the changes in the cells. The plasma membrane of blood cells is a selectively permeable membrane. You should compare all three slides and only discard them once you have seen a difference between the slides.

Caution! Make sure you wear protective gloves while conducting this experiment to avoid any potential transmission of disease.

A Activity

1. Mark each slide with a permanent marker. Slide 1 should read "0.9%." Slide two should be labeled "5%." Slide three should be "0%."
2. Place a drop of fresh mammal blood on slide 1.
3. To this slide add a drop of physiological saline (0.9% sodium chloride) and place a coverslip on the slide.
4. Observe the cells on high power under the microscope and note their shape.

5. Describe this in the space provided.

❓ When red blood cells lose water, they shrivel in a process known as **crenation.** Did any of the cells show crenation?

_____ (21)

❓ Which solution might produce this? _____ (22)

When water moves into a cell at a rapid rate, the cell becomes inflated and sometimes bursts in a process known as **hemolysis.**

❓ Did any of the cells undergo hemolysis?_____ (23)

❓ Which solution might produce this effect?_____ (24)

Examine fig. 4.10 for the various effects of solutions on red blood cells.
Note: In clinical settings, 0.9% saline and 5% dextrose in water (D5W) are isotonic to human red blood cells.

6. Place another drop of blood on slide 2.
7. To this slide add a drop of 5% sodium chloride (NaCl) solution.
8. Place a coverslip on the slide and record your observations.

_____

_____

9. Examine the slide for at least a few minutes or until a change of shape becomes obvious. Describe the shape of the cells in 5% NaCl.

CASE STUDY—WATER BALANCE

Electrolyte imbalance can lead to headache, confusion, and even death. In some cases, participating in a water drinking contest or drinking large amounts of pure water after intensive exercise causes such an imbalance. Water absorbed in the blood increases the flow of water to the brain cells, putting pressure on vital centers, causing them to malfunction.

A 28-year-old female who is in the final stages of running a marathon on a hot day has regularly consumed water along the path. She begins to have a severe headache and shows symptoms of being confused. She has been sweating profusely, losing electrolytes and, to compensate, has been drinking pure water. Discuss with your lab partner the osmotic nature of what is happening in relationship to the movement of water in her blood and brain cells and what should be done to correct the problem.

For slide 3 add a drop of blood and a drop of distilled water. Immediately observe this slide and then continue to look at it for a few minutes. Record your observations.

❓ Describe the shape of cells in distilled water. _____

_____ (20)

Virtual Lab—🧪PhILS

Open the Ph.I.L.S. program (Ph.I.L.S. 4.0) and select the first simulation, "Osmosis and Diffusion 1. Varying Extracellular Concentration." Read the Objectives and Introduction and take the pre-lab quiz to make sure you understand the principles of the exercise. Use the scroll bar on the right-hand side of the screen to control the text. Once you have finished reading the lab, either click "Continue" or click "PRE-LAB QUIZ" at the top of the screen. Once you have finished the pre-lab portion, you can select the "WET LAB" to run the experiment. In the wet lab experiment, the use of transmitted light in a spectrophotometer is the way that

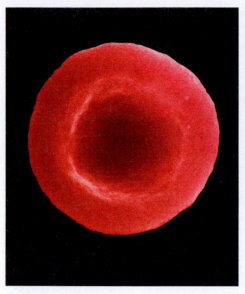
(a) In isotonic solution

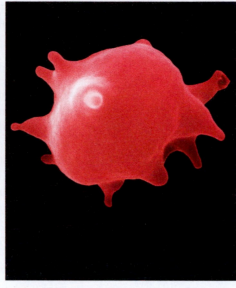
(b) In hypertonic solution

(c) In hypotonic solution

FIGURE 4.10 Stages of Red Blood Cells. (a) Normal; (b) crenated; (c) inflated.
David M. Phillips/Science Source

osmotic relationships are determined between the red blood cell and the solution surrounding the red blood cell. The more light that is blocked, the less the transmission of light. If red blood cells burst, then the transmission of light increases because there are fewer cells absorbing the light.

To begin the wet lab, read and follow the instructions. When you click the power switch on the spectrophotometer, it will turn green. You can then set the spectrophotometer setting by clicking the up or down arrows on the spectrophotometer. In order to load the pipette, you must click and hold the arrow on the pipette. Once it is filled, you can click and drag the pipette to the appropriate test tubes. Once you do this three times, the rest of the tubes will automatically fill.

You must set the spectrophotometer to zero. Click on the "zero" button of the spectrophotometer (to the left of the calibrate term) to open the chamber. Click and drag the zero test tube to the holder and use the up or down arrows to set the calibration to the value of 100.

You must double-click the test tube holder to remove the tube. The values are automatically entered into the journal. Once you have sampled all the tubes, click on "POST-LAB QUIZ AND LAB REPORT." Take the quiz. Follow your instructor's directions for printing the results (if that is what he or she wants) and analyze your results.

Between what values of NaCl concentration is there a drop in transmittance? _____

How does this relate to hemolysis of red blood cells?

_____

Filtration

The process of **filtration** is important in certain cells of the body and results as the pressure of a fluid forces particles through a filtering membrane. Filtration is a major component of kidney function. Hydrostatic pressure from the blood forces urea, ions, sugars, and other materials from the blood through the holes in the membranes and spaces between the cells of the kidney. In this way, small particles in the blood are forced into kidney tubules, while larger particles, such as proteins, remain in the blood. In this exercise, you learn the basic principles of filtration, such as the **selectivity** of the filtration membrane and the **filtration rate** of a system.

Ⓐ Activity

1. Fold a piece of filter paper in half and then fold it in half again so it forms a cone (fig. 4.11).
2. Place the cone in a funnel mounted on a ring stand over a beaker.
3. Into this cone filter add the filtration solution, a mixture of blue food coloring, powdered charcoal, and starch in water.
4. Fill the funnel to near the top of the filter paper cone and let the filtrate (the material passing through the filter) collect in the beaker.
5. When the funnel is approximately half full, place a 10 mL graduated cylinder under it and record the time it takes to fill the cylinder to the 2 mL mark.

❓ Number of seconds to produce 2 mL: _____ (25)

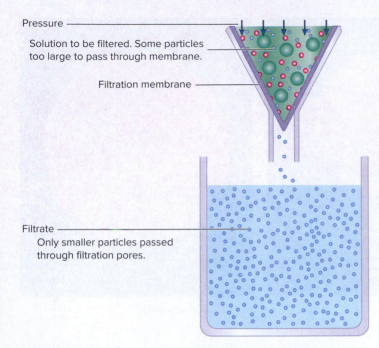

Pressure

Solution to be filtered. Some particles too large to pass through membrane.

Filtration membrane

Filtrate

Only smaller particles passed through filtration pores.

FIGURE 4.11 Filtration. A filtration membrane has pores that larger particles cannot pass through. The filter paper inside your funnel allows some particles to pass through it, while preventing others.

6. Calculate the filtration rate expressed as mL/minute.
7. Record this value.

❓ Filtration rate: _____ min/mL (26)

You should be able to determine what material passes through the membrane by the following method. Remove the funnel from the beaker. If the solution in the beaker has a blue cast to it, then blue food coloring passed through the membrane. If black particles are found in the filtrate, then charcoal passed through the membrane. If you add a few drops of iodine to the beaker and the solution turns black, then starch passed through the membrane. Record what material passed through the membrane in the space provided.

❓ Substance	Yes	No	
Blue food coloring	_____	_____	(27a)
Activated charcoal	_____	_____	(27b)
Uncooked starch	_____	_____	(27c)

The force that drives filtration in the funnel is the force of gravity on the liquid. In the kidney, the force that drives filtration is blood pressure.

REVIEW SECTION

Cell Structure and Function

Name _____   Date _____

Lab Section _____   Time _____

❓ Chapter Summary Data

Use this section to record your results from questions within the exercise.

1. _____   10. _____

2. _____   _____

3. _____   11a. _____

4. _____   11b. _____

_____   11c. _____

5a. _____   11d. _____

5b. _____   12a. _____

6a. _____   12b. _____

6b. _____   12c. _____

7. _____   12d. _____

_____   12e. _____

8. _____   12f. _____

_____   12g. _____

9. _____   12h. _____

_____   12i. _____

12j. _____   18. _____

12k. _____   19. _____

12l. _____   20. _____

12m. _____   21. _____

12n. _____   22. _____

12o. _____   23. _____

12p. _____   24. _____

13. _____   25. _____

14. _____   26. _____

15. _____   27a. _____

16. _____   27b. _____

17. _____   27c. _____

Review Questions

1. Cells in the body have a fluid surrounding them. What is the name of this fluid? _____

2. The cytoplasm has a liquid portion. What is it called? _____

3. Name the major parts of the cell. _____

4. What structure in a cell is composed mostly of a phospholipid bilayer? _____

5. Which organelle is responsible for ATP production? _____

6. Which organelle makes protein for use outside the cell? _____

7. Which organelles in the cell produce lipids? _____

8. Which organelle directs the activity of the cell and contains most of the genetic information of the cell? _____
_____

9. What cellular structure is responsible for ribosome production? _____
_____

10. What are the three stages of the cell cycle? _____
_____
_____

11. Of the three stages of the cell cycle, in which one is DNA duplicated? _____

12. In what part of the cell cycle do chromosomes first split apart? _____

13. What part of the cell cycle involves the division of the cytoplasm? _____

14. Describe the four phases of nuclear (mitotic) division and what occurs during those phases. _____

_____

_____

_____

15. Label the illustration using the terms provided.

   1. Centriole   4. Nucleolus   7. Rough endoplasmic reticulum
   2. Golgi complex   5. Plasma membrane   8. Smooth endoplasmic reticulum
   3. Mitochondrion   6. Ribosomes

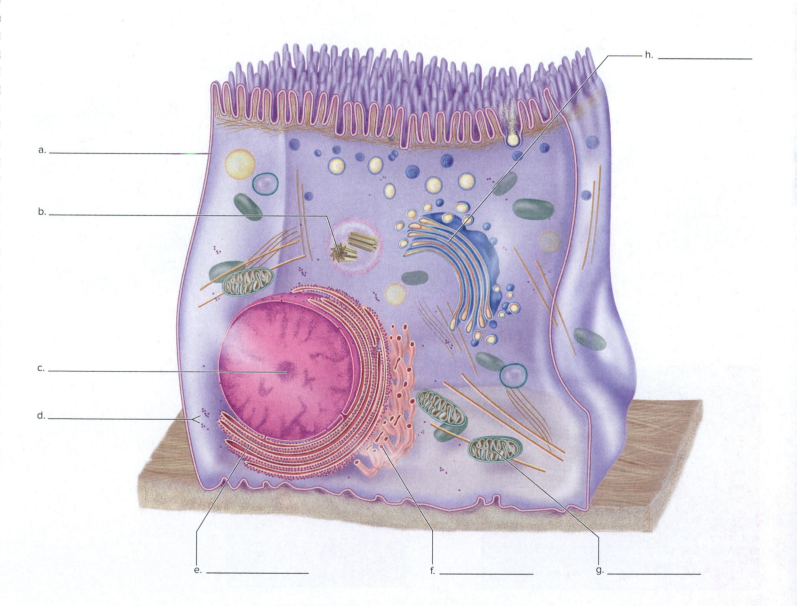

16. Name the phases of interphase and describe what happens during those phases. _____

_____

_____

17. Particles that are visible in the light microscope and move erratically by molecular collision represent what process?

_____

18. Particles in a gas or liquid moving from a region of higher concentration to a region of lower concentration represent what

process? _____

19. Name the phases of the cell cycle, as illustrated. Use the terms provided.

Anaphase   Prophase

Interphase   Telophase

Metaphase

(a) _____

(b) _____

(c) _____

(d) _____

(e) _____

(a–e) ©Eric Wise

20. What is a concentration gradient? _____

_____

21. Define *solute*. _____

_____

22. When particles from solutions of different concentrations are evenly dispersed or distributed in solution, what condition is the solution said to be in? _____

_____

23. What process drives Brownian motion? _____

24. Based on your experimentation with Brownian motion, what do you predict would happen to the movement of the particles you examined if you cooled the slide? _____

Why? _____

_____

25. What particle diffused the farthest, potassium permanganate or methylene blue? _____

What was the factor in the difference in diffusion rates? _____

_____

26. Placing the cold petri dish in a refrigerator may not produce the same results as placing the dish on ice. Refrigerators vibrate some during the cooling cycles. What physical phenomenon might affect the diffusion rate if this is the case?_____

_____

27. Which osmosis bag in your experiment gained the most weight? _____

28. Was the bag that gained the most weight hypertonic, hypotonic, or isotonic to the solution in the beaker?

_____

29. Which bag lost weight in your experiment? _____

What caused the weight loss? _____

30. Why should you not inject a patient with a 10% saline solution? _____

_____ What osmotic effect would that cause? _____

_____

31. What process occurs as water moves from regions of higher water concentration to regions of lower water concentration across a selectively permeable membrane? _____

_____

32. A solution containing less solute is (hypertonic/hypotonic/isotonic) to a solution containing more solute. _____

_____

33. A solution with 5% sugar is (isotonic/hypertonic/hypotonic) to a 3% sugar solution. _____

_____

34. If the two solutions in question 33 were separated by a selectively permeable membrane, which solution would lose water?

_____

35. The pressure needed to stop osmosis is called the _____.

36. Define *hemolysis*. _____

_____

37. Is selectivity needed for filtration to occur? Why or why not? _____

_____

38. What would happen to the filtration rate if you applied pressure to the filtration system? _____

_____

What might happen to the filtration membrane if the pressure is too high?_____

_____

Why might this be of concern to people who have both kidney disease and high blood pressure? _____

_____

Tissues

INTRODUCTION

The study of tissues, called **histology,** is microscopic anatomy. Individual tissues consist of cells and extracellular material that have a particular function. Organs of the body are formed by two or more tissues. The study of histology is important because many diseases of the human body are diagnosed at the tissue level. Surgical specimens are routinely sent to pathology labs so that accurate assessment of the health of the tissue, and consequently the health of the individual, can be made.

Stem cells are unspecialized, or undifferentiated, cells that have great potential for regeneration. Stem cells may divide to produce more stem cells, or they may **differentiate** to become other cells of the body, such as those in neural tissue or blood.

There are four main tissue types found in the human body: **epithelial tissue, connective tissue, muscular tissue,** and **nervous tissue.** These topics are discussed in chapter 5, "The Human Tissues," in the Saladin text. In this exercise you examine numerous slides of organ tissue and begin an introduction to histology. In later exercises you revisit histology as you examine various organ systems.

OBJECTIVES

At the end of this exercise, you should be able to

1. examine a slide under the microscope or a picture of a tissue and name the cell type or specific tissue represented;
2. associate a particular tissue type with an organ, such as kidney or bone and recall that many tissues exist together to form an organ;
3. list the three parts of a neuron;
4. describe the muscle cell types according to location and structure.

MATERIALS

Microscope

Colored pencils

Epithelial Tissue Slides

Simple squamous epithelium

Simple cuboidal epithelium

Simple columnar epithelium

Pseudostratified ciliated columnar epithelium

Stratified squamous epithelium

Urothelium (transitional epithelium)

Muscular Tissue Slides

Skeletal muscle

Cardiac muscle

Smooth muscle

All three muscle types

Nervous Tissue Slides

Spinal cord smear

Connective Tissue Slides

Dense regular connective tissue

Dense irregular connective tissue

Elastic tissue

Reticular tissue

Areolar tissue

Adipose tissue

Ground bone

Hyaline cartilage

Fibrocartilage

Elastic cartilage

Blood

PROCEDURE

Before you begin this exercise, you should be thoroughly familiar with the microscopes in your lab. If you are not familiar with the use of the microscope, review Exercise 3. The **extracellular material** in tissues is known as **matrix.** Tissues are stained so that the details become visible. The most common stain is hematoxylin and eosin (H&E) stain. Hematoxylin stains the nucleus purple, and eosin colors the cytoplasm pink.

Epithelial Tissue

Epithelial tissue is mostly composed of cells, with little matrix. It covers or lines surfaces of the body (such as the surface of the skin and on the inside of the digestive tract) and is found in glandular tissue, such as the sweat glands and the pancreas. In most cases, epithelial tissue adheres to the underlying layers by way of a **basement membrane,** a noncellular adhesive layer. In the skin, the epidermis is made of epithelial tissue and the basement membrane connects the epidermis to the underlying dermis. Epithelial tissue is classified according to the shape of the cells and the number of layers present. The cell shapes are **squamous** (flattened), **cuboidal,** and **columnar.** The number of layers is **simple** (cells in a single layer) or **stratified** (cells stacked in more than one layer). Examine tables 5.1 and 5.2 for an overview of epithelium. Epithelium is listed by the tissue type in the following discussion. The edge of the cell touching the basement membrane is the **basal surface,** and the upper edge is the **apical surface.**

TABLE 5.2	Stratified Epithelia
Stratified Squamous Epithelium	
Microscopic appearance: many layers, cells cuboidal but flatten toward surface	
Significant locations: epidermis, oral cavity, esophagus, and vagina	
Function: resists abrasion, prevents microbial infection, retards water loss in skin	
Urothelium	
Microscopic appearance: many layers with teardrop-shaped cells that do not flatten toward surface	
Significant locations: urinary bladder	
Function: allows stretching of urinary bladder	

TABLE 5.1	Simple Epithelia
Simple Squamous Epithelium	
Microscopic appearance: single layer of flat cells	
Significant locations: lungs, inside of heart and blood vessels	
Function: diffusion and smooth lining, secretion of serous fluid	
Simple Cuboidal Epithelium	
Microscopic appearance: small cubes or wedge-shaped cells in single layer	
Significant locations: kidney tubules and liver	
Function: absorption and secretion	
Simple Columnar Epithelium	
Microscopic appearance: tall cells in one layer with nuclei typically in basal part of cell	
Significant locations: from stomach through intestines, uterine tube	
Function: absorption and secretion	
Pseudostratified Columnar Epithelium	
Microscopic appearance: looks stratified, but all cells arise from basement membrane, often ciliated	
Significant locations: respiratory passages	
Function: secretes mucus and traps dust particles, moving them away from lung	

STUDY HINT

As you examine various tissues, look for distinguishing features that will identify the tissue. It is a good idea to examine more than one slide of a particular tissue so that you see a range of samples of that tissue. Frequently, the material you see in the lab is from a slice of an organ and as such includes more than one tissue type. For example, a sample of cartilage taken from the trachea contains epithelial tissue, adipose tissue, and other connective tissues in addition to the cartilage you want to study. If you are looking for smooth muscle from the digestive tract, you may also find both epithelial tissue and connective tissue in the slide. Use the figures in this exercise to help you locate tissues on the prepared slides. Examine each slide by holding the slide up to the light and visually locating the sample. Then put the slide on the microscope and examine it on low power, scanning around the slide. Move to progressively higher powers after you have identified the tissue. If you cannot identify the tissue after some searching, ask your lab partner or your instructor for help. When you look at epithelial tissue under the microscope, examine the edge of the sample because epithelial tissue is frequently found as a lining.

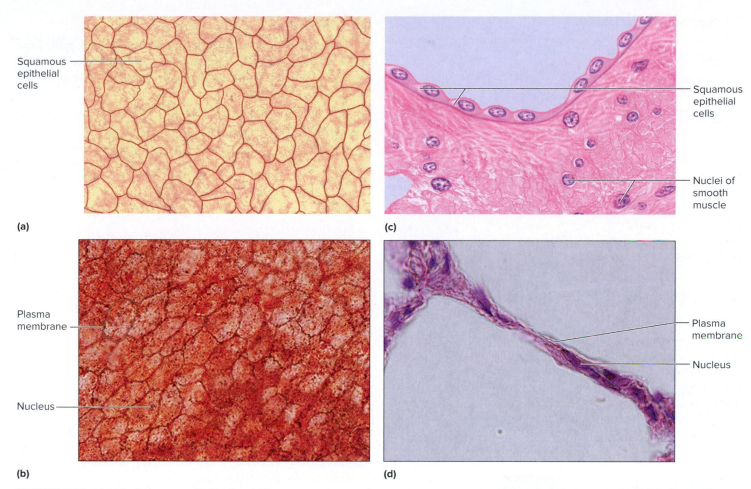

Squamous epithelial cells

(a)

Plasma membrane

Nucleus

(b)

Squamous epithelial cells

Nuclei of smooth muscle

(c)

Plasma membrane

Nucleus

(d)

FIGURE 5.1 Simple Squamous Epithelium. Top view of mesothelium (a) diagram; (b) photomicrograph (400×). Side view of human lung (c) diagram; (d) photomicrograph (1,000×).
(b), (d) ©Eric Wise

Simple Epithelia

Simple epithelium is only one cell layer thick. The cells are located on the basement membrane. Each epithelial type is further classified according to shape.

Simple Squamous (SKWAY-mus) ***Epithelium***   This epithelium is composed of thin, flat cells that lie on the basement membrane like floor tiles. If these cells are seen from a side view, they look flat and their resemblance to floor tiles becomes apparent.

Ⓐ **Activity**

Examine a prepared slide of simple squamous epithelium, usually seen as either a surface view or a side view. Simple squamous epithelium occurs in the air sacs of lungs and in the lining of blood vessels (where it is called **endothelium).** It can also be found as the surface layer of many membranes, where it is called **mesothelium.** It provides a smooth lining (as found on the inside of blood vessels), filters, or allows diffusion (as in the lungs). Compare your slide to fig. 5.1.

Draw what you see under the microscope in the space provided. Note whether your slide shows the cell as a surface or side view.

Illustration of simple squamous epithelium:

Kidney tubule   Cuboidal epithelial cells

Basement membrane

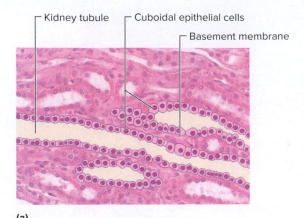

(a)

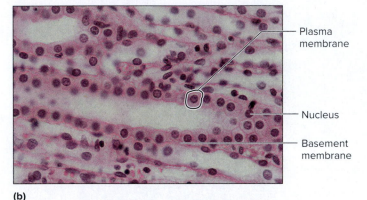

Plasma
membrane

Nucleus

Basement
membrane

(b)

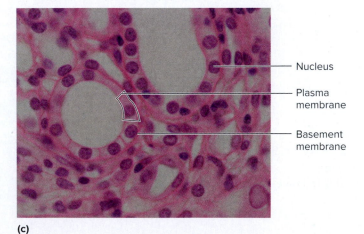

Nucleus

Plasma
membrane

Basement
membrane

(c)

FIGURE 5.2 Simple Cuboidal Epithelium—Kidney. Long
section (a) diagram; (b) photomicrograph (400×). Cross
section (c) photomicrograph (800×).
(b), (c) ©Eric Wise

Ⓐ Activity

Examine a prepared slide of simple cuboidal epithelium
and compare it to fig. 5.2. In the space provided, draw what you
see under the microscope and label the nucleus of the cell and the
basement membrane.

Illustration of simple cuboidal epithelium:

Simple Columnar Epithelium   This epithelium resembles tall
columns anchored at one end to the basement membrane. The
nuclei are frequently aligned in a row. Nonciliated simple columnar epithelium lines the inner portion of the digestive tract and
provides an absorptive area for digested food. Mucus-secreting
goblet cells are frequently found with it. Ciliated simple columnar
epithelium lines the uterine tubes.

Ⓐ Activity

Examine a prepared slide of simple columnar epithelium and
compare it to fig. 5.3. Draw a representation in the space provided.
Illustration of simple columnar epithelium:

Simple Cuboidal Epithelium   Simple cuboidal epithelium consists of cubelike or wedge-shaped cells mostly uniform in size.
These cells form many of the major glands and glandular organs of
the body and are the major cell type of the kidney. Simple cuboidal epithelium is frequently found lining tubules. Simple cuboidal
epithelium often is involved in the secretion of fluids (oil) or in
reabsorption (as in the kidneys).

Pseudostratified Ciliated Columnar Epithelium   These cells
appear as if they occur in a few layers, but all of the cells rest on the
basement membrane; thus, it is a simple epithelium. The nuclei are
at different levels in the tissue. Pseudostratified ciliated columnar
epithelium lines some portions of the respiratory passages, where
it protects the lungs by trapping dust particles in a mucous sheet

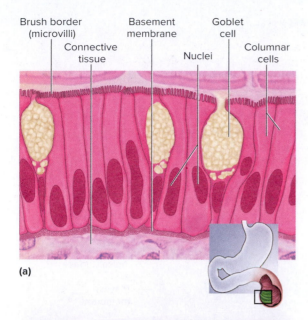

Brush border (microvilli) · Connective tissue · Basement membrane · Nuclei · Goblet cell · Columnar cells

(a)

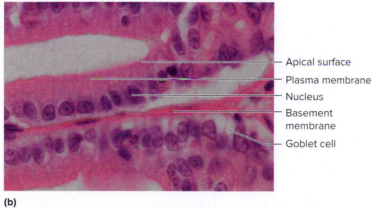

Apical surface
Plasma membrane
Nucleus
Basement membrane
Goblet cell

(b)

FIGURE 5.3 Simple Columnar Epithelium—Gastrointestinal Tract. (a) Diagram; (b) photomicrograph (400×).
(b) ©Eric Wise

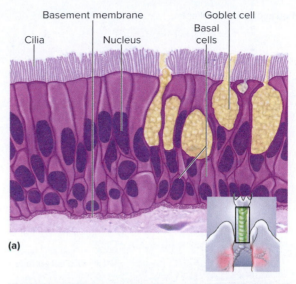

Cilia · Basement membrane · Nucleus · Basal cells · Goblet cell

(a)

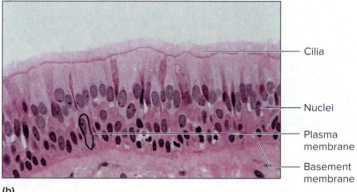

Cilia
Nuclei
Plasma membrane
Basement membrane

(b)

FIGURE 5.4 Pseudostratified Ciliated Columnar Epithelium—Trachea. (a) Diagram; (b) photomicrograph (400×).
(b) ©Eric Wise

and by moving the particles away from them. Mucus is secreted by goblet cells.

Ⓐ Activity

Examine a prepared slide of pseudostratified ciliated columnar epithelium and compare it to fig. 5.4. Draw a representative sample in the space provided.

Illustration of pseudostratified ciliated columnar epithelium:

Stratified Epithelia

Stratified epithelium is so named because many epithelial cells occur in layers on a basement membrane. The term *stratified* comes from the word *strata* and refers to the layering of these cells. You will examine two common types belonging to this group.

Stratified Squamous Epithelium Stratified squamous epithelium is an epithelium that makes up the outermost layer of skin. It also lines the vaginal canal, oral cavity, and esophagus. The multiple layers of this cell type protect the underlying tissue from mechanical abrasion. This epithelium may have cuboidal-shaped cells at the basement layer, but it derives its name from the cell shape at the free surface. Stratified squamous epithelium comes in two distinct types, keratinized and nonkeratinized (keratin is a tough protein that hardens cells in the outer layer of the skin).

Ⓐ Activity

Examine a prepared slide of stratified squamous epithelium and locate the basement membrane. Compare your slide to fig. 5.5 and draw what you see under the microscope in the space provided.

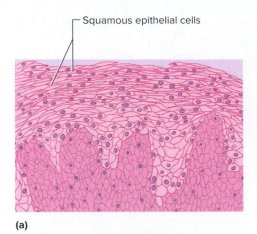

Squamous epithelial cells

(a)

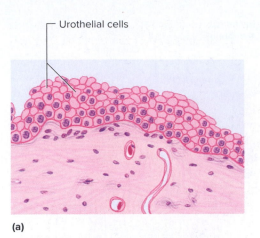

Urothelial cells

(a)

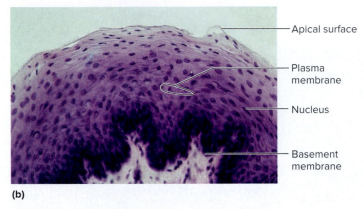

Apical surface

Plasma membrane

Nucleus

Basement membrane

(b)

FIGURE 5.5 Stratified Squamous Epithelium—Esophagus. (a) Diagram; (b) photomicrograph (400×).
(b) Ed Reschke/Getty Images

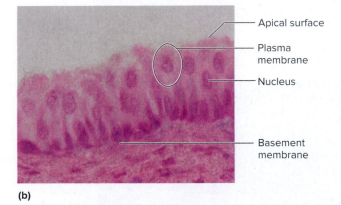

Apical surface

Plasma membrane

Nucleus

Basement membrane

(b)

FIGURE 5.6 Urothelium—Bladder. (a) Diagram; (b) photomicrograph (400×).
(b) ©Eric Wise

Locate the basement membrane, and count the layers of cells that occur between it and the surface of this cell type.

Illustration of stratified squamous epithelium:

(A) Activity

Examine a prepared slide of urothelium. Look at fig. 5.6 and draw the cell type in the space provided.

Illustration of urothelium:

Record the numbers of layers of cells: _____

Urothelium   **Urothelium (transitional epithelium)** is unusual in that it has some remarkable stretching capabilities. Urothelium lines the ureter, urinary bladder, part of the kidney, and proximal urethra and allows the bladder to expand as it fills with urine.

Muscular Tissue

Like epithelial tissue, muscular tissue is a cellular tissue with the tissue having mostly cells and little matrix. These cells are contractile and shorten due to the sliding of protein filaments across one another. There are three types of muscle: skeletal, cardiac, and smooth muscle.

Skeletal Muscle

The cells of skeletal muscle are called **fibers** or **myocytes.** When you refer to the muscles of your body, you are referring to organs made mostly of skeletal muscle. Skeletal muscle is sometimes labeled as striated muscle on some microscope slides because it has obvious **striations** (stry-A-shuns) (they look like stripes) in the fiber. Cardiac muscle has striations as well. Skeletal muscle is **voluntary** because you have conscious control over this type of muscle in your body. The individual muscle cells have many nuclei and are thus called **multinucleate** or **syncytial.** The nuclei are elongated and occur on the periphery of the cell.

(A) Activity

Examine a prepared slide of skeletal muscle under high power. Compare this slide to fig. 5.7. In the space provided, draw a representation of the muscle. How many widths of muscle cells fit across the diameter of your microscope when viewed at high power? You can use this information as one way to distinguish different muscle types. Approximately how many muscle fiber widths do you count? _____

Illustration of skeletal muscle:

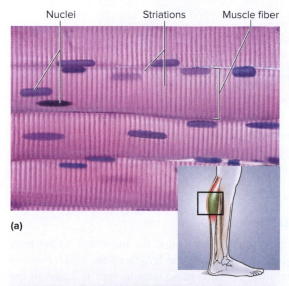

Nuclei   Striations   Muscle fiber

(a)

Peripheral nucleus

Striation

Width of fiber

(b)

FIGURE 5.7 Skeletal Muscle—Longitudinal Section.
(a) Diagram; (b) photomicrograph (400×).
(b) ©Eric Wise

> **STUDY HINT**
>
> By counting the cell widths in the slide, you should be able to see yet another way that you can distinguish among skeletal muscle, cardiac muscle, and smooth muscle.

Cardiac Muscle

Cardiac muscle is found only in the heart and is the main tissue making up that organ. Cardiac muscle is somewhat similar to skeletal muscle in that it is striated, but the striations are much less obvious, and the individual cell diameter is less than that of skeletal muscle cells. Cardiac muscle cells are called **cardiomyocytes.** Another difference between cardiac muscle and skeletal muscle is that cardiomyocytes are branched.

(A) Activity

Place a prepared slide of cardiac muscle under the microscope, and count the number of widths of cardiomyocytes across the diameter of the microscope under high power.

Record the number of cell widths that occupy the diameter of your field of view at high power: _____

Cardiac muscle contracts on its own and is therefore called **involuntary muscle.** The cells are mostly uninucleate with the nucleus centrally located and oval in shape. Cardiomyocytes are joined together by **intercalated discs,** which facilitate the transmission of the electrical impulses in the heart. These intercalated discs have connections between adjacent cells called **gap junctions.** When one muscle cell receives an impulse, it sends it on to the next cell.

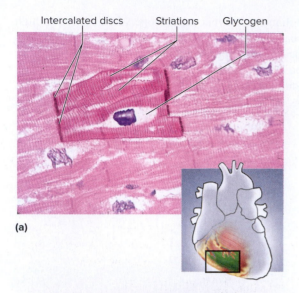

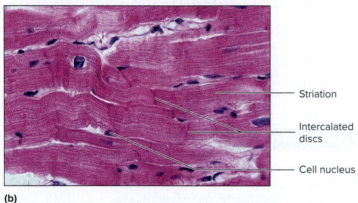

FIGURE 5.8 Cardiac Muscle—Longitudinal Section.
(a) Diagram; (b) photomicrograph (1,000×).
(b) ©Eric Wise

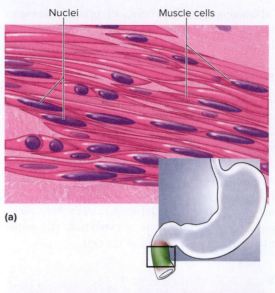

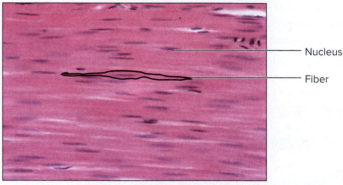

FIGURE 5.9 Smooth Muscle—Small Intestine, Longitudinal
Section. (a) Diagram; (b) photomicrograph (400×).
(b) ©Eric Wise

Look at a prepared slide of cardiac muscle and compare it to fig. 5.8, noting the striations, intercalated discs, and nuclei of the cells. In the space provided, draw the cells.

Illustration of cardiac muscle:

Smooth Muscle

Smooth muscle is **nonstriated** in that the myocytes do not have crosshatchings perpendicular to the length of the fiber. Smooth muscle is **involuntary,** like cardiac muscle, and is found in the gastrointestinal tract, where it propels food along by a process known as peristalsis and segmentation. The uterine contractions during labor are produced from smooth muscle. Smooth muscle is also found in other areas involved in contraction, such as hair follicles and some sphincters. The cells of smooth muscle are spindle-shaped and uninucleate, and the nucleus is centrally located and elongated in shape when the muscle is relaxed. When the muscle is contracted the nucleus appears corkscrew-shaped.

(A) Activity

Examine a prepared slide of smooth muscle and note the narrow diameter of the fibers. How many fiber widths do you count across the diameter of the field of view of your microscope under high power? Compare your slide to fig. 5.9, and in the space provided, draw an illustration of smooth muscle.

Illustration of smooth muscle:

Approximate number of cell widths under high power: _____

A **Activity**

Examine a prepared slide of all three muscle types, if available. Work with your lab partner and identify each muscle cell. Fill in the chart comparing the various muscle types and their characteristics. Review muscles in table 5.3.

Muscle Characteristics			
Muscle Type	Number of Nuclei	Location Found	Voluntary/Involuntary
	Many		
	One		
	Mostly One		

TABLE 5.3	Muscular Tissue

Skeletal Muscle

Microscopic appearance: large cell with many nuclei and obvious striations

Significant locations: skeletal muscles (biceps brachii, rectus abdominis, etc.)

Function: voluntary contractions

Cardiac Muscle

Microscopic appearance: smaller branched cell with mostly one nucleus, intercalated discs, and less obvious striations

Significant locations: heart

Function: rhythmic contraction of heart

Smooth Muscle

Microscopic appearance: small, slender cell with one central nucleus and no striations

Significant locations: digestive tract, blood vessels, and uterus

Function: sustained contractions, propulsion of food or delivery of infant

Nervous Tissue

Nervous tissue, as with epithelial and muscular tissues, is a cellular tissue. It is found in the brain, the spinal cord, the clusters of nerve cell bodies called ganglia, and the peripheral nerves of the body. The conductive cell of this tissue is the **neuron,** which receives and transmits electrochemical impulses. A neuron is a specialized cell with three major regions, the **dendrites,** the **cell body,** and the **axon.** Examine fig. 5.10 for the regions of a typical neuron; also see table 5.4.

A **Activity**

Examine the neurons of a smear of the spinal cord. Look for purple star-shaped structures, the cell body. Compare them to fig. 5.10. Draw what you see in the space provided.

Illustration of nervous tissue:

Special cells of the nervous system are called **glial cells** or **neuroglia.** These cells guide developing neurons to synapses, remove some neurotransmitters from the synapse, and perform numerous other functions. You will study glial cells in greater detail in Exercise 16.

Connective Tissue

The tissues that you have seen so far, epithelial, muscular, and nervous, are composed mostly of cells with very little extracellular matrix (ECM). This is not the case with connective tissue. There is usually more ECM than cells in connective tissue. This matrix usually consists of fibers and fluid, gel, or solid ground substance. Because of the abundance of ECM, connective tissue is classified by *specific tissue.*

Connective tissues have diverse appearances and functions, and can be divided into several subgroups for easier identification. These subgroups are **fibrous connective tissue, adipose tissue, cartilage, bone,** and **blood.** They are described in more detail next and are outlined in table 5.5.

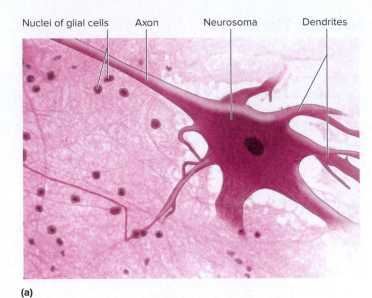

Nuclei of glial cells   Axon   Neurosoma   Dendrites

(a)

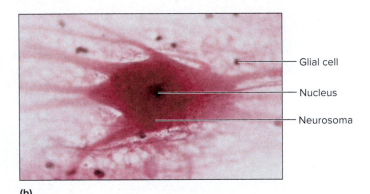

— Glial cell

— Nucleus

— Neurosoma

(b)

FIGURE 5.10 Typical Neuron. (a) Diagram; (b) photograph (400×).

(b) ©Eric Wise

TABLE 5.4	Nervous Tissue
Microscopic appearance: large, star-shaped cells (neurons in brain and spinal cord) with smaller cells (glial cells) nearby	
Significant locations: brain and spinal cord (nerves and ganglia)	
Function: multiple, including transmission and storage of information, learning	

Fibrous Connective Tissue

The fibers in **dense regular connective tissue** are parallel. The fibers in **dense irregular connective tissue** run in many directions. Dense regular connective tissue is found in tendons and ligaments. Dense irregular connective tissue is found in the deep layers of the skin and the white of the eye. The fibers in these tissues are made of a protein called **collagen** and are called **collagenous fibers. Fibroblasts** are found in the tissue in addition to the collagenous fibers. Fibroblasts secrete fibers and then mature into **fibrocytes.**

TABLE 5.5	Connective Tissue
Fibrous Connective Tissue	
Dense connective tissue	
Dense regular connective tissue	
Dense irregular connective tissue	
Elastic tissue	
Loose connective tissue	
Reticular tissue	
Areolar tissue	
Adipose Tissue	
Cartilage	
Hyaline cartilage	
Fibrocartilage	
Elastic cartilage	
Bone	
Blood	

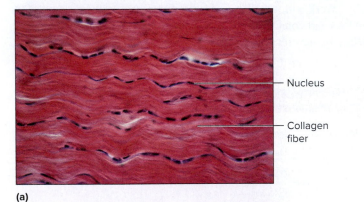

— Nucleus

— Collagen fiber

(a)

— Collagen fiber

(b)

FIGURE 5.11 Dense Connective Tissue. (a) Dense regular connective tissue—tendon (400×); (b) dense irregular connective tissue—skin (100×).

(a) Ed Reschke/Getty Images; *(b)* ©Eric Wise

(A) Activity

Examine a slide of dense connective tissue and compare it to fig. 5.11. In the space provided, draw a section of dense regular or dense irregular connective tissue.

Illustration of dense connective tissue:

TABLE 5.6	Dense Connective Tissue

Dense Regular Connective Tissue

Microscopic appearance: closely packed, wavy collagen fibers (or elastic fibers, in the case of elastic tissue)

Significant locations: tendons and ligaments (vocal cords in elastic tissue)

Function: binds bones together or muscle to bone (provides elastic structure)

Dense Irregular Connective Tissue

Microscopic appearance: randomly appearing collection of densely clustered collagen fibers

Significant locations: dermis and sheaths around cartilage and bone

Function: provides strength and resists stress and strain against tearing

Elastic Tissue   This tissue has a cellular component of fibroblasts. The fibers in this tissue are made of collagenous and **elastic fibers.** Elastic fibers are made of the protein **elastin.** Elastic tissue occurs in the walls of arteries and can be identified as dark, squiggly lines. These are sheets of elastic tissue. Elastic tissue can also be found in the vocal cords. Examine fig. 5.12 as you look at a prepared slide of elastic tissue. Then, in the space provided, draw the specific tissue. Review these tissues in table 5.6.

Illustration of elastic tissue:

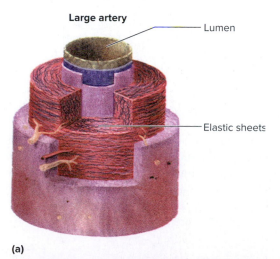

Large artery
— Lumen

— Elastic sheets

(a)

Reticular Tissue   Reticular tissue has fibroblasts with **reticular fibers.** Reticular fibers are made of a different type of collagen than collagenous fibers. Reticular tissue is found in soft internal organs, such as the liver, spleen, and lymph nodes. This tissue provides an internal framework for these organs. If your slide is stained with a silver stain, the reticular fibers appear black. If your slide is stained with Masson stain, they appear blue. In either case, look for fibers that appear branched among small, round cells. The other cells in the prepared slide are the cells of the organ that the reticular tissue is holding together.

Ⓐ Activity

Examine a prepared slide of reticular tissue and compare it to fig. 5.13. Compare what you see with table 5.7. In the space provided, draw a sample of what you see.

Illustration of reticular tissue:

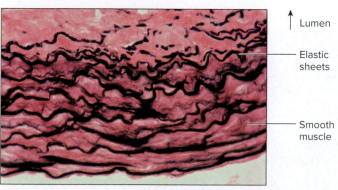

↑ Lumen

— Elastic sheets

— Smooth muscle

(b)

FIGURE 5.12 Elastic Tissue. (a) Overview of artery; (b) photomicrograph (400×).
(b) ©Eric Wise

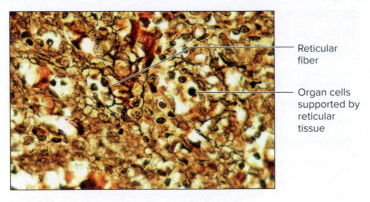

FIGURE 5.13 Reticular Tissue (400×).
©Eric Wise

Reticular fiber

Organ cells supported by reticular tissue

TABLE 5.7	**Loose Connective Tissue**

Reticular Tissue

Microscopic appearance: reticular fibers forming meshwork around organ cells

Significant locations: spleen, thymus, lymph nodes, and liver

Function: internal skeleton (framework) for soft organs

Areolar Tissue

Microscopic appearance: scattered arrangement of collagenous fibers with elastic and reticular fibers along with many cells (many with protective immune functions)

Significant locations: attaches epithelia to lower layers and around many internal organs

Function: binds epithelia to lower layers, insulates organs from infections

Adipose Tissue

Microscopic appearance: large, pale, open cells with nuclei near periphery of cell

Significant locations: under skin, breast tissue, and outside of heart and kidney

Function: energy storage, physical protection

Areolar Tissue   Areolar tissue is a complex collection of fibers and cells that has a distinctive look. Areolar tissue is found as a wrapping around organs, as sheets of tissue between muscles, and in many areas of the body where two different tissues meet (such as the boundary layer between fat and muscle). Areolar tissue consists of large, pink-stained **collagenous fibers;** smaller, dark **elastic** and **reticular fibers;** and a collection of cells that includes **fibroblasts, mast cells,** and **macrophages.**

(A) Activity

Examine a prepared slide of areolar tissue and compare it to fig. 5.14. Make a drawing of it in the space provided.

Illustration of areolar tissue:

Adipose Tissue

Adipose tissue (fat) is unusual as a connective tissue in that it is highly cellular. The cells that make up adipose tissue are called **adipocytes.** Adipocytes store lipids in large inclusions, while the nucleus and cytoplasm remain on the outer part of the cell near the cell membrane. In prepared slides of adipose tissue, the fat typically has been dissolved in the preparation process.

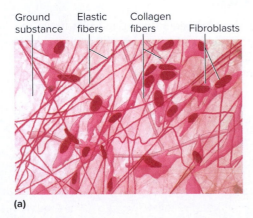

Ground substance   Elastic fibers   Collagen fibers   Fibroblasts

(a)

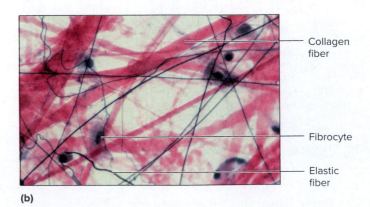

Collagen fiber

Fibrocyte

Elastic fiber

(b)

FIGURE 5.14 Areolar Tissue. (a) Diagram; (b) photomicrograph (400×).
(b) ©Eric Wise

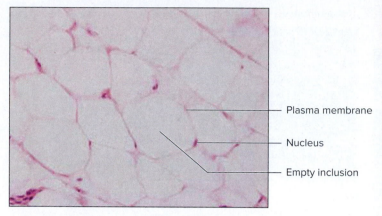

Plasma membrane

Nucleus

Empty inclusion

FIGURE 5.15 Adipose Tissue (400×).
©Eric Wise

TABLE 5.8	Cartilage
Hyaline Cartilage	
Microscopic appearance: usually light blue- or pink-stained matrix, frequently with pairs of cells appearing like "eyes"	
Significant locations: ends of long bones, ribs, larynx, and trachea	
Function: reduces friction at joints, keeps air passages open	
Fibrocartilage	
Microscopic appearance: numerous collagen fibers; cells frequently in rows of four or five	
Significant locations: pubic symphysis and intervertebral discs	
Function: protects from wear and tear at weight-bearing or stressed joints	
Elastic Cartilage	
Microscopic appearance: netlike pattern of fibers around chondrocytes	
Significant locations: external ear and epiglottis	
Function: provides flexible framework	

Ⓐ Activity

Locate the large, empty cells, with the nuclei on the edge of some of the cells. Compare what you see in the microscope to fig. 5.15 and make a drawing in the space provided. You may need to close down the iris diaphragm or reduce the amount of light shining on the specimen to best see this tissue.

Illustration of adipose tissue:

Cartilage (Cartilage and Perichondrium)

Cartilage is a type of connective tissue in which the matrix is composed of a pliable material that allows for some degree of movement. The matrix of cartilage has abundant amounts of chondroitin sulfate and forms a semisolid gel in which fibers and cells are found. The cells that occur in hyaline cartilage are called **chondrocytes** and in prepared slides, sometimes a cell has come out of the matrix and left a cavity. This cavity is known as a **lacuna** and is reasonably diagnostic of cartilage tissue. The **perichondrium** is dense irregular connective tissue that is found as a thin layer on the surface of some cartilages. There are three types of cartilage in the human body, and they vary by the number and type of protein fibers. These three types are hyaline cartilage, fibrocartilage, and elastic cartilage. Compare their features in table 5.8.

Hyaline Cartilage   The most common cartilage in the body is hyaline cartilage, found at the apex of the nose, at the ends of many long bones, between the ribs and the sternum, in the respiratory passages, and in other locations as well. Hyaline cartilage is clear and glassy in fresh tissue but will probably appear light pink or blue in prepared slides. The fibers in hyaline cartilage are **collagenous fibers.**

Ⓐ Activity

Find the chondrocytes in a prepared slide of hyaline cartilage. The fibers will not appear distinct because they blend into the color of the matrix and are relatively small in diameter. The chondrocytes of hyaline cartilage frequently occur in pairs. Some people say that they look like pairs of eyes staring back at you as you look at them in the microscope. Look around the edge of the sample of hyaline cartilage and locate the perichondrium. Compare the material under the microscope to fig. 5.16. In the space provided, draw what you see.

Illustration of hyaline cartilage:

Fibrocartilage   Fibrocartilage is similar to hyaline cartilage in that it has chondrocytes and collagenous fibers, but it differs from hyaline cartilage in that it has many more collagenous fibers. These can be seen in the prepared slides because the fibers are coarse and

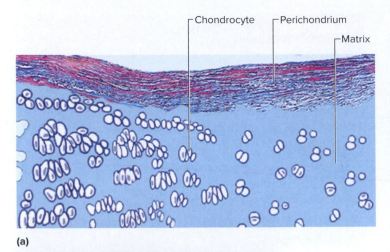

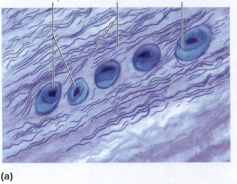

(a)

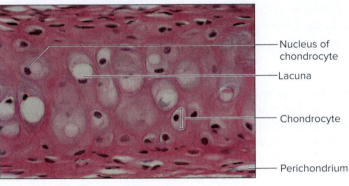

(b)

FIGURE 5.16 Hyaline Cartilage—Trachea. (a) Overview;
(b) photomicrograph (400×).
(b) ©Eric Wise

(b)

FIGURE 5.17 Fibrocartilage. (a) Diagram; (b) photomicrograph (400×).
(b) Victor P. Eroschenko

in larger bundles than those in hyaline cartilage. Fibrocartilage is found in areas where more stress is placed on the cartilage, such as in the intervertebral discs, the pubic symphysis, and the menisci of each knee. One characteristic of fibrocartilage is that the chondrocytes are frequently found in rows of four or five in the tissue.

Ⓐ **Activity**

Examine a prepared slide of fibrocartilage, and locate the collagenous fibers and chondrocytes. Compare the slide to fig. 5.17 and make a drawing of fibrocartilage in the space provided.
Illustration of fibrocartilage:

Elastic Cartilage   Elastic cartilage is unique in that it contains **elastic fibers,** which give the tissue its flexible nature. Elastic cartilage is found in the external ear and in the part of the larynx called the epiglottis. If the prepared slide is stained with a silver stain, the elastic fibers appear black. If the slide is not stained in this way, the fibers may be hard to distinguish.

Ⓐ **Activity**

Examine a prepared slide of elastic tissue and look for the chondrocytes and elastic fibers. Compare your slide to fig. 5.18 and make a drawing of elastic cartilage in the space provided.
Illustration of elastic cartilage:

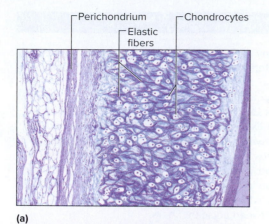

Perichondrium ─ ─Chondrocytes

─Elastic
fibers

(a)

─Elastic fiber

─Chondrocyte

(b)

FIGURE 5.18 **Elastic Cartilage.** (a) Diagram; (b) photomicrograph (400×).
(b) Victor P. Eroschenko

Bone

Bone is a type of connective tissue in which the extracellular matrix contains a calcium phosphate mineral called **hydroxyapatite** and collagenous fibers. Hydroxyapatite and **collagenous fibers** occur in the tissue along with osteocytes or bone cells. Look at a prepared slide of ground bone and compare it to fig. 5.19. Locate the large, circular regions called **osteons** (which resemble cross sections of tree trunks with the growth rings). An osteon is a long, thin unit of bone with space in the middle of it known as the **central canal**

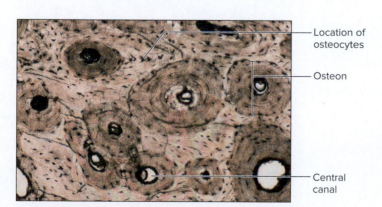

─Location of osteocytes

─Osteon

─Central canal

FIGURE 5.19 **Ground Bone (100×).**
©Eric Wise

TABLE 5.9	Bone
Microscopic appearance: cells enclosed in calcium salts in concentric circles	
Significant locations: skeleton	
Function: protection of soft organs, locomotion along with muscles, support	

and the dark dots around the central canal are small spaces where **osteocytes** (bone cells) occur in bone tissue. A more detailed examination of bone appears in Exercise 7.

A **Activity**

Review the features of bone in table 5.9. After you study the slide, record your observations in the space provided.

Illustration of ground bone:

Blood

Blood is a connective tissue in which the extracellular matrix is fluid and no fibers are visible. Fibers occur in blood as blood clots. The extracellular matrix of the blood is called **plasma,** and the cells of blood are either **erythrocytes** (eh-RITH-ro-sites) (red blood cells) or **leukocytes** (white blood cells). Small cellular fragments known as **platelets** are also present.

A **Activity**

Examine a prepared slide of blood and look for the numerous pink discs in the sample. These are the erythrocytes. You may find larger cells with lobed, purple nuclei. These are the leukocytes. Examine table 5.10. The details of blood are studied

TABLE 5.10	Blood
Microscopic appearance: many cells, most with no nucleus, suspended in plasma	
Significant locations: heart and blood vessels	
Function: transports gases, nutrients, hormones, water, and other material throughout the body; immunity	

in Exercise 24. Examine the blood slide under high power and compare it to fig. 5.20. Make a drawing of blood in the space provided.

Illustration of blood:

Review the types of connective tissue in table 5.5.

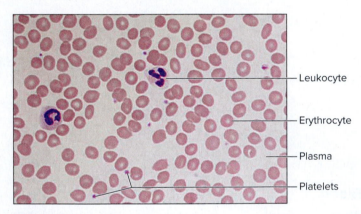

— Leukocyte

— Erythrocyte

— Plasma

— Platelets

FIGURE 5.20 Blood (400×).
©Eric Wise

REVIEW SECTION

Tissues

Name _____ *Date* _____

Lab Section _____ *Time* _____

Review Questions

1. List the four main tissues of the body. _____

   _____

2. What is the name of the noncellular layer that attaches epithelial tissue to other layers?

   _____

3. How many layers of cells are in simple epithelium? _____

4. Name the three general cell shapes of epithelial tissue. _____

   _____

5. A single, flattened layer of cells represents what type of epithelium? _____

6. What functional problem could occur if the stomach and small intestine were lined with stratified squamous epithelium instead of
   simple columnar epithelium? _____

   _____

7. Multiple layers of flattened epithelial cells represent what cell type? _____

8. What tissue type lines the inside of the urinary bladder? _____

9. Can you make the hairs on your arm "stand on end"? _____
   Based on your results, what type of muscle is responsible for doing this?_____

10. What muscle cell types have the following features?

    a. striated and voluntary _____

    b. striated and involuntary _____

    c. nonstriated and involuntary _____

11. What muscle cell type has intercalated discs? _____

12. The intestinal wall contains what kind of muscle? _____

13. The heart is composed primarily of what type of muscle? _____

14. The muscles of your arm are primarily composed of what cell type? _____

15. What cell type is responsible for the transmission of electrochemical impulses? _____

16. What is the extracellular matrix in blood called? _____

17. Name the three types of fibers in areolar tissue.

    a. _____

    b. _____

    c. _____

18. What type of connective tissue is springlike and found in the middle walls of arteries? _____

19. What is the cell type in adipose tissue? _____

20. Name the outer connective tissue layer that envelops some hyaline cartilage. _____

21. What types of fibers are in fibrocartilage? _____

22. Describe the functional difference between fibrocartilage and hyaline cartilage. _____

_____

23. What is another name for calcium and phosphate minerals in bone? _____

24. How do you distinguish generally between epithelium and connective tissue? _____

_____

25. Label the following photomicrographs by tissue type and by using the terms provided.

    a. collagenous fibers, elastic fibers, reticular fibers

1. _____

2. _____

3. _____

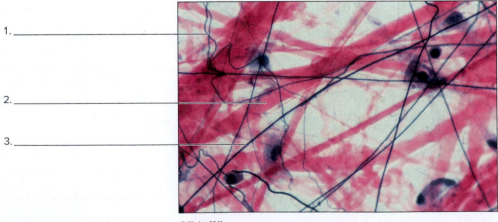

©Eric Wise

tissue name (400×) _____

b. basement membrane, cilia, nucleus

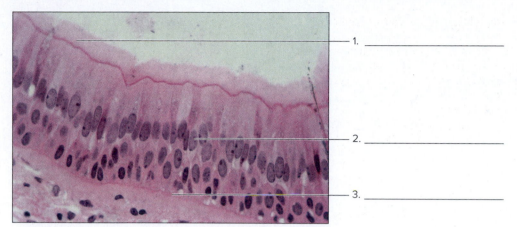

1. _____

2. _____

3. _____

©Eric Wise

tissue name (400×) _____

c. intercalated discs, nucleus, striations

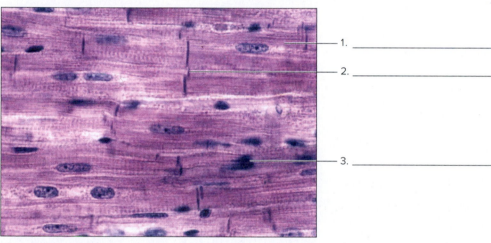

1. _____

2. _____

3. _____

Victor P. Eroschenko

tissue name (1,000×) _____

d. glial cells, neurosoma

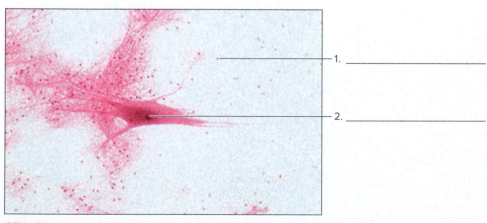

1. _____

2. _____

©Eric Wise

tissue name (100×) _____

26. Draw side views of the following tissues. Label the apical surface, basement membrane, cell membrane, and nucleus.

    a. simple columnar epithelium

    b. stratified squamous epithelium

Integumentary System

INTRODUCTION

The integumentary system consists of the skin and associated structures, such as hair, nails, and several glands. The integument consists primarily of a cutaneous membrane composed of a superficial epidermis and a deeper dermis. Below these two layers is an underlying layer, called the hypodermis, which anchors the integument to deeper structures. The skin is the largest organ of the body and is vital for temperature regulation and for protection against water loss and invasion from microorganisms. These topics are discussed in chapter 6, "The Integumentary System," in the Saladin text.

OBJECTIVES

At the end of this exercise, you should be able to

1. list the two layers of the integument;
2. list all the layers of the epidermis from superficial to deep;
3. describe the structure and function of sweat glands and sebaceous glands;
4. draw a hair in a follicle in longitudinal and cross section;
5. discuss the protective nature of melanin;
6. describe the hypodermis and its relationship to the integument.

MATERIALS

Models and charts of the integumentary system

Prepared Microscope Slides

Thick skin (with Pacinian corpuscles)

Hair follicles

Pigmented and nonpigmented skin

Microscopes

PROCEDURE

Overview of the Integument

(A) **Activity**

Examine models or charts of the **integumentary system** in the lab and locate the two major regions of the integument: the **epidermis** and the **dermis.** Most of the epidermis is made of **keratinocytes,** which are cells that produce the protective protein **keratin.** The epidermis is the most superficial layer and can be determined by the epithelial cells of that layer. You can look for cells that have obvious nuclei. The dermis consists primarily of collagenous fibers. Also locate the **hypodermis,** which is not a layer of the integument but anchors the integument to underlying bone or muscle. The hypodermis can be distinguished by the abundant adipose tissue. Compare the models in the lab to fig. 6.1.

Locate the **hair follicles** in the dermis, **sebaceous (oil) glands,** and **sweat glands.** The sweat glands are connected to the surface by **sweat ducts.**

Microscopic Examination of the Integument

(A) **Activity**

Examine a prepared slide of thick skin under low power. Locate the hypodermis, dermis, and epidermis. The slide will probably show three layers—one relatively clear due to a significant amount of adipose tissue, a pink layer, and a multicolored layer. The clear layer is the hypodermis, the pink layer is the dermis, and the multicolored layer is the epidermis. The epidermis may be determined by the epithelial cells of that layer. Look for cells with obvious nuclei. The colors may vary in the slides used in your lab. Compare these layers in the prepared slide to fig. 6.2.

Draw and label the overview of the integument in the following space, locating the epidermis, dermis, and hypodermis.

Illustration of integument:

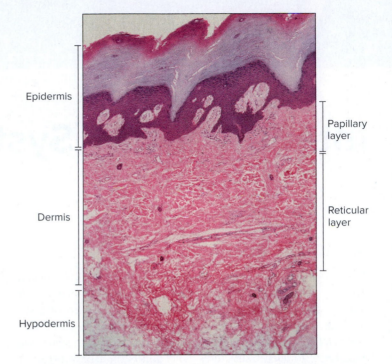

FIGURE 6.2 **Photomicrograph of the Integument (40×).**
©Eric Wise

Epidermis

The keratinocytes of the epidermis are composed of keratinized **stratified squamous epithelium** toughened with the protein keratin. Keratin is a tough material that protects the lower layers of the epidermis from wear and tear. The epidermis has a number of layers known as **strata** (singular, *stratum*). The epidermis does

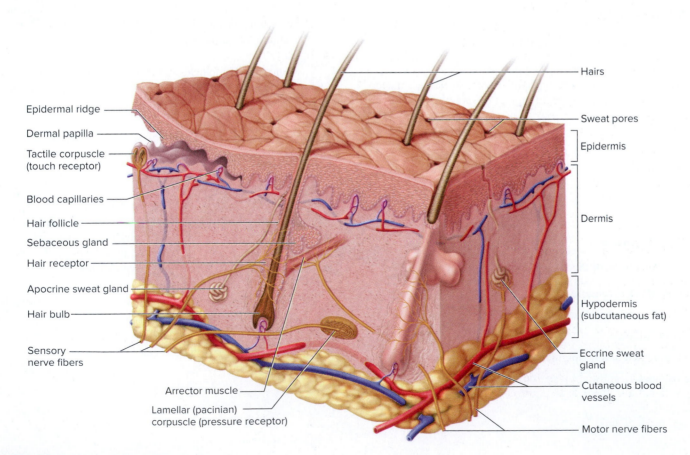

FIGURE 6.1 **Diagram of the Integument.**

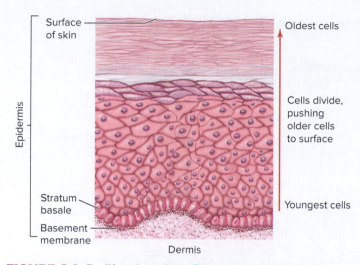

Surface of skin

Oldest cells

Cells divide, pushing older cells to surface

Epidermis

Stratum basale

Youngest cells

Basement membrane

Dermis

FIGURE 6.3 Proliferation of the Epidermis.

(*lucid* = light) because this layer appears translucent in fresh specimens. This layer generally appears clear or pink in prepared slides stained with hematoxylin and eosin (H&E) stain.

The most superficial layer of the integument is the **stratum corneum,** a tough layer consisting of dead cells flattened at the surface. Cells of the stratum corneum have been toughened by complete keratinization; thus, the epidermis is said to be made of **keratinized stratified squamous epithelium.**

Ⓐ **Activity**

Examine a prepared slide of thick skin and locate the various strata of the epidermis, as illustrated in fig. 6.4. Draw an example of the epidermis in the following space.

Illustration of epidermis:

not have blood vessels between the cells but is nourished by the vessels in the dermis.

Strata of the Epidermis

The deepest layer of the epidermis is known as the **stratum basale.** This single layer of cells is attached to the dermis by the **basement membrane,** a noncellular layer. Cells in the stratum basale divide repeatedly and push up toward the superficial layers as the cells increase in age. This process is illustrated in fig. 6.3.

In the stratum basale are specialized cells known as **melanocytes** (fig. 6.4). These cells produce a brown pigment called **melanin.** Melanin protects the stratum basale from the damaging effects of ultraviolet radiation. The number of melanocytes is relatively constant among people of all skin colors, but the amount of melanin produced by the melanocytes is variable. People of darker pigmentation produce more melanin, while people of lighter pigmentation have melanocytes that produce less melanin. People of lighter pigmentation are more susceptible to the effects of ultraviolet radiation and skin cancer.

Ⓐ **Activity**

Locate the stratum basale in a section of skin.

The layer of numerous cells superficial to the stratum basale is the **stratum spinosum.** The stratum basale and stratum spinosum are frequently classified as the stratum germinativum.

A layer of granular cells is present in the next layer. This is known as the **stratum granulosum.** The cells have purple-staining keratohyalin granules in their cytoplasm. The granules are precursors of keratin found in the outermost layer of the epidermis.

Superficial to this layer is the **stratum lucidum.** It is found only in the palms of the hand and soles of the feet. It is so named

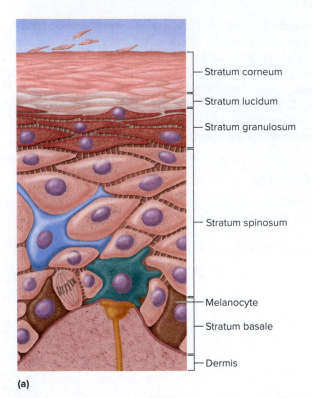

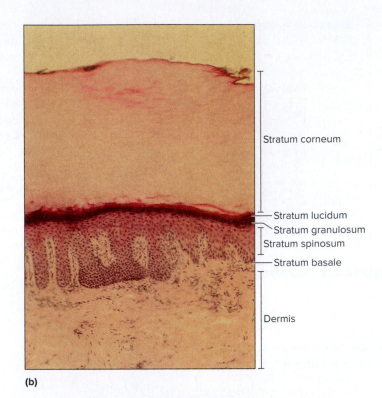

(a)

(b)

FIGURE 6.4 Strata of the Epidermis. (a) Diagram; (b) photomicrograph (100×).
(b) ©Eric Wise

Cells in the epidermis go through three major phases. In the deepest layer, the stratum basale, the cells are involved in rapid cell division. More superficial to this layer, cells undergo a process of producing precursor molecules that leads to the general waterproofing of the cells by keratinization. The final phase is the completion of the waterproofing process. Cells eventually die, providing a tough barrier. The most superficial cells of the epidermis are tough. They protect the skin from microorganisms and desiccation and eventually slough off. The epidermis renews itself about every 5–6 weeks.

Dermis

The dermis is responsible for the structural integrity of the integumentary system. It is composed primarily of closely apposed fibers along with blood vessels, nerves, sensory receptors, hair follicles, and glands. The dermis consists of two major regions, the superficial **papillary layer** and a deeper **reticular layer.** The papillary layer is so named for the **papillae** (bumps) that interdigitate with the epidermis, providing good adhesion between the two layers. The reticular layer is the largest layer of the dermis. Examine a slide of thick skin and locate the two layers of the dermis, as seen in fig. 6.5.

The majority of the fibers of the dermis are irregularly arranged **collagenous** along with lesser numbers of **elastic** and **reticular fibers.** The collagenous fibers provide strength and flexibility. The elastic fibers, as their name implies, provide elasticity to the skin.

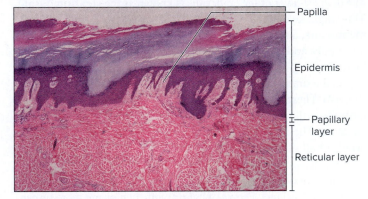

FIGURE 6.5 Layers of the Dermis (40×).
©Eric Wise

 Activity

Draw a representation of the dermis in the following space. Illustration of dermis:

Blood vessels occur in the dermis, bringing nutrients to both the dermis and epidermis. The blood vessels do not penetrate the epidermis, but the nutrients and oxygen these vessels carry diffuse from the dermis into the nearby cells of the epidermis. Dermal blood vessels are important in that they release heat to the external environment. As the internal temperature of the body core rises, the blood vessels of the dermis vasodilate and allow heat to radiate from the skin. In cold weather the vessels constrict, decreasing the blood flow to the dermis, which conserves heat in the body core.

Nerves in the dermis receive sensory inputs and transmit impulses to the central nervous system. Sensory information is received by numerous structures, such as **light touch receptors** of the dermis. Two of these are the **tactile (Meissner) corpuscles,** in the upper portion of the dermis, and the **tactile (Merkel) discs,** located in the upper dermis and lower epidermis. These two receptors allow for the perception of very slight touch stimuli (such as a fly lightly walking over your cheek). In addition, deep touch or pressure receptors are found in the dermis, farther away from the epidermis. **Lamellar (pacinian) corpuscles** sense pressure, such as when you lean against a wall or feel a vibration. Other receptors in the skin are **warm receptors** and **cool receptors.** When you are at a comfortable temperature, both of these receptors are firing. If you increase or decrease the skin temperature beyond the range of these receptors, **pain receptors** are stimulated. Pain receptors are naked nerve endings in the dermis that respond to numerous environmental stimuli.

Ⓐ Activity

Locate the tactile corpuscles and lamellar corpuscles in a prepared slide of skin and compare them to fig. 6.6. Draw an example of a tactile corpuscle and a lamellar corpuscle in the following spaces.

Illustration of tactile corpuscle:

Illustration of lamellar corpuscle:

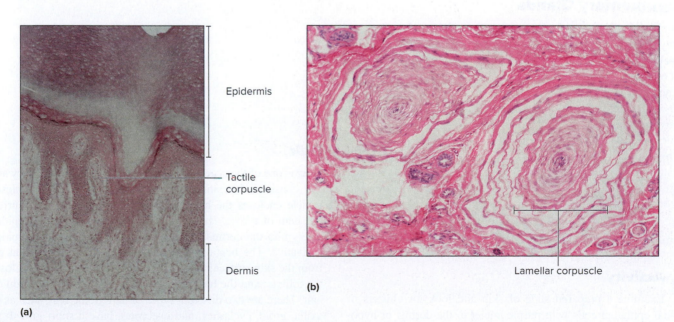

Epidermis

Tactile corpuscle

Dermis

(a)

Lamellar corpuscle

(b)

FIGURE 6.6 Skin Receptors (100×). (a) Tactile corpuscle; (b) lamellar corpuscle.
©Eric Wise

The Hypodermis

The hypodermis (figs. 6.1 and 6.2) is also known as **subcutaneous tissue.** It is a layer deep to the dermis consisting of dense irregular connective tissue, adipose tissue, nerves, blood vessels, and sense receptors.

The hypodermis attaches to the dermis by collagenous and elastic fibers. In some locations, such as the front of the neck, the hypodermis is loosely attached to the dermis and underlying fascia. In other areas, such as the soles of the feet, it is dense and strongly attached to the adjacent tissues.

The adipocytes in the hypodermis are clustered in lobules and can mobilize fatty acids to the venous system when the body needs energy.

Ⓐ Activity

Draw the hypodermis in the following space and label the adipocytes and collagenous fibers.

Illustration of hypodermis:

Integumentary Glands

A number of glands occur in the dermis and hypodermis. Two broadly distributed glands are **sweat glands** (figs. 6.7a and 6.7b) and **sebaceous (oil) glands** (figs. 6.7a and 6.7c). The secretory parts of sweat glands are composed of simple cuboidal epithelium and connective tissue. There are two different types of sweat glands. **Eccrine (merocrine) glands** are the most common, are sensitive to temperature, and produce normal body perspiration. Perspiration reduces the temperature of the body by **evaporative cooling. Apocrine sweat glands** secrete water and a higher concentration of organic acids than eccrine glands. The organic acids, along with bacterial action, produce the characteristic pungent body odor. These glands are concentrated in the region of the axilla and groin and are associated with hair follicles. Two more localized glands are **mammary glands** and **ceruminous (earwax) glands**.

Ⓐ Activity

Examine a prepared slide of skin and look for clusters of cuboidal epithelial cells with purple nuclei in the dermis or hypodermis. The glands are tubules, and when these are cut in section on your slide, you are mostly seeing the tubules cut in cross section.

These are the sweat glands illustrated in fig. 6.7a and 6.7c. Draw a sweat gland.

Illustration of sweat gland:

Sebaceous glands are typically associated with hair follicles and are larger than sweat glands (fig. 6.7a and 6.7c). You may not see a hair follicle with the sebaceous gland if the section was made through the gland and not through the gland and follicle. These glands secrete an oily material called **sebum,** which lubricates the hair and decreases the wetting action of water on the skin.

Ⓐ Activity

Examine a slide of hair follicles and compare the slide to fig. 6.7. Draw a sebaceous gland.

Illustration of sebaceous gland:

Hair

Hair is one of the accessory structures of the integumentary system and consists of keratinized cells produced in **hair follicles.** The follicle encloses the hair in the same way a bud vase surrounds the stem of a rose. Hair follicles are projections of the epidermal layers into the dermis. Hair can thus be considered an epidermal derivative. The hair consists of the **shaft,** the portion that erupts from the skin surface; the **root,** a deeper portion that is enclosed by the follicle; and the **hair bulb,** the actively growing portion of the hair. There are two different types of hair. Some hair, such as in the axilla, groin, eyelashes, and eyebrows, have a short growth phase and a long rest phase. Other hair, such as scalp hair and beard hair, have a longer growth phase.

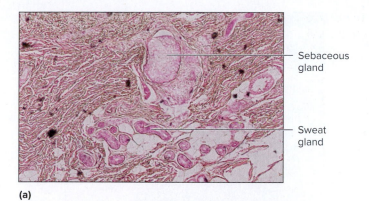

(a)

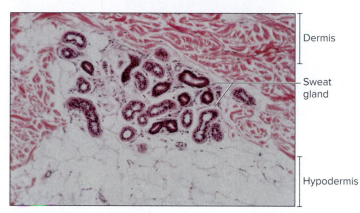

(b)

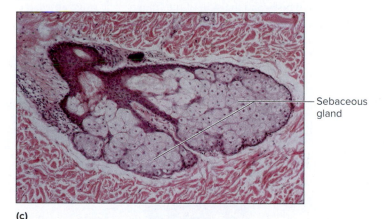

(c)

FIGURE 6.7 **Integumentary Glands.** (a) Overview of glands (20×); (b) sweat gland (100×); (c) sebaceous gland (100×). ©Eric Wise

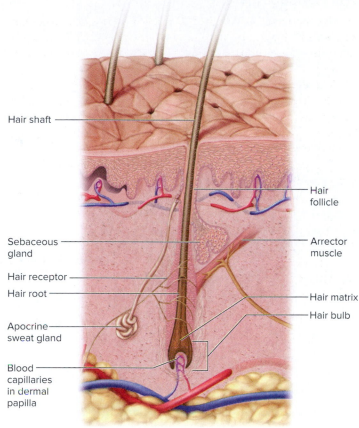

FIGURE 6.8 Diagram of Hair in a Follicle.

Illustration of hair:

(A) Activity

Examine models and charts of hair in the lab and compare them to fig. 6.8.

(A) Activity

Look at a prepared slide of hair and locate the root, follicle, and hair bulb. At the center of the bulb is a small structure known as the **dermal papilla,** a region with blood vessels and nerves that reach the hair bulb. The follicle is composed of an outer dermal layer of connective tissue and an inner epithelial layer. These form the **root sheath** of the hair. Locate these structures and compare them to fig. 6.9. Draw a longitudinal section of hair.

Arrector Muscles

The hair "stands on end" by the contraction of **arrector muscles.** An arrector muscle is a cluster of parallel smooth muscle fibers that connects the hair follicle to the upper regions of the dermis. These muscles are controlled by the autonomic nervous system. In response to cold conditions the arrector muscles contract, producing goose bumps. The arrector muscles also contract when a person is suddenly frightened.

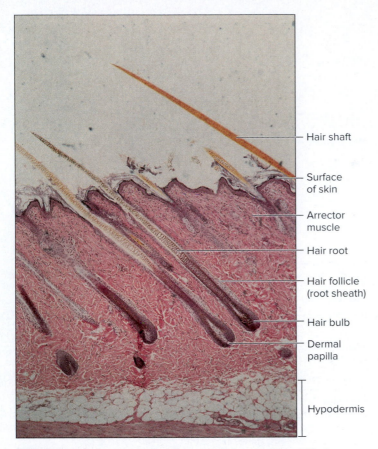

FIGURE 6.9 Photomicrograph of Hair in a Follicle (40×).

©Eric Wise

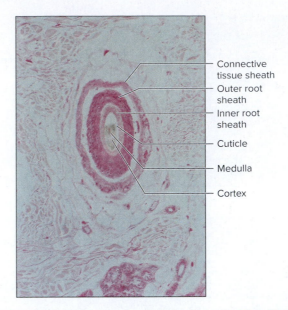

FIGURE 6.10 Hair Root and a Follicle, Cross Section (100×).

©Eric Wise

(A) Activity

Examine fig. 6.11 and compare your fingernails to the diagram.

On each side of the nail is the **nail groove,** a depression between the nail body and the skin of the fingers. The ridge above the groove is the **nail fold.**

(A) Activity

Find an arrector muscle in a prepared slide and compare it to figs. 6.8 and 6.9. If you have trouble finding an arrector muscle, look for slender slips of smooth muscle that angle from the follicle to the superficial regions of the dermis. You may have to look at more than one slide to find them.

Cross Section of Hair

The central portion of the hair is known as the **medulla,** enclosed by an outer **cortex.** The cortex may contain a number of pigments, which give the hair its color. The brown and black pigments are due to varying types and amounts of **melanin.** Iron-containing melanin pigments produce red hair. Gray hair is the result of a lack of pigment in the cortex and air in the medulla. The layer superficial to the cortex is the **cuticle** of the hair and looks rough in microscopic sections. Figure 6.10 shows a cross section of hair.

Nails

Nails are another accessory structure of the integument. They are keratinized cells of the stratum corneum of the skin that occur on the distal ends of the fingers and toes. Most of the nail consists of the **nail body,** and as it grows, the **free edge** of the nail is the part you clip. The cuticle of the nail is known as the **eponychium,** and the **nail root** is underneath the eponychium and is the area that generates the rest of the nail body. The **nail bed** is the layer deep to the nail body, while the **lunule** is the small white crescent that occurs at the base of the nail. The **hyponychium** is the region under the free edge of the nail.

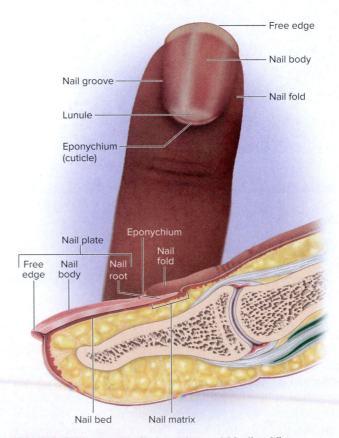

FIGURE 6.11 Fingernail, Superior and Median View.

REVIEW SECTION

Integumentary System

Name _____ *Date* _____

Lab Section _____ *Time* _____

Review Questions

1. The main organ of the integumentary system is the skin. Name three associated structures in the integumentary system.

   _____

   _____

2. The dermis has two main layers. Which one of these is the most superficial? _____

3. What is the most common connective tissue fiber found in the dermis? _____

4. The release of heat from the body by blood vessels occurs in what main layer of the integument? _____

5. The most superficial layer of skin, made of keratinocytes, is dead and provides a tough, protective barrier. If the keratinocytes were alive at the surface, how could that compromise the protective function of the integument? _____

   _____

   _____

6. Which type of sweat gland (eccrine or apocrine) is involved in evaporative cooling? _____

7. Distinguish among lamellar corpuscles, tactile corpuscles, and pain receptors in the skin. _____

   _____

   _____

   _____

8. Electrolysis is the process of hair removal by using electric current. Explain how this might destroy the process of hair growth in relation to the hair bulb. _____

   _____

   _____

9. Since hair color is determined by pigment in the cortex and the hair shaft is dead, explain the fallacy of claims of a person's hair turning white overnight. _____

   _____

10. Tattoos consist of ink injected into the skin. Do you think the ink is injected into the epidermis or the dermis? _____ What support do you give for your answer? _____
_____
_____

11. Approximately how long does it take for the epidermis to renew itself? _____

12. What cell type produces a pigment that darkens the skin? _____

13. Which one of the three layers (epidermis, dermis, hypodermis) is not part of the integument? Circle the correct answer.

14. A subcutaneous injection is inserted in the hypodermal or subcutaneous tissue. Based on structures found in the hypodermis and not in the epidermis, why is this a preferred area for administering an injection? _____
_____

15. What integumentary gland secretes an oil-like substance? _____

16. What part of the hair is found on the outside of the skin? _____

17. What does an arrector muscle do? _____

18. Circle the correct answer. The outermost portion of the hair is known as the

    a. cuticle.   b. lunula.   c. root.   d. medulla.

19. Label the following illustration using the terms provided.

dermis   hair root   arrector muscle   stratum corneum
epidermis   hair shaft   sebaceous gland   sweat gland
hair follicle   hypodermis

Introduction to the Skeletal System

INTRODUCTION

The skeletal and muscular systems are intimately related. The skeletal system has numerous functions, including supporting the body, protecting soft tissues (such as the brain and lungs), providing a structure for locomotion in the interplay between the skeletal and muscular systems, storing calcium, and forming blood (hematopoiesis). These topics are covered further in the Saladin text in chapter 7, "Bone Tissue," and chapter 8, "The Skeletal System."

The interaction between the skeletal and muscular systems is one in which the skeletal system provides the lever system on which the muscles can work, resulting in movement. This will be discussed in greater detail in Exercise 10. In this exercise, you examine the structure, composition, and development of the skeletal system.

OBJECTIVES

At the end of this exercise, you should be able to

1. describe the composition of bone tissue;
2. discuss how the skeletal system provides for efficient movement;
3. list the two major types of bone material found in long bones;
4. draw or describe the microscopic structure of cortical (compact) bone;
5. relate the anatomy of the bone to its growth.

MATERIALS

Chart of the skeletal system

Cut sections of bone showing cortical and spongy bone

Articulated human skeleton

Articulated cat skeleton (if available)

Disarticulated human skeleton

Prepared slide of cortical (ground) bone

Prepared slide of decalcified bone

Bone heated in oven at 350°F for 3 hours or more

Bone placed in vinegar for several weeks or 1 N nitric acid for a few days

Microscopes

PROCEDURE

Divisions of the Skeletal System

The skeletal system can be sorted into two main divisions according to location. The **axial skeleton** is so named because it is in the vertical axis of the body. It consists of the skull, auditory ossicles, hyoid bone, vertebral column, ribs, and sternum. The **appendicular skeleton** is that part of the skeleton "appended" (attached) to the axial skeleton. It can be divided further into the **pectoral girdle,** the **upper limb,** the **hip bones,** and the **lower limb.**

A Activity

Locate the major bones in lab and compare them to fig. 7.1 and table 7.1.

Major Bones of the Body

Figure 7.1 shows the major bones or bone regions of the body. You should be familiar with them before you study the specifics of the individual bones in later exercises. It helps to take a bone from a disarticulated skeleton and compare it to an articulated skeleton. Locate the major bones in the figure as you examine the bones in the lab.

The typical young adult has around 206 bones, with more found in younger individuals and fewer in older individuals. Many bones fuse as a person ages, forming larger bones, thus reducing the overall number.

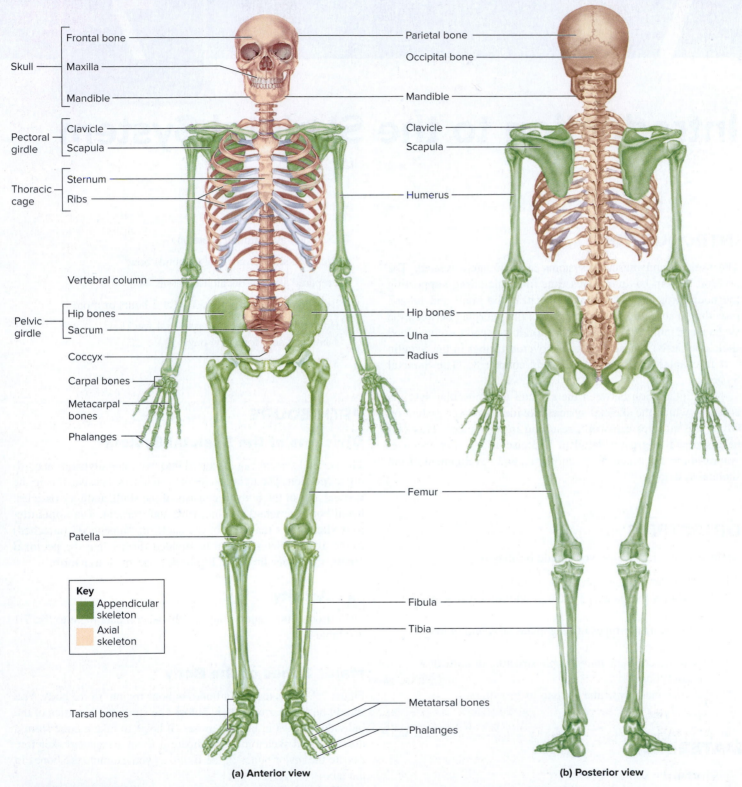

FIGURE 7.1 **Major Bones of the Body.** Axial skeleton in beige, appendicular skeleton in green.

TABLE 7.1	Axial and Appendicular Skeletons
Axial Skeleton	
Skull	Ribs
Hyoid bone	Sternum
Vertebral column	Auditory ossicles
Appendicular Skeleton	
Pectoral girdle	Hip bone
Clavicle	
Scapula	Lower limb
Upper limb	Femur
Humerus	Patella
Radius	Tibia
Ulna	Fibula
Carpals	Tarsals
Metacarpals	Metatarsals
Phalanges	Phalanges

Bone Features

There are specific terms that describe bumps, hollows, or spaces of bones. These are grouped by projections, depressions, or holes in the bone. **Projections** arise from the surface of bones, **depressions** occur on the surface of bones, and **cavities** are enclosed spaces in bones and holes that pass through bones. Projections on bones reflect different functions. Some projections, such as condyles and heads, are surfaces where the bone articulates with another. Other projections, such as tubercles, spines, and trochanters, are attachment points for muscles or ligaments.

(A) Activity

Holes in bones, such as foramina, canals, and fissures, are often passageways for blood vessels and/or nerves. Study the terms in table 7.2 and find them on bones in lab.

Composition of Bone Tissue

Bone consists of **organic material** and **inorganic material.** The organic portion is composed mostly of cells and collagenous fibers and makes up about one-third of the bone by weight. The inorganic portion

TABLE 7.2	Bone Features
Projections from the Bone Surface	**Example**
Process—a general term for a projection from the surface of the bone	Styloid process of ulna
Tubercle—a relatively small bump on a bone	Greater tubercle of humerus
Tuberosity—a relatively large, rough area on a bone	Deltoid tuberosity of humerus
Spine—a short, sharp projection	Vertebral spine
Condyle—a smooth surface that articulates with another bone	Lateral condyle of femur
Epicondyle—a projection above a condyle	Medial epicondyle of humerus
Head—a terminal projection that articulates with another bone	Head of femur
Neck—a constriction below the head	Neck of rib
Crest—an elevated ridge of bone	Crest of ilium
Line—a smaller elevation than a crest	Gluteal line of ilium
Facet—a smooth, flat face	Articular facets of vertebrae
Trochanter—a large bump (on femur)	Greater and lesser trochanter of femur
Ramus—a branch	Ramus of mandible
Depressions, Passageways, and Cavities in Bones	**Example**
Foramen—a shallow hole	Foramen magnum of occipital bone
Sinus—a cavity	Maxillary sinus
Meatus or canal—a deeper hole	External auditory meatus of temporal bone
Fossa—a depression in a bone	Iliac fossa
Notch—a deep cut-out	Greater sciatic notch of ilium
Groove or sulcus—an elongated depression	Intertubercular groove of humerus
Fissure—a long, deep cleft	Inferior orbital fissure

makes up the remaining two-thirds of the bone and mostly consists of **hydroxyapatite,** a complex salt of calcium phosphate. There is also a lesser amount of calcium carbonate in the inorganic portion of bone.

(A) Activity

Examine bones that have been soaked in acid. The acid has dissolved the mineral salts, leaving the collagenous fibers as a flexible framework. Your instructor may also show bones baked in an oven for a few hours. In this preparation, the structural characteristics of proteins have been destroyed, and though the bones remain hard because of the mineral salts, they are brittle due to the destruction of the protein fibers by the heat. Your instructor may wish to demonstrate the fragility of specially prepared bones.

Bone Morphology

The general structure of a long bone consists of the proximal and distal ends of the bone, called the **epiphyses** (eh-PIF-uh-seez), and the shaft of the bone, called the **diaphysis** (die-AH-fuh-sis) (fig. 7.2). The epiphyses of bones have **articular cartilage** at the

ends of the bone. The articular cartilage is composed of hyaline cartilage and helps reduce friction as the joint moves.

In individuals who are growing in height, there is a plate of hyaline cartilage between the epiphysis and the diaphysis. This is the **epiphyseal** (EP-ih-FIZZ-ee-ul) **plate,** which increases in thickness by division of the chondrocytes. As the cartilage grows, the individual increases in height. During growth, the cartilage away from the epiphyseal plate is replaced by bone. When a person reaches adult height, all the cartilage is replaced by bone and the remnants of the plate are seen as the **epiphyseal line.** Examine long bones cut lengthwise for the epiphyseal line. In the cut sections of bone, you can also see the hard, **cortical (compact) bone** on the outside of the bone, and the inner **spongy (cancellous) bone.** The spongy bone is made of **trabeculae,** thin rods or plates of bone that run in the same direction as the stress applied to the bone. Stress on the bone may be in the form of gravity, or it may occur due to a common force applied to the limb. Trabeculae make an internal framework that strengthens the bone.

The innermost section of bone is hollow and is called the **marrow** (or **medullary** (MED-you-lerr-ee)) **cavity.** Marrow is

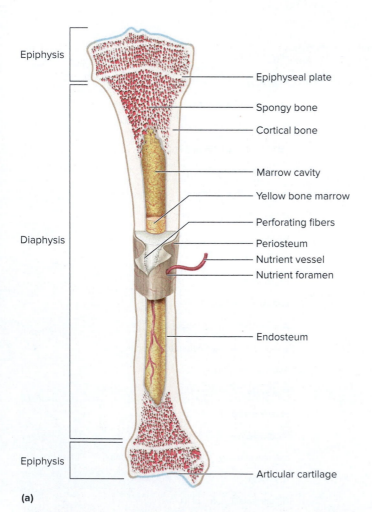

Epiphysis

Epiphysis

Diaphysis

Epiphysis

Epiphysis

Epiphyseal plate

Spongy bone

Cortical bone

Marrow cavity

Yellow bone marrow

Perforating fibers

Periosteum

Nutrient vessel

Nutrient foramen

Endosteum

Articular cartilage

(a)

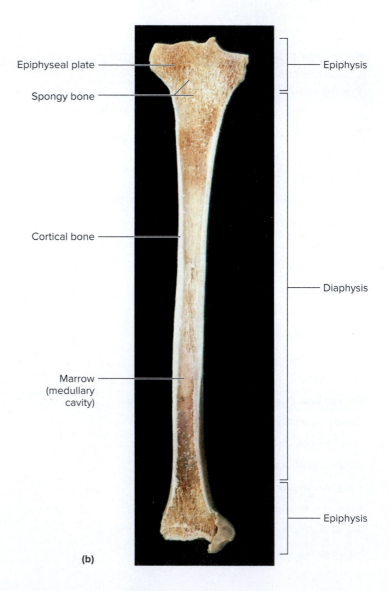

Epiphyseal plate

Spongy bone

Cortical bone

Marrow (medullary cavity)

Epiphysis

Diaphysis

Epiphysis

(b)

FIGURE 7.2 Long Bone, Longitudinal Section. (a) Diagram; (b) photograph.
(b) ©Eric Wise

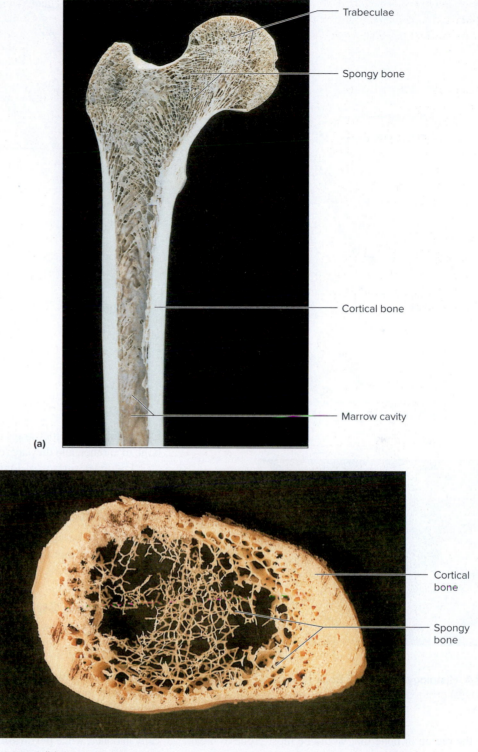

FIGURE 7.3 Cut Bone. (a) Long section; (b) cross section.
(*a*), (*b*) ©Eric Wise

the material that occupies this cavity and can be of two types: a hematopoietic (blood-forming) **red marrow** or an adipose tissue-containing **yellow marrow.** In flat bones, there is an outer and an inner layer of cortical bone with spongy bone sandwiched between them. In flat bones, the spongy bone found in cranial bones is called **diploe** (DIP-low-ee).

Ⓐ Activity

Compare the cut bones in the lab to figs. 7.2 and 7.3 and note the features listed in the discussion. In a long bone, such as the femur, the cortical bone is thicker in the middle of the bone than at the ends. Many long bones have this feature and resemble an archery bow in this regard. This reinforcement reduces the risk of

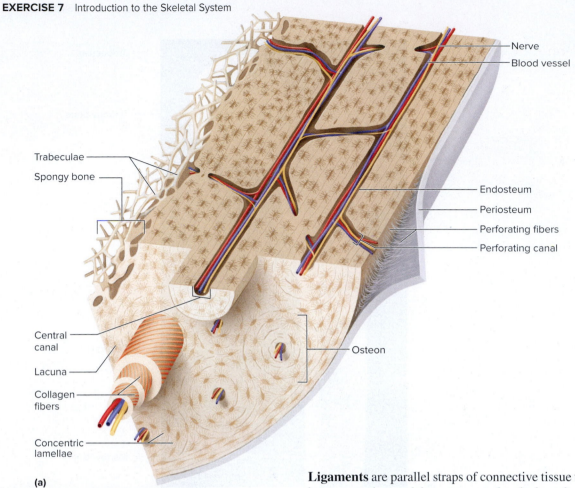

Nerve

Blood vessel

Trabeculae

Spongy bone

Endosteum

Periosteum

Perforating fibers

Perforating canal

Central canal

Osteon

Lacuna

Collagen fibers

Concentric lamellae

(a)

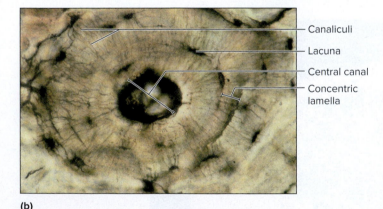

Canaliculi

Lacuna

Central canal

Concentric lamella

(b)

FIGURE 7.4 **Histology of Cortical Bone (Osteon).**
(a) Diagram; (b) cross section (400×).
(b) ©Eric Wise

breaking at the region of bone that would be most prone to break, the middle.

On the surface of the diaphysis, locate the **nutrient foramina** (for-AM-ih-nuh). These are small holes that allow for the passage of blood vessels into and out of the bone. The nutrient foramina lead to **perforating (Volkmann) canals** that pass through cortical bone.

The outer surface of the bone is covered with a dense connective tissue sheath called the **periosteum** (PEAR-ee-OS-tee-um), where nerves and blood vessels are located. The periosteum is an anchoring point for **tendons and ligaments.** Tendons attach muscles to bone at the periosteum, and this attachment is strengthened by **perforating fibers** that penetrate the cortical bone.

Ligaments are parallel straps of connective tissue that connect one bone to another, and they are secured to the bone by the periosteum.

The inner surface of long bones, near the marrow cavity, is lined with a layer known as the **endosteum** (end-OS-tee-um).

Bone Location

In addition to being part of either the axial or the appendicular skeleton, bones more specifically belong to the upper limb, lower limb, pelvic girdle, skull, or vertebral column.

Microscopic Structure of Bone

Cortical Bone

(A) Activity

Examine a prepared slide of cortical (ground) bone and locate the modular units of bone called **osteons.** Each osteon has a hole in the middle called a **central canal,** which houses blood vessels and nerves in the dense bone tissue. The central canals typically run vertically in bones, while **perforating canals** carry nutrients horizontally. Around the central canal are rings of bone tissue known as **lamellae.** Also in the slide are dark spots. These are holes that held **osteocytes** in living tissue but fill with bone dust when the bones are thinly ground for slide preparation. These spaces are the **lacunae** (singular, *lacuna*). The lacunae are connected by thin tubes called **canaliculi** (CAN-uh-LIC-you-lie). The role of canaliculi is to provide a passageway through the dense bone material. Osteocytes are living and need to receive oxygen and nutrients and remove wastes. They do this by extending cellular strands (**dendrites**) through the canaliculi between osteocytes. The canaliculi connect to the central canal, forming rings called **concentric lamellae.** Compare what you see to fig. 7.4.

Decalcified Bone

Ⓐ **Activity**

Examine a slide of decalcified bone. In this preparation the bone salts have been removed and the remaining section of bone has been sliced, mounted, and stained. The periosteum is visible on the surface of the bone, with the cortical bone in the middle and the endosteum on the inside of the bone (fig. 7.5). Look for as many bone cells as you can find. These cells are described in the following section and illustrated in fig. 7.6.

Bone Cells

There are three main types of bone cells: **osteoblasts, osteocytes,** and **osteoclasts.** Osteoblasts originate from stem cells called **osteogenic cells** (fig. 7.6), which occur in the periosteum, endosteum, and central canals of osteons. These osteogenic cells maintain mitotic activity and produce osteoblasts, which secrete collagen fibers and calcium salts. Osteoblasts are most numerous in the endosteum and periosteum.

Osteocytes are mature bone cells; they respond to stresses placed on bone and remodel the bone in response to those stresses. For example, an increase in weight-bearing activity increases the density of bone. This is seen in the bones of athletes, which are denser than those of people who do not exercise. Osteocytes deposit or reabsorb bone matrix, regulating bone density, and control calcium and phosphate balance. Osteocytes also repair microfractures that frequently

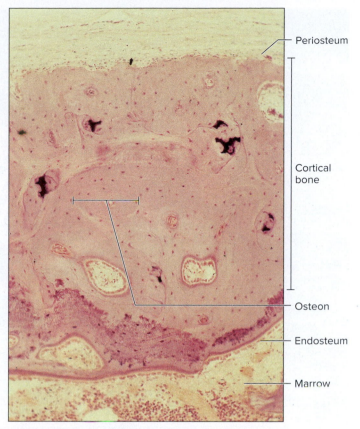

FIGURE 7.5 Decalcified Bone (100×).
©Eric Wise

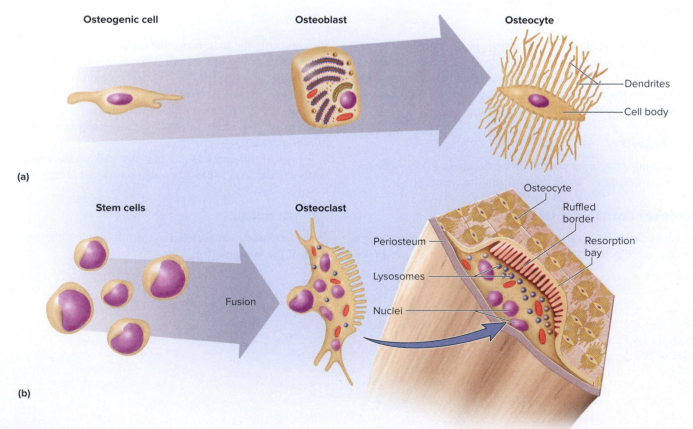

FIGURE 7.6 Bone Cells. (a) Osteogenic cells produce osteoblasts that develop into osteocytes. (b) Stem cells in bone marrow fuse to form osteoclasts.

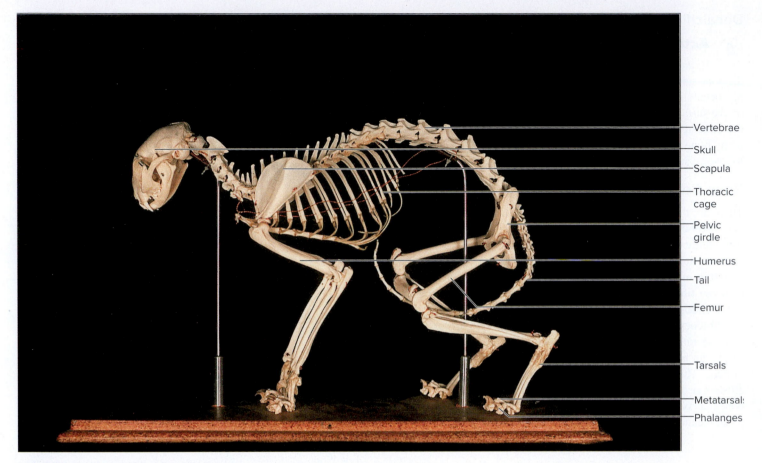

FIGURE 7.7 Cat Skeleton, Lateral View.
©Eric Wise

occur in bone. Osteocytes have extensions of the cytoplasm known as dendrites that connect with other osteocytes via the canaliculi. Examine the slide of decalcified bone and find the osteocytes. **Osteoclasts** are cells that remove collagen and remove minerals with HCl. They arise by the fusion of **stem cells.**

Ⓐ Activity

Examine the inner portion of the decalcified bone slide and locate the osteoclasts. These cells are multinucleate.

Skeletal Anatomy of the Cat

The cat skeleton differs from the human skeleton in several ways. Humans are bipedal, and this is reflected in the nature of the vertebral column. In humans, the bodies of the vertebrae become much larger as you move from superior to inferior. Cats are quadrupeds, so the weight of the vertebral column is distributed more evenly, and their vertebrae are more evenly matched in size. In humans, the absence of a tail is reflected in a corresponding absence of tail vertebrae. In cats, the tail vertebrae are present and fully functional.

Another difference is that humans walk on the entire inferior surface of the foot, while cats walk on just part of their foot.

Ⓐ Activity

Examine a cat skeleton in the lab and note the general features of the skeleton (fig. 7.7).

REVIEW SECTION

Introduction to the Skeletal System

Name _____ *Date* _____

Lab Section _____ *Time* _____

Review Questions

1. Circle the correct answer. The hyoid bone belongs to the

   a. appendicular skeleton.   b. axial skeleton.   c. upper limb.   d. skull.

2. Circle the correct answer. The clavicle belongs to the

   a. axial skeleton.   b. pectoral girdle.   c. hip bones.   d. upper limb.

3. In the disease osteoporosis, there is a significant loss of spongy bone. Explain how the loss of this specific bone material can weaken a bone. _____

   _____

   _____

4. Label the following illustration using the numbers for the terms provided.

   1. canaliculi
   2. central canal
   3. concentric lamella
   4. lacuna
   5. osteon

   a. _____
   b. _____
   c. _____
   (the thin lines)
   d. _____
   e. _____
   (the dark spot)

   ©Eric Wise

5. The ends of a long bone are known as the _____.

6. Name two bones of the forearm. _____

7. Circle the correct answer. The ribs are part of which skeletal division?

   a. axial   b. appendicular

8. Name the bone found between the femur and the tibia. _____

9. A young adult has how many bones, on average? _____

10. The inorganic portion of bone tissue is made of what complex mineral salt (consisting of calcium phosphate)?

   _____

11. What is an osteon? _____

   _____

12. Synthetic bone material known as hydroxyapatite is sometimes molded into the shape of bone. This procedure is done
   to replace diseased, damaged, or surgically removed bone. Cells in the body remodel the synthetic material and produce new bone. What bone cells do you predict to be involved in this remodeling, and what role do these cells have?

   _____

   _____

   _____

13. What role do osteocytes have in bone tissue? _____

   _____

   _____

14. How does the shape of a long bone resist breaking when put under stress? _____

   _____

15. Describe the nature of decalcified bone. What was removed in the process of decalcification, and what impact did this have on the
   bone structure? _____

   _____

16. How does the central canal differ from a lacuna in terms of location and the material found in each respective space?

   _____

   _____

17. What is another name for the shoulder blade, and what two bones attach to it? _____

   _____

18. Label the illustration of the skeletal system using the numbers for the terms provided.

1. carpals	6. humerus	11. phalanges (feet)	16. sternum
2. clavicle	7. mandible	12. phalanges (hand)	17. tibia
3. femur	8. metacarpals	13. radius	18. ulna
4. fibula	9. metatarsals	14. rib	19. vertebra
5. hip bone	10. patella	15. scapula	

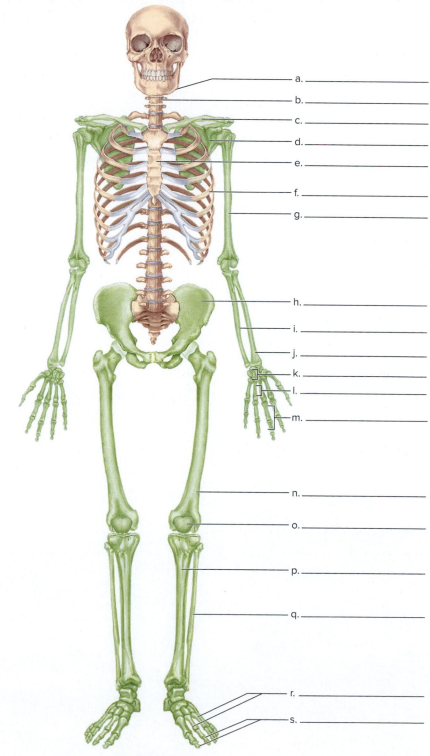

a. _____

b. _____

c. _____

d. _____

e. _____

f. _____

g. _____

h. _____

i. _____

j. _____

k. _____

l. _____

m. _____

n. _____

o. _____

p. _____

q. _____

r. _____

s. _____

Anterior view

Notes

Axial Skeleton

INTRODUCTION

The axial skeleton consists of 80 bones, including the skull, middle-ear bones (auditory ossicles), hyoid bone, vertebrae, sternum, and ribs. The skull consists of two major regions, the **cranium** and the **face.** The skull is the most complex region of the skeletal system and not only houses the brain but also contains a significant number of the sense organs. The cranium consists mostly of flat bones, while the face is composed of irregular bones. The middle-ear bones are covered in Exercise 22, Ear, Hearing, and Equilibrium.

There are five regions of the vertebral column. These are the cervical (SERVE-ik-ul), thoracic (thor-AH-sic), lumbar vertebrae, and vertebrae that make up the sacrum and the coccyx. There are usually 33 individual vertebrae, some of which fuse into larger structures, such as the sacrum.

The thoracic cage consists of the 12 pairs of ribs and the sternum, protecting the lungs and heart yet providing flexibility during breathing. The hyoid bone is a small, floating bone between the floor of the mouth and the upper anterior neck. These topics are discussed further in the Saladin text in chapter 8, "The Skeletal System."

OBJECTIVES

At the end of this exercise, you should be able to

1. list all the bones of the cranium and the major features of those bones;
2. list all the bones of the face and the major features of those bones;
3. name all the bones that occur singly or in pairs in the skull;
4. locate the major foramina of the skull;
5. name the major sutures of the skull;
6. list all the fontanelles of the fetal skull;
7. find and name a specific vertebra (such as T6) on an articulated vertebral column;
8. name the bony features of an individual vertebra;
9. describe the features that determine cervical, thoracic, lumbar, sacral, and coccygeal vertebrae and identify the atlas and the axis;
10. describe the normal and abnormal spinal curvatures;
11. distinguish between the different types of ribs, the markings of the individual ribs, and whether a rib comes from the left or right side of the body;
12. identify the bony features of the sternum.

MATERIALS

Articulated skeleton or plastic casts of a skeleton

Disarticulated skeleton or plastic casts of bones

Disarticulated skull, if available

Articulated skulls with the calvaria cut

Fetal skulls

Charts of the skeletal system

Plastic straws cut on a bias or pipe cleaners for pointer tips

Foam pads of various sizes (to protect bone from hard countertops)

PROCEDURE

As you locate the cranial and facial bones on a skull in the lab, make sure you do not poke pencils, pens, or fingers into the delicate regions of the eyes or nasal cavity. You may break the delicate structures in these regions. Be very careful handling skulls. Once you have found the major bones of the skull, use the descriptions and illustrations in this exercise to find the structures of the skull. The skull is described from a series of views, and you should locate the structures listed for those views.

When you examine bones in the lab, do not use your pen or pencil to locate a structure. These leave marks on the bones. Use a cut plastic straw or a pipe cleaner to point out structures. Your instructor may want you to place real bone material on foam pads to cushion the bone from the tabletop.

Overview of the Skull

(A) Activity

Familiarize yourself with the bones of the skull in the cranium and the face. The skull bones in the cranium are listed with a number after the name of the bone indicating whether the bone occurs *singly* (1) or as a *pair* (2). Take a skull to your lab table and locate the bones listed in fig. 8.1.

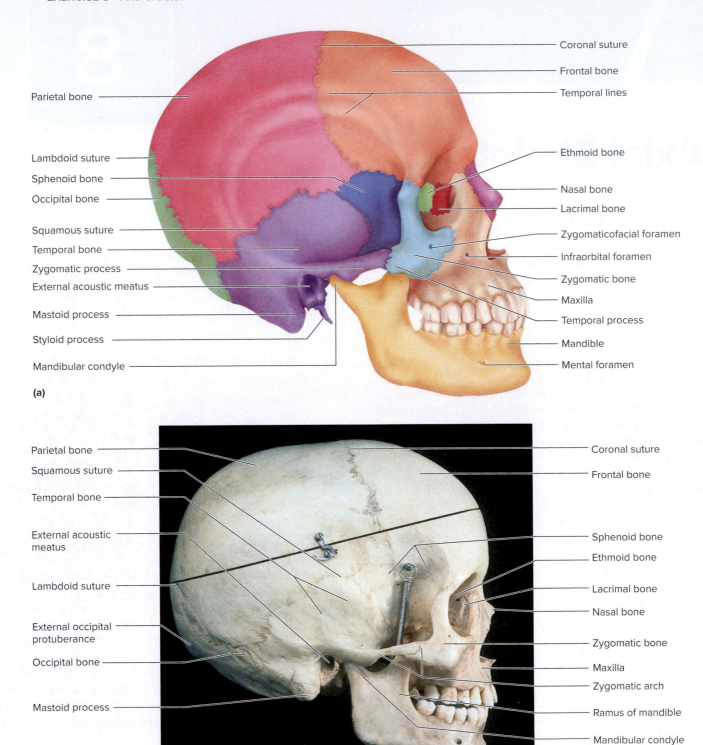

Coronal suture

Frontal bone

Temporal lines

Parietal bone

Lambdoid suture

Sphenoid bone

Occipital bone

Squamous suture

Temporal bone

Zygomatic process

External acoustic meatus

Mastoid process

Styloid process

Mandibular condyle

Ethmoid bone

Nasal bone

Lacrimal bone

Zygomaticofacial foramen

Infraorbital foramen

Zygomatic bone

Maxilla

Temporal process

Mandible

Mental foramen

(a)

Parietal bone

Squamous suture

Temporal bone

External acoustic
meatus

Lambdoid suture

External occipital
protuberance

Occipital bone

Mastoid process

Coronal suture

Frontal bone

Sphenoid bone

Ethmoid bone

Lacrimal bone

Nasal bone

Zygomatic bone

Maxilla

Zygomatic arch

Ramus of mandible

Mandibular condyle

Body of mandible

(b)

FIGURE 8.1 **Skull, Lateral View.** (a) Diagram; (b) photograph.
(b) ©Eric Wise

Bones of the Cranium

frontal (1)   parietal (2)

occipital (1)   temporal (2)

sphenoid (1)   ethmoid (1)

You can remember the bones of the cranium with the mnemonic "of pets." Each letter represents a cranial bone (*o* for *occipital, f* for *frontal,* etc.).

The bones of the face are listed next. Locate these bones on the skull in the lab, as shown in figs. 8.1, 8.2, and 8.3.

Bones of the Face

maxilla (2)   nasal (2)

mandible (1)   palatine (2)

vomer (1)   zygomatic (2)

lacrimal (2)   inferior nasal concha (plural, *conchae*) (2)

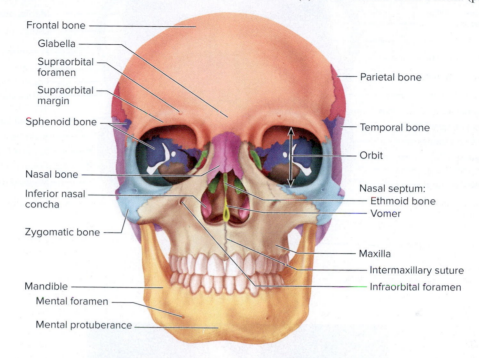

(a)

(b)

FIGURE 8.2 Skull, Anterior View. (a) Diagram; (b) photograph.

(b) ©Eric Wise

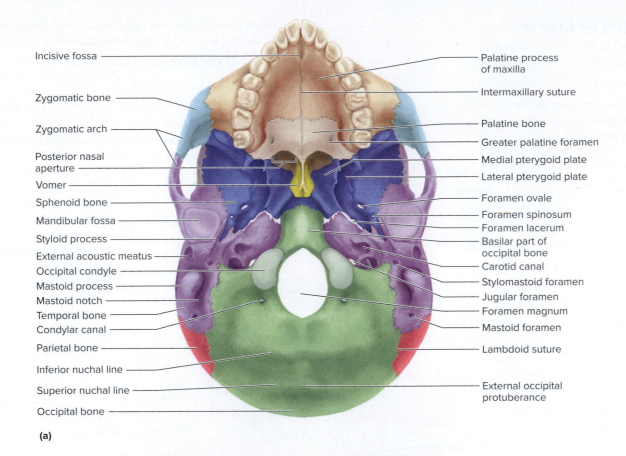

Incisive fossa

Zygomatic bone

Zygomatic arch

Posterior nasal aperture

Vomer

Sphenoid bone

Mandibular fossa

Styloid process

External acoustic meatus

Occipital condyle

Mastoid process

Mastoid notch

Temporal bone

Condylar canal

Parietal bone

Inferior nuchal line

Superior nuchal line

Occipital bone

Palatine process of maxilla

Intermaxillary suture

Palatine bone

Greater palatine foramen

Medial pterygoid plate

Lateral pterygoid plate

Foramen ovale

Foramen spinosum

Foramen lacerum

Basilar part of occipital bone

Carotid canal

Stylomastoid foramen

Jugular foramen

Foramen magnum

Mastoid foramen

Lambdoid suture

External occipital protuberance

(a)

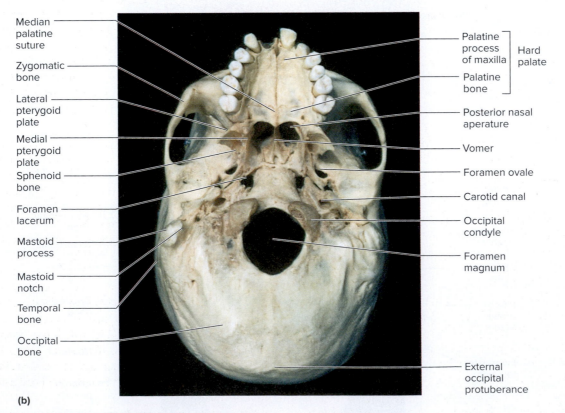

Median palatine suture

Zygomatic bone

Lateral pterygoid plate

Medial pterygoid plate

Sphenoid bone

Foramen lacerum

Mastoid process

Mastoid notch

Temporal bone

Occipital bone

Palatine process of maxilla

Palatine bone

} Hard palate

Posterior nasal aperature

Vomer

Foramen ovale

Carotid canal

Occipital condyle

Foramen magnum

External occipital protuberance

(b)

FIGURE 8.3 Skull, Inferior View. (a) Diagram; (b) photograph.

(b) ©Eric Wise

The facial bones can be remembered by this mnemonic device: "Monkeys in New Zealand like peaches very much." In the mnemonic device, all the bones except the last two (vomer and mandible) are paired bones.

Lateral View

A Activity

Locate the cranial bones from a lateral view. These are the **frontal, parietal, occipital, temporal, sphenoid,** and **ethmoid** bones. Find the coronal and lambdoid sutures, as well as the **squamous** (SKWAY-mus) **suture,** which separates the **temporal bone** from the **parietal** (pa-RY-eh-tul) **bone.** You may be able to see a bump on the posterior of the occipital bone from this angle. This is known as the **external occipital** (oc-SIP-ih-tul) **protuberance.** The hole on the lateral aspect of the head where the ears attach is the **external acoustic meatus.** The large process posterior and inferior to this opening is the **mastoid process.** A long, thin spine medial to the mastoid process is the **styloid process,** which has muscles that connect it to the hyoid bone, tongue, and larynx. The sphenoid bone can also be seen from this view just anterior to the temporal bone. A bony process is found lateral to the sphenoid bone. This is the **zygomatic arch,** which makes up the superior part of the cheek. The zygomatic arch is composed of part of the temporal bone, part of the zygomatic bone, and part of the maxilla. The zygomatic bone forms part of the lateral wall of the orbit. On the medial wall of the orbit are the ethmoid bone, lacrimal bone, maxilla, and nasal bone. Examine these structures in fig. 8.1.

Anterior View

A Activity

Examine the skull from the anterior. The large bone that makes up the forehead is the **frontal bone.** It is a single bone (fused from two bones in the fetus) that makes up the superior portion of the **orbits** (the eye sockets). The frontal bone has two ridges above the eyes called **supraorbital margins** (these are deep to the eyebrows). The hole in each of these ridges is called the **supraorbital foramen** and it allows the nerves and arteries to reach the face. Most of the bones inferior to the frontal bone are facial bones. The bony part of the nose is composed of the paired **nasal bones,** which join with the cartilage that forms the tip of the nose. Posterior and lateral to the nasal bones are the superior portions of the **maxillae** (mac-SILL-ee). The maxillae form the inferior, medial portion of the orbit, extend lateral and inferior to the nose, and hold the upper teeth. On the medial side of the orbits are the thin **lacrimal** (LACK-rih-mul) **bones,** which contain the **nasolacrimal duct,** a tube that drains tears into the nose. Locate these bones in fig. 8.2. Posterior to the lacrimal bones is the **ethmoid** (ETH-moyd) **bone.** It is very delicate and frequently broken on mishandled skulls. Posterior to the ethmoid is the **sphenoid bone,** which forms the posterior wall of the orbit and contains not only the **optic canal** (a passageway for

the optic nerve) but also the **superior orbital fissure** and the **inferior orbital fissure.** Note the bones on the lateral side of the orbit. These are the **zygomatic** (ZY-go-MAT-ic) **bones.**

The major bone inferior to the orbit is the maxilla, commonly known as the upper jaw, which also makes up the floor of the orbit. The two maxillae have sockets called **alveoli** (singular, *alveolus*), which contain the upper teeth. Extensions of bone between each pair of sockets are called **alveolar processes.** The **infraorbital foramen** is a small hole inferior to the eye in the maxilla and is a passageway for nerves and blood vessels. The most inferior bone of the face is the **mandible,** commonly known as the lower jaw. The mandible also has alveoli and alveolar processes and holds the lower teeth (fig. 8.2). The mandible begins as two bones in utero and fuses at the midline of the chin. This fusion of the **mental symphysis** joins the two mandible bones into one. On the lateral aspects of the mandible are the **mental foramina,** which conduct nerves and blood vessels to the tissue anterior to the jaw.

Inferior View

A Activity

Place the skull in front of you with the mandible removed (fig. 8.3). The largest hole in the skull, the **foramen magnum,** should be close to you. The foramen magnum is in the occipital bone and is where the brain joins the spinal cord. Lateral to the foramen magnum are the **occipital condyles,** processes that articulate with the first cervical vertebra. A small bump at the posterior part of the occipital bone is the external occipital protuberance, an attachment site for muscles. At the junction of the occipital bone and the temporal bone is the **jugular foramen,** a hole where the internal jugular vein begins. If you carefully insert a pipe cleaner into the jugular foramen and turn the skull over, you will see that the jugular foramen leads to the posterior part of the skull.

You can see the mastoid process of the temporal bone and a depression just medial to the process known as the **mastoid** (MAS-toyd) **notch.** Find the styloid processes in your specimen; they may be hard to locate because they frequently get broken in lab specimens. The styloid process is an attachment point for muscles that move the tongue, larynx, or hyoid. Medial to the styloid process is the **carotid canal,** which encloses the internal carotid artery, a vessel that takes blood to the brain. If you *carefully* insert a pipe cleaner into the carotid canal of a real skull, you will notice that the canal bends at about a 90-degree angle; if you turn the skull over, you will see that the opening occurs in the middle of the skull. You cannot do this with most plastic casts of skulls. At the junction of the temporal bone and the sphenoid bone is the **foramen lacerum** (lah-SER-um), which is next to the carotid canal. The temporal bone also has a **mandibular fossa,** the articulation site of the mandible. The zygomatic process of the temporal bone can also be seen from this view.

The sphenoid bone runs from one side of the skull to the other. The part of the sphenoid seen from the lateral view is one of the **greater wings** of the sphenoid. Two pairs of flattened processes can also be seen in the view, the **lateral pterygoid** (TERR-ih-goyd) plate and the **medial pterygoid plate** (*pterygoid* means winglike). These are attachments for muscles that extend from the sphenoid to the mandible. Just posterior to the pterygoid processes is the **foramen ovale,** which conducts one of the branches of the trigeminal nerve to the mandible.

In the midline of the skull and sometimes confused as part of the sphenoid is the **vomer.** This is a single bone of the face that forms part of the **nasal septum.** The holes on either side of the vomer are the **posterior nasal apertures.** Connected to the vomer and forming part of the **hard palate** are the **palatine bones.** The palatine bones are L-shaped bones with a horizontal plate and a vertical plate. The horizontal plates normally join at the **median palatine suture.** If this suture does not fuse completely at birth (along with the **intermaxillary suture,** which occurs between the two maxillae), an individual has a cleft palate. The rest of the hard palate consists of horizontal shelves of the maxillae. These form the **palatine processes of the maxilla.**

The major openings of the skull are presented in table 8.1. Locate the openings and note the number of structures that pass through these holes.

Superior View

Ⓐ Activity

Locate the major suture lines of the skull from the superior view (fig. 8.4). The frontal bone is separated from the pair of parietal bones by the **coronal suture.** The parietal bones are separated from each other by the **sagittal suture.** The parietal bones are separated from the occipital bone by the **lambdoid** (LAM-doyd)

suture (named after the Greek letter lambda), which looks similar to an upside-down Y. There may be small bones between the occipital bone and the parietal bones (or between other skull bones), and these are known as **sutural,** or **Wormian, bones.**

TABLE 8.1	Openings of the Skull
Opening	**Function or Structure in Opening**
Foramen magnum	Spinal cord vertebral arteries
Jugular foramen	Internal jugular vein, vagus, and other nerves
Carotid canal	Internal carotid artery
Stylomastoid canal	Facial nerve exits skull
Foramen lacerum	Closed by cartilage
Foramen ovale	Mandibular branch of trigeminal nerve
Foramen spinosum	Meningeal blood vessels
Foramen rotundum	Maxillary branch of trigeminal nerve
Optic canal	Optic nerve
Superior orbital fissure	Nerves to the eye and face
Inferior orbital fissure	Maxillary branch of trigeminal nerve
Mandibular foramen	Mandibular branch of trigeminal nerve
Mental foramen	Mental nerve and blood vessels
Supraorbital foramen	Supraorbital nerve and artery for face
Infraorbital foramen	Infraorbital nerve and artery for face
External acoustic meatus	Opening for sound transmission
Internal acoustic meatus	Vestibulocochlear nerve and facial nerve

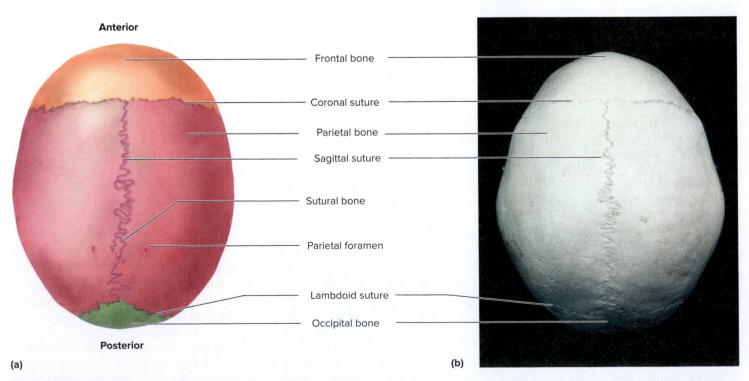

Anterior

Frontal bone

Coronal suture

Parietal bone

Sagittal suture

Sutural bone

Parietal foramen

Lambdoid suture

Occipital bone

Posterior

(a)

(b)

FIGURE 8.4 Skull, Superior View. (a) Diagram; (b) photograph.
(b) ©Eric Wise

Interior of the Cranium

Ⓐ **Activity**

With the superior portion of the skull removed, examine the cranial cavity, which is divided into three major regions. These are the **anterior cranial fossa,** a depression anterior to the lesser wings of the sphenoid; the **middle cranial fossa,** which lies between the lesser wings of the sphenoid and the petrous part of the temporal bone; and the **posterior cranial fossa,** posterior to the petrous part of the temporal bone. Examine fig. 8.5 for a view of the interior of the cranium.

Beginning with the anterior cranial fossa, you should find the centrally located **ethmoid bone.** A sharp ridge known as the **crista galli** projects from the main portion of this bone.

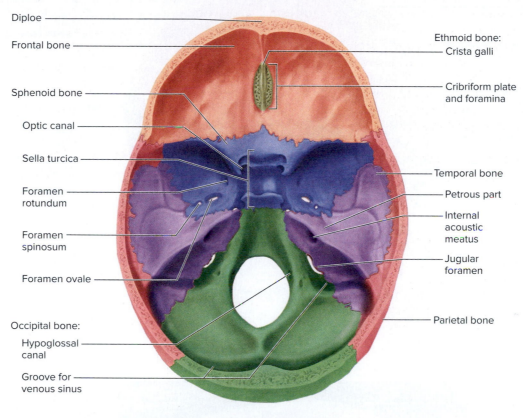

Diploe

Frontal bone

Sphenoid bone

Optic canal

Sella turcica

Foramen rotundum

Foramen spinosum

Foramen ovale

Occipital bone:
Hypoglossal canal

Groove for venous sinus

Ethmoid bone:
Crista galli

Cribriform plate and foramina

Temporal bone

Petrous part

Internal acoustic meatus

Jugular foramen

Parietal bone

(a) Superior view of cranial floor

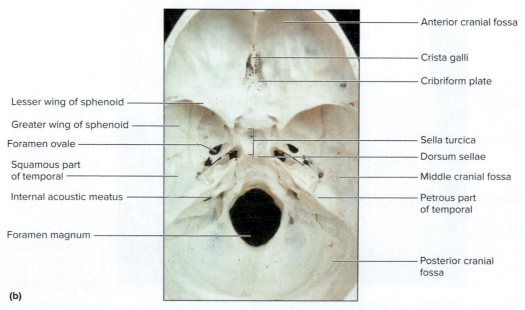

Lesser wing of sphenoid

Greater wing of sphenoid

Foramen ovale

Squamous part of temporal

Internal acoustic meatus

Foramen magnum

Anterior cranial fossa

Crista galli

Cribriform plate

Sella turcica

Dorsum sellae

Middle cranial fossa

Petrous part of temporal

Posterior cranial fossa

(b)

FIGURE 8.5 Interior of the Cranium. (a) Diagram; (b) photograph. *(b)* ©Eric Wise

The small, horizontal plate of bone with numerous holes lateral to the crista galli is the **cribriform** (CRIB-rih-form) **plate.** The holes in this plate are called the **olfactory foramina.** The nerves in these holes transmit the sense of smell from the nose to the brain. The nerves synapse in the olfactory bulbs, which are parts of the nervous system and are superior to the cribriform plate. If the skull was cut close to the orbit, you can see the **frontal sinus,** a hollow space, in the anterior portion of the frontal bone.

The dividing line between the anterior and middle cranial fossae is the sphenoid bone. Locate the lesser wings of the sphenoid

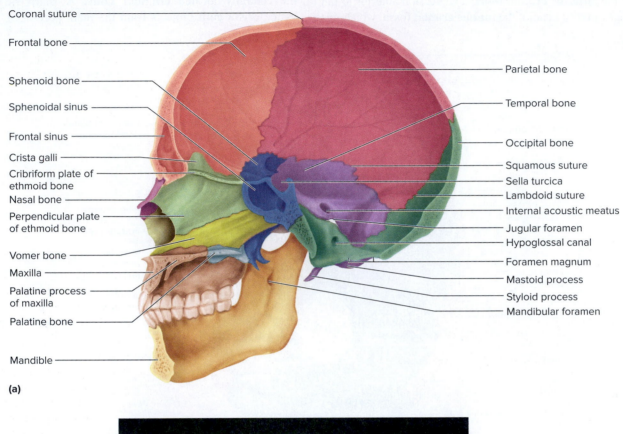

(a)

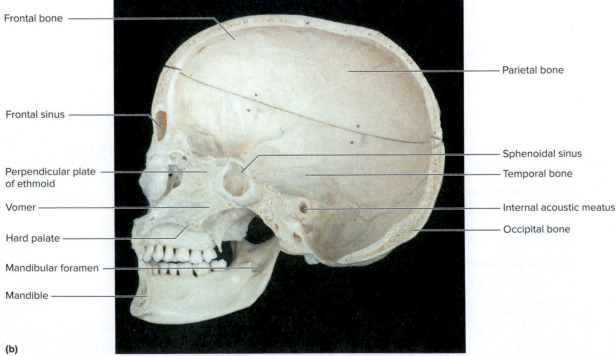

(b)

FIGURE 8.6 Skull, Midsagittal Section. (a) Diagram; (b) photograph.
(b) ©Eric Wise

and the **sella turcica** (SEL-la TUR-sih-ca) just posterior to it. The sella turcica has a small depression in which the pituitary gland sits. The posterior, raised part of the sella turcica is known as the **dorsum sellae.** The greater wings of the sphenoid are more inferior than the lesser wings, and each greater wing contains a **foramen rotundum,** which encloses a branch of the trigeminal nerve. The **foramen ovale** can be seen from this view as well.

Behind the sphenoid bone is the temporal bone, which has a flattened lateral section known as the **squamous part** and a heavier mass of bone known as the **petrous part.** The petrous part divides the middle and posterior cranial fossae. The petrous part has a hole in the posterior surface, the **internal acoustic meatus.** This is a passageway for the nerves of the inner ear.

The posterior cranial fossa is located dorsal to the petrous part of the temporal bone. It contains the foramen magnum and the jugular foramina. Two small openings are found near the foramen magnum: the **hypoglossal canals,** which allow the passage of the hypoglossal nerve. Most of this fossa is formed by the occipital bone.

Median Section of the Skull

Ⓐ Activity

Be extremely careful with the median section of the skull because many of the internal structures are fragile. Use fig. 8.6 as a guide.

Locate the **nasal septum,** composed of the **vomer,** the **perpendicular plate** of the ethmoid bone, and the **nasal cartilage** (absent in skull preparations). If the nasal septum is removed, locate the **superior nasal concha** of the ethmoid bone and the **middle nasal concha** also of the ethmoid bone (fig. 8.10). Below these is the **inferior nasal concha,** a separate, distinct bone. Look for the junction between the palatine bone and the palatine process of the maxilla. These two bony plates make up the hard palate.

If the mandible is present, locate the **mandibular foramen** on the inner aspect of the mandible. It transmits branches of the trigeminal nerve and blood vessels to the mandible.

Sinuses

There are numerous sinuses and air cells in the skull. These spaces provide shape to the skull while decreasing its weight and some add resonance to the voice. The **paranasal sinuses** occur around the region of the nose and are named for the bones in which they are found. They include the **frontal sinus,** the **maxillary sinus,** the **ethmoidal sinus** (ethmoidal cells), and the **sphenoidal sinus.** These sinuses may fill with fluid when a person has a cold and harbor bacteria causing secondary infections.

Ⓐ Activity

Locate the sinuses in skulls in the lab and compare them to fig. 8.7.

Fontanelles

The **fontanelles** are the "soft spots" of an infant's skull. There are four types of fontanelles, and some of them allow for the passage of the skull through the birth canal by enabling the bones of the cranium to slide over one another. After birth, the fontanelles allow for further expansion of the skull. The **anterior (frontal) fontanelle** is an area between the frontal bone and the parietal bones. The **posterior (occipital) fontanelle** is between the occipital bone and the parietal bones. The **sphenoid (anterolateral) fontanelles** are paired structures on each side of the skull and are located superior to the **sphenoid** bone, and the **mastoid (posterolateral) fontanelles** are paired structures posterior to the temporal bone. Most fontanelles fuse before 1 year of age, although the frontal fontanelle may fuse as late as age 2.

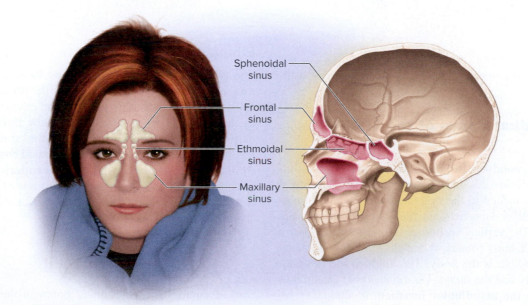

Sphenoidal sinus

Frontal sinus

Ethmoidal sinus

Maxillary sinus

FIGURE 8.7 Sinuses of the Skull.

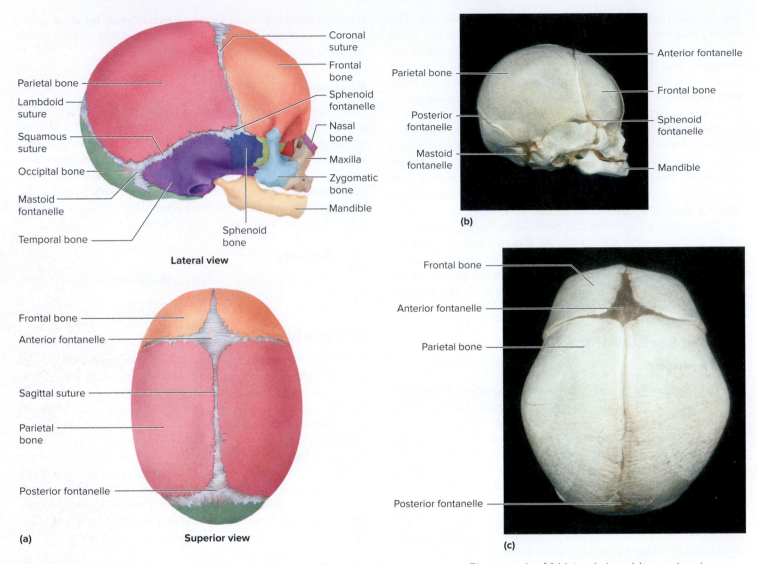

FIGURE 8.8 **Fetal Skull and Fontanelles.** (a) Diagram of lateral and superior views. Photograph of (b) lateral view; (c) superior view. *(b), (c)* ©Eric Wise

ⓐ Activity

Locate these structures in fig. 8.8 and on the material available in the lab.

Select Individual Bones of the Skull

Mandible

The mandible has a **condylar process** with a terminal **mandibular condyle,** which articulates with the temporal bone; a **coronoid process,** which lies medial to the zygomatic arch; and the **mandibular notch,** a depression between the condylar process and coronoid process. The vertical section of the mandible is known as the **ramus** (RAY-mus) (*ramus* = branch), and the horizontal portion of the mandible is the **body.** The **angle** is the posterior junction of the body and the ramus. On the inside of each ramus of the mandible is the **mandibular foramen,** a conduit for an artery, a vein, and a nerve. The parts of the mandible can also be identified in figs. 8.1 and 8.9.

Ethmoid

The **ethmoid bone** is a cranial bone located in the middle of the skull. It is a delicate bone due to the presence of ethmoidal cells. The **perpendicular plate** of the ethmoid can be seen in median view or from the anterior view through the external nares. The **orbital plate** is the part of the ethmoid that lines the medial wall of the orbit. The **middle nasal conchae** can be seen from the nasal cavity as well, but the **superior nasal conchae** are best seen by looking at an inferior view of the skull through the posterior nasal aperture or at a median view with the nasal septum removed.

ⓐ Activity

Examine isolated ethmoid bones in the lab and find the **crista galli, cribriform plate,** and other structures, as shown in fig. 8.10.

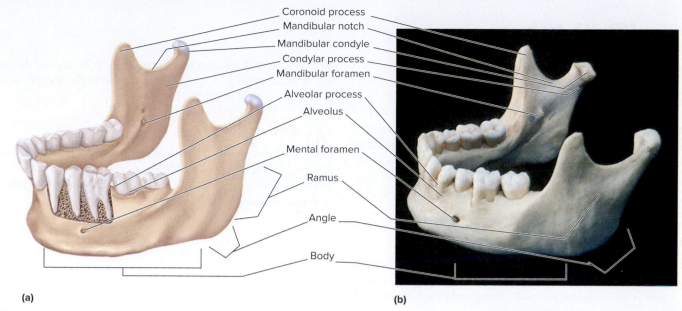

Coronoid process
Mandibular notch
Mandibular condyle
Condylar process
Mandibular foramen
Alveolar process
Alveolus
Mental foramen
Ramus
Angle
Body

(a)

(b)

FIGURE 8.9 **Mandible, Lateral View.** (a) Diagram; (b) photograph.
(b) ©Eric Wise

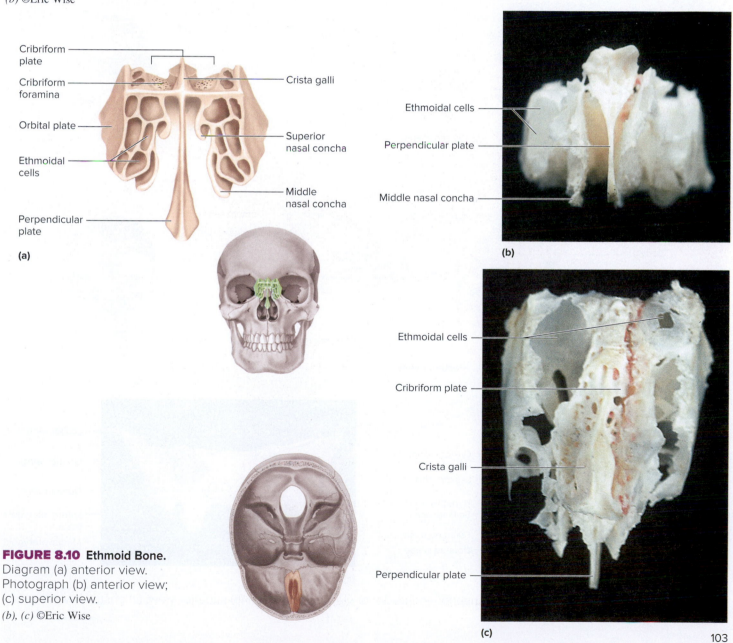

Cribriform plate
Cribriform foramina
Orbital plate
Ethmoidal cells
Perpendicular plate

Crista galli
Superior nasal concha
Middle nasal concha

(a)

Ethmoidal cells
Perpendicular plate
Middle nasal concha

(b)

Ethmoidal cells
Cribriform plate
Crista galli
Perpendicular plate

(c)

FIGURE 8.10 **Ethmoid Bone.**
Diagram (a) anterior view.
Photograph (b) anterior view;
(c) superior view.
(b), (c) ©Eric Wise

103

Sphenoid

(A) Activity

The **sphenoid** (SFEE-noyd) **bone** is seen in fig. 8.11. Examine an isolated sphenoid bone in the lab and locate the **greater wings,** the **lesser wings,** the **medial** and **lateral pterygoid plates,** the sella turcica, the **dorsum sellae,** and other features, as seen in fig. 8.11. The sella turcica has a small depression in it called the **hypophyseal fossa** in which the pituitary gland sits.

Temporal

The temporal bone is a paired cranial bone that has a squamous part that is the lateral part of the bone and forms part of the cranial vault. There is also a medial part of the temporal called the petrous

part. The petrous part contains the middle-ear bones and the opening of the internal acoustic meatus, as seen in the medial view of the temporal bone in fig. 8.12.

(A) Activity

Examine an isolated temporal bone and locate the zygomatic process, which articulates with the zygomatic bone, the mastoid process, which can be palpated (felt) as a bump posterior to the ear, and other features in fig. 8.12.

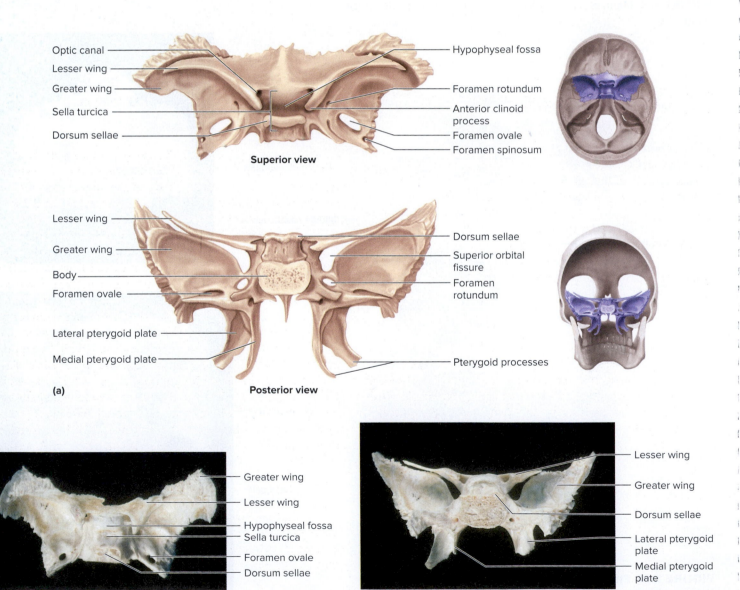

FIGURE 8.11 Sphenoid Bone. Diagram (a) superior and posterior views. Photograph (b) superior view; (c) posterior view.
(b), (c) ©Eric Wise

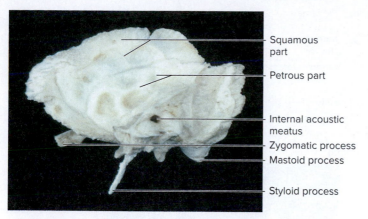

FIGURE 8.12 Right Temporal Bone, Medial View.
©Eric Wise

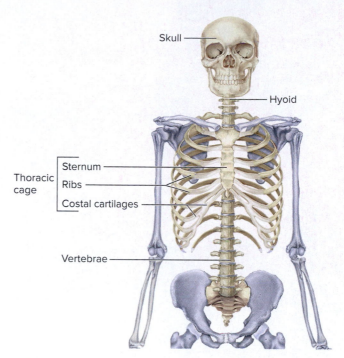

FIGURE 8.13 Major Bones of the Axial Skeleton.

Overview of the Vertebral Column

(A) Activity

Locate the bones of the axial skeleton in the lab and on fig. 8.13. Examine both the isolated bones and the articulated skeleton. Hold each bone up to the skeleton to see how it is positioned in relation to the other bones of the body.

Vertebral Column

The **vertebral column** in humans is significantly different from other mammals in that we are the only habitually bipedal mammal. In humans the vertebrae increase in size from the cervical to the lumbar vertebrae. This is due to the increase in weight on the lower vertebrae. The diameter of the vertebral canal increases from the sacral region to the cervical region. This reflects the greater number of neural fibers in the spinal cord as sensory information is transmitted to the brain and motor information is transmitted from the brain.

Spinal Curvatures

The spine has four curvatures, which alternate from superior to inferior. The **cervical lordosis (curvature)** is convex (bowed forward) when seen from the anatomical position. The **thoracic kyphosis (curvature)** is concave, while the **lumbar lordosis (curvature)** is convex, and the **sacral kyphosis (curvature)** is concave. These allow for balance in an upright posture.

(A) Activity

Locate the curvatures in fig. 8.14. In addition to the spinal curvatures described, there are abnormal spinal curvatures. **Scoliosis** is a lateral curvature. **Hyperkyphosis** (HI-per KY-foh-sis) is an exaggerated thoracic curvature, and **hyperlordosis** is an exaggerated lumbar curvature commonly seen during pregnancy (fig. 8.14).

Typical Vertebra

(A) Activity

Between the vertebrae are fibrocartilaginous pads known as **intervertebral discs.** Obtain a vertebra and examine it for the typical features (fig. 8.15). Place the vertebra in front of you so you can see through the large hole known as the **vertebral foramen.** Note that the large **body** of the vertebra supports the weight of the vertebral column and is in contact with the intervertebral discs. Note also the vertebral foramen, a hole where the spinal cord is located, and the **vertebral arch,** which consists of two **pedicles** (PED-ik-uls) and two **laminae** (LAM-i-nee). The pedicles are the parts of the arch that extend from the body of the vertebra to the two lateral projections, called the **transverse processes.** Each lamina (LAM-in-uh) is a broad, flat structure between the transverse process and the dorsal **spinous process.**

If you rotate the vertebra as illustrated in fig. 8.15, you should be able to see the vertebral body in lateral view, with the **superior articular process** and the **superior articular facet** visible. There is also an **inferior articular process** and an **inferior articular facet.** The inferior articular facet of a superior vertebra joins with the superior articular facet of the next inferior vertebra. If you put two adjacent vertebrae together, you can see the **intervertebral foramina,** holes where spinal nerves are located. These features are seen in fig. 8.16.

(A) Activity

After examining a representative vertebra, look at the vertebrae from each region of the spinal column and note the specific features of each region. Compare these to an articulated vertebral column (fig. 8.17).

Cervical Vertebrae

There are seven **cervical vertebrae** (listed as C1–C7), and these can be distinguished from all other vertebrae in that

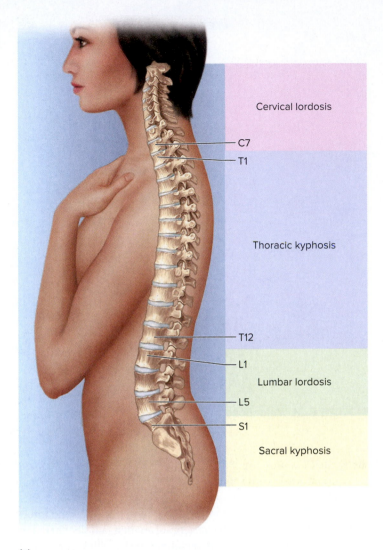

Cervical lordosis

C7
T1

Thoracic kyphosis

T12
L1

Lumbar lordosis

L5
S1

Sacral kyphosis

(a)

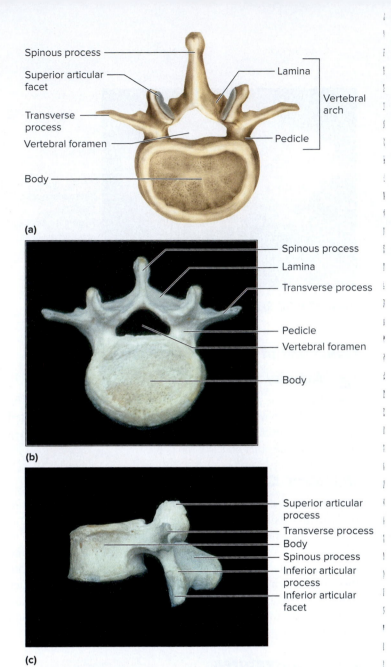

Spinous process
Superior articular facet
Lamina
Transverse process
Vertebral arch
Pedicle
Vertebral foramen
Body

(a)

Spinous process
Lamina
Transverse process
Pedicle
Vertebral foramen
Body

(b)

Superior articular process
Transverse process
Body
Spinous process
Inferior articular process
Inferior articular facet

(c)

FIGURE 8.15 Features of a Typical Vertebra. (a) Diagram, superior view. Photograph (b) superior view; (c) lateral view. *(b), (c)* ©Eric Wise

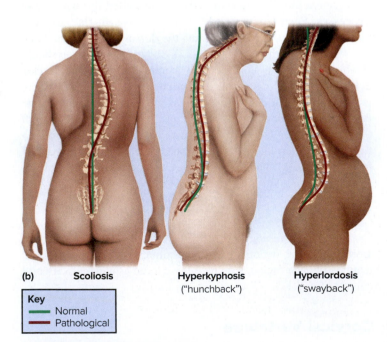

(b)   **Scoliosis**   **Hyperkyphosis** ("hunchback")   **Hyperlordosis** ("swayback")

Key
— Normal
— Pathological

FIGURE 8.14 Spinal Curvatures. (a) Lateral view; (b) abnormal spinal curvatures.

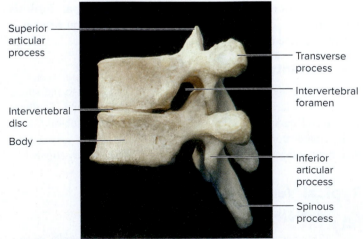

Superior articular process
Transverse process
Intervertebral foramen
Intervertebral disc
Body
Inferior articular process
Spinous process

FIGURE 8.16 Articulated Vertebrae, Lateral View.
©Eric Wise

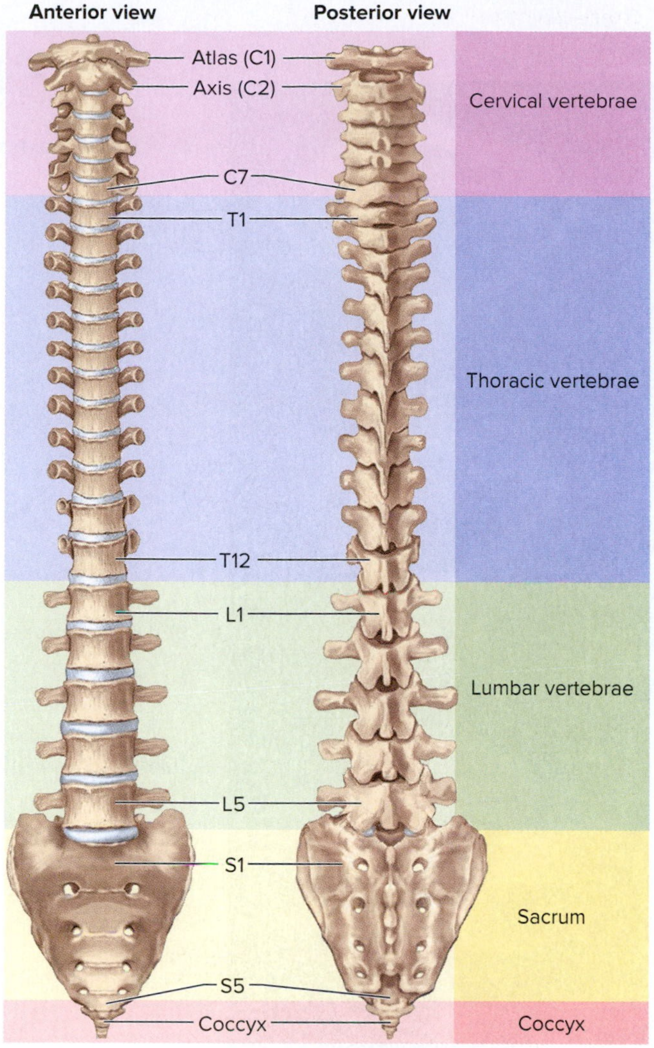

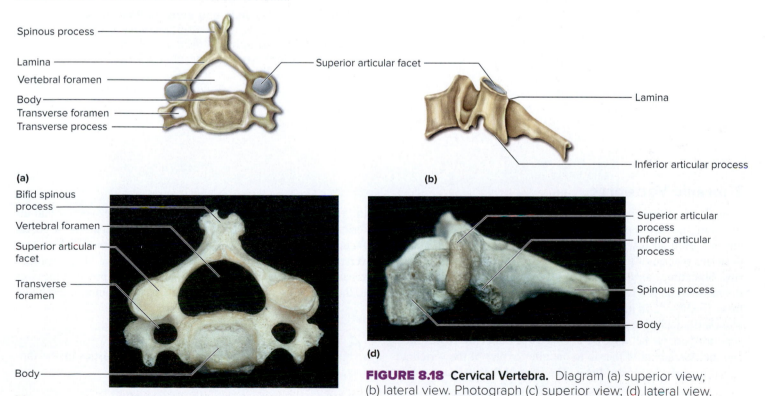

Anterior view   **Posterior view**

Atlas (C1)
Axis (C2)
Cervical vertebrae

C7
T1

Thoracic vertebrae

T12
L1

Lumbar vertebrae

L5
S1

Sacrum

S5
Coccyx   Coccyx

FIGURE 8.17 Overview of the Vertebral Column.

each cervical vertebra has two **transverse foramina** in addition to the vertebral foramen. The transverse foramina house the vertebral arteries and vertebral veins. Some of the cervical vertebrae (C2–C6) have **bifid** (BY-fid) **spinous processes** (meaning split in two), and the bodies of the cervical vertebrae are less massive than those of the inferior vertebrae.

A Activity

Examine the isolated cervical vertebrae in the lab and compare them to fig. 8.18.

The first cervical vertebra (C1) is known as the **atlas** and is the only cervical vertebra without a body. The atlas carries the weight of the head and was named after the Greek mythological figure Atlas, a giant who carried the heavens on his shoulders. The atlas joins with the head and allows you to nod your head to indicate "yes." The second cervical vertebra (C2) is the **axis,** and it has a unique process called the **dens** or **odontoid process,** that runs superiorly through the atlas. The odontoid process allows the atlas to rotate on the axis and allows you to rotate your head to indicate "no." The seventh cervical vertebra (C7) is known as the **vertebra prominens.** It has a **spinous process** that projects sharply in a posterior direction and can be palpated (felt) as a significant bump at the posterior base of the neck. Look at an atlas, an axis, and a vertebra prominens in the lab and compare them to fig. 8.19.

Spinous process
Lamina
Vertebral foramen
Body
Transverse foramen
Transverse process
Superior articular facet
Lamina
Inferior articular process
(a)   (b)

Bifid spinous process
Vertebral foramen
Superior articular facet
Transverse foramen
Body
Superior articular process
Inferior articular process
Spinous process
Body
(c)   (d)

FIGURE 8.18 Cervical Vertebra. Diagram (a) superior view; (b) lateral view. Photograph (c) superior view; (d) lateral view. *(c), (d)* ©Eric Wise

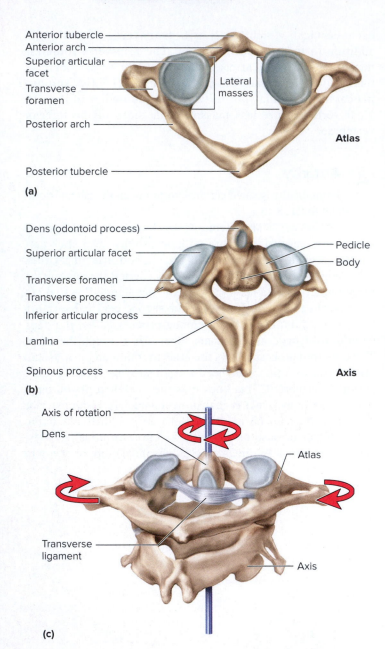

(a) Atlas

- Anterior tubercle
- Anterior arch
- Superior articular facet
- Transverse foramen
- Lateral masses
- Posterior arch
- Posterior tubercle

(b) Axis

- Dens (odontoid process)
- Superior articular facet
- Transverse foramen
- Transverse process
- Inferior articular process
- Lamina
- Spinous process
- Pedicle
- Body

(c)

- Axis of rotation
- Dens
- Atlas
- Transverse ligament
- Axis

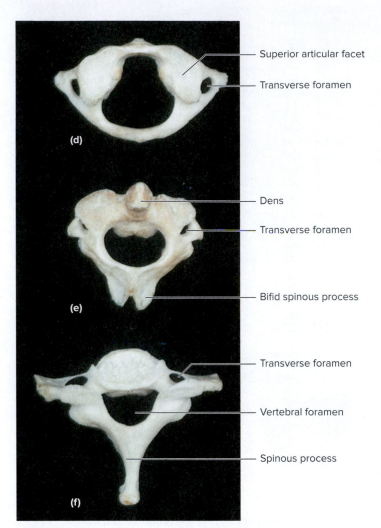

(d)
- Superior articular facet
- Transverse foramen

(e)
- Dens
- Transverse foramen
- Bifid spinous process

(f)
- Transverse foramen
- Vertebral foramen
- Spinous process

FIGURE 8.19 Atlas, Axis, and Vertebra Prominens. (a) Atlas, superior view; (b) axis, supero-posterior view; (c) atlas and axis joined. Photographs of three vertebrae, superior view (d) atlas; (e) axis; (f) vertebra prominens.
(d), (e), (f) ©Eric Wise

Thoracic Vertebrae

There are 12 **thoracic vertebrae** (listed as T1–T12 with T1 being the most superior thoracic vertebra). These vertebrae are distinguished from all other vertebrae by their markings on the lateral posterior surface of the body, which are attachment points for the ribs. Sometimes a rib attaches to just one vertebra, in which case the marking on the vertebral body is known as a **complete costal facet** (FASS-et). In most sections of the thoracic region, the head of a rib is attached to two vertebrae. The point of attachment at the superior part of the vertebra is called the **superior costal facet,** and the attachment of the rib to the inferior part of the vertebra is called the **inferior costal facet.** The head of the rib spans both of the vertebrae and articulates with the facet of the superior and inferior vertebrae. Vertebrae T1 through T10 have **transverse costal facets** for rib attachments on the terminal portions of the transverse processes. The thoracic vertebrae also have longer spinous processes than the cervical vertebrae, and the spinous processes of the thoracic vertebrae tend to angle in a more inferior direction.

(A) Activity

Examine fig. 8.20 and note the characteristics of the thoracic vertebrae as seen in the lab.

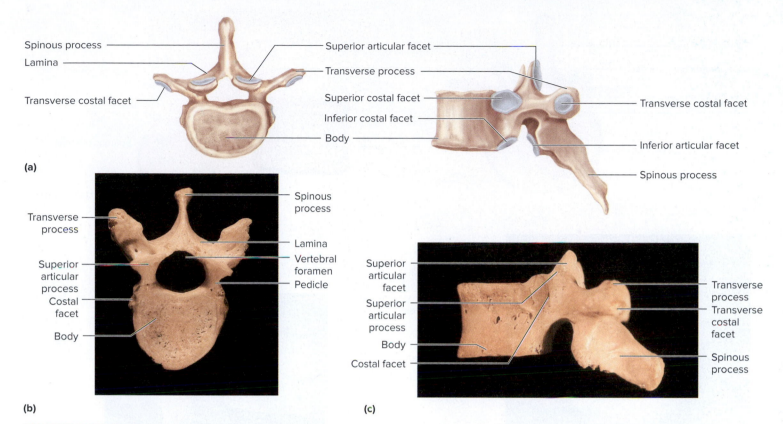

FIGURE 8.20 Thoracic Vertebra. Diagram (a) superior and lateral views. Photograph (b) superior view; (c) lateral view. *(b), (c)* ©Eric Wise

Lumbar Vertebrae

There are five **lumbar vertebrae** (listed as L1–L5). The lumbar vertebrae are distinguished from the other vertebrae by having neither transverse foramina nor rib facets. The spinous processes of the lumbar vertebrae tend to be more horizontal than those of the thoracic vertebrae, and the bodies of the lumbar vertebrae are larger, since they carry more weight than the thoracic vertebrae; yet the twelfth thoracic and the first lumbar vertebrae are remarkably similar. The last thoracic vertebra has rib facets on it, and the first lumbar does not.

Activity

Look at the lumbar vertebrae in the lab and compare them to fig. 8.21.

Sacrum

The **sacrum** (SAY-krum, SACK-rum) is a large, wedge-shaped bone composed of five fused vertebrae. The lines of fusion are called **transverse lines,** and these may be seen on both the anterior and the posterior sides. The sacrum is shaped like a shallow, triangular bowl. If you place the sacrum in front of you as you would a cereal bowl, the shallow depression is the **anterior surface.** The two rows of holes you see are the **anterior sacral foramina.** The holes on the back are known as the **posterior sacral foramina.**

Activity

Compare a bone in the lab to fig. 8.22.

The **sacral promontory** is a rim on the anterior superior part of the sacrum, and the **alae** (AIL-ee) (singular *ala*) are two expanded regions of the sacrum lateral to the promontory. The roughened areas on the lateral surfaces of the alae are the **auricular** (aw-RIC-you-lur) **surfaces** of the sacrum; each joins with the ilium to form the **sacroiliac** (SACK-ro-ILL-ee-ac) **joint.** As additional force is applied to the sacrum, it wedges itself between the iliac bones. If you examine the posterior surface of the sacrum, you will notice the posterior sacral foramina, the **median** and **lateral sacral crests,** and the superior and inferior openings of the **sacral canal.** The **sacral hiatus** is the inferior opening of the sacrum. As with the other vertebrae, the **superior articular process** and the **superior articular facets** join with the next most superior vertebra (the fifth lumbar vertebra). These features can be seen in fig. 8.22. The sacrum in humans is wedge-shaped; this allows the upper body to carry more weight than if the sacrum were box-shaped, as it is in many quadrupeds. Hold a block of wood or a box between your hands, as illustrated in fig. 8.23a. Have your lab partner push down on the box and see how much force is required for the box to slip through your hands. This would be similar to the forces that would act on a rectangular sacrum. Turn the box or block of wood so that it

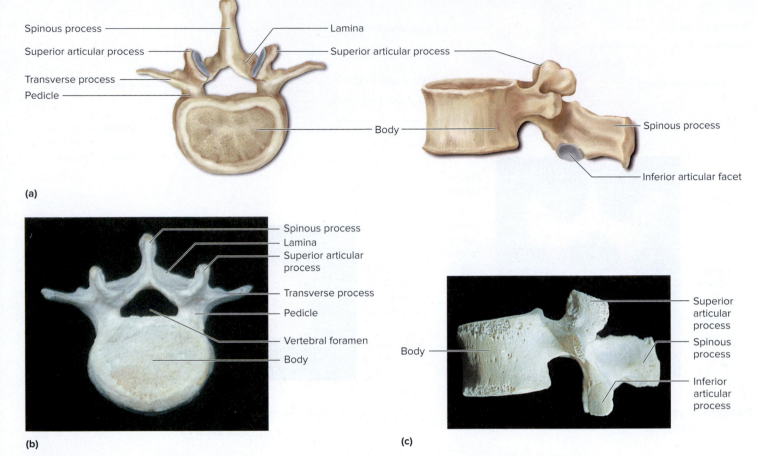

Spinous process — Lamina
Superior articular process — Superior articular process
Transverse process
Pedicle
Body — Spinous process
Inferior articular facet

(a)

Spinous process
Lamina
Superior articular process
Transverse process
Pedicle
Vertebral foramen
Body

(b)

Body
Superior articular process
Spinous process
Inferior articular process

(c)

FIGURE 8.21 Lumbar Vertebra. Diagram (a) superior and lateral views. Photograph (b) superior view; (c) lateral view. *(b), (c) ©Eric Wise*

forms a wedge in your hands, as illustrated in fig. 8.23b. Have your lab partner push down on the box. Is it easier or harder to dislodge the "sacrum" in this way?

Coccyx

The **coccyx** is the terminal portion of the vertebral column, and it consists of usually four fused vertebrae. The coccyx is fused with the sacrum in some individuals, especially females who disarticulated the joint during childbirth.

 Activity

Examine the coccyx in the lab and compare it to fig. 8.22.

Ribs

There are 12 pairs of **ribs** in humans, and these, along with the sternum and thoracic vertebrae, make up the thoracic cage. Each rib has a **head** that articulates with the body of one or two vertebrae. On the head is one or two **facets**, sites of articulation with the vertebral body. A narrow region near the head is the **neck** of

the rib. Locate this on fig. 8.24. The process near the neck on ribs 1–10 is the **tubercle** of the rib. This tubercle articulates with the transverse process of the vertebra.

 Activity

Examine the articulated skeleton in the lab and notice that the **shaft** of the ribs bends at about the same distance from the midline of the body as the inferior angle of the scapula. This can be seen on ribs 2 through 10 and is known as the **angle of the rib** or the **costal angle**. Some ribs have a flat **sternal end** that attaches to cartilage prior to joining the sternum. The superior edge of the rib is more rounded than the inferior edge. The depression that runs along the inferior side of each rib is known as the **costal groove**. Blood vessels and nerves are nestled in the space provided by the costal groove. With the costal groove in an inferior position and the flat sternal end toward the midline, determine whether you are looking at a left rib or a right rib.

The first seven pairs of ribs are **true ribs.** A true rib is one that attaches to the sternum by its own cartilage. These are known also as **vertebrosternal ribs,** as they attach to

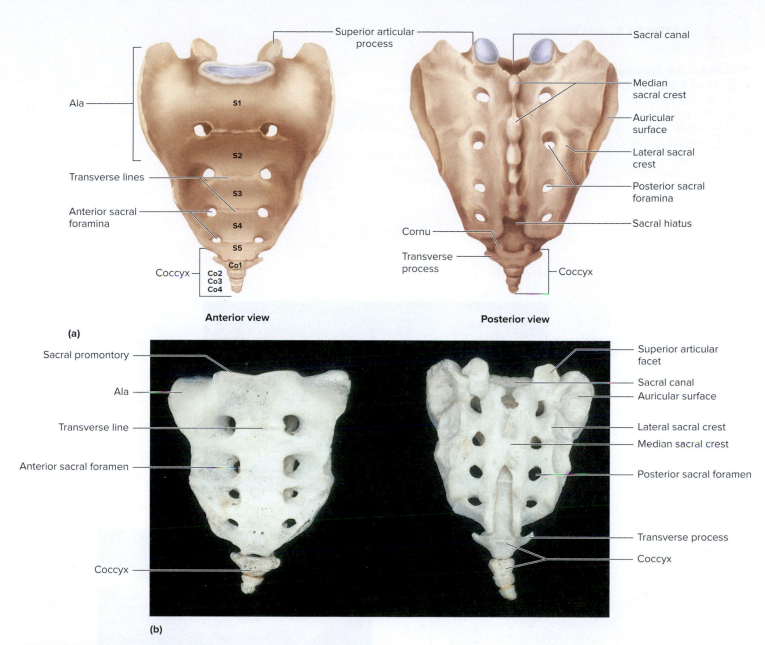

FIGURE 8.22 Sacrum and Coccyx. (a) Diagram; (b) photograph.
(b) ©Eric Wise

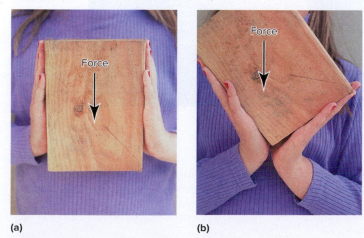

FIGURE 8.23 Forces Acting on the Sacrum. (a) Rectangular sacrum; (b) wedge-shaped sacrum.
(a), (b) ©Eric Wise

the vertebra posteriorly and to the sternum anteriorly. Ribs 8 through 12 are **false ribs,** because they do not attach to the sternum by their own cartilage. Ribs 8 through 10 in most people attach to the sternum by way of the cartilage of rib 7. These ribs are also called **vertebrochondral ribs** for their posterior attachment to the vertebrae and their anterior attachment to the cartilage of rib 7. Ribs 11 and 12 do not attach to the sternum at all and are known as **floating ribs** (a specific type of false rib) or **vertebral ribs.** Examine the articulated skeleton in the lab and compare it to fig. 8.25.

Sternum

The **sternum** is composed of three fused bones. The superior segment is the **manubrium** (ma-NOO-bree-um), and the depression at the top of the manubrium is the median **suprasternal notch** (**jugular notch**). The lateral indentations are the

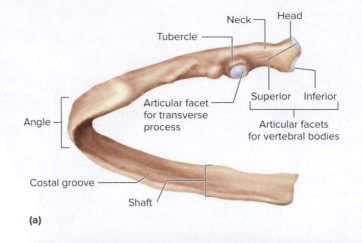

(a)

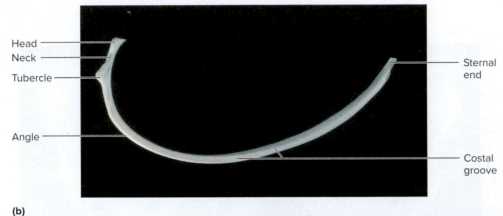

(b)

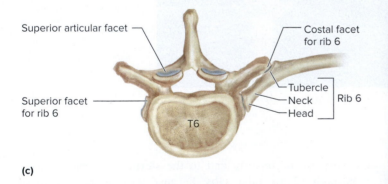

(c)

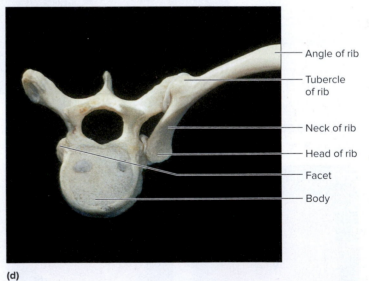

(d)

clavicular notches. The main portion of the sternum is the **body,** and between the body and the manubrium is the **sternal angle.** The sternal angle is a landmark for finding the second rib when using a stethoscope to listen to heart sounds. Locate the **costal notches** on the body of the sternum. These are where the cartilages of the ribs attach. The narrow, bladelike part that is the most inferior segment of the sternum is the **xiphoid** (ZYE-foyd) **process.** Care must be taken when performing cardiopulmonary resuscitation (CPR) so that pressure is applied to

the body of the sternum and not to the xiphoid process. If the force is applied to the xiphoid process, it could be fractured and driven into the liver.

Ⓐ **Activity**

Examine the structures of the sternum on lab specimens and in fig. 8.25.

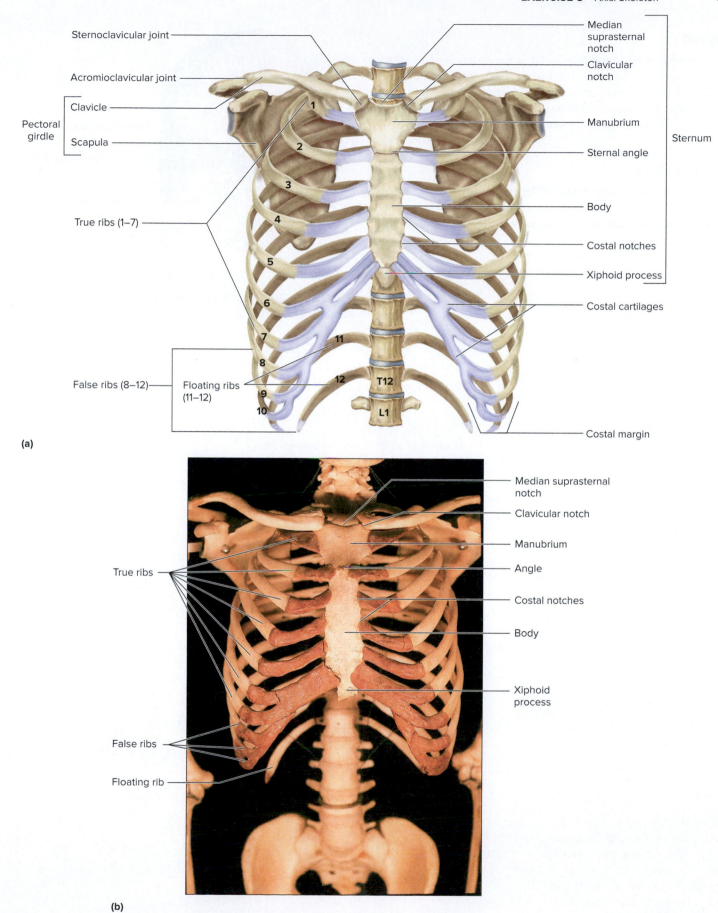

Figure 8.25 Thoracic Cage. (a) Diagram; (b) photograph.
(b) ©Eric Wise

Hyoid

The **hyoid** is a floating bone found at the junction of the floor of the mouth and the neck. The hyoid may be anchored by muscles from the anterior, posterior, or inferior directions; it aids tongue movement and swallowing. The larynx and trachea are suspended from the hyoid by the thyrohyoid ligament.

Ⓐ Activity

Locate the central **body** of the hyoid, the **greater horns (cornua)** (COR-new-uh) (singular, *cornu* [COR-new]), and the **lesser horns (cornua)** on the material in the lab and in fig. 8.26.

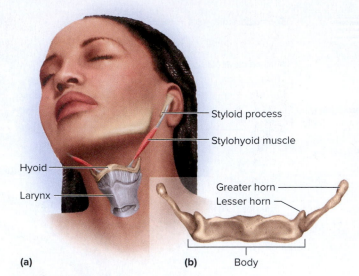

Styloid process

Stylohyoid muscle

Hyoid

Larynx

Greater horn

Lesser horn

(a)

(b)   Body

FIGURE 8.26   Hyoid Bone, Anterior View. (a) In situ; (b) details of hyoid.

REVIEW SECTION

Axial Skeleton

Name _____ *Date* _____

Lab Section _____ *Time* _____

Review Questions

1. The eyebrows are superficial to what bone? _____

2. What is the common name for the zygomatic bone? _____

3. What is the name of the bony process posterior to the earlobe? _____

4. The hard palate is made up of what bones? _____

5. What are the names of the major paranasal sinuses? _____

6. The mandible fits into what part of the temporal bone to form the jaw joint? _____

7. What bone is found just posterior to the ethmoid bone in the orbit? _____

8. The sella turcica is found in what bone? _____

9. What is the name of the bone that makes up most of the temple (the region superior and anterior to the ears)?

_____

10. What are the names of the bones that surround the anterior opening of the nose? _____

_____

11. The upper teeth are held by what bones? _____

12. In what bone would you find the foramen magnum? _____

13. What is the name of the bone that makes up most of the posterior surface of the orbit? _____

14. What are the two bony structures that make up the nasal septum? _____

_____

15. The mastoid process is located on which bone? _____

16. The sagittal suture separates the _____ bone from the _____ bone. (Circle the correct answer.)

  a. sphenoid, ethmoid   b. left parietal, right parietal   c. frontal, parietal   d. parietal, occipital

17. Which bone does *not* occur in the orbit? (Circle the correct answer.)

  a. maxilla   b. zygomatic   c. ethmoid   d. sphenoid   e. temporal

18. Which bone is *not* a paired bone of the skull? (Circle the correct answer.)

  a. zygomatic   b. temporal   c. lacrimal   d. vomer

19. Based on what you know about the maxillary sinus, why would a significant impact to the maxilla create a more difficult situation for healing than would the fracture of a long bone? _____

_____

20. Label the following illustration using the numbers for the terms provided.

  1. carotid canal   5. mastoid notch   9. palatine process of maxilla

  2. external occipital protuberance   6. mastoid process   10. posterior nasal aperture

  3. foramen magnum   7. occipital condyle   11. vomer

  4. jugular foramen   8. palatine bone   12. zygomatic bone

21. Label the following illustration using the numbers for the terms provided.

1. angle   5. mandibular notch

2. body   6. mental foramen

3. condyloid process   7. ramus

4. coronoid process

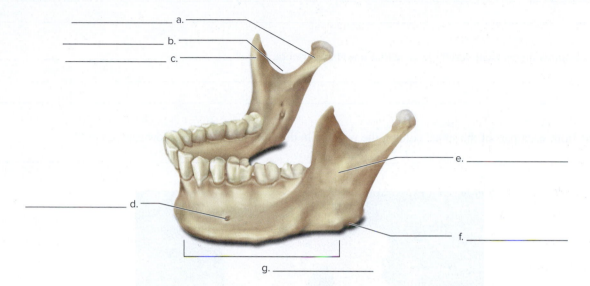

22. Label the following illustration using the numbers for the terms provided.

1. body   3. pedicle   5. transverse process

2. lamina   4. spinous process   6. vertebral arch

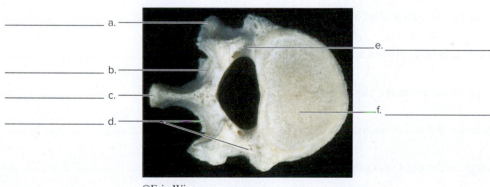

©Eric Wise

23. What part of a rib articulates with the transverse process of a vertebra? _____

24. Circle the correct answer. The most inferior portion of the sternum is the

    a. body.   b. manubrium.   c. angle.   d. xiphoid process.

25. Distinguish between the posterior sacral foramina and the sacral canal. _____

_____

_____

26. What two features do cervical vertebrae have that no other vertebrae have? _____

_____

_____

27. Determine from what part of the spinal column the vertebra in the photograph comes. How can you tell?

_____

©Eric Wise

28. What is the name of the superior portion of the sternum? _____

29. What structures make up the vertebral arch? _____

30. The superior articular facet of a vertebra articulates with what specific structure? _____

31. How many lumbar vertebrae are in the human body? _____

32. What is another name for the second cervical vertebra? _____

33. Which spinal curvature is the most superior? _____

34. Does the hyoid bone have any solid bony attachments? _____

35. A rib that attaches to the sternum by the cartilage of another rib has what name? _____

36. Is the angle of the rib on the anterior or posterior side of the body? _____

Appendicular Skeleton

INTRODUCTION

The appendicular skeleton consists of approximately 126 bones in four major groupings. These are the pectoral girdle, upper limbs, hip bones, and lower limbs. The **pectoral girdle** is composed of the scapula and the clavicle on each side. The **upper limb** consists of the humerus, radius, ulna, and bones of the hand, including the carpal bones, metacarpal bones, and phalanges. The lower portion of the appendicular skeleton is composed of the **hip bones,** each of which consists of three fused bones, the ilium, the ischium, and the pubis. The **lower limb** consists of the femur, patella, tibia, fibula, tarsal bones, metatarsal bones, and phalanges. These topics are discussed further in the Saladin text in chapter 8, "The Skeletal System."

OBJECTIVES

At the end of this exercise, you should be able to

1. locate and name all the bones of the appendicular skeleton;
2. name the significant surface features of the major bones;
3. place a bone into one of the four major groupings;
4. name the individual carpal bones and tarsal bones in articulated hands and feet, respectively;
5. determine whether a selected bone is from the left side or right side of the body;
6. place a clavicle with the sternal end pointing medially;
7. determine the gender of isolated or articulated hip bones.

MATERIALS

Articulated skeleton or plastic cast of articulated skeleton

Disarticulated skeleton or plastic casts of bones

Charts of the skeletal system

Plastic drinking straws cut on a bias or pipe cleaners for pointer tips

Foam pads of various sizes (to protect real bones from hard countertops)

PROCEDURE

General Considerations

(A) Activity

Locate the bones of the appendicular skeleton on the material in the lab and on fig. 9.1. When you study a bone, compare an isolated (or disarticulated) bone to the articulated skeleton (one that is joined together). Hold the individual bone up to the skeleton to see how it is positioned in relation to the other bones of the body. When you examine bones in the lab, do not use your pen or pencil to locate a structure. They can leave marks on the bones. Use a cut plastic straw, wooden applicator stick, or pipe cleaner to point out structures. Your instructor may want you to place real bone material on foam pads to cushion the bone from the tabletop. Use your time in lab to find the material at hand. Once you are at home, draw or trace bone images from your manual and be able to name all the parts.

Pectoral Girdle

The pectoral, or shoulder, girdle consists of the right and left **scapulae** (singular, *scapula*) and the right and left **clavicles.** The pectoral girdle provides a movable yet stable support for the upper limb.

Scapula

The scapula, commonly known as the shoulder blade, is found on the posterior, superior portion of the thorax. Because it has few bony attachments, it provides a great range of motion for the arm. The scapula is roughly triangular in shape and has three borders, a **superior border,** a **medial (vertebral) border,** and a **lateral (axillary) border.** On the superior border of the scapula is an indentation known as the **scapular notch,** which contains a nerve that innervates shoulder muscles. The anterior surface of the scapula is a smooth, hollowed depression known as the **subscapular fossa.** The posterior surface of the scapula is divided by the **scapular spine** into a **supraspinous fossa** and an **infraspinous fossa.** The spine of the scapula is a posterior crest that runs from the medial border to the lateral edge of the scapula, where it expands to

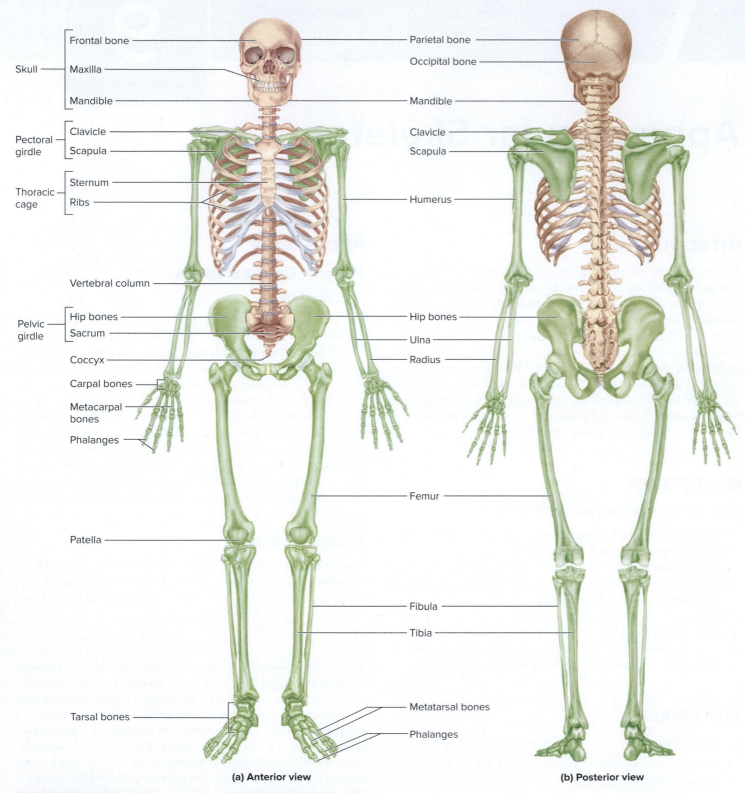

(a) Anterior view

(b) Posterior view

FIGURE 9.1 **Bones of the Appendicular Skeleton, Shaded in Green.**

form the **acromion** (ah-CRO-me-on). Another projection from the scapula is the **coracoid** (COR-uh-coyd) **process,** which projects anteriorly from the scapula. The scapula also has three sharp angles known as the **lateral angle,** the **inferior angle,** and the **superior angle.** The humerus moves in the shallow depression of the scapula known as the **glenoid cavity.** Just below the glenoid cavity is a bump known as the **infraglenoid tubercle,** an attachment point for the triceps brachii muscle. The features of the scapula are important because numerous muscles attach to them. You should also be able to distinguish a right scapula from a left one. The spine of the scapula is posterior, the inferior angle is less than 90 degrees, and the glenoid cavity is lateral for articulation with the humerus.

 Activity

Locate the features of the scapula on fig. 9.2.

Clavicle

Commonly known as the collarbone, the **clavicle** is a small bone between the scapula and the sternum. The blunt end of the clavicle is the **sternal end,** and the horizontally flattened end is the **acromial end.** Note the **conoid** (CON-oyd) **tubercle** on the inferior surface of the clavicle. Examine the clavicles in the lab and compare them to fig. 9.3. The arm transmits force to the clavicle, which is one of the most frequently broken bones in the body.

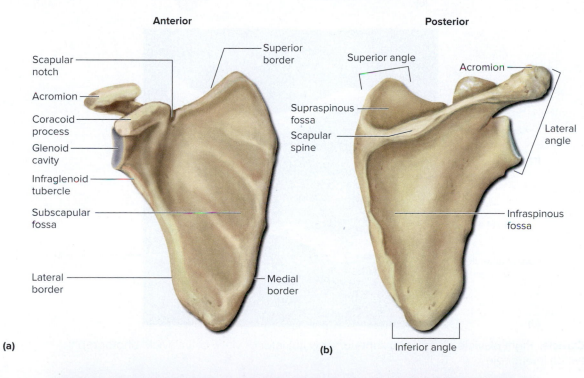

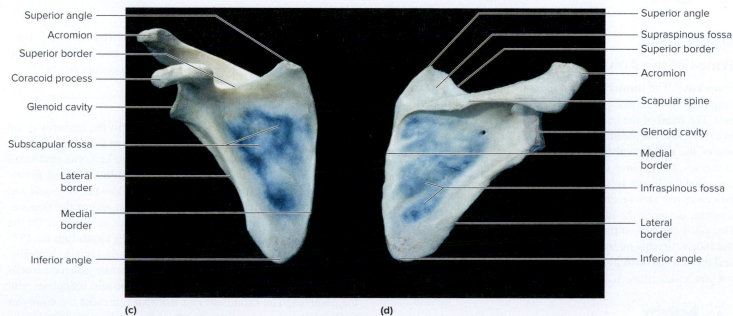

FIGURE 9.2 Right Scapula. Diagram (a) anterior view; (b) posterior view. Photograph (c) anterior view; (d) posterior view.
(c), (d) ©Eric Wise

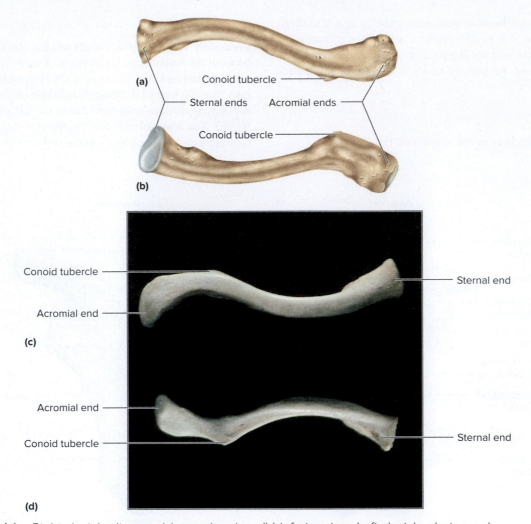

FIGURE 9.3 Clavicle. Right clavicle diagram (a) superior view; (b) inferior view. Left clavicle photograph (c) superior view; (d) inferior view.

(c), (d) ©Eric Wise

Upper Limb

Bones of the Arm

Humerus   The **humerus** is a long bone that articulates with the scapula at its proximal end and with the radius and ulna at its distal end. The **head** of the humerus is a hemispheric structure that ends in a rim known as the **anatomical neck.** The anatomical neck is the site of the **epiphyseal line.** The humerus has two large proximal processes, the **greater tubercle,** the largest and most lateral process, and the **lesser tubercle,** which is medial and smaller. These processes are attachment points for muscles originating from the scapula. Between the tubercles is the **intertubercular sulcus,** an elongated depression through which passes one of the tendons of the biceps brachii muscle. The intertubercular sulcus is anterior, and the head of the humerus is medial. These two features can orient you to determine if you have the right or left humerus.

Ⓐ Activity

Examine fig. 9.4 and locate these features. Below the tubercles of the humerus is a constricted region known as the **surgical neck.**

The surgical neck is so named because it is a frequent site of fracture. Locate the surgical neck on your arm, and in the following space, name a commonly known muscle that occurs nearby.

❓ Muscle near surgical neck: _____ (1)

A roughened area on the lateral surface of the humerus is the **deltoid tuberosity,** named for the attachment of the deltoid muscle. The shaft makes up most of the length of the humerus, and small holes penetrating the shaft are **nutrient foramina.** Nutrient foramina are holes that occur in many bones and allow blood vessels and nerves to enter into these bones. At the distal end of the humerus are the regions of articulation with the bones of the forearm. On the lateral side is a round hemisphere known as the **capitulum** (ca-PIT-you-lum). This is where the head of the radius fits onto the humerus. On the medial side is an hourglass-shaped structure called the **trochlea** (TROCK-lee-uh). The trochlea is where the ulna articulates with the humerus. The capitulum and trochlea represent the **condyles** (KON-dials) of the humerus. To the sides of these condyles are the **epicondyles** (EP-ee-KON-dials). The medial epicondyle is the bump that you can palpate (feel) on the medial side of the elbow. The ridge

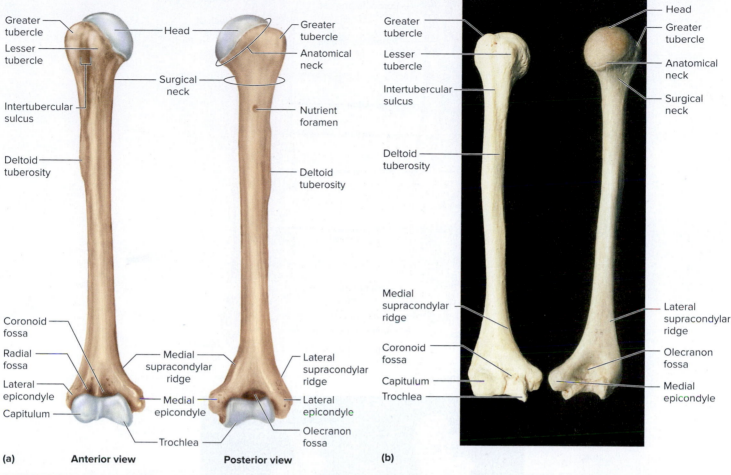

FIGURE 9.4 Right Humerus. (a) Diagram; (b) photograph.
(b) ©Eric Wise

that forms a wing of bone, proximal to this epicondyle, is the **medial supracondylar ridge.** The lateral epicondyle also has a **lateral supracondylar ridge.** The epicondyles and supracondylar ridges are points of attachments for muscles that manipulate the forearm and hand. At the distal end of the humerus are depressions. The ones on the anterior surface are the **coronoid fossa** and the **radial fossa,** and the one on the posterior surface is the **olecranon** (uh-LEC-ruh-non) **fossa.** Locate these structures on specimens in the lab and on fig. 9.4.

Bones of the Forearm

Two bones make up the skeletal portion of the forearm. These are the **radius,** a lateral bone, and the **ulna,** a medial one. The radius has a proximal **head** that looks like a wheel. Distal to this is the **radial tuberosity,** where the biceps brachii muscle inserts, and at the most distal end is the **styloid** (STY-loyd) **process** of the radius. On the distal end of the radius is a medial depression called the **ulnar notch.** This is where the distal part of the ulna joins with the radius.

Ⓐ **Activity**

Examine fig. 9.5 for these structures and compare them to the bones in the lab.

The ulna has a U-shaped depression (some students remember this as "U for ulna") on the proximal portion. This depression is called the **trochlear** (TROCK-lee-ur) **notch.** The most proximal portion of the ulna is the **olecranon,** commonly known as the apex of the elbow. The process distal to the trochlear notch is the **coronoid process,** which provides a relatively tight fit with the humerus and, along with the **tuberosity of the ulna,** provides an area for muscle attachment. The ulna has a *distal* **head,** unlike the radius (the head of the radius is proximal), and a styloid process near the head. Locate these structures in fig. 9.5. On the proximal part of the ulna, where the head of the radius fits into the ulna, is a depression known as the **radial notch.** If you know that the trochlear notch is anterior and the radial notch is lateral, you can determine whether you are looking at a left or right ulna. In the following space, list what side of the body the bones of the forearm you are studying come from.

❓ Side of the body: _____ (2)

Bones of the Hand

Each hand consists of three groups of bones, the **carpal bones,** the **metacarpal bones,** and the **phalanges** (singular, *phalanx*).

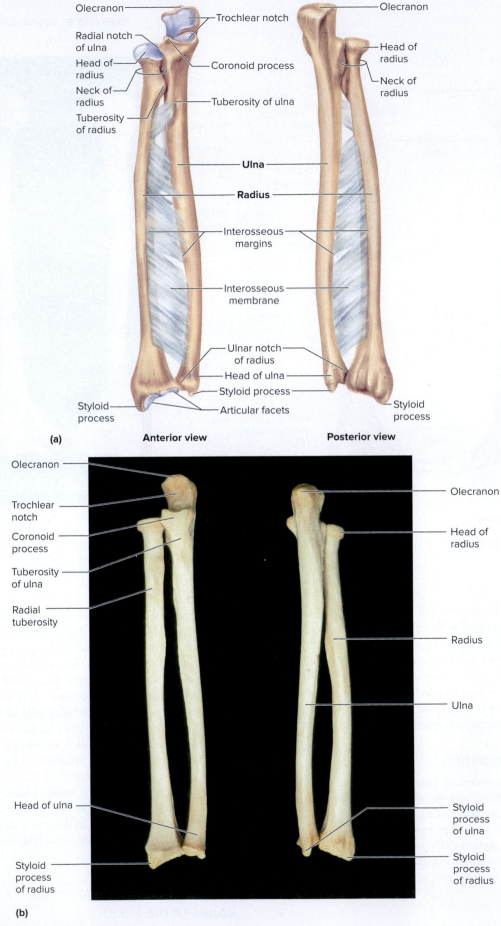

(a)

Olecranon — Trochlear notch

Radial notch of ulna

Head of radius

Neck of radius

Tuberosity of radius

Coronoid process

Tuberosity of ulna

Ulna

Radius

Interosseous margins

Interosseous membrane

Ulnar notch of radius

Head of ulna

Styloid process

Styloid process

Articular facets

Anterior view

Olecranon

Head of radius

Neck of radius

Styloid process

Posterior view

(b)

Olecranon

Trochlear notch

Coronoid process

Tuberosity of ulna

Radial tuberosity

Head of ulna

Styloid process of radius

Olecranon

Head of radius

Radius

Ulna

Styloid process of ulna

Styloid process of radius

FIGURE 9.5 Right Radius and Ulna. (a) Diagram; (b) photograph.

(b) ©Eric Wise

Each human hand has eight carpal bones (fig. 9.6). These bones are aligned in two rows. The proximal row from lateral to medial includes the **scaphoid, lunate, triquetrum** (tri-KWĒ-trum), and **pisiform** (PIE-sih-form). The distal row includes the **trapezium**, the **trapezoid**, the **capitate**, and the **hamate**. On the hamate is a hooklike projection called the **hamulus**. Another way to put the bones in proper order is to use a mnemonic device that uses the first letter of each carpal bone (S for scaphoid, L for lunate, etc.)

in a sentence: "Say Loudly To Pam, Time To Come Home." Most anatomical terms derive from Latin or Greek. The carpal bones are little bones yet they have their own distinctive shape. The scaphoid is the only carpal bone with a Greek name. It is shaped like a little boat (*scaphos* is Greek for boat). The rest of the carpal bones have Latin names. The lunate is shaped like a crescent moon (*luna* means moon), triquetrum is derived from *triangle*, and pisiform is *pea-shaped*. In the distal row, the trapezium (it rhymes with *thumb*

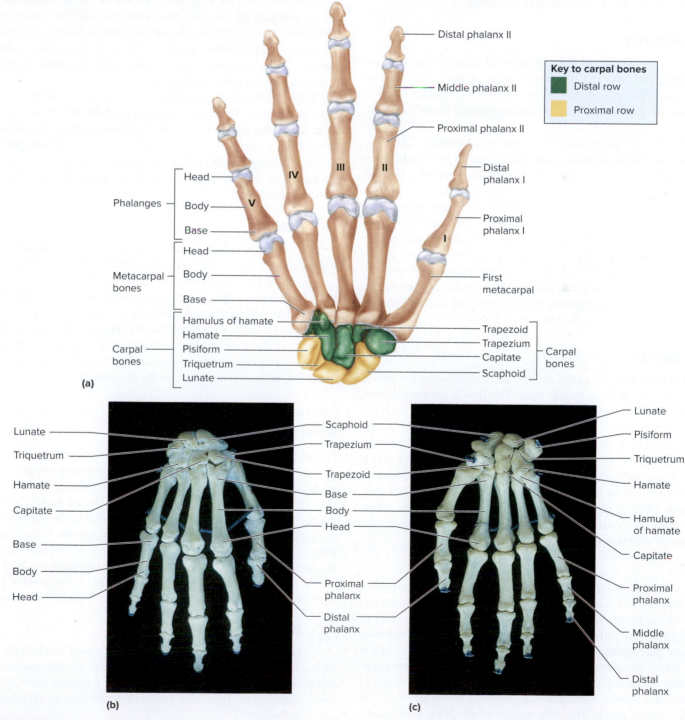

FIGURE 9.6 Bones of the Right Hand. Diagram (a) anterior view. Photograph (b) posterior view; (c) anterior view.
(b), (c) ©Eric Wise

and is proximal to the thumb) and the trapezoid are both named for a *four-sided table*. The capitate refers to it being *head-like* and the hamate refers to *hook*. The carpal bones serve as regions of attachment for forearm muscles and for intrinsic muscles of the hand.

The **metacarpal bones** are located in the palm. Each metacarpal bone consists of a proximal **base**, a **body**, and a distal **head**. The metacarpals are labeled by Roman numerals I to V, with I being proximal to the thumb, and V being proximal to the little finger. Metacarpal bones II to V have little movement, but metacarpal I allows for significant movement of the thumb.

 Activity

Locate metacarpal I on the skeletal material and on your own hand, and note the substantial range of movement in your own hand. This type of movement is covered in greater detail in Exercise 10. Examine the carpal and metacarpal bones in the lab and in fig. 9.6.

Distal to the metacarpal bones are the **phalanges**. There are 14 bones that make up this group. Each finger of the hand, except for the thumb, has three phalanges, a **proximal, middle,** and **distal** phalanx. The thumb has only a proximal and distal phalanx. Identify each bone by noting the digit it comes from and whether it is proximal, middle, or distal. The thumb is the first phalanx and the little finger is the fifth phalanx. Therefore, the tip of the little finger is the distal phalanx V and the base of the middle finger is the proximal phalanx III. Like the metacarpal bones, each phalanx has a **base, body,** and **head,** and also like the metacarpal bones, phalanges are classified as long bones. Locate the various bones of the fingers in material in the lab and in fig. 9.6.

Hip Bones

The **hip bones** (ossa coxae) and the sacrum make up the **pelvic girdle.** Each hip bone results from the fusion of three bones, the **ilium** (ILL-ee-um), the **ischium** (ISS-kee-um), and the **pubis** (PYOU-bis). The hip bones are joined anteriorly by the interpubic disc, a fibrocartilage pad, and this along with the anterior portions of the pubis forms the **pubic symphysis.** Inferior to this is the **subpubic angle** (pubic arch). This angle is greater than 90 degrees in females and less than 90 degrees in males. The large hole at the inferior part of the hip bone is the **obturator** (OB-tur-aye-tur) **foramen.** This space is covered by the **obturator membrane.**

Ilium

The most superior hip bone is the ilium. If you feel the top part of your hip, you are feeling the **iliac crest,** a long, crescent-shaped ridge of the ilium. The two processes that jut from the ilium in the front are the **anterior superior iliac spine** and the **anterior inferior iliac spine.** These are attachment points for some of the thigh flexor muscles. The posterior ilium has the **posterior superior iliac spine** and the **posterior inferior iliac spine.** Inferior to the posterior inferior iliac spine is a large depression known as the **greater sciatic notch.** The outer surface of the ilium is an attachment point for the gluteal muscles. The interior, shallow depression of the ilium is known as the **iliac fossa.**

 Activity

Locate these areas in fig. 9.7.

Ischium and Pubis

The ischium is inferior to the ilium and is the part of the hip bone on which you sit. The **ischial spine** is a sharp projection on the ischium, and the **ischial tuberosity** is an attachment site for the hamstring muscles. Anterior and superior to the ischial tuberosity is a cuplike depression known as the **acetabulum** (a region in which all three hip bones are joined). The acetabulum is the socket into which the femur fits. The ischium connects to the pubis by an elongated portion of bone known as the **ischial ramus.** This ramus connects to the **inferior pubic ramus,** which is posterior to the pubic symphysis.

If the hip bones are articulated, you can easily see the upper basin of the pelvis, known as the **greater pelvis.** The greater pelvis is medial to the iliac fossa. A rim of bone separates the greater pelvis from a deeper, smaller basin known as the **lesser pelvis.** This rim is known as the **pelvic brim** in an intact pelvis or **arcuate line** on an isolated hip bone. The lesser pelvis is medial to the obturator foramen.

 Activity

Examine the disarticulated hip bone in lab. You can determine if you have a left or right bone and the gender of the individual. In the pelvis from a female, the greater sciatic notch has a fairly broad angle, which approximates the arc inscribed by your outstretched thumb and index finger. In the pelvis of a male, the greater sciatic notch is less than 90 degrees and approximates the angle if you were to separate the index finger and the middle finger. As you examine the bones in the lab, look at fig. 9.7 and locate the various terms listed on the illustration.

Lower Limb

Bones of the Thigh

Femur   The **femur** is the longest and heaviest bone in the body, and it is the only bone of the thigh. It articulates proximally with the hip bone and inferiorly with the **tibia.** The femur has a proximal, medial spherical **head,** which inserts into the acetabulum of the hip bone, and an elongated, constricted **neck.** On the head of the femur is a depression known as the **fovea capitis.** It is a region where a ligament attaches from the femur to the acetabulum. Below the neck are two large processes called the **trochanters,** areas of muscle attachment. The larger, anterior process is the **greater trochanter,** and the smaller, posterior process is the **lesser trochanter.**

 Activity

Locate these structures in fig. 9.8. You should also find the **intertrochanteric crest,** a posterior ridge between the two trochanters and the anterior **intertrochanteric line.** In the following space, determine which is more broad, the intertrochanteric crest or the intertrochanteric line.

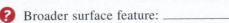 Broader surface feature: _____ (3)

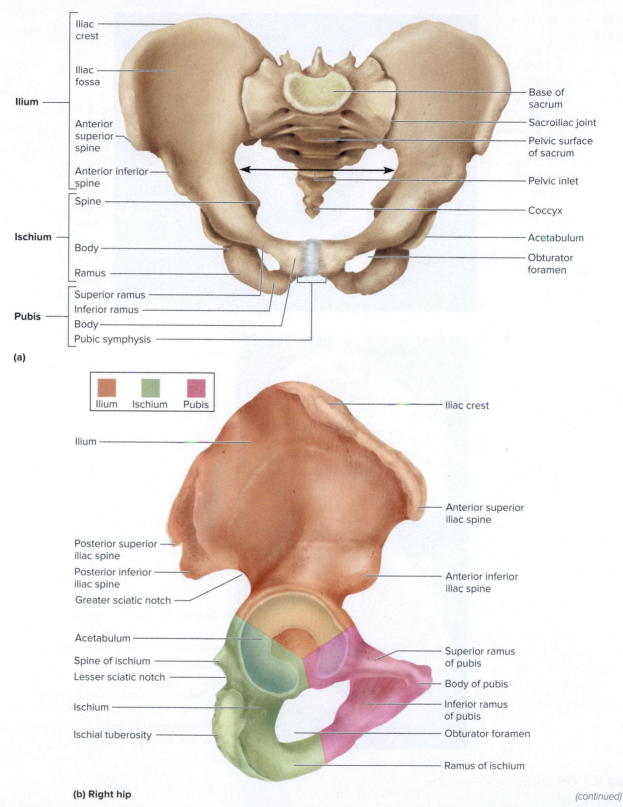

(a)

Ilium
- Iliac crest
- Iliac fossa
- Anterior superior spine
- Anterior inferior spine

Ischium
- Spine
- Body
- Ramus

Pubis
- Superior ramus
- Inferior ramus
- Body
- Pubic symphysis

Base of sacrum
Sacroiliac joint
Pelvic surface of sacrum
Pelvic inlet
Coccyx
Acetabulum
Obturator foramen

Ilium Ischium Pubis

Ilium
Iliac crest
Anterior superior iliac spine
Posterior superior iliac spine
Posterior inferior iliac spine
Greater sciatic notch
Anterior inferior iliac spine
Acetabulum
Spine of ischium
Lesser sciatic notch
Ischium
Ischial tuberosity
Superior ramus of pubis
Body of pubis
Inferior ramus of pubis
Obturator foramen
Ramus of ischium

(b) Right hip

(continued)

FIGURE 9.7 Hip Bones. Diagram (a) anterior view; (b) lateral view. Photograph (c) anterior view; (d) lateral view.

The **shaft** of the femur is curved and bows anteriorly with the posterior shaft marked by the **linea aspera** (meaning *rough line*). At the proximal portion of the linea aspera is a roughened area called the **gluteal tuberosity,** an attachment site for the gluteus maximus muscle. The distal portion of the femur consists of **medial** and **lateral condyles.** These two smooth processes come into contact with the condyles of the tibia. To the sides of the condyles are bulges known as the **epicondyles.** Locate the **lateral**

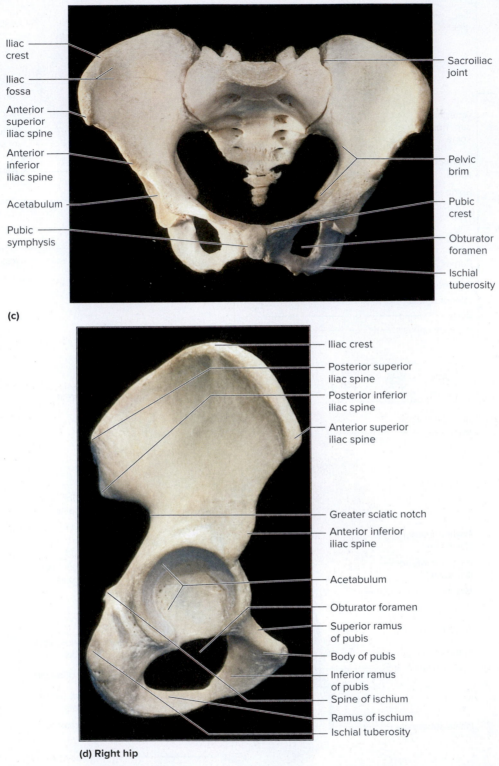

(c)

(d) Right hip

FIGURE 9.7 *Continued.*

(c), (d) ©Eric Wise

epicondyle and the **medial epicondyle** on the femur. Also note the location of the **adductor tubercle,** a small, triangular process proximal to the medial epicondyle. This is one of the attachment points for one of the adductor muscles. The head of the femur is medial, and the linea aspera is posterior. These features help distinguish whether you have a left or right femur.

Patella   The **patella** is a bone formed in the tendon that runs from the quadriceps muscles on the anterior thigh to the tibia. The patella is a specific type of bone known as a **sesamoid bone.** Sesamoid bones develop in tendons. The patella ossifies between ages three and six. When the leg is flexed, as when kneeling, the patella protects the ligaments of the knee joint.

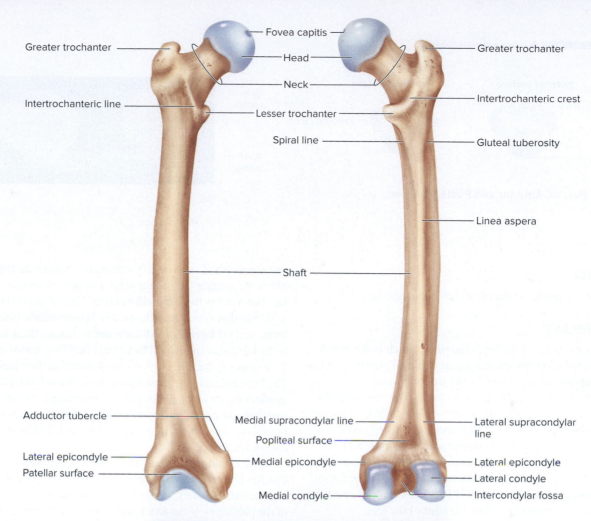

Greater trochanter

Fovea capitis

Head

Neck

Greater trochanter

Intertrochanteric line

Lesser trochanter

Intertrochanteric crest

Spiral line

Gluteal tuberosity

Linea aspera

Shaft

Adductor tubercle

Medial supracondylar line

Lateral supracondylar line

Popliteal surface

Lateral epicondyle

Medial epicondyle

Lateral epicondyle

Patellar surface

Lateral condyle

Medial condyle

Intercondylar fossa

(a)

Anterior view

Posterior view

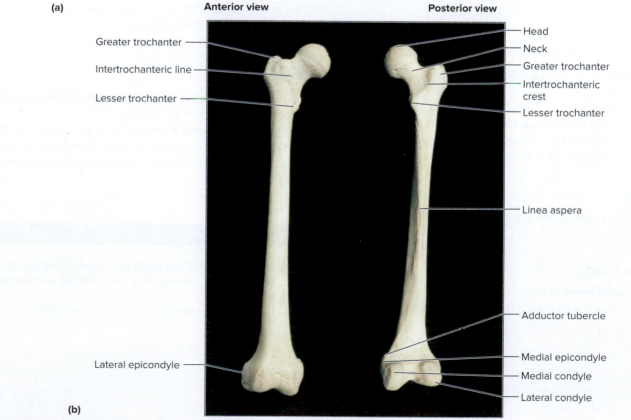

Greater trochanter

Head

Neck

Intertrochanteric line

Greater trochanter

Intertrochanteric crest

Lesser trochanter

Lesser trochanter

Linea aspera

Adductor tubercle

Lateral epicondyle

Medial epicondyle

Medial condyle

Lateral condyle

(b)

FIGURE 9.8 Right Femur. (a) Diagram; (b) photograph.

(b) ©Eric Wise

Anterior surface   **Posterior surface**

Base of patella

Apex of patella

Articular facets

(a)

Base of patella

Apex of patella

Base

Articular facets

Apex

(b)

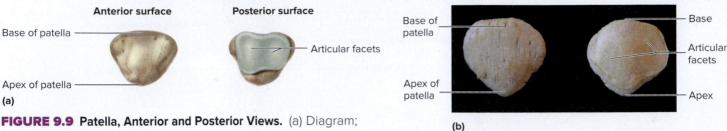

FIGURE 9.9 Patella, Anterior and Posterior Views. (a) Diagram; (b) photograph.

(b) ©Eric Wise

 Activity

Examine the patella in the lab and compare it to fig. 9.9.

Bones of the Leg

Tibia   The largest bone of the leg is the **tibia,** which is the weight-bearing bone of the leg inferior to the femur. It is medial to the fibula. The **tibial condyles** are separated by the **intercondylar eminence,** and they articulate with the condyles of the femur. The tibia is roughly triangular in cross section. The ridge on the anterior tibia is known as the **anterior border.** This is the ridge that hurts when you hit your shin. There is a rough area on the anterior surface of the tibia known as the **tibial tuberosity.** This is the attachment point for the patellar ligament. At the distal end of the tibia is an extension of bone known as the **medial malleolus** (MAH-lee-OH-lus). This bump is one part of the ankle joint as it articulates with the talus of the foot.

 Activity

Locate the structures of the tibia in the lab and in fig. 9.10.

Fibula   The **fibula** is smaller than the tibia (you can remember that it is smaller than the tibia because you "tell a little fib"). The fibula is lateral to the tibia. The fibula has a proximal **head** and a distal process called the **lateral malleolus.** The lateral malleolus is the other part of the leg that forms a joint with the talus. The fibula is not a weight-bearing bone but is an attachment point for muscles.

 Activity

Examine the bones in the lab and compare the fibula to fig. 9.10.

Bones of the Foot

There are seven **tarsal bones** of the foot, including the **talus,** which is the bone of articulation with the leg. Directly below the talus is the

calcaneus (cal-CAY-nee-us), commonly known as the heel bone. The bone anterior to both the talus and the calcaneus is the **navicular.** The foot has three **cuneiform** (cue-NEE-ih-form) bones, known as the **medial** (first) **cuneiform,** the **intermediate** (second) **cuneiform,** and the **lateral** (third) **cuneiform.** Locate these on specimens in the lab and in fig. 9.11. The lateral cuneiform bone is *not* the most lateral bone in the foot but is the most lateral of the cuneiform bones. The bone lateral to the third cuneiform is the **cuboid.** One way to remember the tarsal bones is with the mnemonic "children that never march in line cry." The beginning letter of each word is the same as a tarsal bone (calcaneus, talus, navicular, medial, intermediate, lateral cuneiform, and cuboid).

The **metatarsal bones** are similar to the metacarpal bones in that the first metatarsal bone is under the largest digit, the big toe, and the fifth metatarsal bone is under the smallest digit. The pattern of the phalanges of the foot is the same as in the hand, with toes II to V having a **proximal, middle,** and **distal phalanx** each, and the big toe having a proximal and distal phalanx.

 Activity

Examine the material in the lab and compare the metatarsal bones and phalanges to fig. 9.11.

When you have finished locating the major features of the bones of the appendicular skeleton, test yourself and your lab partner by pointing to various structures and seeing how accurately you identify the structures.

STUDY HINT

Use your time in lab to *find* the material at hand. Once you are at home, draw or trace material from your manual and name all of the parts.

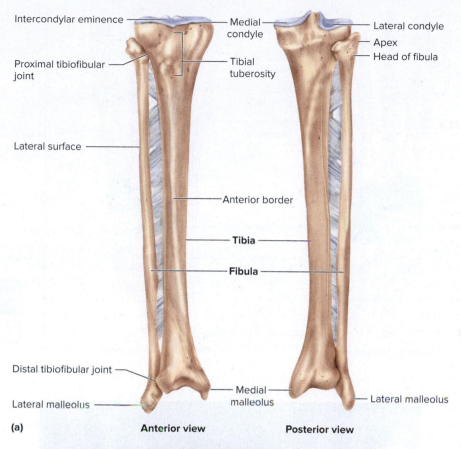

Intercondylar eminence

Medial condyle

Lateral condyle

Apex

Head of fibula

Proximal tibiofibular joint

Tibial tuberosity

Lateral surface

Anterior border

Tibia

Fibula

Distal tibiofibular joint

Medial malleolus

Lateral malleolus

Lateral malleolus

(a) **Anterior view** **Posterior view**

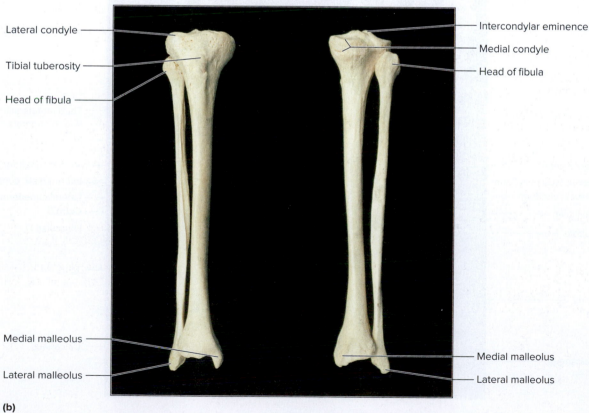

Lateral condyle

Intercondylar eminence

Tibial tuberosity

Medial condyle

Head of fibula

Head of fibula

Medial malleolus

Lateral malleolus

Medial malleolus

Lateral malleolus

(b)

FIGURE 9.10 Right Tibia and Fibula. (a) Diagram; (b) photograph.

(b) ©Eric Wise

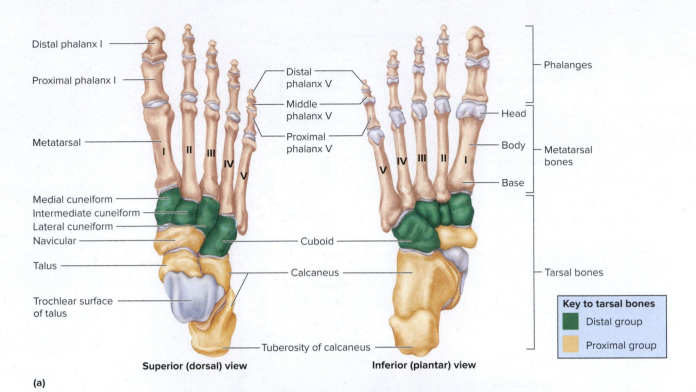

(a)

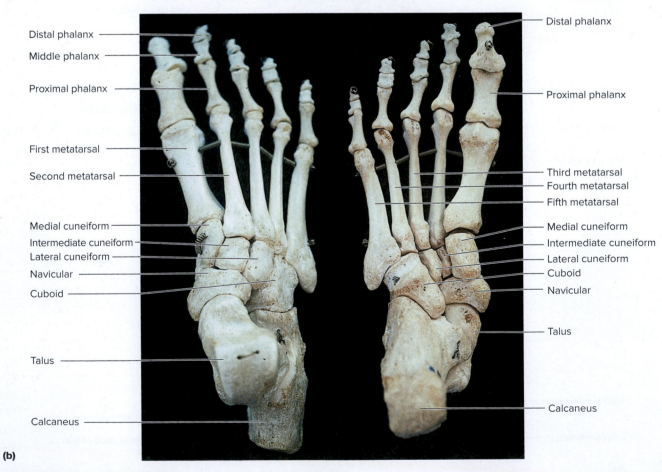

(b)

FIGURE 9.11 Bones of the Right Foot. (a) Diagram; (b) photograph.

(b) ©Eric Wise

REVIEW SECTION

Appendicular Skeleton

Name _____   Date _____

Lab Section _____   Time _____

Chapter Summary Data

Use this section to record your results from questions within the exercise.

1. _____

2. _____

3. _____

Review Questions

1. Name the anterior depression on the scapula. _____

2. The humerus fits into what specific part of the scapula? _____

3. What specific part of the clavicle attaches to the scapula? _____

4. Frequently, the clavicle is broken when the arms are extended to brace a fall. Explain how hitting the ground with your hands can

   fracture the clavicle. _____

   _____

   _____

5. What is another name for the epiphyseal line of the humerus? _____

6. The condyles of the humerus have specific names. What are they? _____

7. Name the ridge of bony tissue proximal to the lateral condyle of the humerus. _____

8. What is the name of the lateral bone of the forearm? _____

9. Name the depression on the ulna into which the humerus inserts. _____

10. Name the bony process that extends distally from the head of the ulna. _____

11. How does the head of the ulna differ in position from the head of the radius? _____

_____

12. Each metacarpal bone consists of three major regions. What are they? _____

_____

13. Name the carpal bone at the base of the thumb. _____

14. The _____ joins the two hip bones at the anterior junction of these bones.

15. What curved area occurs directly on the hip bone below the posterior inferior iliac spine? _____

16. What is the cuplike depression of the hip bone into which the head of the femur fits? _____

17. Explain the difference between the greater pelvis and the lesser pelvis. _____

_____

18. Name the roughened line that runs along the length of the posterior femur. _____

19. Name the sesamoid bone that forms in the quadriceps tendon between the femur and the tibia. _____

20. What is the name of the process on the distal portion of the tibia? _____

21. What is the function of the fibula? _____

_____

22. What is another name for the heel bone? _____

23. What is the name of the bone of the foot that joins with the tibia and fibula? _____

24. Match the bone with the region it comes from.

    ____a. clavicle   1. foot

    ____b. talus   2. hand

    ____c. ilium   3. hip bone

    ____d. radius   4. pectoral girdle

    ____e. scaphoid   5. forearm

25. How many ankle bones are there versus the number of wrist bones? _____

26. Label the parts of the scapula in the following illustration. Use the numbers for the terms provided.

   1. acromion   5. lateral border
   2. coracoid process   6. medial border
   3. inferior angle   7. scapular spine
   4. infraspinous fossa   8. supraspinous fossa

©Eric Wise

27. Label the parts of the ulna as illustrated, using the numbers for the terms provided. Determine if the bone is from the left or right side of the body. _____

   1. coronoid process   4. radial notch
   2. head   5. styloid process
   3. olecranon   6. trochlear notch

©Eric Wise

28. Name the bone at the tip of the middle finger. _____

29. Label the following illustration of the foot, using the numbers for the terms provided.

   1. calcaneus   5. medial cuneiform
   2. cuboid   6. navicular
   3. intermediate cuneiform   7. talus
   4. lateral cuneiform

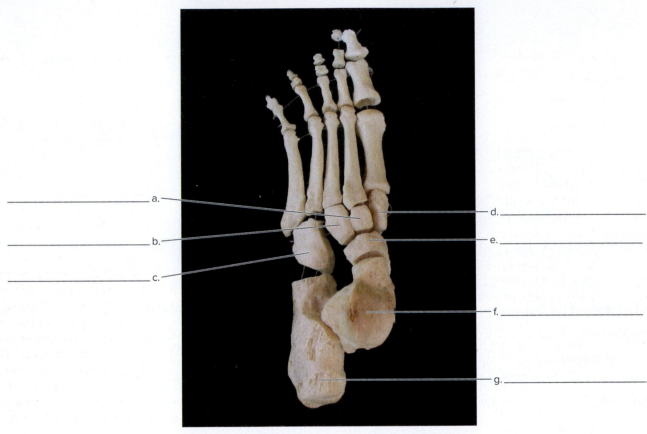

©Eric Wise

30. Label the following illustration of the hand, using numbers for the terms provided.

   1. capitate   4. pisiform
   2. hamate   5. trapezoid
   3. lunate

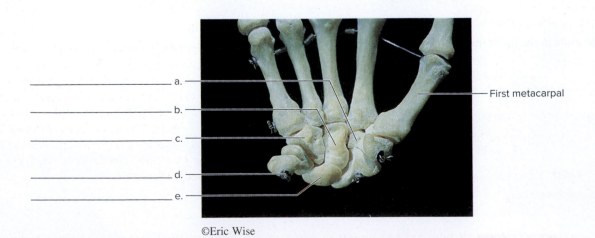

©Eric Wise

Joints

INTRODUCTION

The study of **joints** or **articulations** is known as **arthrology.** Joints are important because they are the points of movement of the body and they can suffer significant trauma or disease. Diseases such as arthritis affect millions of people worldwide. The development of artificial joints is important to replace joints that can no longer provide either a comfortable range of movement or the stability required for support. In this exercise, you study the various types of joints in the body in terms of their structure. Joints can be classified according to their physical composition, or they can be classified according to their range of movement. In terms of composition, joints are classified as bony, fibrous, cartilaginous, or synovial. In terms of range of movement, joints are immovable, semimovable, or freely movable. As you study the joints in the lab, use the joints in your body as reference and perform the action of a specific articulation. These topics are covered in the Saladin text in chapter 9, "Joints."

OBJECTIVES

At the end of this exercise, you should be able to

1. distinguish among fibrous, cartilaginous, and synovial joints;
2. discuss the nature of a synovial joint;
3. locate the fibrous capsule, synovial membrane, synovial fluid, and articular cartilage in a dissected synovial joint;
4. list five types of synovial joints;
5. explain the structure of the knee, hip, jaw, ankle, and shoulder.

MATERIALS

Vertebrate joint with intact synovial capsule

Dissection tray with scalpel, blunt probe, and protective gloves

Waste container

Model or chart of joints, including those of the shoulder, knee, hip, ankle, and jaw

Articulated skeleton

PROCEDURE

Types of Joints

The four major groups of joints classified according to composition are bony, fibrous, cartilaginous, and synovial joints. **Bony joints** occur when two bones fuse to form one. **Fibrous joints** are typically composed of connective tissue fibers between two bones. They permit little movement. **Cartilaginous joints** consist of cartilage between two bones and are generally more movable than fibrous joints, although some cartilaginous joints may have no movement at all. **Synovial joints** have the most complex structure, including a joint capsule, an inner membrane, and synovial fluid, and they are the most movable of the joints (table 10.1).

Bony Joints

A bony joint, or **synostosis** (SIN-oss-TOE-sis), occurs when two separate bones fuse to become one. The jaw starts out developmentally as two bones but fuses early in life as a single bone. This is an example of a synostosis. These joints are immovable joints.

Fibrous Joints

Fibrous joints connect one bone to another with collagenous fibers. If the fibers are short between the bones, the joint does not allow for much, if any, movement. One example of this type of joint is called a suture, and it occurs between adjacent bones in the cranium. In these joints, the bones of the cranium are tightly bound together by dense, fibrous connective tissue. These bones do not move much if at all. Longer fibers, such as those of the interosseous membranes between the radius and ulna or tibia and fibula allow for more movement.

TABLE 10.1	Classification of Joints
Bony Joints—Joints Composed of Two Fused Bones	**Examples**
Synostosis	Fusion of two frontal bones into one bone
Fibrous Joints—Joints Held Together by Collagenous Fibers	
Suture	Frontal and parietal bones
Gomphosis	Teeth and mandible
Syndesmosis	Interosseus membrane between tibia and fibula
Cartilaginous Joints—Joints Held Together by Cartilage	
Synchondrosis	Epiphyseal plate, humeral head and shaft
	Costal cartilages, first rib and sternum
Symphysis	Joint between two vertebral bodies
Synovial Joints—Joints Enclosed by Synovial Capsule	
Plane	Between carpal bones
Hinge	Humerus and ulna
Pivot	Atlas and axis
Condylar	Radius and scaphoid
Saddle	Trapezium and first metacarpal
Ball-and-socket	Acetabulum and femur

Ⓐ Activity

Examine a skull in the lab and locate the sutures. Examine the different types of sutures in fig. 10.1.

Another type of fibrous joint is a **gomphosis** (gom-FOE-sis), which is the joint of the teeth in the sockets of the maxilla and the mandible. This type of joint is like a peg in a socket. A gomphosis connects the bone of the jaw to the tooth by fibrous connective tissue called a **periodontal ligament.**

Ⓐ Activity

Examine a jaw or complete skull in the lab and locate a gomphosis. Compare your specimen to fig. 10.2.

Another type of fibrous joint is called a **syndesmosis** (SIN-dez-MO-sis). The fibrous connective tissue is longer in these joints than in sutures or gomphoses. An example of a syndesmosis is the connection between the distal radius and ulna and the interosseus membrane of the tibia and fibula.

Ⓐ Activity

Examine an articulated skeleton in the lab and compare the fibrous joints with fig. 10.2.

Cartilaginous Joints

If bones are held together by cartilage, such as the joint between two vertebrae, the articulation is known as a cartilaginous joint.

A cartilaginous joint is known as a **synchondrosis** (SIN-kon-DRO-sis) (plural *synchondroses*). It consists of hyaline cartilage and a **symphysis** if it is composed of fibrocartilage. If the cartilage is long, there is more movement in the joint. In one case, the cartilaginous joint is a phase in the development of the skeleton. This is seen in the **epiphyseal** (EP-ih-FIZ-ee-ul) **plate,** which is illustrated in fig. 10.3. This joint eventually fuses to form a single bone.

Another example of a synchondrosis is the articulation between the first rib and the sternum by the **costal cartilage.** In this joint, hyaline cartilage binds the rib to the sternum yet lets it move somewhat.

A **symphysis** (SIM-fih-sis) is a **fibrocartilaginous** pad and the two bones connecting it (for example, the pubic symphysis or the intervertebral discs). These joints also allow for some movement. The physical stress on a symphysis is greater than the stress on the costal cartilages, and the joint reflects this in its fibrocartilage composition. Fibrocartilage endures much greater stress than hyaline cartilage.

Ⓐ Activity

Locate a symphysis in lab and compare it to fig. 10.4, which compares symphyses to costal cartilages.

Synovial Joints

These joints consist of two bones in a capsule that contains synovial fluid such as the knee joint. Synovial joints allow for

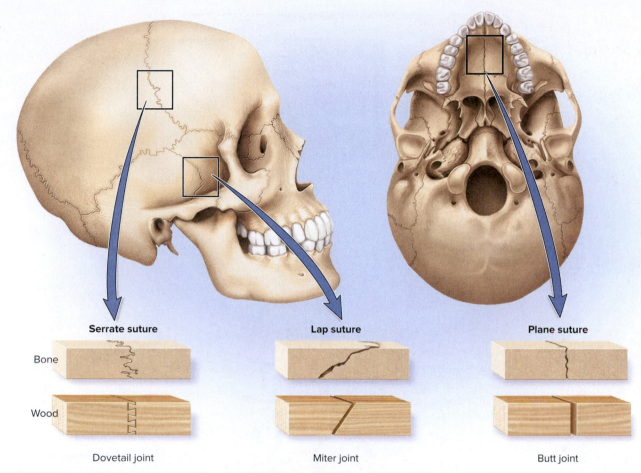

Serrate suture   Lap suture   Plane suture

Bone

Wood

Dovetail joint   Miter joint   Butt joint

FIGURE 10.1 Sutures. Examples and locations in the skull, along with cross sections and analogous wood joints.

extensive movement. The outer part of the synovial joint is the **joint capsule,** which is made of an outer **fibrous capsule** and an inner **synovial membrane.** The synovial membrane secretes **synovial fluid,** a lubricating liquid that reduces friction inside the joint. The space inside the joint is called the **synovial cavity,** and each bone of the joint ends in a hyaline cartilage cap called the **articular cartilage.** There are other structures in synovial joints that are characteristic of specific joints, and these are discussed later in this exercise. The shape of the bones determines the type of movement at the articulation, and in general the more movable a joint is, the less stable it is. The stability of a joint depends on the number and types of ligaments, tendons, and muscles and on the way the bones fit together.

Activity

Compare the features of the joint in fig. 10.5 to models or charts in the lab.

Dissection of a Synovial Joint   To understand the structure of a synovial joint, it is beneficial to examine a fresh or recently thawed vertebrate joint such as a beef or chicken joint.

Activity

Wear protective gloves as you dissect the joint, and wash your hands thoroughly with soap and water after the dissection. Place the joint in front of you on a dissecting tray, and cut into the joint capsule with a scalpel.

Note the tough, white material that surrounds the joint. This is the joint capsule, and it may be fused with ligaments (parallel fibers) that bind the bones of the joint together. Find any trace of muscle and follow it back to its bony attachment. Muscles frequently provide stability to a joint. Notice the synovial fluid, which is a slippery substance that provides a slick feel to the inside of the capsule. Once you cut into the joint, examine the articular cartilage found on the ends of the bones. *Carefully* cut into this cartilage with a scalpel, and notice how the material chips away from the bone. When you have finished with the dissection, make sure to remove the scalpel blade, wash the dissection equipment, dispose of the joint in the appropriate animal waste container, and wash your hands.

Other Synovial Structures   Bursae and tendon sheaths are synovial structures. **Bursae** (singular, *bursa*) are small synovial sacs

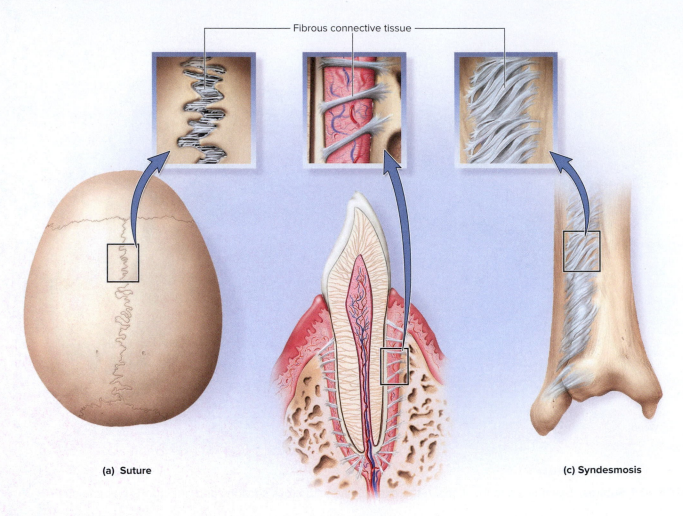

Fibrous connective tissue

(a) Suture

(b) Gomphosis

(c) Syndesmosis

FIGURE 10.2 Fibrous Joints. (a) Suture; (b) gomphosis; (c) syndesmosis.

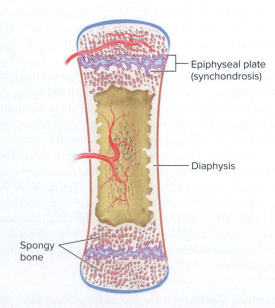

Epiphyseal plate (synchondrosis)

Diaphysis

Spongy bone

FIGURE 10.3 Synchondrosis. Immovable epiphyseal plate.

between tendons and bones or other structures. The bursae cushion the tendons as they pass over the other structures. **Tendon sheaths** are synovial structures, such as those encircling the tendons that pass through the palm of the hand. Tendon sheaths protect the tendons as they slide past one another. Examine fig. 10.6 for these structures.

Ⓐ Activity

Examine the joints on an articulated skeleton in lab and compare them to fig. 10.7.

1. **Plane joints** allow for movement between two flat surfaces, such as between the superior and inferior facets of adjacent vertebrae or intertarsal or intercarpal joints.
2. **Hinge joints** allow for angular movement, such as in the elbow, in the knee, or between the phalanges of the fingers. You can increase or decrease the angle of the two bones with this joint.
3. **Pivot joints** allow for rotational movement between two bones, such as in the movement of the atlas and the axis when a person is moving the head to indicate "no." They also occur at the proximal radius and ulna.

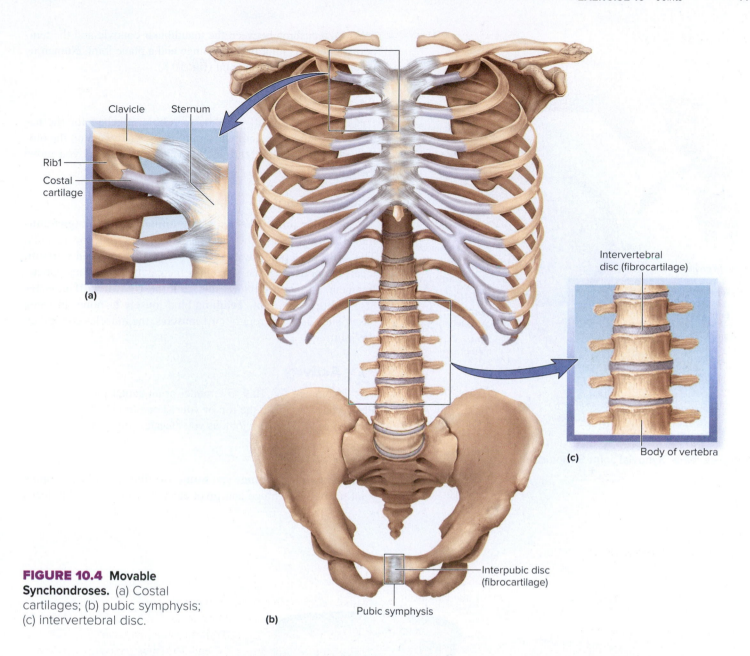

Clavicle   Sternum

Rib 1

Costal
cartilage

(a)

Intervertebral
disc (fibrocartilage)

Body of vertebra

(c)

Interpubic disc
(fibrocartilage)

Pubic symphysis

(b)

**FIGURE 10.4 Movable
Synchondroses.** (a) Costal
cartilages; (b) pubic symphysis;
(c) intervertebral disc.

4. **Condylar** or **ellipsoid joints** allow significant movement
   generally in two planes. Condylar joints consist of a convex
   surface paired with a concave surface. The junction between
   the radius and scaphoid bone is a good example of a condylar joint. Others are found between the metacarpals and
   phalanges of digits II through V.
5. **Saddle joints** have two concave surfaces that articulate with
   each other. An example of a saddle joint is that between the
   trapezium and the first metacarpal of the thumb. This provides for greater movement in the thumb than the condylar
   joint of the wrist.
6. **Ball-and-socket joints** consist of a spherical head in a
   round concavity, such as in the shoulder and the hip.
   There is extensive movement in these joints, yet they are
   inherently less stable due to the freedom of movement that
   they afford.

Monaxial joints move only in one plane. Hinge and pivot
joints are monaxial joints. **Biaxial joints** move in two planes.
Condylar and saddle joints are examples of biaxial joints.
Multiaxial joints move in many planes. Ball-and-socket joints are
multiaxial joints. These are represented in figure 10.7.

Specific Joints of the Body

There are several types of specific joints in the body; some are
primarily hinge joints, such as the jaw, elbow, and knee, while others are ball-and-socket joints, such as the shoulder and hip joints.

Jaw Joint

The **temporomandibular joint (TMJ)** is the only diarthrotic
joint of the skull. The **articular disc** is a pad of fibrocartilage that

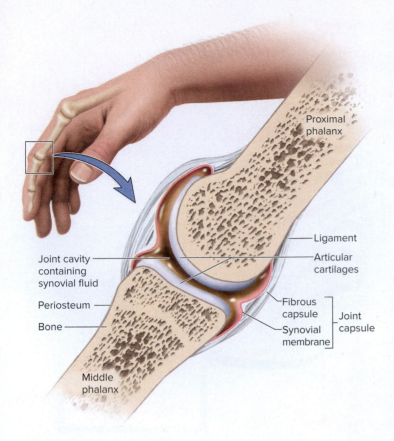

FIGURE 10.5 Synovial Joint Structure.

provides a cushion between the mandibular condyle and the temporal bone. This joint is both a hinge and a plane joint. Numerous ligaments strengthen this joint (fig. 10.8).

Ⓐ Activity

Examine a skull with the mandible attached for the nature of the temporomandibular joint. Put your fingers on the outside of your jaw joint and palpate (feel) the joint as you move your jaw.

Shoulder Joint

The **shoulder joint** is known as the **humeroscapular (glenohumeral) joint** because the humerus articulates with the scapula. The shallow glenoid cavity is deepened by the **glenoid labrum,** a cartilaginous ring that surrounds the cavity. Numerous bursae, ligaments, a tough joint capsule, and the **rotator cuff muscles** also stabilize the joint. **Tendons** bind muscle to bone. In some cases, such as in the rotator cuff muscles, the muscles strengthen the joint.

Ⓐ Activity

Compare fig. 10.9 to a model or an actual joint in the lab. Place your fingers on the top of your shoulder and determine the location of the shoulder joint as you elevate your arm.

Elbow Joint

The **elbow (humeroulnar** and **humeroradial) joint** is a complex joint having both hinge and pivot characteristics. The ulna locks

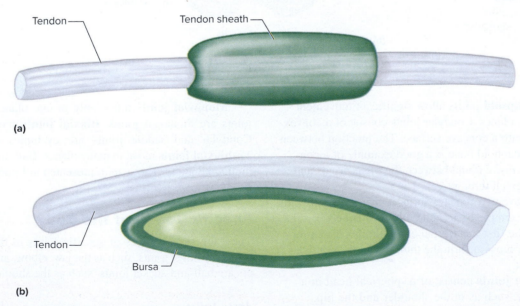

FIGURE 10.6 **Other Synovial Structures.** (a) Tendon sheath; (b) bursa.

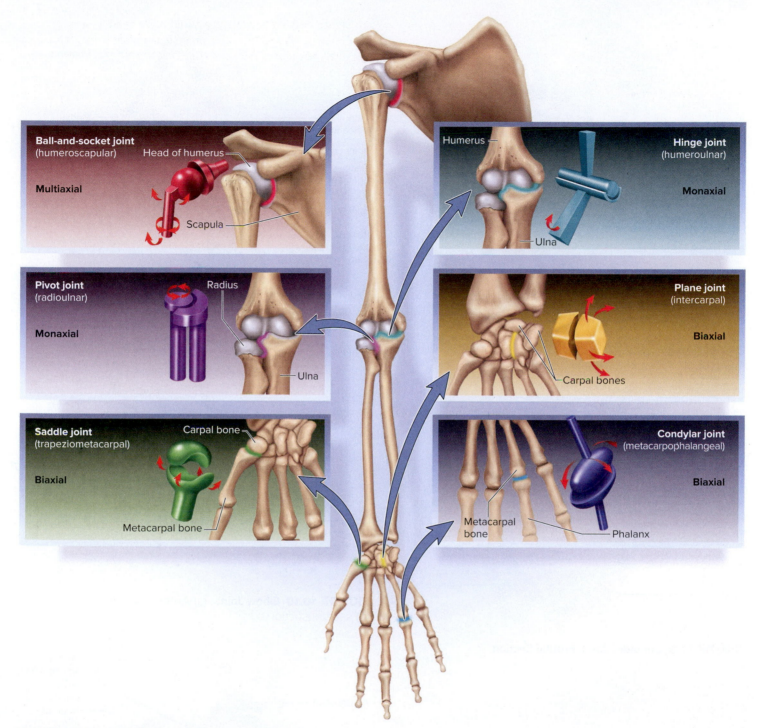

FIGURE 10.7 Six Major Types of Synovial Joints.

into the humerus tightly, but the **anular ligament** that wraps around the radius is commonly torn from the radial head when the forearm is abruptly pulled (fig. 10.10).

 Activity

Rotate your forearm and locate the anular ligament.

Hip Joint

As with the shoulder joint, the **acetabular labrum** deepens the hip socket. Numerous ligaments, including the iliofemoral, pubofemoral, and ischiofemoral, bind the femur to the hip bone. The **round ligament** is a band of dense connective tissue that attaches the acetabulum to the **fovea capitis** of the femur.

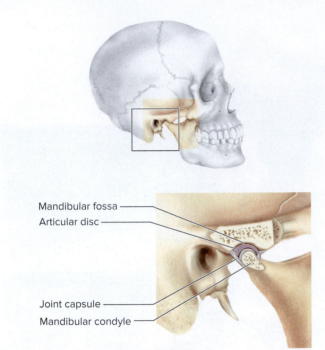

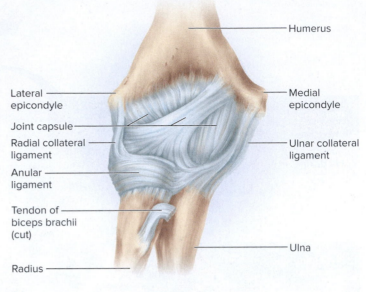

Humerus

Lateral epicondyle

Joint capsule

Radial collateral ligament

Anular ligament

Tendon of biceps brachii (cut)

Radius

Medial epicondyle

Ulnar collateral ligament

Ulna

(a)

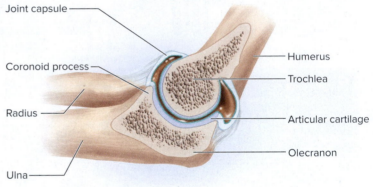

Joint capsule

Coronoid process

Radius

Ulna

Humerus

Trochlea

Articular cartilage

Olecranon

(b)

FIGURE 10.10 Elbow Joint. (a) Anterior view; (b) sagittal section.

FIGURE 10.8 Temporomandibular Joint.

Mandibular fossa

Articular disc

Joint capsule

Mandibular condyle

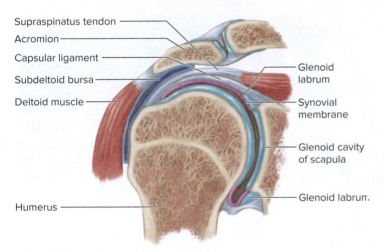

Supraspinatus tendon

Acromion

Capsular ligament

Subdeltoid bursa

Deltoid muscle

Humerus

Glenoid labrum

Synovial membrane

Glenoid cavity of scapula

Glenoid labrum

FIGURE 10.9 Shoulder Joint, Frontal Section.

 Activity

Compare the models, charts, or specimens in the lab to fig. 10.11.

Knee Joint

The **tibiofemoral joint,** also known as the **knee joint,** is the largest, most complex joint of the body. It has several major ligaments, including the **tibial (medial) collateral ligament,** the

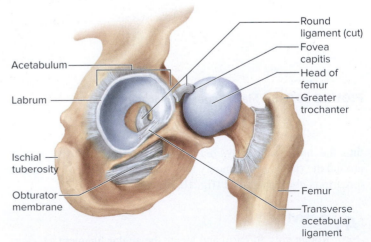

Acetabulum

Labrum

Ischial tuberosity

Obturator membrane

Round ligament (cut)

Fovea capitis

Head of femur

Greater trochanter

Femur

Transverse acetabular ligament

FIGURE 10.11 Hip Joint.

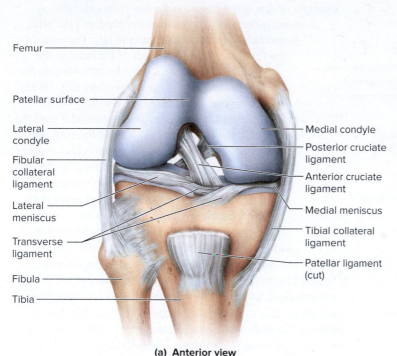

Femur
Patellar surface
Lateral condyle
Fibular collateral ligament
Lateral meniscus
Transverse ligament
Fibula
Tibia

Medial condyle
Posterior cruciate ligament
Anterior cruciate ligament
Medial meniscus
Tibial collateral ligament
Patellar ligament (cut)

(a) Anterior view

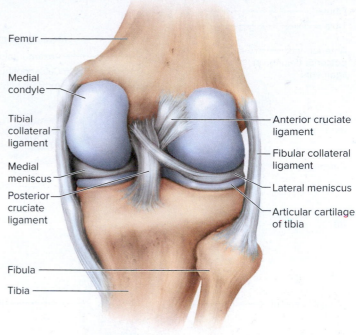

Femur
Medial condyle
Tibial collateral ligament
Medial meniscus
Posterior cruciate ligament
Fibula
Tibia

Anterior cruciate ligament
Fibular collateral ligament
Lateral meniscus
Articular cartilage of tibia

(b) Posterior view

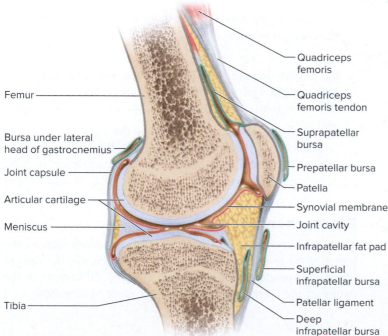

Femur
Bursa under lateral head of gastrocnemius
Joint capsule
Articular cartilage
Meniscus
Tibia

Quadriceps femoris
Quadriceps femoris tendon
Suprapatellar bursa
Prepatellar bursa
Patella
Synovial membrane
Joint cavity
Infrapatellar fat pad
Superficial infrapatellar bursa
Patellar ligament
Deep infrapatellar bursa

(c) Sagittal section

FIGURE 10.12 Knee Joint. (a) Anterior; (b) posterior; (c) sagittal.

fibular (lateral) collateral ligament, the **anterior cruciate ligament,** and the **posterior cruciate ligament.** Another important structure is the **patellar tendon,** which runs from the quadriceps femoris muscle to the patella. The **patellar ligament** connects the patella and the tibial tuberosity. The **medial** and **lateral menisci** are wedge-shaped pads of fibrocartilage that provide a cushion between the femur and tibia. The knee is primarily a hinge joint with a little lateral movement allowed.

(A) Activity

Compare specimens in the lab to fig. 10.12. Palpate the tibial collateral ligament, the fibular collateral ligament, and the patellar tendon on yourself.

Ankle Joint

The **talocrural joint,** or ankle joint, is formed from the tibia, the fibula, and the talus. The lateral malleolus of the distal fibula and the medial malleolus of the distal tibia are on each side of the talus, and all three bones share a joint capsule. The tibia and fibula are connected by an **anterior tibiofibular ligament** and a **posterior tibiofibular ligament.** The tibia is connected to the calcaneus, talus, and navicular by fibers of the **medial ligament,** and the fibula is connected to the calcaneus and talus by fibers of the **lateral ligament.** A sprained ankle is the stretching of these fibers; it frequently occurs when the foot is excessively inverted.

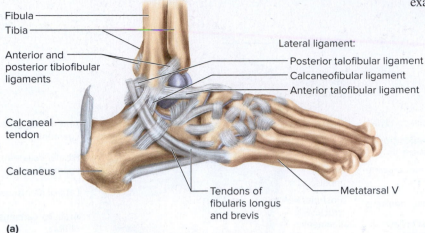

(a)

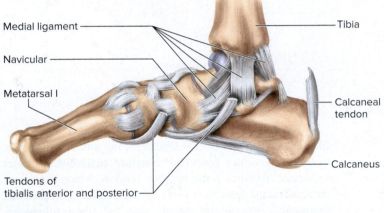

(b)

FIGURE 10.13 **Ankle Joint.** (a) Lateral view; (b) medial view.

Ⓐ **Activity**

Examine fig. 10.13 and models in lab, or a dissected ankle in a cadaver, and find the bones and major ligaments of the ankle.

Lever Systems

Lever systems are mechanical devices that increase power or speed with a rigid lever arm pivoting on a fulcrum. In the study of body mechanics, the musculoskeletal system can be viewed as various lever systems that provide a mechanical advantage for the body. There are three parts to a lever: the fulcrum, the resistance arm, and the effort arm. The fulcrum is the point of movement or rotation of the lever. The resistance arm is the part of the lever that is to be moved, and the effort arm is that part to which an action is applied to move the lever (fig. 10.14). Bones are the lever arms, joints are the fulcrum, the weight is the resistance, and muscles produce the effort. There are three classes of levers and these are explained next.

1. A **first-class lever** is one where the effort is on one side of the fulcrum and the resistance is on the other side. A classic

example of this is a see-saw, and a common example in humans is the junction between the head and the atlas. Examine an articulated skeleton in lab. The atlas is the fulcrum, the anterior part of the head is the resistance, and the neck muscles attaching to the back of the head produce the effort (fig. 10.15).

2. A **second-class lever** has the fulcrum at one end, the resistance in the middle, and the effort at the other end. A wheelbarrow is a good representative of a second-class lever. The wheel is the fulcrum, the load in the basin of the wheelbarrow is the resistance, and the effort is produced by your arms pulling up on the handles of the wheelbarrow (fig. 10.15). If you are sitting down and lift your thigh and leg, the hip joint is the fulcrum, the resistance is the weight of the thigh and leg, and the effort is caused by the pulling of the quadriceps muscle.

3. A **third-class lever** has the fulcrum at one end, the effort in the middle, and the resistance at the other end. Most of the joints of the body involve third-class levers, and a common example is flexing the forearm. The fulcrum is the elbow, the effort is caused by the biceps brachii muscle, and the resistance is the weight of the forearm. Examine this joint on an articulated skeleton (fig. 10.15).

Answer the following questions to see if you understand the concepts of levers.

❓   When you are lying on your back with your thighs and legs straight and you elevate your lower limbs, what kind of lever is this? _____ (1)

❓   The triceps brachii muscle on the back of your arm is attached to the tip of your elbow. When you use this muscle to straighten your arm, what kind of lever is this? _____ (2)

Biomechanics

The efficiency and strength of levers vary, depending on the length of the lever arm, the amount of resistance, and the amount of effort expended. These can have dramatic differences in the nature of the lever. If you examine one lever system, such as a

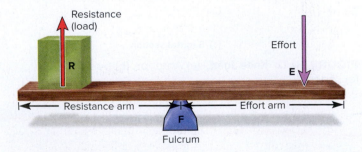

FIGURE 10.14 **Lever System.**

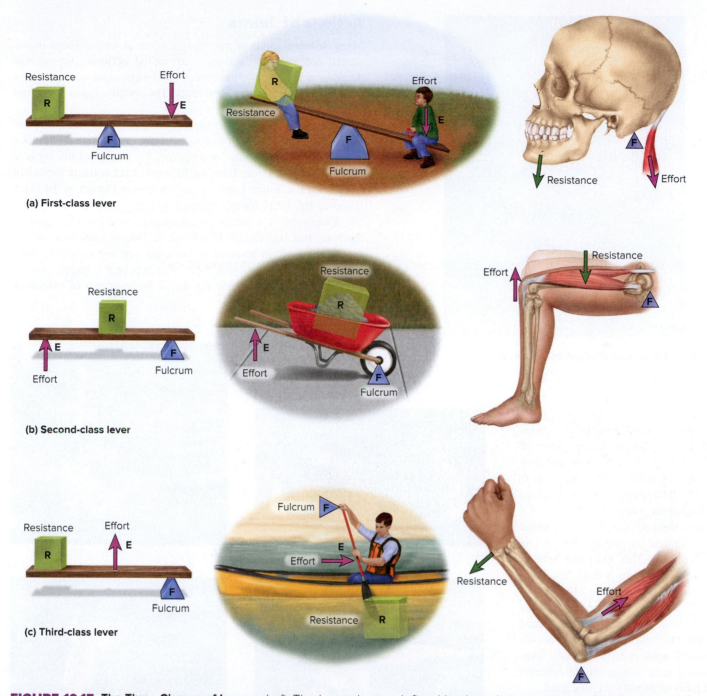

FIGURE 10.15 The Three Classes of Levers. *Left:* The lever classes defined by the relative positions of the resistance (load), fulcrum, and effort. *Center:* Mechanical examples. *Right:* Anatomical examples. (a) Muscles of the back of the neck pull down on the occipital bone to oppose the tendency of the head to drop forward. The fulcrum is the occipital condyles. (b) The quadriceps muscle of the anterior thigh elevates the knee. The fulcrum is the hip joint. (c) In flexing the elbow, the biceps brachii muscle exerts an effort on the radius. Resistance is provided by the weight of the forearm or anything held in the hand. The fulcrum is the elbow joint.

third-class lever, you can see the impact of changing just one variable. Figure 10.16 shows a demonstration of a lever system. It is a simple demonstration where there are three variations in the distance between the effort and the fulcrum. If you pull on string "A" it is difficult to raise the resistance, but small amounts of movement by the effort result in a significant motion in the lever arm. If you pull on string "B" it takes less effort, but the lever arm moves less. If you pull on string "C," it is easy to move the lever arm, but you have to pull the string over a significant distance in order to raise the lever arm. We can explain this by measuring the mechanical advantage (MA) of each system. The actual distances are not important in this discussion, but the relative

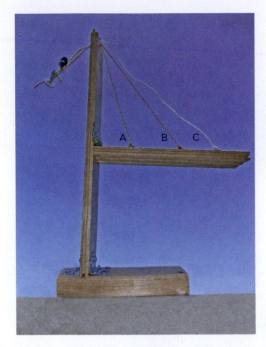

FIGURE 10.16 Demonstration of a Lever System.

Eric Wise

Actions at Joints

Many different kinds of movements occur at joints. These movements, controlled by muscles, are called **actions.** Actions can decrease a joint angle, increase a joint angle, and cause rotation at a joint, among other movements. The specific types of actions are as follows:

Flexion is a decrease in the joint angle from anatomical position. If you bend your elbow, you are flexing your forearm. Flexion of the thigh is in the anterior direction, yet flexion of the leg is in the *posterior* direction. Bending forward at the waist is flexion of the vertebral column. Looking at your toes is flexion of the head. Examine fig. 10.17 for examples of flexion.

Extension is a return to anatomical position of a part of the body that was flexed. If you are looking at your toes and lift your head back to anatomical position, you are extending your head. If you straighten your knee after it is bent (flexed), you are extending the leg. Examine fig. 10.17 for examples of extension.

distances are critical. The distance from the fulcrum to the attachment point of string "A" is 1.25 units. The distance from the fulcrum to the attachment point "B" is 3 units. The distance to the attachment point of "C" is 4.5 units, and the total length is 5.5 units. The mechanical advantage for "A" is 1.25/5.5 or .23. This does not help in multiplying force, but it does help multiply distance. For "B," the MA is 3/5.5 or .55. This means that it is easier to move the lever arm but it doesn't move as far. For "C," the MA is 4.5/5.5 or .82. This is the easiest to move the lever arm, but a muscle that would contract would have to shorten significantly.

Which one of these three variations represents the elbow joint with the biceps brachii providing the effort? Use a ruler and an articulated skeleton. Measure the distance from the pivot point of the elbow to the attachment of the biceps brachii muscle (it is the radial tuberosity) and the distance to the end of the fingers. Calculate the mechanical advantage for this lever and record it in the space below.

❓ MA of the biceps brachii, forearm lever: _____ (3)

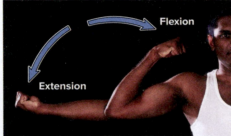

(a)

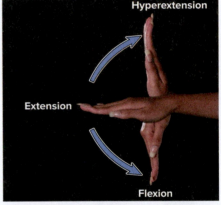

(b)

(c)

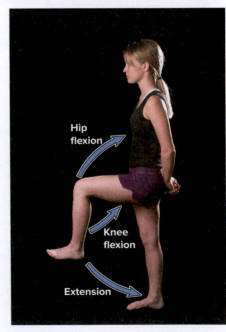

(d)

FIGURE 10.17 Flexion and Extension.
(a) Flexion and extension of the elbow.
(b) Flexion, extension, and hyperextension of the wrist. (c) Flexion and hyperextension of the shoulder. (d) Flexion and extension of the hip and knee.

Timothy L. Vacula/Ken Saladin

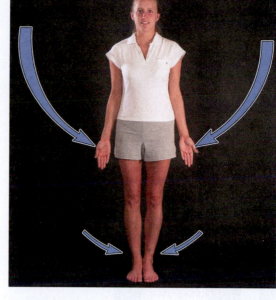

(a) Abduction

(b) Adduction

FIGURE 10.18 Abduction and Adduction.

Timothy L. Vacula/Ken Saladin

Extension of the part of the body beyond anatomical position is known as **hyperextension.** When you are about to roll a bowling ball and your arm reaches the very back of the arc, you are hyperextending your arm.

Abduction is movement of the limbs in the coronal plane away from the body (*abduct* = to take away). Abduction is taking away a part of the body in a lateral direction. This can be seen in fig. 10.18.

Adduction is the return of the part of the body to anatomical position after abduction. In doing "jumping jacks," you are abducting and adducting in series. Think of adduction as "adding" a limb back to the body, as seen in fig. 10.18.

Rotation is the movement of a part of the body around an axis. **Lateral rotation** of a limb moves the anterior surface of the limb toward the lateral side of the body. **Medial rotation** of a limb turns the anterior surface of the limb toward the midline.

Supination is lateral rotation of the forearm and consequently the hand. The hands are supinated when the body is in anatomical position.

Pronation is medial rotation of the forearm and consequently the hand as well. When you turn your palms posteriorly from anatomical position, you are pronating your hands.

Ⓐ Activity

Rotate various joints of the body and compare them to fig. 10.19.

Protraction is a horizontal movement in the anterior direction, as in jutting the chin forward.

Retraction is the reverse of protraction. The jaw that moves from anterior to posterior is retracted. These two actions are illustrated in fig. 10.20.

Elevation means to move in a superior direction. Elevation of the shoulders occurs when you shrug your shoulders.

Depression is the opposite of elevation; it is movement in the inferior direction. Elevation and depression are seen in fig. 10.20.

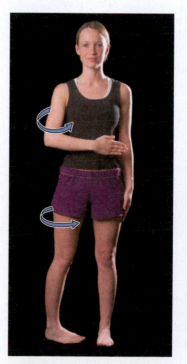
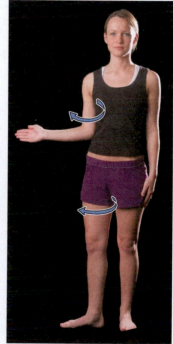

(a) Medial (internal) rotation

(b) Lateral (external) rotation

FIGURE 10.19 Medial (Internal) and Lateral (External) Rotation. Shows rotation of both the humerus and femur.

Timothy L. Vacula/Ken Saladin

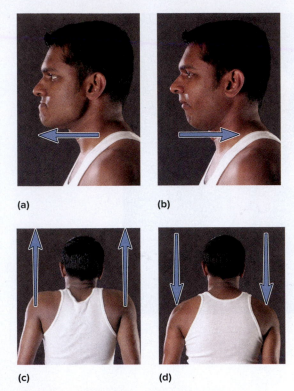

(a)   (b)

(c)   (d)

FIGURE 10.20 **Protraction, Retraction, Elevation, and Depression.** (a) Protraction of the jaw; (b) retraction of the jaw; (c) elevation of the scapulae; (d) depression of the scapulae.

Timothy L. Vacula/Ken Saladin

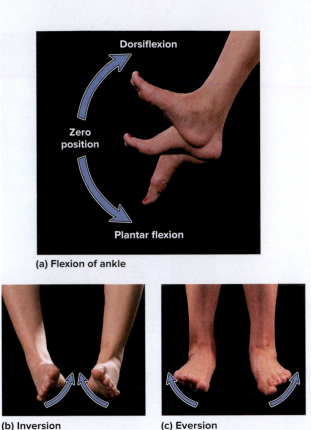

(a) Flexion of ankle

(b) Inversion   (c) Eversion

FIGURE 10.21 **Movements of the Foot.**

Timothy L. Vacula/Ken Saladin

Inversion (supination) refers to movement of the feet. Turning the soles of the feet medially so they face each other is known as inversion.

Eversion (pronation) means turning the soles of the feet laterally. Inversion and eversion can be seen in fig. 10.21.

Fixing a muscle prevents motion in either direction. This is done with opposing muscles contracting simultaneously. The muscle that has the main force on a joint is called the **prime mover.** Muscles that assist with the prime mover are **synergists,** while those that oppose the muscle are **antagonists.**

REVIEW SECTION

Joints

Name _____ *Date* _____

Lab Section _____ *Time* _____

❓ Chapter Summary Data

Use this section to record your results from questions within the exercise.

1. _____
2. _____
3. _____

Review Questions

1. Which one of the following joints has the greatest range of movement? Circle the correct answer.

   a. gomphosis

   b. suture

   c. synchondrosis

   d. hinge

2. In which of these joints would you find a meniscus? Circle the correct answer.

   a. cartilaginous

   b. fibrous

   c. synovial

3. Label the following illustration using the numbers for the terms provided.

   1. articular cartilage

   2. bone

   3. joint capsule

   4. synovial cavity

   5. synovial membrane

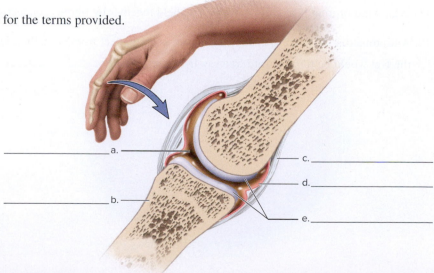

_____ a. _____

_____ b. _____

c. _____

d. _____

e. _____

4. Match the joint in the left column with the type of joint in the right column.

_____ acetabulofemoral   a. ball-and-socket

_____ intertarsal   b. condylar

_____ radiocarpal   c. hinge

_____ tibiofemoral   d. plane

5. Rank the following joints from least movable to most movable, with 1 being the least movable and 5 being the most movable.

ball-and-socket _____   plane _____   saddle _____   suture _____   syndesmosis _____

6. What is the function of the meniscus in the knee? _____

_____

_____

7. What is the function of the labrum in the humeroscapular joint? _____

_____

_____

8. Bones held together by cartilage are known as _____ joints.

9. A class of joint with great movement is known as a _____.

10. The teeth are held in the jaw by what specific type of joint? _____

11. What is the name of a joint held together by a joint capsule? _____

12. The joint between the femur and the tibia is what specific type of joint? _____

13. Synovial fluid is secreted by what structure? _____

14. A skull suture is what type of joint, in terms of movement? _____

15. What type of joint is found at the wrist (between the radius and the scaphoid bones)? _____

16. The joint between the first metacarpal and the proximal phalanx is what type of joint? _____

17. When two bones fuse into a single bone, this union is called a(n) _____.

18. What kind of joint is slightly movable and held together by fibrous connective tissue? _____

19. Antipronation shoes are specialized shoes to keep the feet from pronating. Does this mean that by using the shoes, the bottom of the feet would be tilted toward or away from the midline? _____

Axial Muscles 1: Muscles of the Head and Neck

INTRODUCTION

The skeletal muscles are covered in Exercises 11 through 14. There are more than 600 of these muscles in the human body, with most of them occurring as pairs. The muscles are grouped in this lab manual according to their location. The attachment points, action, and innervation are listed for about 100 muscles in this lab manual, and your instructor may wish to customize the list, so that you learn specific muscles or specific things about each muscle. The study of muscles can be both fun and challenging. Begin your study of muscles early and do not wait to learn them the night before the lab exam!

Skeletal muscle is voluntary muscle. It contracts when consciously stimulated by specific nerves. Muscles pull on bones or other muscles. They do not push. The muscular system functions in movement, maintenance of posture, generation of heat (shivering), compression of the abdomen, and storage of glycogen, among other functions. In this exercise, you are introduced to some muscles of the axial skeleton, starting with the muscles of the head and neck.

The study of human musculature is seen as a daunting task by some students, while other students recall the muscle section of the course as their favorite part of the study of human anatomy. Studying muscles is more satisfying if attainable goals are set. This is best accomplished by choosing small numbers of muscles to study at a sitting and using flash cards as study aids. It is essential that you know the bones and bony markings before studying muscles. Look over the lab exercises on the skeletal system if you need a review. You may also want to make a quick sketch of bones and how muscles attach to them to help you visualize points of attachment. A thorough study of muscles involves knowing not only the name of the muscle but also the attachment points, action, and innervation. Traditionally muscles have been listed by both their **origin,** which generally is the most stable part of the muscle, and the **insertion,** the more movable attachment of the muscle. Since some muscles have two movable attachment points (depending on what part of the body is stable), some authors are moving away from the concept of the origin and the insertion. For the purpose of this lab manual, the origins of the muscle are also listed as *attachment point 1,* and the insertions are also listed as *attachment point 2,* so either can be used as per the wishes of your instructor.

Every muscle has an **action,** which is the effect the muscle has on a part of the body (such as **flexion** of the head). There are two schools of thought on action. One is that the action has an impact on a joint, such as flexion of the wrist. The other is that an action moves the region beyond a joint, such as flexion of the hand. This lab manual will mostly follow the former. Finally, the **innervation** of a muscle is the specific nerve that controls it. Nerves are important in that, if the motor portion of a nerve is damaged, the muscle innervated by that nerve cannot function.

The head muscles can be subdivided into muscles of mastication (chewing), muscles of facial expression, and muscles that close the eyes and mouth. The neck muscles are those that move the neck or head or those that move the hyoid, larynx, or tongue during speech or swallowing. These muscles are discussed in the Saladin text in chapter 10, "The Muscular System."

OBJECTIVES

At the end of this exercise, you should be able to

1. locate the muscles of the head and neck on a torso model, chart, cadaver, or cat;
2. list the attachment points, and action of each muscle presented;
3. describe what nerve controls each muscle;
4. name all the muscles that have a specific action at a joint, such as all the muscles that flex the head;
5. reproduce the actions of select muscles on your head or neck.

MATERIALS

Human torso model or head and neck model

Human muscle charts

Articulated skeleton

Cadaver

Cat

Materials for Cat Dissection

Dissection trays   Blunt probe

Scalpel   Pins

Protective gloves

PROCEDURE

A Activity

Review the actions at joints as outlined in Exercise 10. The details of the muscles are listed in table 11.1. Examine a skull or an articulated skeleton and review the bony markings as you study the attachment points of the muscles in this exercise. Once you know them, the actions should be more comprehensible.

Muscles of the Neck

The trapezius is a muscle that has an action on the neck. It is covered in Exercise 13.

A Activity

Examine the following muscles on models, charts, or cadavers in lab as you read the following descriptions. Locate the muscles on your body as you study them. The **levator scapulae** (le-VAY-tur SCAP-u-lay) is named for what it does: It elevates the scapula (figs. 11.1 and 11.4). The levator scapulae attaches to the lateral side of the neck and also on the upper scapula. If the neck is fixed, the levator scapulae raises the scapula, as in shrugging. If the scapula is fixed, it rotates or abducts the neck.

The **scalenes** (SCALE-eens) are a group of muscles found on the lateral side of the neck. They are deep to the sternocleidomastoid in the front and anterior to the levator scapulae in the back. They rotate the neck or elevate ribs 1 and 2. Examine fig. 11.1 for these muscles.

The **sternocleidomastoid** (STER-no-KLY-doh-MAS-toyd) has two inferior attachments, one to the sternum and the other to the clavicle. It has a superior attachment to the mastoid process of the temporal bone (figs. 11.1, 11.2, and 11.4). It rotates the head in a unique way. Place your hand on the right sternocleidomastoid and turn your head to the right. Notice how the muscle does not contract. Now turn your head to the left, and you can feel the muscle contract. The right sternocleidomastoid turns the head to the left, and the left sternocleidomastoid turns the head to the right.

The **sternohyoid, sternothyroid,** and **omohyoid** are named for their attachments. In the case of the sternohyoid and sternothyroid, the sternum anchors the stable attachment of the muscle (the origin), and

the hyoid bone and thyroid cartilage of the larynx move when the respective muscles contract. In the case of the omohyoid, the term *omo* means shoulder, and the scapula anchors the stable part of the muscle. The hyoid is depressed when the omohyoid contracts. Examine fig. 11.2 for muscles of the anterior neck.

The **platysma** (plah-TIS-mah) is a broad, thin muscle that has a soft attachment (on the fascia of the pectoral and deltoid muscles). It also attaches to the mandible and skin of the lips and can be seen if you elevate your chin and subsequently pout. The thin wings that stick out on the side of your neck are the edges of the platysma muscle. It also elevates the skin of the anterior neck. Examine the platysma in fig. 11.3 and in the models or on the cadaver in the lab. There are numerous muscles of the neck involved in moving the head. The splenius capitis (see fig. 11.1) and the semispinalis capitis, described in Exercise 12 (fig. 12.4), are both deep to the trapezius and rotate and extend the head.

Muscles of the Head

The **digastric** (di-GAS-trik) is so named because it has two bellies. Few muscles open the mandible. The digastric is one that does. In addition, the digastric has significant action on the hyoid, which is important in tongue movement for speech and swallowing. Deep to the digastric is the **mylohyoid,** a broad muscle of the floor of the mouth that aids in pushing the tongue superiorly when swallowing. These muscles can be seen in figs. 11.2 and 11.3.

The **occipitofrontalis** (AUK-sip-ih-TOE fron-TAL-is) muscle attaches to the **galea aponeurotica** (GALE-ee-uh AP-oh-nu-ROT-ih-KAH). Galea means helmet and aponeurotica is a wide, thin, tendinous sheet on the upper aspect of the skull. The **frontal belly** of the occipitofrontalis attaches to the subcutaneous tissue deep to the eyebrows. The **occipital belly** attaches to the posterior part of the galea aponeurotica. If both muscles contract, the eyebrows are raised. If just the frontal belly contracts, then the scalp is brought forward as in frowning. Examine the occipitofrontalis in fig. 11.4.

The **masseter** (MASS-ih-tur) is a large muscle of the head. It elevates the mandible. If you place your fingers on the ramus of the mandible and clench your teeth, you can feel the masseter tighten (fig. 11.4).

The **temporalis** is a powerful muscle that elevates the mandible and is a synergist to the masseter. It is a chewing muscle, or a muscle of *mastication*. Its origin is on the temporal fossa and is so named because it is a depression medial to the zygomatic arch. The temporalis muscle is deep to the zygomatic arch and inserts on the coronoid process and on the superior and medial surface of the ramus of the mandible (figs. 11.4 and 11.5).

The **pterygoid** (TARE-ih-goyd) muscles are deep muscles that have one attachment point on the sphenoid bone and the other on the medial side of the mandible. They pull the jaw horizontally, which helps in rotatory chewing, characteristic of a person chewing gum. These muscles can be seen in fig. 11.5.

The **orbicularis oculi** and the **orbicularis oris** are muscles that close the eyes and the mouth, respectively. The orbicularis

TABLE 11.1	Muscles of the Neck and Head			
Name	**Attachment 1 (Origin)**	**Attachment 2 (Insertion)**	**Action**	**Innervation**
Neck Muscles				
Levator scapulae	C1–4	Upper vertebral border of scapula	Elevates scapula, abducts and rotates neck	Spinal nerves C3–5 and posterior scapular nerve
Scalenes (anterior, middle, and posterior)	Transverse process of cervical vertebrae	Ribs 1 and 2	Flexes and rotates neck, elevates ribs 1 and 2	Spinal nerves C3–8
Sternocleidomastoid	Sternum, clavicle	Mastoid process of temporal bone	Abducts, rotates, and flexes head	Spinal nerves C2–4, accessory nerve (XI)
Sternohyoid	Manubrium of sternum	Hyoid bone	Depresses hyoid bone	Spinal nerves C1–3
Sternothyroid	Manubrium of sternum	Thyroid cartilage of larynx	Depresses thyroid cartilage	Spinal nerves C1–3
Omohyoid	Superior border of scapula	Hyoid bone	Depresses hyoid bone	Spinal nerves C1–3
Platysma	Fascia covering pectoralis major and deltoid	Mandible and skin of inferior region of face	Depresses inferior lip, opens jaw	Facial nerve (VII)
Digastric	Inferior, distal margin of mandible, mastoid notch of temporal	Hyoid bone	Elevates hyoid bone and mandible	Trigeminal (V) and facial (VII) nerves
Mylohyoid	Inferior margin of mandible	Hyoid bone	Elevates hyoid bone and tongue	Trigeminal nerve (V)
Head Muscles				
Frontal belly of occipitofrontalis	Galea aponeurotica	Subcutaneous tissue of eyebrows	Raises eyebrows, draws scalp anteriorly	Facial nerve (VII)
Occipital belly of occipitofrontalis	Occipital and temporal bone	Galea aponeurotica	Retracts scalp	Facial nerve (VII)
Temporalis	Temporal fossa	Coronoid process and ramus of mandible	Elevates (closes) mandible	Trigeminal nerve (V)
Masseter	Zygomatic arch	Angle and ramus of mandible	Elevates (closes) mandible	Trigeminal nerve (V)
Pterygoids (medial and lateral)	Pterygoid processes of sphenoid bone	Medial ramus of mandible	Medial excursion of mandible (for chewing)	Trigeminal nerve (V)
Orbicularis oculi	Frontal and maxilla on medial margin of orbit	Skin of eyelid	Closes eyelid	Facial nerve (VII)
Orbicularis oris	Modiolus of mouth	Skin near lips	Closes lips	Facial nerve (VII)
Corrugator supercilii	Medial portion of frontal bone	Skin of eyebrows	Adducts eyebrows (pulls them medially)	Facial nerve (VII)
Risorius	Fascia of masseter, zygomatic bone	Modiolus at angle of mouth	Abducts corner of mouth (draws edge of mouth lateral)	Facial nerve (VII)
Mentalis	Anterior mandible	Skin of chin below lower lip	Protrudes lower lip	Facial nerve (VII)
Buccinator	Maxilla and mandible near molar teeth	Orbicularis oris	Compresses cheek	Facial nerve (VII)
Zygomaticus (major and minor)	Zygomatic bone	Superolateral angle of mouth and muscles of upper lip	Elevates corners of mouth (in smiling and laughing)	Facial nerve (VII)
Depressor labii inferioris	Mandible	Muscles and skin of inferior lip	Depresses inferior lip	Facial nerve (VII)
Levator labii superioris	Maxilla and zygomatic bones	Muscle of superior lip	Elevates superior lip, flares nostril	Facial nerve (VII)

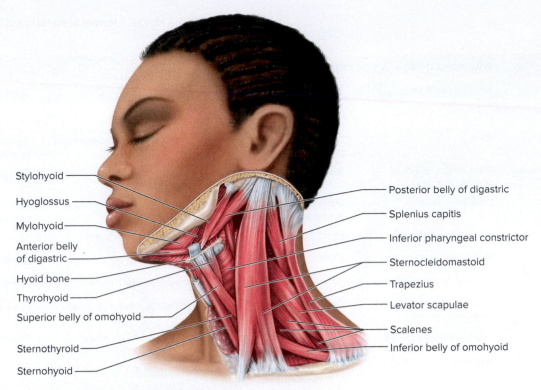

Stylohyoid

Hyoglossus

Mylohyoid

Anterior belly
of digastric

Hyoid bone

Thyrohyoid

Superior belly of omohyoid

Sternothyroid

Sternohyoid

Posterior belly of digastric

Splenius capitis

Inferior pharyngeal constrictor

Sternocleidomastoid

Trapezius

Levator scapulae

Scalenes

Inferior belly of omohyoid

FIGURE 11.1 Muscles of the Neck, Lateral View. (Platysma is removed).

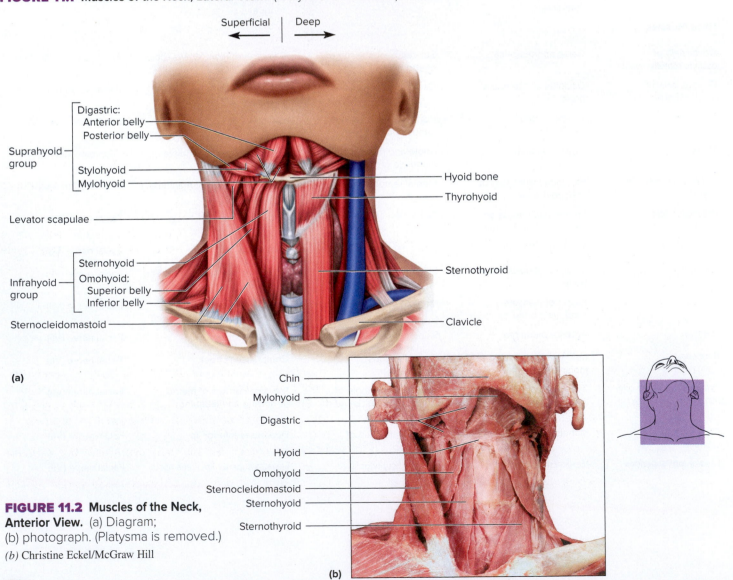

Superficial ← | → Deep

Digastric:
  Anterior belly
  Posterior belly

Suprahyoid
group

Stylohyoid
Mylohyoid

Levator scapulae

Sternohyoid

Omohyoid:
  Superior belly
  Inferior belly

Infrahyoid
group

Sternocleidomastoid

Hyoid bone

Thyrohyoid

Sternothyroid

Clavicle

(a)

Chin

Mylohyoid

Digastric

Hyoid

Omohyoid

Sternocleidomastoid

Sternohyoid

Sternothyroid

**FIGURE 11.2 Muscles of the Neck,
Anterior View.** (a) Diagram;
(b) photograph. (Platysma is removed.)
(b) Christine Eckel/McGraw Hill

(b)

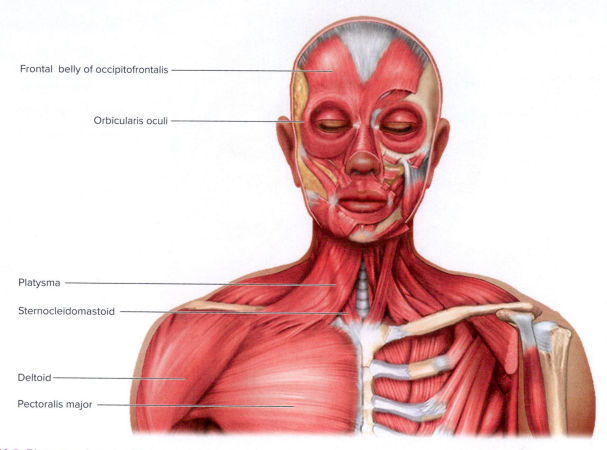

Frontal belly of occipitofrontalis ————

Orbicularis oculi ————

Platysma ————

Sternocleidomastoid ————

Deltoid————

Pectoralis major ————

FIGURE 11.3 Platysma, Anterior View.

oculi has a medial, bony attachment and the other attachment on the eyelid. The muscles ring the eyes and close the eyelids. A muscle that opens the eye is the levator palpebrae superioris and is seen in Exercise 21. The orbicularis oris attaches to the corners of the mouth in a complex arrangement of muscles and connective tissue known as the **modiolus.** The orbicularis oris attaches to the skin of the lips, thus closing the lips. Examine the facial muscles in fig. 11.6.

The **corrugator supercilii** (SOUP-ur-SIL-ee-ee) muscle (fig. 11.6) has a medial point of attachment between the eyebrows and another attachment point laterally. It furrows the eyebrows.

The **zygomaticus** (ZY-goh-MAT-ih-cus) **major** and the **zygomaticus minor** elevate the corners of the mouth by pulling them superiorly and laterally, as in smiling or laughing. They are named for their attachment on the zygomatic bone. The **risorius** (rise-OH-ree-us) is known as the laughing muscle because it pulls the lips laterally. It does not have a bony attachment but rather connects to the fascia of the masseter (see fig. 11.6).

The **mentalis** attaches to the chin (anterior mandible) and is another of the pouting muscles. The **buccinator** (BUK-sin-ay-tur) muscle of the cheek (fig. 11.5) runs in a horizontal

direction. It tightens the cheeks and pushes food toward the molars in chewing.

The **depressor labii** (LABE-ee-ee) **inferioris** pulls the lower lip inferiorly, as when pouting. It is named for its action (depressing the lower lip), while the **levator labii superioris** muscle raises the skin of the upper lip and expands the nostrils, as in the expression of showing extreme disgust. These muscles are seen in fig. 11.6.

Ⓐ **Activity**

1. Turn your head to the left. Which sternocleidomastoid muscle contracts? _____

2. Raise your head so that your chin is elevated and pout your lower lip. What muscle forms a thin membrane along the anterolateral neck? _____

3. Purse your lips and feel the buccinator muscle as it contracts in the cheek.

4. Clench your teeth and palpate the temporalis muscle and the masseter.

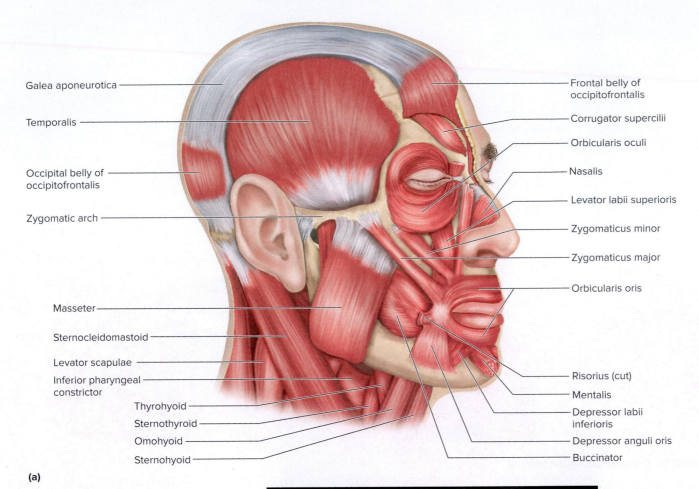

Galea aponeurotica

Temporalis

Occipital belly of occipitofrontalis

Zygomatic arch

Masseter

Sternocleidomastoid

Levator scapulae

Inferior pharyngeal constrictor

Thyrohyoid

Sternothyroid

Omohyoid

Sternohyoid

Frontal belly of occipitofrontalis

Corrugator supercilii

Orbicularis oculi

Nasalis

Levator labii superioris

Zygomaticus minor

Zygomaticus major

Orbicularis oris

Risorius (cut)

Mentalis

Depressor labii inferioris

Depressor anguli oris

Buccinator

(a)

FIGURE 11.4 Muscles of the Head, Lateral View. (a) Diagram; (b) photograph. (Platysma is removed.)
(b) ©Eric Wise

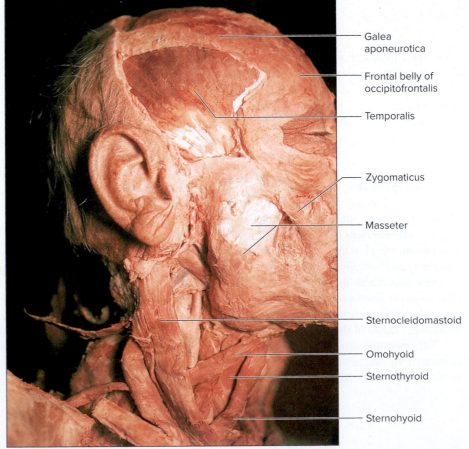

Galea aponeurotica

Frontal belly of occipitofrontalis

Temporalis

Zygomaticus

Masseter

Sternocleidomastoid

Omohyoid

Sternothyroid

Sternohyoid

(b)

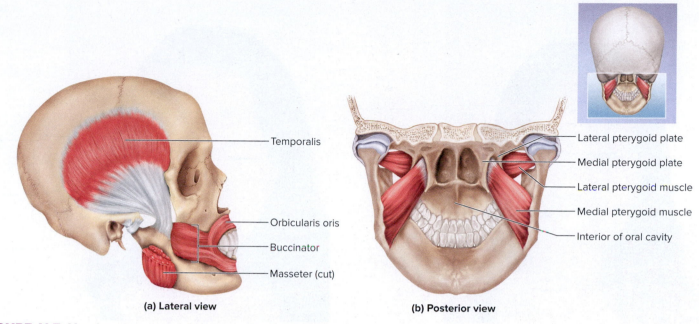

(a) Lateral view

Temporalis

Orbicularis oris

Buccinator

Masseter (cut)

(b) Posterior view

Lateral pterygoid plate

Medial pterygoid plate

Lateral pterygoid muscle

Medial pterygoid muscle

Interior of oral cavity

FIGURE 11.5 Muscles of the Head, Lateral and Posterior Views.

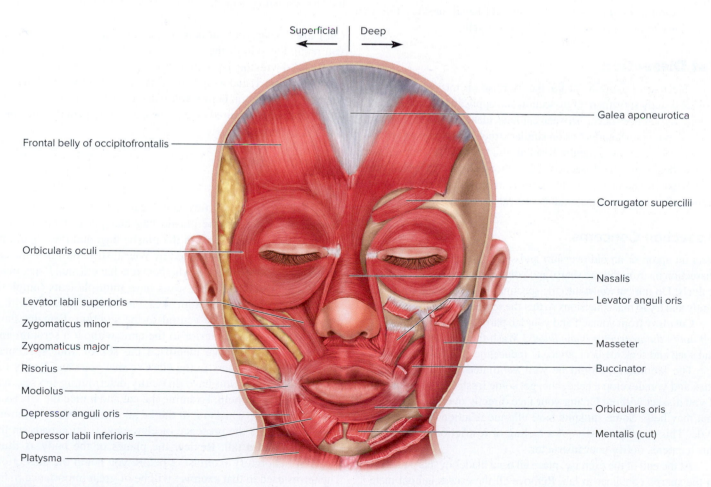

Superficial | Deep

Frontal belly of occipitofrontalis

Orbicularis oculi

Levator labii superioris

Zygomaticus minor

Zygomaticus major

Risorius

Modiolus

Depressor anguli oris

Depressor labii inferioris

Platysma

Galea aponeurotica

Corrugator supercilii

Nasalis

Levator anguli oris

Masseter

Buccinator

Orbicularis oris

Mentalis (cut)

FIGURE 11.6 Muscles of the Face, Anterior View.

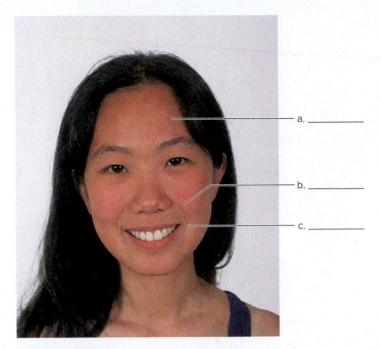

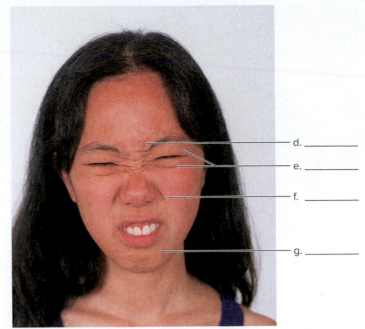

FIGURE 11.7 Muscles of Facial Expression. Label the muscles illustrated.

©Eric Wise

(A) Activity

Examine fig. 11.7 for surface views of facial muscles. Fill in which muscles are represented in the photographs.

Cat Dissection

The objective in using a cat for the exercise on musculature is to provide a study specimen for dissection and application to the human system. Differences occur between cat musculature and human musculature, but you should focus on similar structures to gain an appreciation of human musculature. Read all the material in the exercise prior to beginning the dissection. Dissection is a skill in which you try to separate the overlying structures from those underneath while keeping intact as much material as you can.

Dissection Concerns

Wear an apron or an old overshirt and safety glasses as you dissect. Dissection instruments are sharp, and you must be careful when using scalpels. Do not cut *down* into the specimen but rather lift structures gently and try to make incisions so that the scalpel blade cuts *laterally*.

Cut *away* from yourself and your lab partners. *If you do cut yourself notify the instructor immediately!* Wash the cut with antimicrobial soap and seek medical advice to reduce the chance of infection.

The laboratory should be well ventilated, and if your eyes burn and you develop a headache, get some fresh air for a moment. If you dissect without having your face directly over the specimen, that may help. Some students have allergic reactions to formaldehyde. This usually consists of a feeling of restriction of breath. If this happens, notify your instructor.

At the end of the exercise, place all used blades or sharp material in the sharps container in lab. Remove all the excess animal material on the dissection trays and put it in the appropriate animal waste container. *Do not dump excess animal material in the lab sinks!* Wash the dissection trays and place them in the appropriate area to dry.

Cat Care   Keep your cat in a plastic bag. It is recommended that you retain the skin of the cat as a protective wrapping when you are finished with the day's dissection or that you take a small towel (or an old T-shirt) and wrap the specimen in the cloth, soaking it in a cat wetting solution before you place it back in the bag. Usually, one lab period is required to skin the cat and prepare the specimen for further study.

External Features

(A) Activity

Take a dissection tray and a cat specimen to your table. Remove the cat from the plastic bag and place it on the tray. You may have excess fluid in the plastic bag, and this should be disposed of properly as directed by your instructor. Place the cat on its back and determine whether you have a male specimen or a female specimen. Both sexes have multiple teats (nipples), yet the males have a scrotum near the base of the tail with a prepuce (penile foreskin) ventral to the scrotum. Females have a urogenital opening anterior to the anus without a scrotum and prepuce. Once you have identified the sex of your specimen, compare yours to others in the class so that you can identify the sexes externally. If you have difficulty determining the sex, ask your instructor for help. Examine the cat and notice the vibrissae (vib-RISS-ee), or whiskers, in the facial region. Other variations from humans are the presence of claws and friction pads on the extremities and a tail. Review the planes of the body as illustrated in Laboratory Exercise 2 before you begin the dissection. The terms used in that exercise will be of great importance in the dissection procedures.

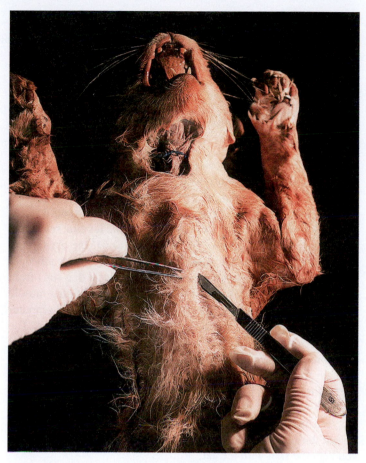

FIGURE 11.8 Lateral Cutting with a Scalpel.
©Eric Wise

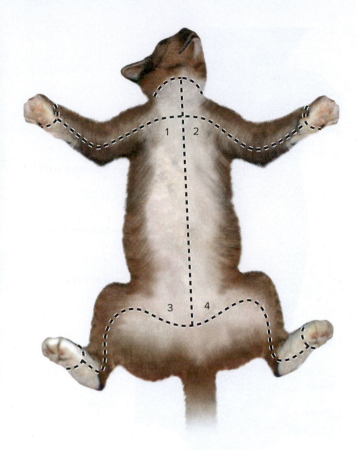

FIGURE 11.9 Removal of the Skin of the Cat. After cutting down the midline, make incisions into the limbs (follow the numbers) and gently remove the skin from underlying structures.

Removal of the Skin Begin your dissection by lifting the skin in the pectoral region with a forceps and making a small cut with a scalpel or sharp scissors in the midline. Work a blunt probe gently into the cut so that you somewhat free the skin from the underlying fascia and muscle. Be careful—the muscles are close to the skin. Insert your scalpel, blade side up, and make small incisions in the skin, cutting away from the underlying muscle as illustrated in fig. 11.8.

Make a cut that runs up the midline of the sternal region until you reach the neck. Likewise, cut posteriorly until you reach an area anterior to the genital region. Leave the genitals intact and carefully make an incision that runs perpendicular to your first cut (see numbers 3 and 4 in fig. 11.9). Likewise, make a perpendicular incision along the upper thoracic region (fig. 11.9, numbers 1 and 2). Continue your caudal incision along the medial aspect of the thigh. Watch out for superficial blood vessels in this area (the great saphenous vein) and stop when you reach the knee. Cut the skin around the knee and begin removing it as a layer from the dorsal side of the cat. Cut the skin from the base of the tail and remove the skin from the back.

Once you reach the shoulders, turn the cat back to the ventral side and remove the skin on the medial side of the arm until you reach the elbow. Stop at this level and remove the skin from the lateral aspect of the arm, working back toward the shoulder. Be careful not to damage the superficial veins on the lateral side of the forearm.

Continue your removal of the skin from the ventral body region. If your cat is a female, locate the mammary glands, elongated, beige, lobular tissue, on each side of the midline on the ventral side (fig. 11.10).

Make an incision on the ventral side of the neck along the midline. Be careful—there are numerous blood vessels along the neck. Do *not* cut through these blood vessels. Continue up into the face and look for beige lumps of glandular tissue in the region of the mandible. These are the salivary glands. Cut the skin carefully from the face and remove it from the head by cutting around the ears (fig. 11.11). You should be able to remove all of the skin from the cat at this time. Examine the undersurface of the skin and note the superficial muscles that cause the skin to move.

These are the cutaneous (que-TAY-nee-us) maximus and the platysma (plah-TISS-mah). Remove as much fat overlying the muscles as you can. Subcutaneous fat is variable from cat to cat, and your specimen may have little or may have significant amounts. The muscle is also covered by fascia, a connective tissue wrapping. Remove the fascia from the muscle so that the fiber direction is apparent. Remove the skin from the forelimb. This can be accomplished by making a longitudinal incision along the length of the forelimb. Be careful not to cut the tendons, blood vessels, or nerves. Pull the skin off gently as if you were removing a pair of knee socks.

As you get to the tips of the digits, cut the skin from them. Pay particular attention and keep the tendons intact. It is a good

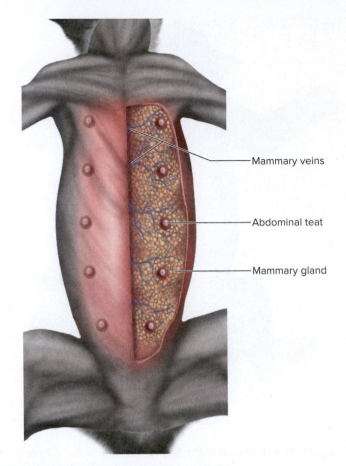

FIGURE 11.10   Mammary Glands of the Cat.

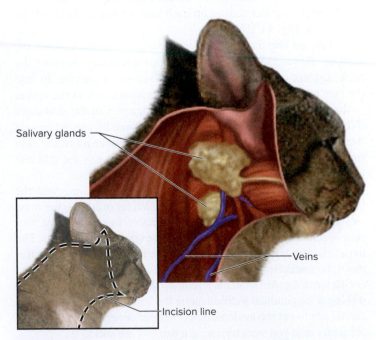

FIGURE 11.11   Removal of Skin from the Head of a Cat.

procedure to dissect only one side of the cat at a time. If you make an error on one side, you will have the other side to dissect. As you cut through the pad on the ventral side of the paw of the cat, make sure that you do not cut through the tendons of the flexor digitorum muscles. As you dissect, leave the muscles of the forelimb intact. Deeper muscles can be seen in future exercises by moving more superficial muscles to the side.

Ⓐ Activity

The head and neck dissection of the cat is beneficial, but you must take care to not cut through the digestive structures, such as the salivary glands, or the circulatory structures, such as the veins and arteries. You will study these structures in later exercises. A dorsal neck muscle of the cat is the **levator scapulae,** found deep to the trapezius. You will need to cut the clavotrapezius to see the levator scapulae. The clavotrapezius is equivalent to part of the trapezius in humans as it is a muscle that runs from the clavicle to the vertebrae. In cats there is an additional muscle called the levator scapulae ventralis, which inserts on the scapular spine. Examine fig. 11.12 for the levator scapulae muscles. The **scalenes** can be dissected by reflecting the pectoralis minor. Notice how the scalenes are composed of separate slips of muscle that run from the ribs to the neck. In humans the muscle is more lateral than in cats. Compare your specimen to fig. 11.2.

The remainder of the muscles you will study in this exercise are seen from a ventral aspect. The **platysma** in the cat was removed during the skinning process, and it will not be seen unless you kept the skin with the cat.

In the cat the sternocleidomastoid consists of two muscles, the **sternomastoid** and the **cleidomastoid.** The sternomastoid extends from the sternum to the mastoid process of the skull, and the cleidomastoid runs from the clavicle to the mastoid process. Underneath the sternomastoid is the most medial muscle of the neck group, the **sternohyoid.** The **sternothyroid** is deeper and more lateral than the sternohyoid. These muscles are seen in fig. 11.13.

The **digastric** muscle runs parallel to the lower edge of the mandible and underneath the submandibular gland. Dissect only one side of the head, leaving the structures on the other side intact for study of the digestive system. Deep to the digastric is the **mylohyoid,** a broad muscle. Notice how the muscle fibers of the mylohyoid run perpendicular to the direction of the digastric. Compare your dissection to fig. 11.13.

The **temporalis** is located more dorsally than the masseter and can be dissected by removing the skin and fascia anterior to the ear. The **masseter** is a large, well-developed muscle in the cat that attaches to the zygomatic arch and has another attachment on the lateral surface of the mandible. Examine these muscles in fig. 11.14.

The **pterygoids** are usually not dissected because you have to cut through the ramus of the mandible to examine them. The muscles of facial expression are generally not studied in the cat. These muscles are small and are frequently removed with the skin. You should study these muscles on human models or a cadaver (if available).

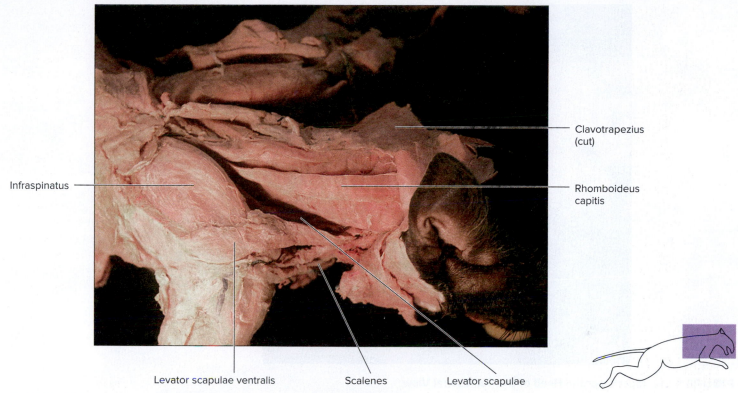

Clavotrapezius
(cut)

Infraspinatus

Rhomboideus
capitis

Levator scapulae ventralis   Scalenes   Levator scapulae

FIGURE 11.12 **Muscles of the Neck of the Cat, Dorsolateral View.**

©Eric Wise

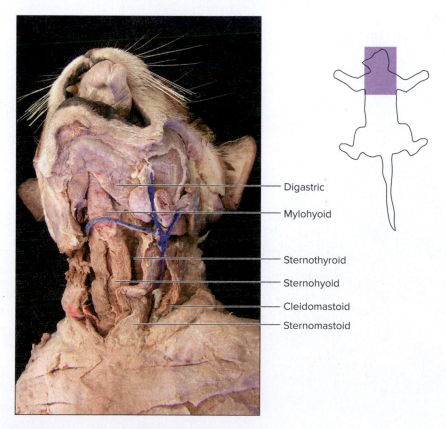

Digastric

Mylohyoid

Sternothyroid

Sternohyoid

Cleidomastoid

Sternomastoid

FIGURE 11.13 **Muscles of the Neck of the Cat, Ventral View.**

McGraw Hill/Cordi Smith, photographer

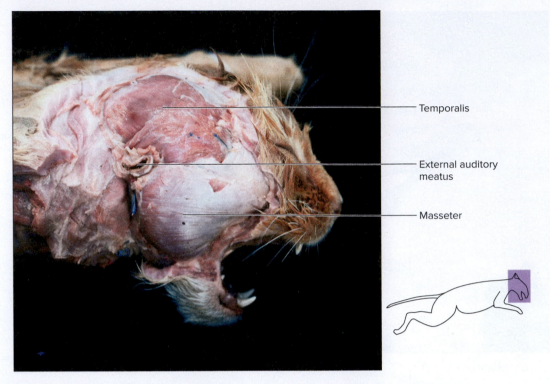

Temporalis

External auditory meatus

Masseter

FIGURE 11.14 **Muscles of the Head of the Cat, Lateral View.**

©Eric Wise

REVIEW SECTION

Axial Muscles 1: Muscles of the Head and Neck

Name _____ *Date* _____

Lab Section _____ *Time* _____

Review Questions

1. What is the superior attachment point of the masseter muscle? _____

2. What is the action of the risorius? _____

_____

3. Where is the inferior attachment point of the levator scapulae muscle? _____

4. What is the action of the orbicularis oculi and orbicularis oris muscles? _____

_____

5. Label the muscles in the following illustration.

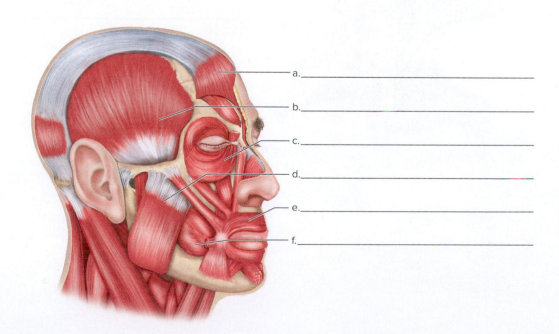

a. _____

b. _____

c. _____

d. _____

e. _____

f. _____

6. What muscle has a stable attachment point on the temporal fossa? _____

7. Name two muscles that close the jaw. _____

8. Where does the sternocleidomastoid muscle attach superiorly? _____

9. What muscle compresses the cheeks? _____

10. Where does the orbicularis oculi attach laterally? _____

11. What is the inferior attachment of the temporalis? _____

12. What muscle elevates the corners of the mouth as in smiling and laughing? _____

13. What is the action of the sternocleidomastoid? _____

14. Label the muscles in the following illustration.

a. _____

b. _____

c. _____

d. _____

e. _____

f. _____

Axial Muscles 2: Muscles of the Trunk

INTRODUCTION

The muscles of the trunk can be grouped into a few functional areas. These are the abdominal muscles, which tighten the abdomen; the respiratory muscles, which assist in breathing; the postural muscles of the back; and the muscles that act on the scapula or on the head and neck. Some trunk muscles are described in Exercise 13 if they act on the shoulder. These muscles are discussed in the Saladin text in chapter 10, "The Muscular System."

OBJECTIVES

At the end of this exercise, you should be able to

1. locate the muscles of the trunk on a torso model, chart, cadaver, or cat;
2. list the attachment points, and action of each muscle;
3. describe what nerve controls each muscle;
4. name all the muscles that have a particular action on the torso, such as all the muscles that compress the abdomen.

MATERIALS

Human torso model

Human muscle charts

Articulated skeleton

Cadaver

Cat

Materials for Cat Dissection

Dissection trays

Scalpel

Protective gloves

Blunt probe

Pins

PROCEDURE

Ⓐ Activity

Review the muscle nomenclature and the actions as outlined in Exercises 10 and 11. Examine a torso model or chart in the lab and locate the muscles described next and in table 12.1. Correlate the shape or fiber direction of the muscle with the name and visualize the muscles as you study the models or charts. Look at an articulated skeleton as you study the attachment points of the muscles so you can better see them. The descriptions in the text portion of this exercise can help you understand the nature of the muscle, while table 12.1 gives you specific information about the muscle.

Once you have learned the attachment points, you should be able to understand the action of the muscle. This is done, in part, by visualizing how one end of the muscle is pulled toward the other when the muscle contracts.

Anterior Muscles

The muscles of the abdomen compress the viscera, which aid in breathing and food regurgitation. These muscles are also involved in the **valsalva maneuver,** which consists of taking a deep breath and contracting these muscles. This aids in a bowel movement, urination, and childbirth. The **external oblique** is a broad, superficial muscle of the abdomen with fibers that run from a superior direction to an inferior, medial direction when viewed from the anterior aspect. The **internal oblique** is deep to the external oblique and has fiber directions that run perpendicular to the external oblique. Locate the abdominal muscles on models in the lab and compare them to fig. 12.1.

TABLE 12.1	Muscles of the Trunk			
Name	**Attachment Point 1 (Origin)**	**Attachment Point 2 (Insertion)**	**Action**	**Innervation**
Muscles of Thorax, Abdomen, and Pelvis				
External oblique	Ribs 5–12	Iliac crest, pubis	Compresses abdominal wall, laterally rotates waist	Intercostal nerves T7–12
Internal oblique	Inguinal ligament, iliac crest	Pubis, ribs 10–12	Compresses abdominal wall, laterally rotates waist	Intercostal nerves T7–12, spinal nerve L1
Transverse abdominal	Inguinal ligament, iliac crest, costal cartilages 7–12	Linea alba, pubis	Compresses abdominal wall, laterally rotates waist	Intercostal nerves T7–12, spinal nerve L1
Rectus abdominis	Crest of pubis, pubic symphysis	Costal cartilages 5–7, xiphoid process	Flexes spine, compresses abdominal wall	Intercostal nerves T6–12
Serratus anterior	Ribs 1–8	Vertebral border and inferior angle of scapula	Abducts scapula (moves scapula away from spinal column)	Long thoracic nerve
External intercostals	Inferior border of ribs 1–11	Superior border of rib below	Elevates ribs (increases volume in thorax)	Intercostal nerves
Internal intercostals	Superior border of a rib	Inferior border of next superior rib (2–12)	Elevates and depresses ribs	Intercostal nerves
Diaphragm	Xiphoid process, ribs 7–12, superior lumbar vertebrae	Central tendon	Inspiration	Phrenic nerve
Deep Muscles of the Back				
Erector spinae				
Iliocostalis *Longissimus* *Spinalis*	Nuchal ligament, thoracic and lumbar vertebrae, sacrum, ilium, ribs 3–12	All ribs, all thoracic vertebrae, temporal bones	Extends and rotates spine and head	Dorsal rami of spinal nerves
Multifidus	Vertebrae C4–L5, sacrum, ilium	Spine	Extends and rotates spine	Dorsal rami of spinal nerves
Quadratus lumborum	Iliac crest	L1–4, rib 12	Extends and abducts spine	T12, L1–4
Rhomboid major	Spines of T2–5	Medial border of scapula	Adducts scapula	Dorsal scapular nerve
Rhomboid minor	Nuchal ligament, spines C7–T1	Medial border of scapula	Adducts scapula	Dorsal scapular nerve
Splenius	Nuchal ligament, C7–T6	C2–4, occipital and temporal bone	Extends and rotates head	Middle and lower cervical nerves
Semispinalis	C4–T10	Occipital bone, spinous process of C2–C5	Extends and rotates head and spine	Cervical and thoracic spinal nerves

The deepest of the abdominal muscles is the **transverse abdominal,** which has fibers running in a horizontal direction. The muscle fibers of the **rectus abdominis** (*rectus* means straight) are oriented in a superior/inferior direction. The rectus abdominis has small connective tissue bands called **tendinous intersections** located horizontally across the muscle, dividing it into small segments. If the abdominal fat is minimal and the muscles are well developed, the "washboard stomach," or "six-pack," is apparent due to the muscle fibers increasing in girth while the tendinous intersections remain undeveloped.

The **serratus anterior** (figs. 12.1, 12.3, and 12.5) is a broad, fan-shaped muscle that has slips of muscle attached to the superior ribs. The anterior part of the muscle is serrated like the cutting edge of a saw or bread knife. These slips of muscles unite and attach to the medial border of the scapula and the inferior angle of the scapula. As the muscle contracts, it pulls the scapula toward the anterior aspect of the ribs.

The **intercostal** muscles and the diaphragm are respiratory muscles. Normally, the **diaphragm** is responsible for about two thirds of the resting breath volume, while the external intercostals contribute to the remaining volume. The diaphragm is a domed muscle that has a peripheral attachment. The other attachment of the diaphragm is central at the base of the mediastinum. If you think of the diaphragm as a trampoline, the outer springs represent the stable attachment while the center (where you jump) represents the moving attachment.

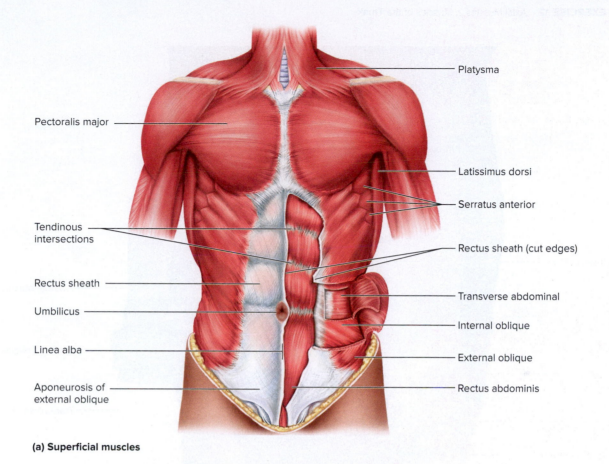

Platysma

Pectoralis major

Latissimus dorsi

Serratus anterior

Tendinous
intersections

Rectus sheath (cut edges)

Rectus sheath

Transverse abdominal

Umbilicus

Internal oblique

Linea alba

External oblique

Aponeurosis of
external oblique

Rectus abdominis

(a) Superficial muscles

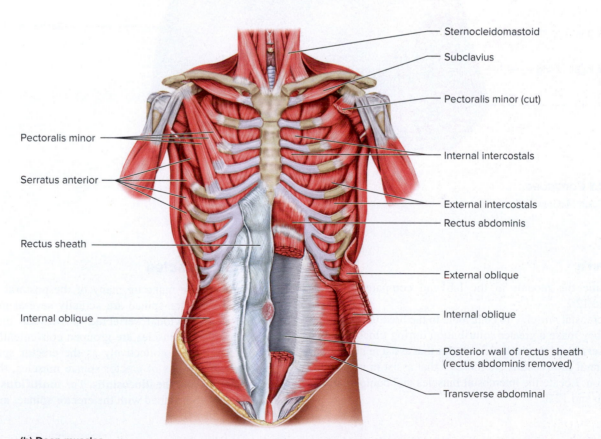

Sternocleidomastoid

Subclavius

Pectoralis minor (cut)

Pectoralis minor

Internal intercostals

Serratus anterior

External intercostals

Rectus abdominis

Rectus sheath

External oblique

Internal oblique

Internal oblique

Posterior wall of rectus sheath
(rectus abdominis removed)

Transverse abdominal

(b) Deep muscles

FIGURE 12.1 Superficial and Deep Muscles of the Abdomen, Anterior View. Diagram (a) superficial, (b) deep; (c) photograph.

169

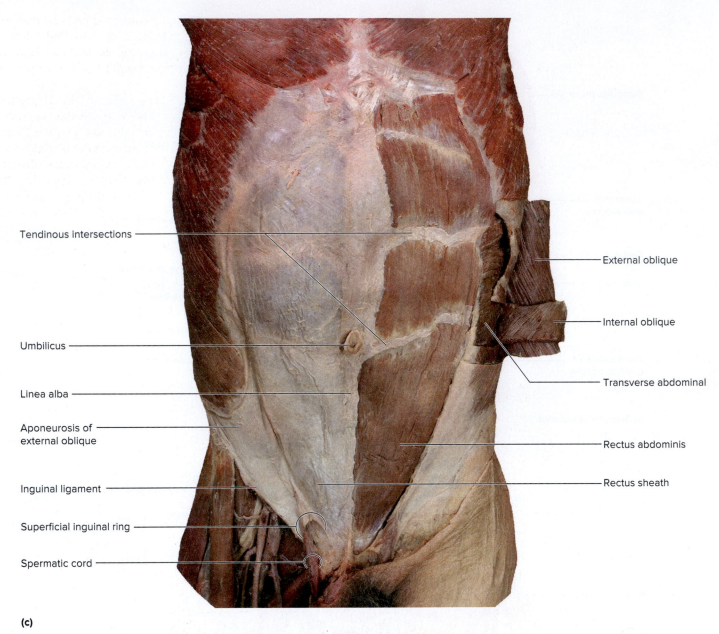

Tendinous intersections

External oblique

Umbilicus

Internal oblique

Linea alba

Transverse abdominal

Aponeurosis of
external oblique

Rectus abdominis

Inguinal ligament

Rectus sheath

Superficial inguinal ring

Spermatic cord

(c)

FIGURE 12.1 *Continued.*

(c) Christine Eckel/McGraw Hill

Ⓐ Activity

Examine the models in the lab and compare them to figs. 12.1 and 12.2.

The intercostal muscles do contribute to the breathing volume at rest, but they make a greater contribution during times of exercise. The **external intercostals** are muscles that assist in inhalation and the **internal intercostals** are muscles that assist in inhalation and exhalation. Locate the intercostal muscles and compare them to figs. 12.1b and 12.3.

Posterior Muscles

The **erector spinae** make up many of the postural muscles of the back. The erector spinae are actually several muscles that occur between individual vertebrae or between the vertebrae and the ribs. These muscles are grouped conveniently into long strap muscles known collectively as the erector spinae. There are three major groups of erector spinae muscles, the **spinalis,** the **longissimus,** and the **iliocostalis.** The **multifidus** is a related muscle and, when grouped with the erector spinae, makes up the

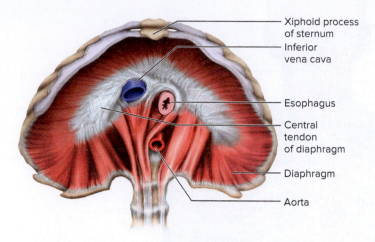

Xiphoid process
of sternum

Inferior
vena cava

Esophagus

Central
tendon
of diaphragm

Diaphragm

Aorta

FIGURE 12.2 Muscles of the Thorax. Diagram, inferior view.

s.l.i.m. muscles as you move from medial to lateral and then inferior. Another muscle that extends the spine (vertebral column) is the **quadratus lumborum,** a square muscle that runs from the iliac crest to the lower vertebrae and twelfth rib. These muscles are illustrated in fig. 12.4.

The **rhomboid muscles** occur deep to the trapezius. Reflection of the trapezius is necessary to see the rhomboid muscles. These muscles attach to the spine and to a tough ligament on the posterior neck known as the **nuchal ligament**. Their moving attachment is on the medial border of the scapula. As they contract they pull the scapulae together, adducting the scapulae. As the name implies, the **rhomboid major** is larger than the **rhomboid minor.** Other deep muscles of the back are the **splenius** and the **semispinalis** muscles. These extend and rotate the head and spine.

(A) **Activity**

Compare the material in lab to fig. 12.5.

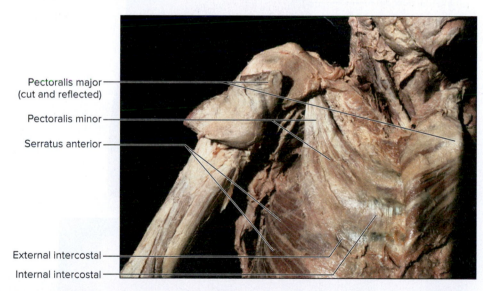

Pectoralis major
(cut and reflected)

Pectoralis minor

Serratus anterior

External intercostal

Internal intercostal

FIGURE 12.3 Muscles of the Thorax. Serratus anterior and intercostal muscles, anterior view.
©Eric Wise

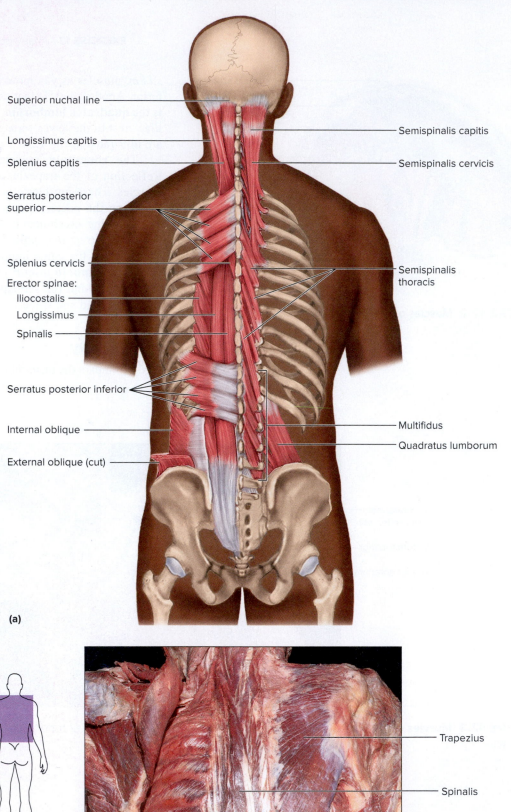

Superior nuchal line

Longissimus capitis

Splenius capitis

Serratus posterior superior

Splenius cervicis

Erector spinae:
  Iliocostalis
  Longissimus
  Spinalis

Serratus posterior inferior

Internal oblique

External oblique (cut)

Semispinalis capitis

Semispinalis cervicis

Semispinalis thoracis

Multifidus

Quadratus lumborum

(a)

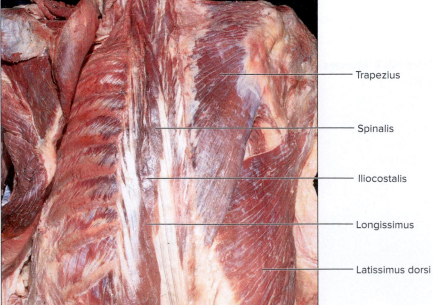

Trapezius

Spinalis

Iliocostalis

Longissimus

Latissimus dorsi

(b)

FIGURE 12.4 Deep Muscles of the Back, Posterior View.
(a) Diagram; (b) photograph.
(b) Rebecca Gray/Don Kincaid/ McGraw Hill

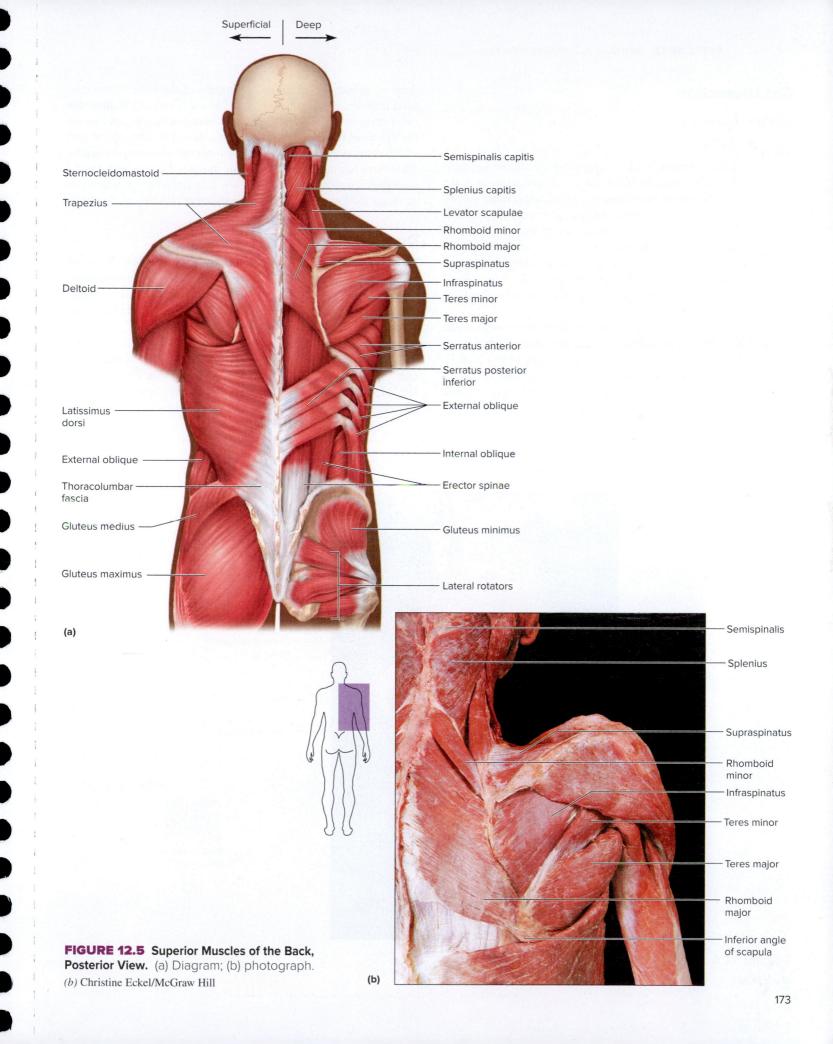

Superficial | Deep

Sternocleidomastoid

Trapezius

Deltoid

Latissimus dorsi

External oblique

Thoracolumbar fascia

Gluteus medius

Gluteus maximus

(a)

Semispinalis capitis

Splenius capitis

Levator scapulae

Rhomboid minor

Rhomboid major

Supraspinatus

Infraspinatus

Teres minor

Teres major

Serratus anterior

Serratus posterior inferior

External oblique

Internal oblique

Erector spinae

Gluteus minimus

Lateral rotators

Semispinalis

Splenius

Supraspinatus

Rhomboid minor

Infraspinatus

Teres minor

Teres major

Rhomboid major

Inferior angle of scapula

(b)

FIGURE 12.5 **Superior Muscles of the Back, Posterior View.** (a) Diagram; (b) photograph.
(b) Christine Eckel/McGraw Hill

Cat Dissection

Ventral Muscles

Ⓐ Activity

Place the cat on its back and examine the abdominal muscles. The abdominal muscles in the cat are similar to those in the human in that the **external oblique** is a broad, superficial muscle on the ventral abdomen. Carefully cut through the external oblique to reveal the **internal oblique,** as illustrated in fig. 12.6. Deep to this is the **transverse abdominal,** and it can be seen by carefully dissecting the internal oblique. If you cut too deeply, you will enter the abdominal cavity, so be careful in this part of the dissection. The **rectus abdominis** is a muscle that runs from the pubic region to the sternum.

Move to the thoracic region and examine the muscle of the lateral thorax. This is the **serratus anterior** (serratus ventralis in the cat), and you should see the scalloped edges of the muscle. Carefully separate this muscle from the others and locate its attachment to the scapula. To see the intercostal muscles, you will have to bisect the superficial chest muscles. If you have not

done so already, cut through the middle of the belly of the pectoral muscles, exposing the ribs of the cat. Carefully remove the outer layer of fascia from the muscle between the ribs and locate the **external intercostal** muscle. You should be able to cut part of this muscle away and expose the **internal intercostal** muscle. The fibers run perpendicular to one another. Do not look for the diaphragm at this time. You can see it in Exercise 33, when you examine the lungs. Examine these thoracic muscles in fig. 12.7.

Dorsal Muscles

Place the cat so you can examine the dorsal surface. You will need to dissect the trapezius carefully to see the **rhomboid** muscles. These muscles attach to the spine and also on the scapula. If you examine the muscles of the neck and head, you should see the **splenius** muscle. These are shown in fig. 12.8.

You will need to bisect the latissimus dorsi and the posterior portion of the external oblique to see the **erector spinae** muscles. The relative position of the cat erector spinae can be seen in fig. 12.9. Compare this figure to your dissection. Locate the **iliocostalis, longissimus, spinalis,** and **multifidus** in the cat.

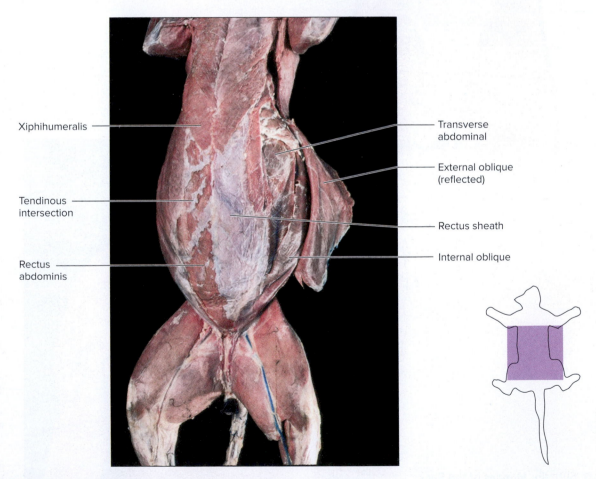

Xiphihumeralis

Tendinous intersection

Rectus abdominis

Transverse abdominal

External oblique (reflected)

Rectus sheath

Internal oblique

FIGURE 12.6 Muscles of the Abdomen of the Cat, Ventral View.

©Eric Wise

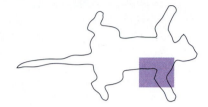

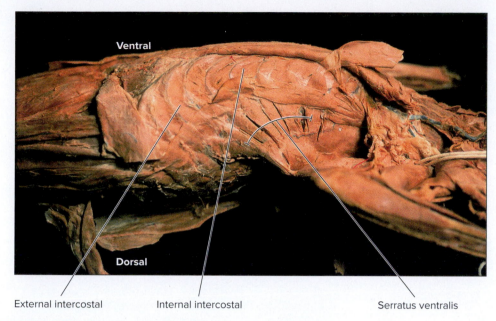

External intercostal   Internal intercostal   Serratus ventralis

FIGURE 12.7 Muscles of the Thorax of the Cat, Lateral View.

©Eric Wise

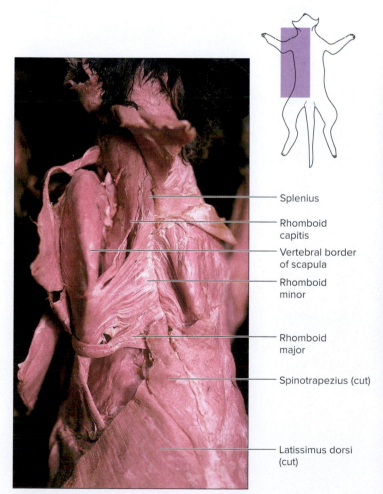

Splenius

Rhomboid
capitis

Vertebral border
of scapula

Rhomboid
minor

Rhomboid
major

Spinotrapezius (cut)

Latissimus dorsi
(cut)

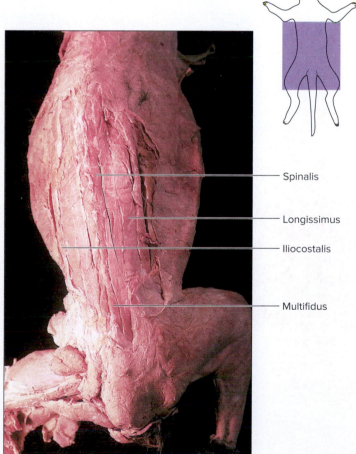

Spinalis

Longissimus

Iliocostalis

Multifidus

FIGURE 12.8 Anterior, Deep Muscles of the Back of the Cat, Dorsal View.

©Eric Wise

FIGURE 12.9 Posterior, Deep Muscles of the Back of the Cat, Dorsal View.

©Eric Wise

Notes

REVIEW SECTION

Axial Muscles 2: Muscles of the Trunk

Name _____ *Date* _____

Lab Section _____ *Time* _____

Review Questions

1. Compression of the abdominal wall occurs by what four muscles? _____
_____

2. Which is the deepest anterior abdominal muscle? _____

3. The tendinous intersections are found in what muscle? _____

4. How does the action of the rectus abdominis differ from that of the other abdominal muscles?
_____
_____

5. What is the action of the serratus anterior muscle? _____
_____

6. What is the physical relationship of the intercostal muscles to each other? _____
_____
_____

7. How does the serratus anterior function as an antagonist to the rhomboid muscles? _____
_____
_____

8. What is the action of the intercostal muscles? _____
_____

9. What muscle attaches to the central tendon? _____

10. Name *five* muscles that extend the spine. _____
_____

11. What is the action of the quadratus lumborum? _____
_____
_____

12. Reviewing *all* the muscles you have studied so far, which ones allow you to look up at the ceiling when you are standing?
_____
_____

13. Which is the more superior muscle, the rhomboid major or the rhomboid minor? _____

14. What is the medial attachment of the rhomboid major muscle? _____
_____

15. Label the muscles in the following illustration. Color the origin (attachment 1) in red and the insertion (attachment 2) in blue for each of the indicated muscles where the attachment points are visible.

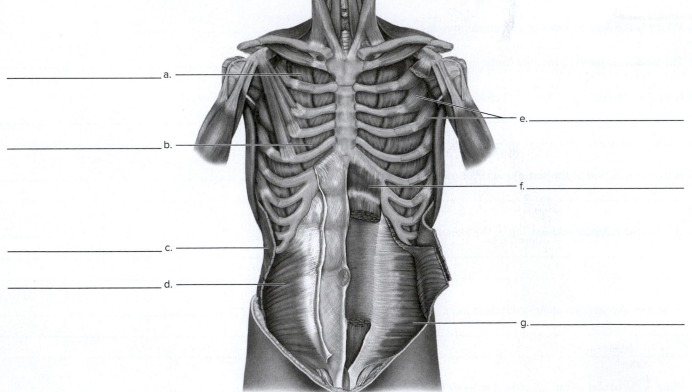

Anterior view

16. In the cat, how do the abdominal muscles compare to those in the human in terms of relative position?
_____
_____
_____

17. Bending over, like doing an abdominal crunch, uses what muscle? _____

_____

_____

_____

18. Label the muscles in the following illustration. Color the origin (attachment 1) in red and the insertion (attachment 2) in blue for each of the indicated muscles where the attachment points are visible.

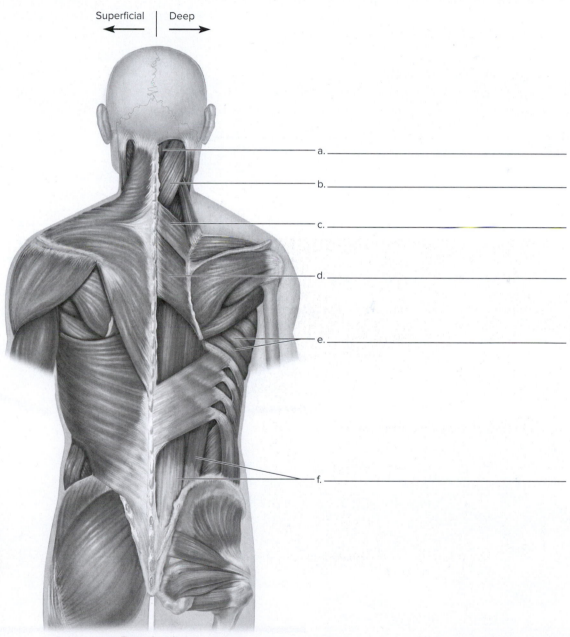

Superficial | Deep

a. _____

b. _____

c. _____

d. _____

e. _____

f. _____

Posterior view

Notes

Appendicular Muscles 1: Muscles of the Upper Limb

INTRODUCTION

Examination of Muscles of the Shoulder and Arm

In this study, the muscles of the arm are treated here for their action on the arm or forearm. Some of these muscles originate on the scapula, which has a small bony attachment to the clavicle. Most of the scapula is held in place by muscles, some of which are covered here, and some that were covered in earlier exercises. The forearm and hand muscles are important, as seen in rehabilitating limbs after trauma or after surgery to relieve carpal tunnel syndrome.

You should learn the attachment points (origins and insertions) as you learn the muscles of the forearm, because many of these muscles look similar. By knowing where a muscle attaches, you will not easily mistake it for another muscle. If your institution has an articulated skeleton that is painted red and blue, the red is one attachment point (the origin) and blue is another attachment point (the insertion).

OBJECTIVES

At the end of this exercise, you should be able to

1. locate the muscles of the shoulder and upper limb on a torso model, chart, cadaver, or cat;
2. list the attchment points and action of each muscle presented;
3. describe what nerve controls each muscle;
4. list what muscles function as synergists or antagonists to the prime mover;
5. name all the muscles that have an action on a joint, such as all the muscles that flex the arm;
6. apply your knowledge of cat musculature to human muscles.

MATERIALS

Human torso model

Human arm models

Human muscle charts

Articulated skeleton

Cadaver (if available)

Cat

Cat wetting solution

Materials for Cat Dissection

Dissection trays

Scalpel and two or three extra blades

Protective gloves

Blunt (mall) probe

String and tags

Pins

Forceps and sharp scissors

First aid kit in lab or prep area

Sharps container

Animal waste disposal container

PROCEDURE

(A) Activity

Look at the charts, models, and cadaver in the lab and locate the muscles presented in this exercise. You will find a list of the muscles of the shoulder and arm in table 13.1. Figure 13.1 illustrates the superficial muscles of the back and shoulder. Locate the diamond-shaped **trapezius** (trah-PEE-zee-us) muscle, which has an attachment point on the head, on the midline of the superior vertebral column, and on a tendon in the back of the neck called the **nuchal ligament.** It attaches laterally to the clavicle and scapula. If the scapula is fixed, the head moves, yet if the vertebral column and head are fixed, the trapezius moves the scapula.

Another superficial muscle of the posterior surface is the **latissimus dorsi** (lah-TISS-ih-mus DOR-si) muscle, which arises from the vertebrae by way of a broad, flat thoracolumbar fascia.

TABLE 13.1	Muscles of the Shoulder and Arm			
Name	**Attachment point 1 (Origin)**	**Attachment point 2 (Insertion)**	**Action**	**Innervation**
Superficial Muscles of the Shoulder				
Trapezius	Posterior occipital bone, nuchal ligament, C7–T4	Clavicle, acromion, and spine of scapula	Extends and abducts head, rotates and adducts scapula, fixes scapula	Accessory nerve (XI)
Deltoid	Clavicle, acromion, spine of scapula	Deltoid tuberosity of the humerus	Abducts arm; flexes, extends, medially, and laterally rotates humerus	Axillary nerve
Pectoralis major	Clavicle, sternum, cartilages of ribs 1–7	Crest of greater tubercle of humerus	Flexes, adducts, and medially rotates humerus	Medial and lateral pectoral nerves
Pectoralis minor	Ribs 3–5	Coracoid process of scapula	Depresses glenoid cavity	Medial and lateral pectoral nerve
Latissimus dorsi	T7–12, L1–5, crest of ilium, ribs 10–12	Intertubercular sulcus of humerus	Extends, adducts, and medially rotates humerus; draws shoulder inferiorly	Thoracodorsal nerve
Deep Muscles of the Shoulder				
Supraspinatus	Supraspinous fossa	Greater tubercle of humerus	Abducts arm, helps stabilize shoulder joint	Suprascapular nerve
Infraspinatus	Infraspinous fossa	Greater tubercle of humerus	Laterally rotates arm, stabilizes shoulder joint	Suprascapular nerve
Subscapularis	Subscapular fossa	Lesser tubercle of humerus	Medially rotates arm, stabilizes shoulder joint	Subscapular nerve
Teres minor	Lateral border of scapula	Greater tubercle of humerus	Laterally rotates and adducts humerus, stabilizes shoulder joint	Axillary nerve
Teres major	Inferior angle of scapula	Crest of lesser tubercle of humerus	Extends, and medially rotates humerus	Subscapular nerve
Muscles of the Arm				
Biceps brachii	Long head: superior margin of glenoid cavity; Short head: coracoid process of scapula	Radial tuberosity	Flexes arm, flexes forearm, supinates forearm	Musculocutaneous nerve
Triceps brachii	Infraglenoid tubercle of scapula, lateral and posterior surface of humerus	Olecranon, tuberosity of ulna	Extends and adducts arm, extends forearm	Radial nerve
Coracobrachialis	Coracoid process of scapula	Midmedial shaft of humerus	Flexes and medially rotates humerus	Musculocutaneous nerve
Brachialis	Anterior, distal surface of humerus	Coronoid process of ulna	Flexes forearm	Musculocutaneous and radial nerves
Brachioradialis	Lateral supracondylar ridge of humerus	Near styloid process of lateral radius	Flexes forearm	Radial nerve

C = cervical vertebrae   T = thoracic vertebrae   L = lumbar vertebrae   S = sacral vertebrae

The latissimus dorsi originates on the back, but the insertion is on the anterior aspect of the humerus. It is a powerful extensor of the arm and has the common name of the swimmer's muscle (fig. 13.1).

The **deltoid** is located on top of the shoulder and is a major abductor of the arm. Locate the deltoid on material in the lab and compare it to figs. 13.1 and 13.2. It is a fan-shaped muscle whose

insertion partially covers the insertion of the **pectoralis** (PEC-tur-AL-is) **major** muscle seen in fig. 13.2. The pectoralis major has fibers that run horizontally across the chest region, and it is a superficial muscle of the chest. Deep to the pectoralis major muscle is the **pectoralis minor** muscle, whose fibers run in a more vertical direction.

Deep to the trapezius and the deltoid are muscles that originate on the scapula proper. The **supraspinatus** (SOO-pra-spy-NATE-us)

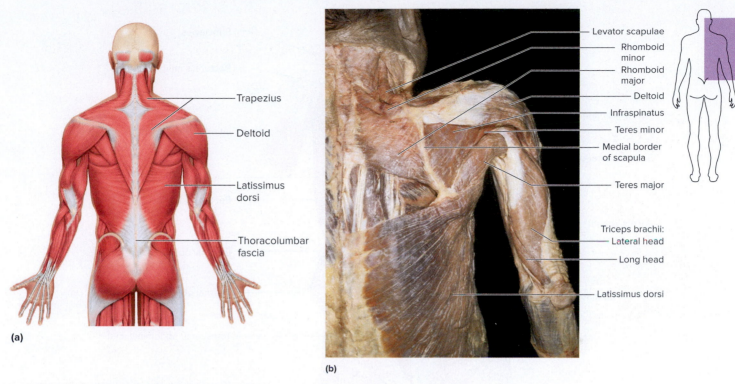

- Trapezius
- Deltoid
- Latissimus dorsi
- Thoracolumbar fascia

(a)

- Levator scapulae
- Rhomboid minor
- Rhomboid major
- Deltoid
- Infraspinatus
- Teres minor
- Medial border of scapula
- Teres major
- Triceps brachii:
  - Lateral head
  - Long head
- Latissimus dorsi

(b)

FIGURE 13.1 Superficial Back Muscles, Posterior View.
(a) Diagram; (b) photograph (trapezius has been removed).

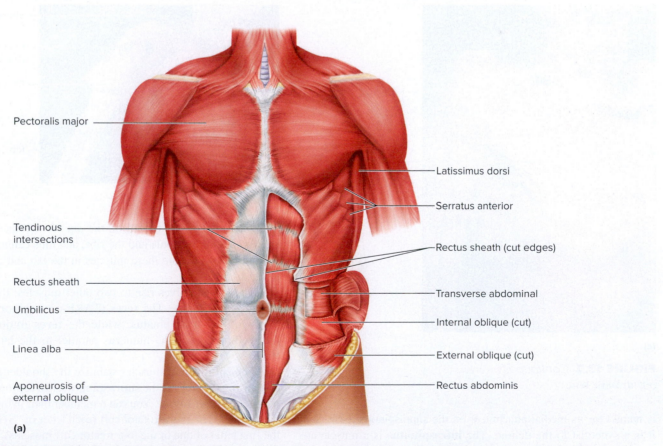

- Pectoralis major
- Tendinous intersections
- Rectus sheath
- Umbilicus
- Linea alba
- Aponeurosis of external oblique
- Latissimus dorsi
- Serratus anterior
- Rectus sheath (cut edges)
- Transverse abdominal
- Internal oblique (cut)
- External oblique (cut)
- Rectus abdominis

(a)

FIGURE 13.2 Thoracic Muscles, Anterior View. Diagram (a) superficial; (b) deep muscles. Photograph (c) superficial muscles; (d) deep muscles.

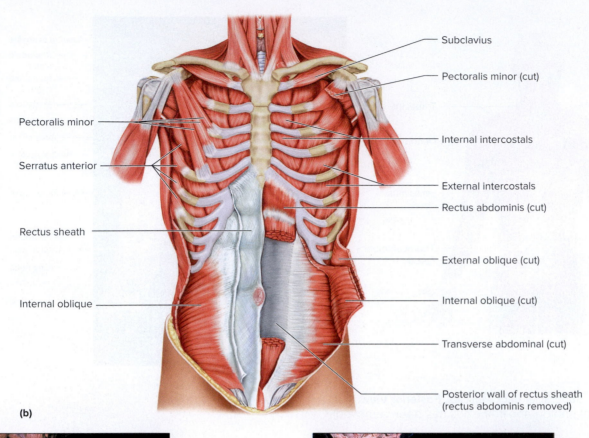

Subclavius

Pectoralis minor (cut)

Pectoralis minor

Serratus anterior

Internal intercostals

External intercostals

Rectus abdominis (cut)

Rectus sheath

External oblique (cut)

Internal oblique (cut)

Internal oblique

Transverse abdominal (cut)

Posterior wall of rectus sheath
(rectus abdominis removed)

(b)

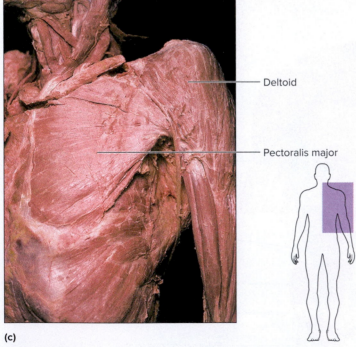

Deltoid

Pectoralis major

(c)

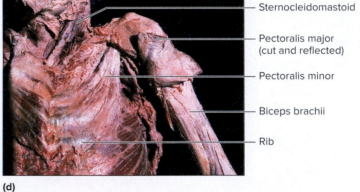

Sternocleidomastoid

Pectoralis major
(cut and reflected)

Pectoralis minor

Biceps brachii

Rib

(d)

FIGURE 13.2 *Continued.*
(c), (d) ©Eric Wise

is named for its medial attachment on the supraspinous fossa, and it is a synergist to the deltoid. The **infraspinatus** is a muscle attaching to the infraspinous fossa and laterally rotates the arm. The **subscapularis** (sub-SCAP-you-LAR-is) is named for its medial attachment to the subscapular fossa, and it medially rotates the arm.

The subscapularis is located on the *anterior* surface of the scapula, between the scapula and the ribs, and thus cannot be seen in a posterior view. Locate these muscles in the lab and compare them to fig. 13.3.

The scapula gives rise to two other muscles: the teres minor and the teres major. The **teres** (TARE-eez) **minor** appears like a slip of the infraspinatus, while the **teres major** crosses on the medial side of the humerus parallel to the latissimus dorsi (fig. 13.3).

The **rotator cuff** muscles stabilize the shoulder joint. Muscles composing the cuff are the **supraspinatus, infraspinatus, subscapularis,** and **teres minor.** You can remember their names because the humerus "SITS" in the rotator cuff (each letter of "SITS" represents the first letter of one of the four rotator cuff muscles). A rotator cuff injury is one where there is damage to any of these four muscles.

The **biceps brachii** (BI-ceps BRAY-key-eye) muscle is a two-headed muscle of the arm. It has an action on the arm, yet

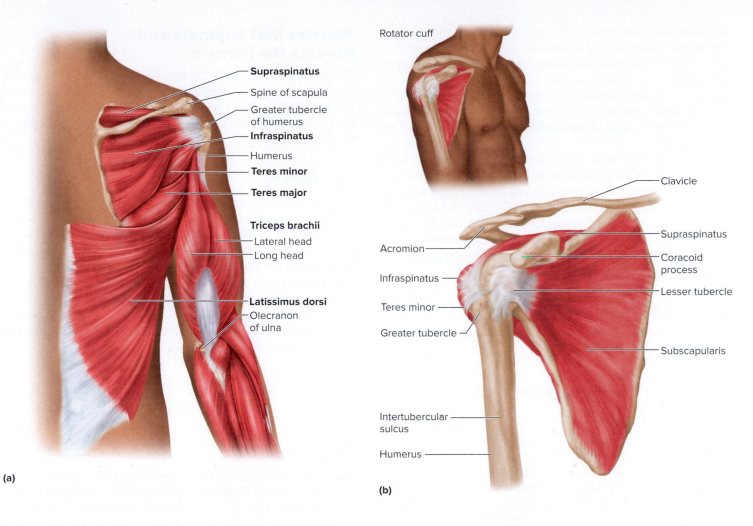

(a)

(b)

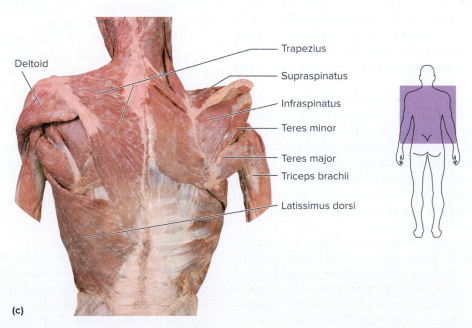

(c)

FIGURE 13.3 Scapular Muscles. Diagram (a) posterior view; (b) anterior view. Photograph (c) posterior view (trapezius and latissimus dorsi are removed on right side).

(c) Christine Eckel/McGraw Hill

the biceps brachii neither originates nor inserts on the humerus. The biceps brachii has one attachment point on the superior margin of the glenoid cavity with a long tendon that runs in the intertubercular sulcus of the humerus. It also has a short tendon that attaches superiorly to the coracoid process. The **triceps** (TRYceps) **brachii** is the only major muscle on the posterior surface of the humerus. The triceps brachii is an antagonist to the biceps brachii.

Along the short head of the biceps brachii is a small slip of muscle known as the **coracobrachialis** (core-AK-oh-BRAY-key-AL-us). This muscle is a short, diagonal muscle of the arm. Underneath the biceps brachii on the anterior surface of the arm is the **brachialis** (BRAY-key-AL-us) muscle. The brachialis only crosses the elbow joint and thus flexes the forearm. The **brachioradialis** (BRAY-key-oh-RAY-dee-AL-us) is a distal muscle of the arm and is the most lateral muscle of the forearm.

(A) **Activity**

Examine fig. 13.4 for these muscles and compare them to descriptions found in table 13.1.

Muscles of the Forearm and Hand

(A) **Activity**

Examine the muscles of the forearm and hand on a model, chart, or cadaver and locate the attachment points of each. Refer to the following descriptions as you examine the muscles. The specific details of these muscles are provided in table 13.2. Review the bones of the hand as well as the bones of the arm and forearm in Exercise 9 to precisely locate the bony origins and insertions.

The muscles of the forearm, due to their similar appearance, provide greater challenges in distinguishing one from another than do the muscles of the shoulder and arm. As you study these muscles, it is important you determine the attachment points, as this will help you determine if you are looking at the correct muscle.

The muscles of the forearm and hand are generally named for their action (*pronate* the forearm or *flex* the digits) or for their insertion (*carpi* for inserting on carpals or metacarpals, *digitorum* for fingers, and *pollicis* for thumb). A few are named for the shape of the muscle (*teres* for round or *quadratus* for square).

Many of the anterior forearm muscles are innervated by the median nerve. As you examine these muscles, look for the median nerve, which is a slender slip of white material resembling a tendon but without coming from any muscle.

Much of the muscle mass for moving the fingers is located on the forearm. In this way the fingers move as if on puppet strings. If all the muscle mass for the hand were located on the hand proper, the hand would look like a softball. By having most of the muscle mass in the forearm, an efficiency of form allows for a powerful grip yet precise movements of the fingers. A connective tissue band known as the **flexor retinaculum** (RET-in-AK-you-lum), anchors the anterior tendons to the wrist and prevents them from pulling away from the wrist when the hand is flexed.

Muscles that Supinate and Pronate the Forearm

The first group of muscles in this study are those muscles that insert on the radius. The **supinator** (SOUP-in-AY-tor) is a muscle that originates on the arm and forearm; it supinates the forearm. It wraps around the radius and is the deepest, proximal muscle of the forearm.

Activity

Examine this muscle in the lab and compare it to fig. 13.5.

The **pronator teres** (PRO-nay-tor TARE-eez) is a round muscle that pronates the forearm. It originates on the medial side of the arm and forearm and inserts on the lateral side of the radius. It is a bit different from the other superficial forearm muscles in that the pronator teres runs at an oblique angle on the forearm, while the other muscles run parallel along the length of the forearm.

The **pronator quadratus** (quah-DRAY-tus) is a square muscle deep to the other forearm muscles on the distal part of the radius and the ulna. It pronates the forearm and hand.

(A) **Activity**

Compare the two pronator muscles in the lab to fig. 13.5.

Flexor Muscles

The next muscles to be studied are grouped by their insertion on the hand. The superficial **palmaris** (pahl-MARE-us) **longus** muscle is absent in about 10% of the population. It is centrally located in the middle, anterior forearm, runs above the flexor retinaculum, and inserts into a broad, flat tendon known as the **palmar aponeurosis.** This aponeurosis has no bony attachment but rather attaches to the fascia of the underlying muscles.

The **flexor carpi** (CAR-pie) **radialis** muscle inserts on the metacarpals on the radial side of the hand. The pulse of the radial artery is taken at the wrist just lateral to the tendon of the flexor carpi radialis muscle. Most of the flexor muscles of the hand attach proximally to the medial epicondyle of the humerus, as seen with the flexor carpi radialis. The tendon of this muscle runs underneath the flexor retinaculum.

The **flexor carpi ulnaris** muscle also has an attachment on the medial epicondyle of the humerus and another on the carpals and a metacarpal of the ulnar side of the hand. The flexor carpi ulnaris is a medial muscle of the forearm.

Activity

Examine these muscles in the lab and in fig. 13.6.

The **flexor digitorum superficialis** (SOUP-er-FISH-ee-AL-is) is a superficial flexor muscle of the digits. It is not the most superficial muscle of the forearm but is actually under three muscles: the palmaris longus, the flexor carpi radialis, and the flexor carpi ulnaris. The flexor digitorum superficialis is so named because it is superficial to the deep digit flexor, discussed next. As with the other flexor muscles, the flexor digitorum superficialis

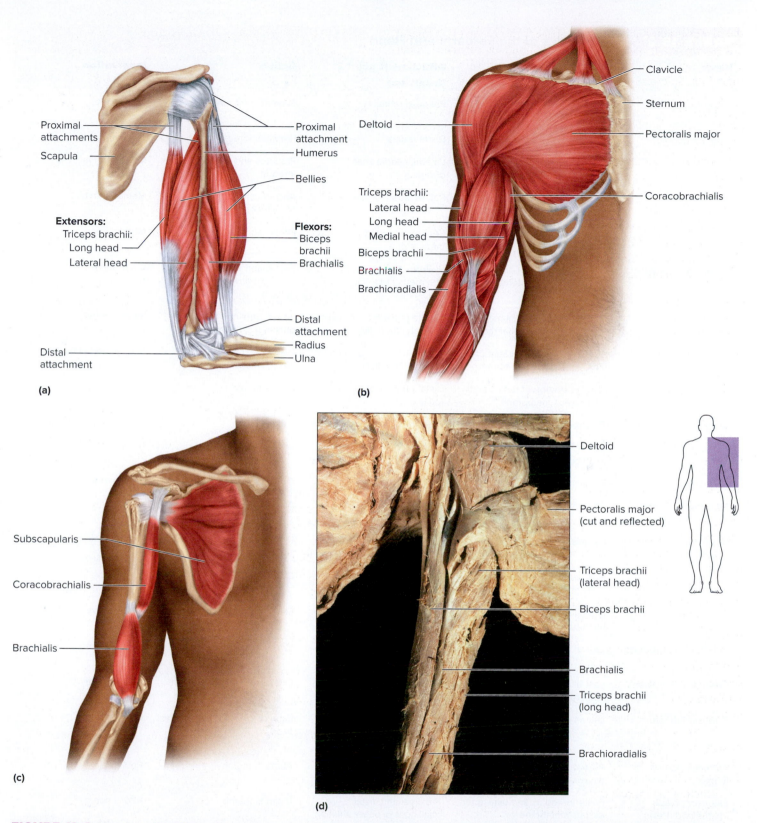

FIGURE 13.4 Muscles of the Arm. Diagram (a) right lateral view; (b) superficial muscles, right anterior view; (c) deep muscles, right anterior view. Photograph (d) left anterior view.
(d) ©Eric Wise

TABLE 13.2	Muscles of the Forearm and Hand			
Name	**Attachment point 1 (Origin)**	**Attachment point 2 (Insertion)**	**Action**	**Innervation**
Supinator	Lateral epicondyle of humerus, anterior ulna	Proximal radius	Supinates forearm	Radial nerve
Pronator quadratus	Distal part of anterior ulna	Distal radius	Pronates forearm and hand	Median nerve
Pronator teres	Medial epicondyle of humerus, coronoid process of ulna	Lateral, middle shaft of radius	Pronates and flexes forearm	Median nerve
Palmaris longus	Medial epicondyle of humerus	Palmar aponeurosis	Anchors skin and fascia of palmar region	Median nerve
Flexor carpi radialis	Medial epicondyle of humerus	Second and third metacarpals	Flexes and abducts wrist	Median nerve
Flexor carpi ulnaris	Medial epicondyle of humerus, olecranon and posterior surface of ulna	Pisiform, hamate, fifth metacarpal	Flexes and adducts wrist	Ulnar nerve
Flexor digitorum superficialis	Medial epicondyle of humerus, proximal ulna, proximal radius	Middle phalanges of second through fifth digits	Flexes proximal and middle phalanges, flexes wrist	Median nerve
Flexor digitorum profundus	Anterior, proximal surface of ulna; interosseus membrane	Distal phalanges of second through fifth digits	Flexes phalanges, flexes wrist	Median and ulnar nerves
Flexor pollicis longus	Anterior portion of radius and interosseus membrane	Distal phalanx of thumb (pollex)	Flexes thumb	Median nerve
Flexor pollicis brevis	Trapezium trapezoid, capitate	Proximal phalanx of first digit	Flexes thumb	Median and ulnar nerves
Abductor pollicis brevis	Scaphoid, trapezium, flexor retinaculum	Proximal phalanx of first digit	Abducts thumb	Median nerve
Opponens pollicis	Trapezium, flexor retinaculum	First metacarpal	Opposes thumb	Median nerve
Abductor digiti minimi	Pisiform	Proximal phalanx of fifth digit	Abducts fifth digit	Ulnar nerve
Flexor digiti minimi brevis	Hamulus of hamate	Proximal phalanx of fifth digit	Flexes fifth digit	Ulnar nerve
Opponens digiti minimi	Hamulus of hamate	Fifth metacarpal	Opposes fifth digit	Ulnar nerve
Extensor carpi radialis longus	Lateral supracondylar ridge of humerus	Second metacarpal	Extends and abducts wrist	Radial nerve
Extensor carpi radialis brevis	Lateral epicondyle of humerus	Third metacarpal	Extends and abducts wrist	Radial nerve
Extensor carpi ulnaris	Lateral epicondyle of humerus, proximal ulna	Fifth metacarpal	Extends and adducts wrist	Radial nerve
Extensor digitorum	Lateral epicondyle of humerus	Middle and distal phalanges of second through fifth digits	Extends phalanges, extends hand	Radial nerve
Abductor pollicis longus	Posterior radius and ulna, interosseus membrane	Trapezium and first metacarpal	Abducts thumb	Radial nerve
Extensor pollicis longus and brevis	Posterior radius and ulna, interosseus membrane	Proximal and distal phalanges of pollex (thumb)	Extends thumb	Radial nerve

has a proximal attachment on the medial epicondyle of the humerus. The distal tendons form a V on the middle phalanges of the second through fifth digits.

The **flexor digitorum profundus** (pro-FUND-us) is a deep muscle of the digits. It is an exception to the other hand flexors in that it does *not* attach to the medial epicondyle of the humerus but on the ulna and the membrane between the radius and the ulna (**interosseus membrane).** It is a deep muscle of the forearm that runs underneath the flexor digitorum superficialis. At the distal attachment point, the tendons of the flexor digitorum

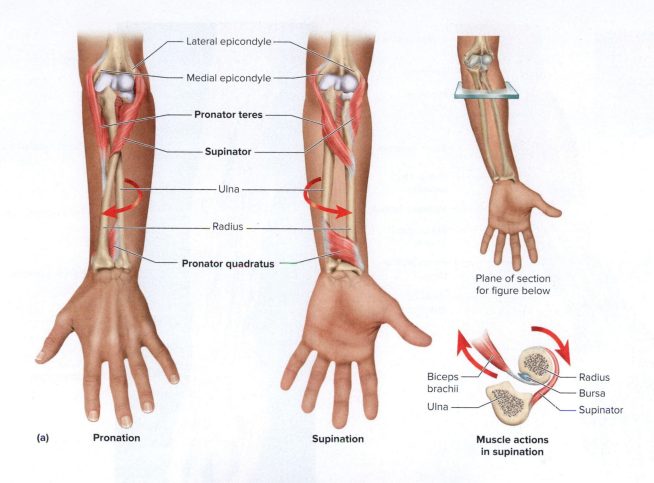

Lateral epicondyle

Medial epicondyle

Pronator teres

Supinator

Ulna

Radius

Pronator quadratus

(a) **Pronation**

Supination

Plane of section
for figure below

Biceps
brachii

Radius

Bursa

Ulna

Supinator

**Muscle actions
in supination**

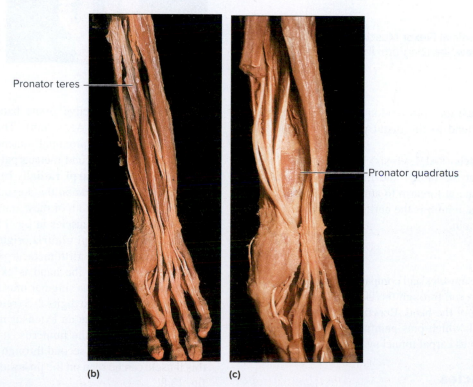

Pronator teres

Pronator quadratus

(b)

(c)

FIGURE 13.5 Supinator and Pronator Muscles of the Right Forearm, Anterior View. (a) Diagram. Photograph (b) superficial muscles; (c) deep muscles.

(b), (c) ©Eric Wise

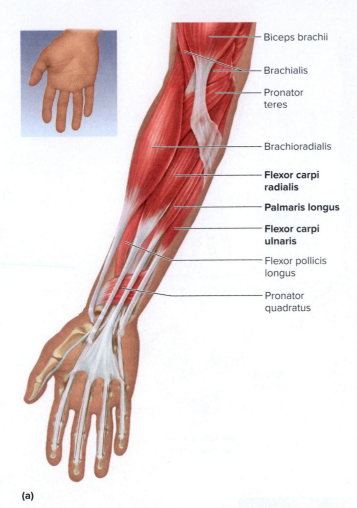

Biceps brachii

Brachialis

Pronator teres

Brachioradialis

Flexor carpi radialis

Palmaris longus

Flexor carpi ulnaris

Flexor pollicis longus

Pronator quadratus

(a)

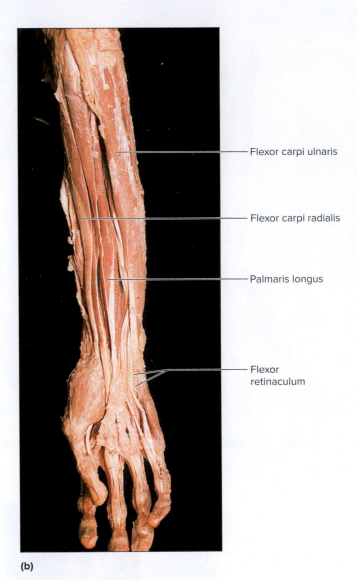

Flexor carpi ulnaris

Flexor carpi radialis

Palmaris longus

Flexor retinaculum

(b)

FIGURE 13.6 Superficial Flexor Muscles of the Right Forearm, Anterior View. (a) Diagram; (b) photograph. *(b) ©Eric Wise*

profundus run through the split tendons of the flexor digitorum superficialis and extend to the distal phalanges of the second through fifth digits.

The **flexor pollicis** (PAUL-eh-sis) **longus** (pollicis refers to thumb) originates on the radius and the interosseus membrane and runs along the lateral forearm to attach to the pad-side of the thumb. Flexion of the thumb is the curling of the thumb as if you were ready to flip a coin.

Ⓐ **Activity**

Examine these muscles and compare them to fig. 13.7. Many of the flexor muscles run through the U-shaped **carpal tunnel** at the proximal portion of the hand. Repetitive use of these muscles causes them to swell, which puts painful pressure on the median nerve. This is known as carpal tunnel syndrome.

Extensor Muscles

As a general rule, most of the extensors originate on the lateral epicondyle or lateral supracondylar ridge of the humerus. The extensor muscle tendons are held to the posterior surface of

the wrist by a connective tissue band known as the **extensor retinaculum** (RET-in-AK-u-lum). The **extensor carpi radialis longus** muscle has a proximal attachment on the humerus and a distal one on the second metacarpal (on the radial side) of the hand. The **extensor carpi radialis brevis** (BREV-is) is deep to the longus, and it inserts on the dorsum (the posterior surface) of the third metacarpal. Both of these muscles extend and abduct the hand. Examine these muscles in fig. 13.8.

The **extensor carpi ulnaris** originates on the lateral epicondyle and inserts on the fifth metacarpal (on the ulnar side) of the hand. As it contracts, the hand is extended and adducted. The **extensor digitorum** is a singular muscle on the posterior forearm whose tendons attach to digits 2–5 (remember there are two flexor digitorum muscles). As an extensor muscle, it originates on the lateral epicondyle of the humerus. It inserts on the middle and distal phalanges of the second through fifth digits. The tendons of this muscle can be seen on the posterior surface of the hand and in fig. 13.8.

The **abductor pollicis longus** muscle inserts on the metacarpal of the thumb. By pulling your thumb ventrally away from the index finger, you are abducting the thumb. There are two extensor

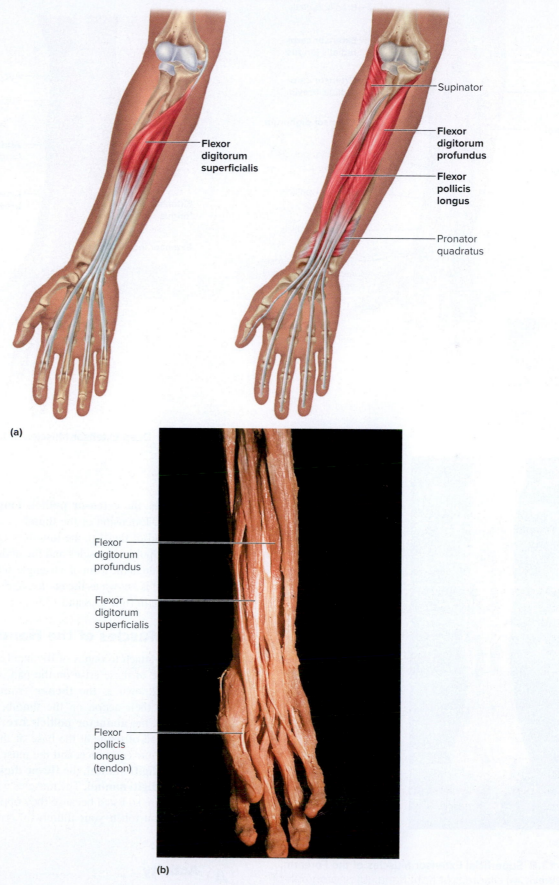

Flexor digitorum superficialis

Supinator

Flexor digitorum profundus

Flexor pollicis longus

Pronator quadratus

(a)

Flexor digitorum profundus

Flexor digitorum superficialis

Flexor pollicis longus (tendon)

(b)

FIGURE 13.7 **Deep Flexor Muscles of the Right Forearm, Anterior View.** (a) Diagram; (b) photograph.
(b) ©Eric Wise

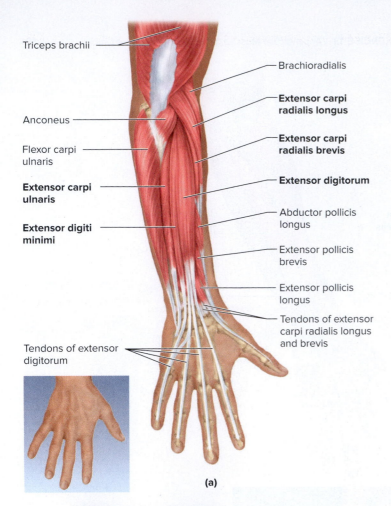

Triceps brachii

Brachioradialis

Extensor carpi radialis longus

Anconeus

Extensor carpi radialis brevis

Flexor carpi ulnaris

Extensor digitorum

Extensor carpi ulnaris

Abductor pollicis longus

Extensor digiti minimi

Extensor pollicis brevis

Extensor pollicis longus

Tendons of extensor carpi radialis longus and brevis

Tendons of extensor digitorum

(a)

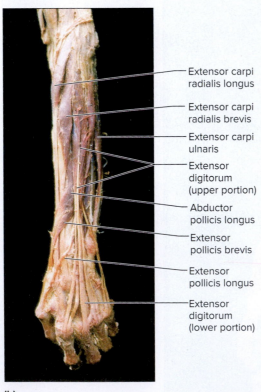

Extensor carpi radialis longus

Extensor carpi radialis brevis

Extensor carpi ulnaris

Extensor digitorum (upper portion)

Abductor pollicis longus

Extensor pollicis brevis

Extensor pollicis longus

Extensor digitorum (lower portion)

(b)

FIGURE 13.8 Superficial Extensor Muscles of the Forearm, **Posterior View.** (a) Diagram of right hand; (b) photograph of left hand.
(b) ©Eric Wise

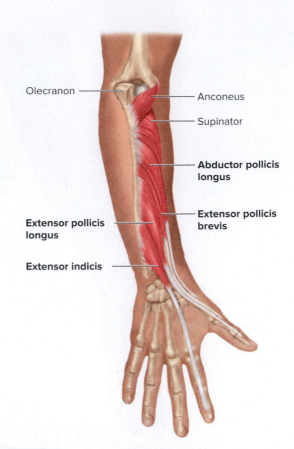

Olecranon

Anconeus

Supinator

Abductor pollicis longus

Extensor pollicis longus

Extensor pollicis brevis

Extensor indicis

FIGURE 13.9 Deep Extensor Muscles of the Right Thumb, **Posterior View.**

pollicis muscles, the **extensor pollicis longus** and the **extensor pollicis brevis.** Extension of the thumb is done when you flip a coin. At the end of the flip, the thumb is extended. The tendons of the extensor pollicis muscles and the abductor pollicis muscles form a depression in the form of a triangle at the base of the thumb. This depression is known as the *anatomical snuff box.* These muscles can be seen in figs. 13.8 and 13.9.

Intrinsic Muscles of the Hand

Many muscles attach to bones of the hand and have an action on the hand. Some of these arise on the pad of muscles at the base of the thumb known as the **thenar eminence.** These muscles are named for their action on the thumb, including the **flexor pollicis brevis,** the **abductor pollicis brevis,** and the **opponens pollicis.** The pad of tissue at the base of the fifth digit is known as the **hypothenar eminence,** and the muscles that arise there are the **abductor digiti minimi,** the **flexor digiti minimi brevis,** and the **opponens digiti minimi.** The muscles with the word *opponens* in their name are so listed because they oppose the thumb or fifth digit. When you touch your thumb to your fifth digit, you are opposing them.

(A) Activity

Find these muscles on models or cadavers in lab and compare them to fig. 13.10.

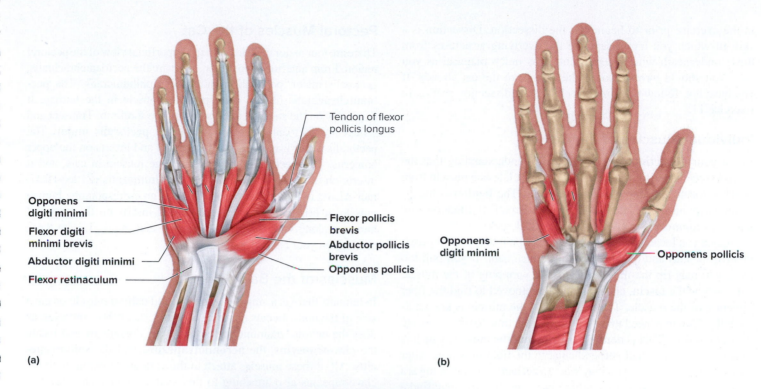

Tendon of flexor
pollicis longus

**Opponens
digiti minimi**

**Flexor digiti
minimi brevis**

Abductor digiti minimi

Flexor retinaculum

**Flexor pollicis
brevis**

**Abductor pollicis
brevis**

Opponens pollicis

**Opponens
digiti minimi**

Opponens pollicis

(a)

(b)

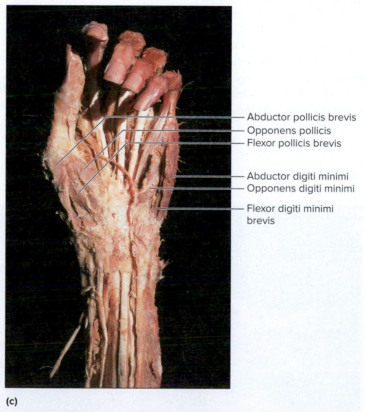

Abductor pollicis brevis
Opponens pollicis
Flexor pollicis brevis

Abductor digiti minimi
Opponens digiti minimi

Flexor digiti minimi
brevis

(c)

FIGURE 13.10 **Intrinsic Muscles of the Right Hand.** Diagram (a) superficial muscles; (b) deep muscles. Photograph
(c) superficial muscles.
(c) ©Eric Wise

Cat Dissection

The objective in using a cat for the exercise on musculature is to provide a study specimen for dissection and application to the human system. Differences occur between cat musculature and human musculature, but you should focus on similar structures to gain an appreciation of human musculature. Read all the material in the exercise prior to beginning the dissection. Dissection is a skill in which you try to separate the overlying structures from those underneath while keeping intact as much material as you can. You should have removed the skin from the cat already. If you have not, follow the directions in the cat dissection section in Exercise 11.

Individual Muscles of the Cat

Begin your **dissection** of the muscles by understanding that the term *dissect* means to separate. When you isolate one muscle from another, locate the tendons of that muscle. The **tendon** is the attachment point of the muscle to a bone. Broad, flat tendons are known as **aponeuroses** (AH-poh-nyeu-ROH-seez).

Once you locate the muscle, tug gently on it to locate its proximal (or medial) and distal (or lateral) attachment. If you pull too hard, you may rip the muscle. The outer wrapping of the muscle is known as the **fascia,** and it should be removed to find the fiber direction of the muscle. The main part of the muscle is known as the **belly.** You may need to cut a muscle from time to time to locate deeper muscles. This is done by **transecting** the muscle, which is to cut the muscle in half perpendicular to the fiber direction. After transecting the muscle, you may want to **reflect** it, or pull it toward its attachment site. When separating two muscles, you may find a cottonlike material between the muscles. This is loose connective tissue that forms part of the fascia.

Once you have removed the skin from your cat and removed the superficial fascia, you should identify the major muscles of the cat. Look for the large **latissimus dorsi** muscle of the back and the **external oblique** muscle. You should also find the **deltoids** and **triceps brachii** muscles of the shoulder region and the **gluteus** and **biceps femoris** muscles of the hip and thigh region. Compare your cat to fig. 13.11.

Pectoral Muscles of the Cat

There are four major muscles seen in a superficial view of the pectoral region. From anterior to posterior, these are the pectoantebrachialis, pectoralis major, pectoralis minor, and xiphihumeralis. The **pectoantebrachialis** has no corresponding muscle in the human. It originates on the sternum and inserts on the forelimb. Transect and reflect the pectoantebrachialis to see the **pectoralis major.** The pectoralis major originates on the sternum and inserts on the upper humerus. The **pectoralis minor** is a large muscle in cats, and it inserts on the upper humerus. The **xiphihumeralis** (ZI-fee-HEU-mur-AL-is) is another cat muscle that has no corresponding human muscle; it originates on the sternum and inserts on the proximal humerus along with the pectoralis major and minor. Locate these muscles on your cat and in fig. 13.12.

Muscles of the Back

In humans there is a singular trapezius and deltoid muscle on each side of the body. In cats the trapezius consists of three muscles, as does the deltoid. Examine the cat from the dorsal side and locate the **clavotrapezius,** the **acromiotrapezius,** and the **spinotrapezius.** All of these muscles attach to the vertebral column, with the clavotrapezius also attaching to the occipital bone. The clavotrapezius goes to the clavicle, the acromiotrapezius to the acromion process of the scapula, and the spinotrapezius to the spine of the scapula. Compare these muscles to fig. 13.13.

The deltoid muscles are also named for their bony attachments. The **clavodeltoid** (clavobrachialis) proximally attaches to the clavicle, the **acromiodeltoid** to the acromion process, and the **spinodeltoid** to the spine of the scapula. The lateral attachments of these muscles are on the arm or forelimb. Locate these muscles on the cat and compare them to fig. 13.13.

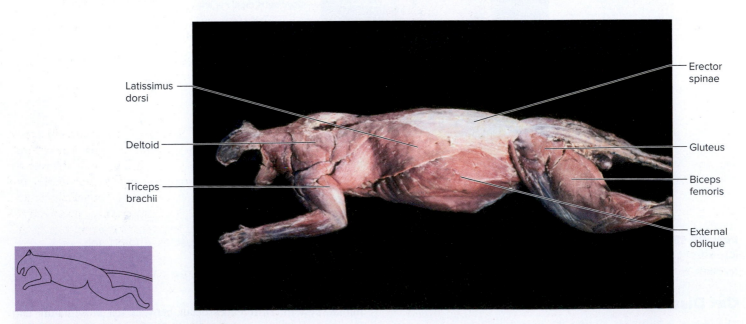

Latissimus dorsi

Deltoid

Triceps brachii

Erector spinae

Gluteus

Biceps femoris

External oblique

FIGURE 13.11 Major Muscles of the Cat.

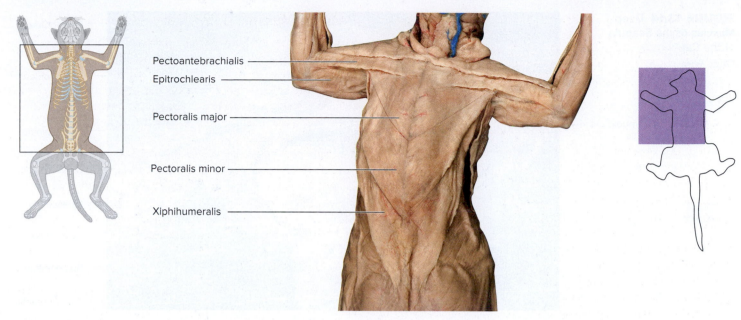

Pectoantebrachialis

Epitrochlearis

Pectoralis major

Pectoralis minor

Xiphihumeralis

FIGURE 13.12 Muscles of the Pectoral Region of the Cat.

FIGURE 13.13 Superficial Muscles of the Shoulder of the Cat.
©Eric Wise

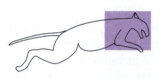

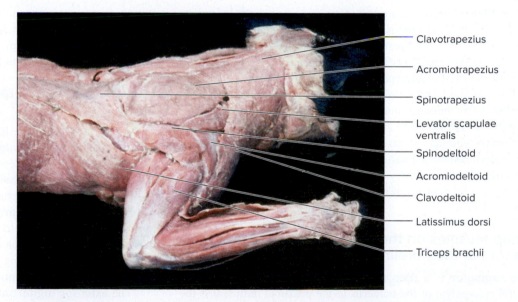

Clavotrapezius

Acromiotrapezius

Spinotrapezius

Levator scapulae ventralis

Spinodeltoid

Acromiodeltoid

Clavodeltoid

Latissimus dorsi

Triceps brachii

Two other muscles of the region are the **latissimus dorsi** and the **levator scapulae ventralis.** These two muscles are similar to those in the human. Locate these muscles on the cat and compare them to fig. 13.13.

The deep muscles of the scapula can be seen by reflecting the overlying muscles. The **supraspinatus, infraspinatus, subscapularis, teres major,** and **teres minor** are roughly equivalent to the same muscles in the human. Locate the supraspinatus, infraspinatus, and teres major and find them in fig. 13.14.

Forelimb Muscles

The muscles that have an action on the forelimb of the cat typically either flex the forelimb or extend the forelimb as their primary action. The **epitrochlearis** is a muscle that does not have a corresponding muscle in humans. The epitrochlearis is on the medial side of the humerus and extends the forelimb. The **biceps brachii** is also a medial muscle, and it flexes the forelimb. The **triceps brachii, anconeus, brachioradialis,** and **brachialis** are lateral or posterior muscles. The triceps brachii and the anconeus extend the forelimb, while the brachialis flexes the forelimb. Find the lateral muscles, using figs. 13.12, 13.15, and 13.16 as a guide.

Superficial Muscles on the Medial Aspect of the Forelimb

Most of the muscles of the forelimb of the cat run parallel to the radius and ulna. An exception to this is the **pronator teres,** a small slip of muscle that runs obliquely down the forelimb. It pronates the forearm.

FIGURE 13.14 Deep Muscles of the Scapula of the Cat.
©Eric Wise

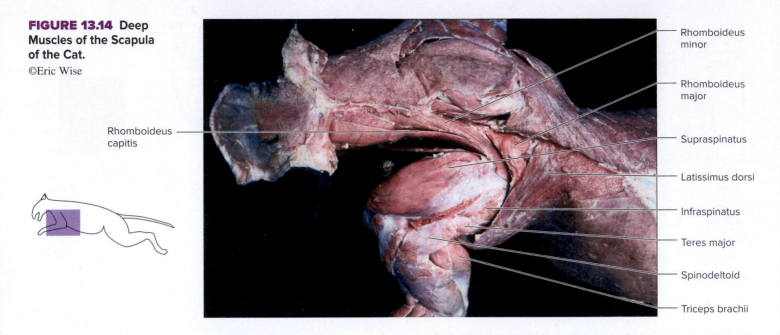

Rhomboideus minor

Rhomboideus major

Supraspinatus

Latissimus dorsi

Infraspinatus

Teres major

Spinodeltoid

Triceps brachii

Rhomboideus capitis

The **palmaris longus** is a broad, flat muscle, superficially located on the forelimb with insertions into the digits. In humans the palmaris longus terminates at the palmar aponeurosis.

The **flexor carpi radialis** is a thin muscle inserting on the second and third metacarpals. It is named for its action, its distal attachment point, and its lateral location. The **flexor carpi ulnaris** attaches to the humerus and ulna and terminates on medial metacarpals and carpals. The **flexor digitorum superficialis** (flexor digitorum sublimis) occurs in the forelimb as a middle-level muscle, underneath the palmaris longus. It originates on fascia of other forelimb muscles (usually muscles originate on bone). Examine the superficial muscles of the medial forelimb and compare them to fig. 13.17.

Deep Muscles on the Medial Aspect of the Forelimb

The **supinator** is a deep muscle that runs diagonally from the lateral epicondyle of the humerus to the proximal radius. It is the deepest of the proximal forelimb muscles. The **flexor digitorum profundus** is an extensive muscle that inserts on the first through fifth digits. It replaces the flexor pollicis longus for the thumb

flexion, because this muscle is absent in cats. The **pronator quadratus** is a square muscle located between the radius and ulna deep to the flexor digitorum profundus. Find the deep muscles in the cat and compare them to fig. 13.18.

Muscles on the Lateral Aspect of the Forelimb

The lateral muscles of the forelimb are the extensor group. The **extensor carpi radialis longus** muscle is deep to the brachioradialis and attaches to the second metacarpal. The **extensor carpi radialis brevis** is underneath the extensor carpi radialis longus and goes to on the third metacarpal.

Locate the **extensor carpi ulnaris,** next to the extensor digitorum lateralis, attaching to the fifth metacarpal. The **extensor digitorum communis** is on the lateral aspect of the forelimb. It is a broad muscle inserting by tendons on the second through fifth digits. Locate these muscles on the cat and in fig. 13.19.

The **extensor digitorum lateralis** is specific to the cat and distally attaches with the tendons of the digitorum communis to the digits. The **extensor pollicis brevis** is a well-developed muscle in the cat, while the abductor pollicis longus is absent in cats.

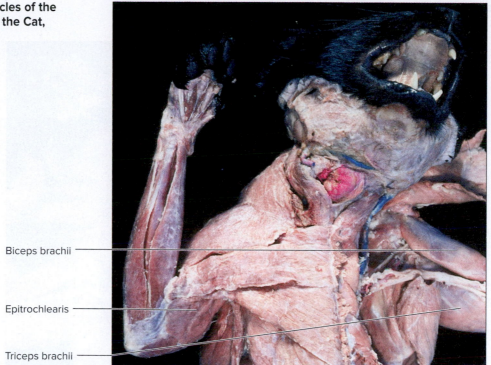

FIGURE 13.15 Muscles of the Proximal Forelimb of the Cat, Ventral View.

©Eric Wise

Biceps brachii

Epitrochlearis

Triceps brachii

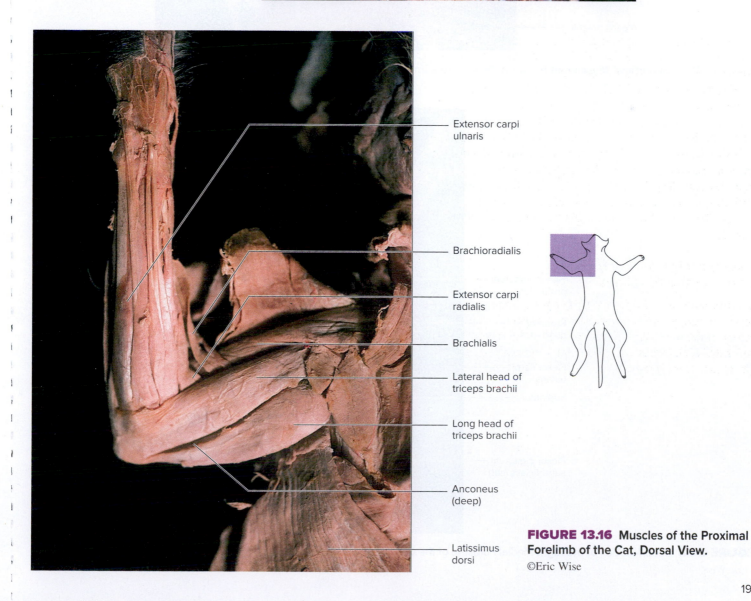

Extensor carpi ulnaris

Brachioradialis

Extensor carpi radialis

Brachialis

Lateral head of triceps brachii

Long head of triceps brachii

Anconeus (deep)

Latissimus dorsi

FIGURE 13.16 Muscles of the Proximal Forelimb of the Cat, Dorsal View.

©Eric Wise

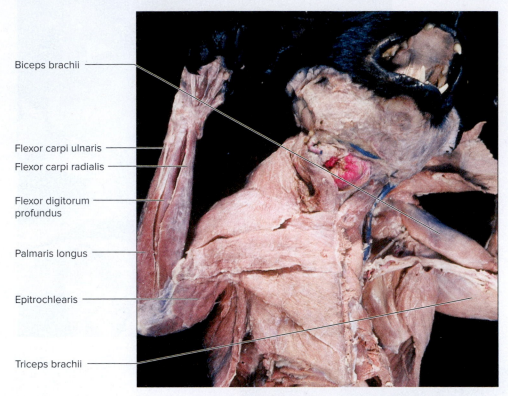

Biceps brachii

Flexor carpi ulnaris

Flexor carpi radialis

Flexor digitorum profundus

Palmaris longus

Epitrochlearis

Triceps brachii

FIGURE 13.17 Superficial Muscles of the Right Medial Forelimb of the Cat.
©Eric Wise

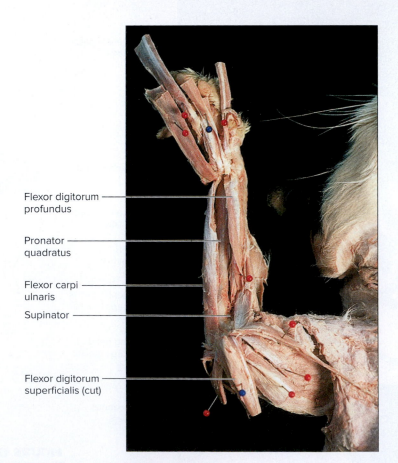

Flexor digitorum profundus

Pronator quadratus

Flexor carpi ulnaris

Supinator

Flexor digitorum superficialis (cut)

FIGURE 13.18 Deep Muscles of the Right Medial Forelimb of the Cat.
©Eric Wise

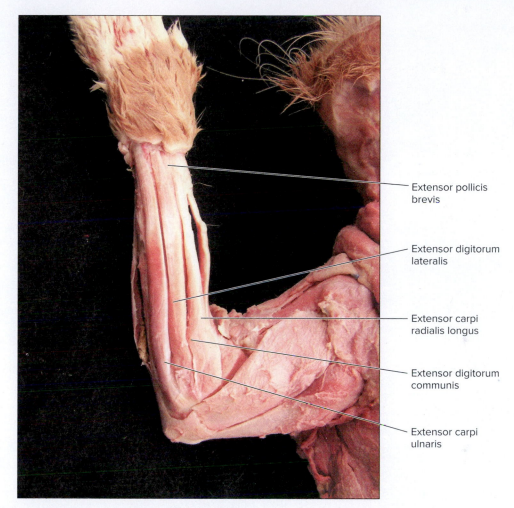

Extensor pollicis brevis

Extensor digitorum lateralis

Extensor carpi radialis longus

Extensor digitorum communis

Extensor carpi ulnaris

FIGURE 13.19 Lateral Muscles of the Left Forelimb of the Cat.

©Eric Wise

Notes

REVIEW SECTION

Appendicular Muscles 1: Muscles of the Upper Limb

Name _____ *Date* _____

Lab Section _____ *Time* _____

Review Questions

1. In human muscles:

   a. What is the action of the deltoid muscle? _____

   b. Name the medial attachment of the supraspinatus muscle. _____

   c. What is the lateral attachment of the trapezius muscle? _____

   d. Does the biceps brachii muscle attach anywhere on the humerus? _____

   e. What is the superior attachment of the pectoralis minor? _____

2. In cat muscles:

   a. What is the action of the epitrochlearis? _____

   b. Circle the muscle that does *not* correspond to a human muscle: biceps brachii, brachialis, xiphihumeralis, latissimus dorsi.

   c. How does the deltoid of the cat differ from the deltoid of the human? _____

   _____

3. Match each term on the left with a description on the right. Refer to Exercises 10 and 11 if you need to.

   1. action   a. what a muscle does

   2. dissect   b. to cut a muscle in half

   3. fixing   c. to stabilize a joint

   4. flexion   d. to separate muscles

   5. reflect   e. to decrease a joint angle

   6. transect   f. to pull back a muscle

4. What muscles are antagonists of the triceps brachii? _____

   _____

5. What is the medial attachment of the trapezius? _____

6. The pectoralis major has what action? _____

7. What is the inferior attachment of the pectoralis minor? _____

8. The latissimus dorsi muscle has what action? _____

9. Name the lateral attachment of the infraspinatus. _____

10. What is the lateral attachment of the subscapularis? _____

11. What is the action of the triceps brachii? _____

12. What is the proximal attachment of the brachialis? _____

13. Name all the muscles that flex the arm. _____

_____

14. As you hit a nail with a hammer, what arm muscle are you using? _____

15. As you look up at the ceiling, what back muscle, from those listed in this exercise, are you using?

_____

16. What is the proximal attachment of the flexor carpi ulnaris in humans? _____

17. What is an antagonist to the supinator muscle? _____

18. Where does the flexor digitorum superficialis of the human distally attach? _____

19. What is the distal attachment of the extensor carpi ulnaris muscle? _____

20. Which muscle is more developed in cats, the flexor digitorum superficialis or the flexor digitorum profundus?

_____

21. What is an antagonist to the extensor pollicis longus and brevis muscles? _____

22. What is the action of the flexor carpi radialis muscles? _____

23. Name all the muscles that extend the hand. _____

_____

24. What muscles flex the hand? _____

25. What muscles extend the thumb? _____

26. Where are the extensor carpi muscles found, on the anterior or posterior side of the forearm? _____

27. Provide proper anatomical names for muscles in this exercise commonly called the

Traps _____.

Lats _____.

Delts _____.

Pecs _____.

28. Identify the muscles in the figure, and color the origins red and the insertions blue.

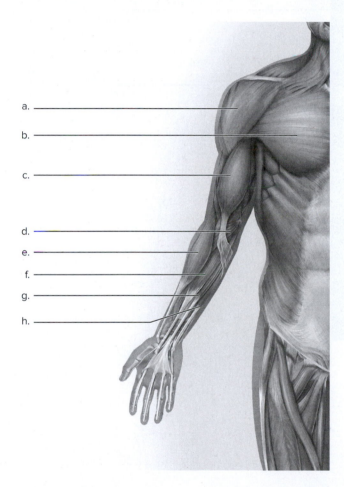

a. _____

b. _____

c. _____

d. _____

e. _____

f. _____

g. _____

h. _____

29. Label the muscles in the following illustration.

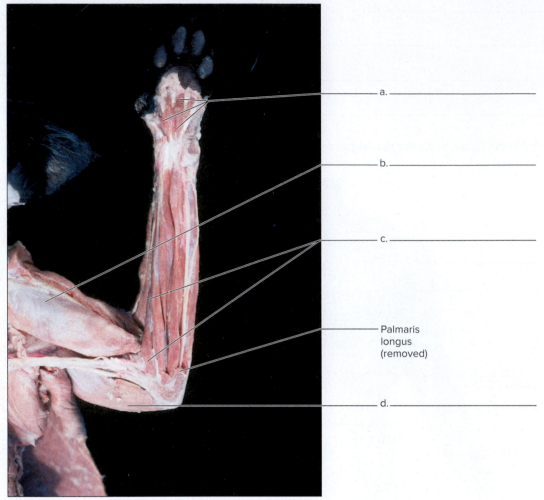

a. _____

b. _____

c. _____

Palmaris
longus
(removed)

d. _____

©Eric Wise

Appendicular Muscles 2: Muscles of the Lower Limb

INTRODUCTION

The muscles of the lower limb are primarily muscles of locomotion. The hip and thigh muscles flex or extend the thigh, adduct or abduct the thigh, and flex or extend the leg. Lateral and medial rotations of the thigh are other actions that occur in muscles of this group. The muscles of the leg and foot dorsiflex the foot (reducing the angle between the shin and the dorsum of the foot), plantar flex the foot (as when you stand on your toes), invert (supinate) the foot (turning the soles medially), or evert (pronate) the foot (turning the soles of the feet laterally). These muscles are discussed in the Saladin text in chapter 10, "The Muscular System."

The size and shape of the muscles of the hip and thigh in humans are different from those of other mammals in that humans are bipedal. This two-legged walking habit is much different from a quadruped's locomotion with respect to balance and forward movement. When a quadruped walks, typically three legs remain on the ground. When a biped walks, the instability of being on one leg makes the support limb that much more important. As you study these muscles, think about what happens to the center of balance when you move a leg to walk or run.

OBJECTIVES

At the end of this exercise, you should be able to

1. locate the muscles of the lower limb on a model, chart, cadaver, or cat;
2. list the attachment points and action of each muscle presented;
3. describe what nerve controls each muscle;
4. list what muscles function as synergists or antagonists to the prime mover;
5. name all the muscles that move a joint, such as all of the muscles that flex the thigh.

MATERIALS

Human torso model

Human leg models

Human muscle charts

Articulated skeleton

Cadaver

Cat

Materials for Cat Dissection

Dissection trays

Scalpel

Protective gloves

Blunt probe

Pins

PROCEDURE

(A) Activity

Review the muscle actions as outlined in Laboratory Exercise 10. Examine muscle models or charts in the lab and locate the muscles described in the following section and in table 14.1. Correlate the shape of the muscle with the name as you study the models or charts. You may want to look at an articulated skeleton as you review the attachments of the muscles so you can better see them. The descriptions in the text portion of this exercise can help you understand the nature of the muscle, while table 14.1 gives you the particular information about the muscle.

Once you have learned the attachment points, you should be able to understand the action of the muscle. This is done, in part, by imagining how the bony points of attachments would come together if the muscle pulled them closer to one another.

TABLE 14.1	Muscles of the Hip and Thigh			
Name	**Attachment Point 1 (Origin)**	**Attachment Point 2 (Insertion)**	**Action**	**Innervation**
Iliopsoas	T12, L1–5, sacrum, iliac crest, iliac fossa	Lesser trochanter of femur	Flexes thigh and lumbar vertebrae	Femoral nerve, spinal nerves L1–3
Sartorius	Anterior superior iliac spine	Proximal, medial tibia	Flexes and laterally rotates thigh, flexes knee	Femoral nerve
Gracilis	Pubis, ischium	Proximal, medial tibia	Flexes and medially rotates leg	Obturator nerve
Pectineus	Pubis	Femur inferior to lesser trochanter	Adducts and flexes thigh	Femoral nerve
Adductor brevis	Pubis	Linea aspera of femur	Adducts thigh	Obturator nerve
Adductor longus	Pubis	Linea aspera of femur	Adducts, flexes, and medially rotates thigh	Obturator nerves
Adductor magnus	Pubis, ischium	Gluteal tuberosity, linea aspera, adductor tubercle of distal femur	Adducts, flexes, extends, and medially rotates thigh	Obturator and tibial nerves
Quadriceps Femoris				
Rectus femoris	Anterior inferior iliac spine, margin of acetabulum	Patella, tibial tuberosity, lateral and medial condyles of tibia	Flexes thigh, extends leg	Femoral nerve
Vastus lateralis	Greater trochanter of femur, linea aspera of femur	Patella, tibial tuberosity, lateral and medial condyles of tibia	Extends leg	Femoral nerve
Vastus intermedius	Proximal, anterior femur	Patella, tibial tuberosity, lateral and medial condyles of tibia	Extends leg	Femoral nerve
Vastus medialis	Linea aspera, medial side	Patella, tibial tuberosity, lateral and medial condyles of tibia	Extends leg	Femoral nerve
Tensor fasciae latae	Iliac crest, anterior iliac surface	Iliotibial tract of fasciae latae to lateral condyle of tibia	Flexes and medially rotates thigh, extends and laterally rotates knee	Superior gluteal nerve
Gluteus maximus	Outer iliac surface, iliac crest, sacrum, coccyx	Gluteal tuberosity of femur, iliotibial tract of fasciae latae	Extends and abducts thigh	Inferior gluteal nerve
Gluteus medius	Lateral surface of ilium	Greater trochanter of femur	Abducts and medially rotates thigh	Superior gluteal nerve
Gluteus minimus	Ilium	Greater trochanter of femur	Abducts and medially rotates thigh	Superior gluteal nerve
Hamstrings				
Biceps femoris	Ischial tuberosity, linea aspera	Head of fibula	Extends thigh, flexes knee	Sciatic nerve
Semitendinosus	Ischial tuberosity	Proximal, medial tibia	Extends thigh, flexes knee	Sciatic nerve
Semimembranosus	Ischial tuberosity	Medial condyle of tibia	Extends thigh, flexes knee	Sciatic nerve

Muscles of the Hip and Thigh

The **iliopsoas** (ILL-ee-oh-SO-az) is actually two muscles, the **iliacus** (ILL-ee-AK-us) and **psoas major.** A companion muscle is the **psoas minor.** In this exercise we treat these muscles as one. The iliopsoas muscle is a major flexor of the thigh. It originates inside the abdominopelvic cavity and crosses over the pelvic brim to insert on the posterior aspect of the femur. Examine the material in the lab and compare it to figs. 14.1 and 14.2.

The **sartorius** (sar-TOR-ee-us) is a slender muscle that runs from a superior, lateral position to an inferior, medial attachment. When tailors hem a pair of pants, they sit cross-legged. This motion mimics the action of the sartorius, named for the Latin word *sartor,* or tailor. The sartorius is the most superficial muscle on the anterior thigh. The most superficial muscle on the medial thigh is the **gracilis** (grah-SIL-us: meaning *gracile* or *slender*). The gracilis is a thin muscle that adducts the thigh. Locate these muscles in the lab and in fig. 14.1.

The muscles in the medial aspect of the thigh are adductor muscles. The gracilis is an example of an adductor muscle, as is the **pectineus** (peck-TIN-ee-us) and the three **adductor muscles** proper. The gracilis is medial, the adductors are more lateral, and the pectineus is more lateral still. The sequence of the first letters of these muscles from medial to lateral spells the word *GAP.* The pectineus is a small, quadrangular muscle, the most proximal of

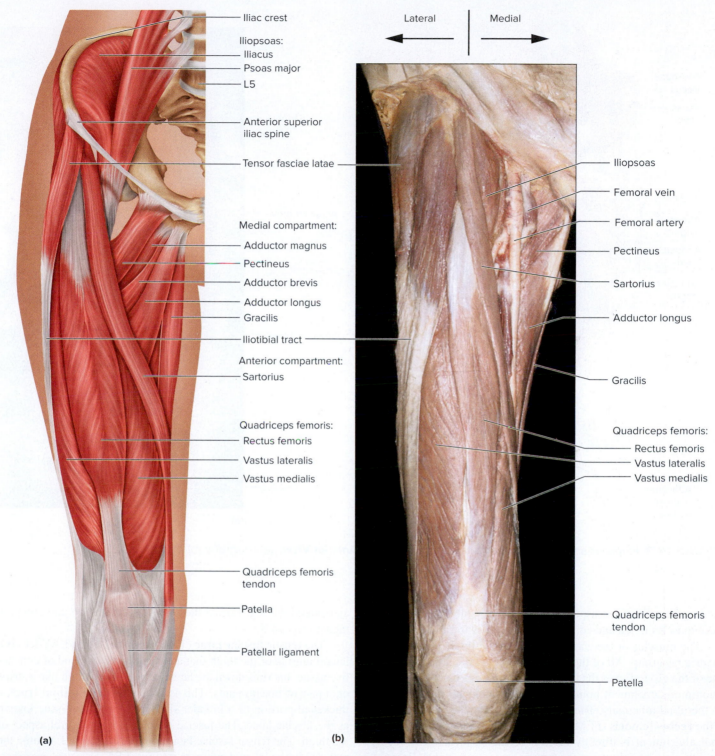

Lateral | Medial

FIGURE 14.1 **Muscles of the Right Thigh, Anterior View.** (a) Diagram: (b) photograph.
(b) Rebecca Gray/Don Kincaid/McGraw Hill

the group. It is superficial to the adductor brevis muscle and proximal to the adductor longus muscle. Examine the material in the lab and compare the pectineus to fig. 14.2.

The **adductor brevis** is the short adductor muscle deep in the medial aspect of the thigh. It can be seen if the pectineus is reflected. The adductor brevis has one attachment on the pubis and

the other on the proximal third of the femur. The **adductor longus** also has a pubic attachment but the other attachment is on the middle third of the femur. The largest of the adductor muscles is the **adductor magnus.** It has a broad attachment along the length of the inferior hip bone (pubis and ischium). The adductor magnus attaches inferiorly to the femur from the gluteal tuberosity, along

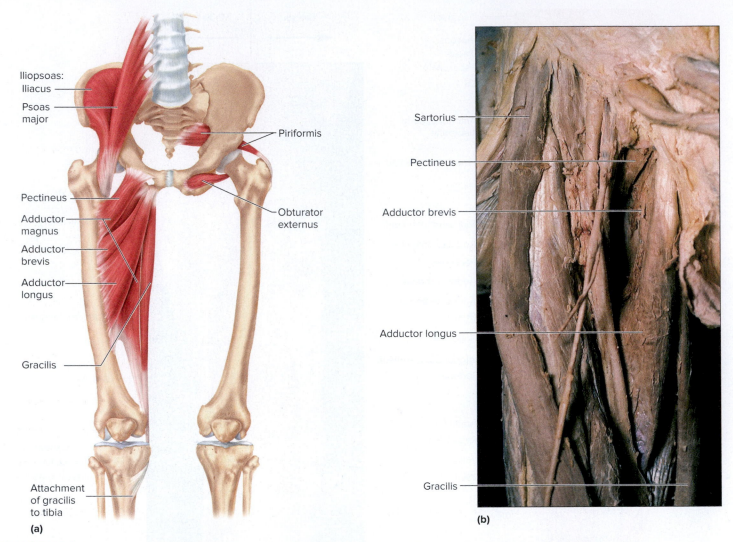

FIGURE 14.2 **Iliopsoas and Adductor Muscles of the Right Thigh, Anterior View.** (a) Diagram; (b) photograph.
(b) ©Eric Wise

the length of the linea aspera to the adductor tubercle. Examine these muscles in lab and compare them to those in fig. 14.2.

The muscles of the anterior aspect of the thigh belong to the quadriceps group. All of these muscles extend the leg, while only one of the group flexes the thigh. The quadriceps muscles all have a common attachment point, the patellar ligament, which inserts on the tibial tuberosity. The most superficial muscle of the group is the **rectus femoris** (FEM-or-us) (figs. 14.1 and 14.3). Its proximal attachment is directly inferior to the sartorius. The rectus femoris runs straight down the femur (*rectus* = straight). Because it crosses the hip joint, the rectus femoris is the only quadriceps muscle to flex the thigh.

The three other quadriceps muscles are the vastus muscles. The **vastus lateralis** is so named because of its lateral position. The **vastus intermedius** is deep to the rectus femoris, and the **vastus medialis** is the most medial of the vastus muscles. These muscles attach to the femur, so they do not have an action on the thigh. They attach to the patella and then to the patellar ligament

and extend the leg. Locate these muscles in the lab and compare them to fig. 14.3.

The **tensor fasciae latae** (TEN-sur FASH-ee-ee LAY-tee) is a lateral muscle of the thigh that attaches to a thick band of connective tissue that runs down the lateral aspect of the thigh like a stripe on a pair of tuxedo pants. This is known as the **iliotibial tract,** a thickened portion of a broader sheet of connective tissue known as the fasciae latae. The fasciae latae covers the lateral aspect of the thigh. The tensor fasciae latae abducts the thigh. Examine the material in the lab and compare it to fig. 14.4.

The gluteus muscles in humans are different from those in other mammals because we are bipedal. The largest of the gluteal muscles is the **gluteus maximus.** It extends the thigh in standing from a sitting position or in actions such as climbing up stairs or pedaling a bicycle. The **gluteus medius** and the **gluteus minimus** both aid in keeping balance during walking. The gluteus medius is superficial to the gluteus minimus, but they both originate on the outer ilium. When you move one leg to

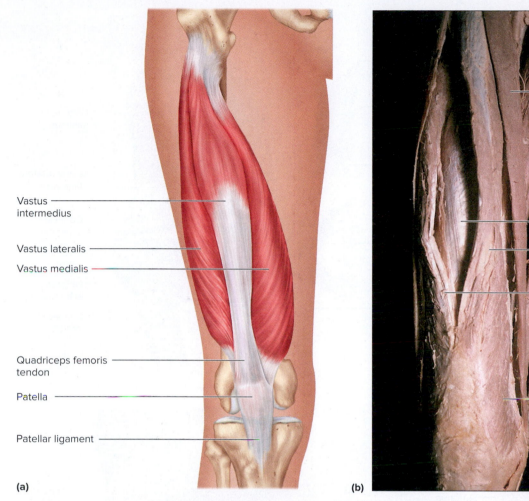

Vastus intermedius

Vastus lateralis

Vastus medialis

Quadriceps femoris tendon

Patella

Patellar ligament

(a)

Sartorius

Adductor longus

Adductor magnus

Vastus intermedius

Rectus femoris

Vastus lateralis

Vastus medialis

(b)

FIGURE 14.3 **Quadriceps Muscles of the Right Thigh, Anterior View.** (a) Diagram (rectus femoris removed); (b) photograph. *(b)* ©Eric Wise

walk, gravity pulls the body in the direction of the lifted leg. The gluteus medius and minimus on the opposite side of the body contract to maintain upright posture. Compare these muscles to fig. 14.4.

Deep to the gluteal muscles are lateral rotators and abductors of the thigh. These are the **piriformis, obturator internus, superior gemellus, inferior gemellus,** and **quadratus femoris.** Find these muscles in lab and compare them with fig. 14.4.

The **hamstring muscles** are so named because these muscles from a pig traditionally were tied together by their tendons (the hamstrings) and hung in a smokehouse to cure. Three muscles make up the hamstrings: the biceps femoris, the semitendinosus, and the semimembranosus. All three of these muscles cross the hip joint, are extensors of the thigh, and flexors of the leg.

The **biceps femoris** is the "two-headed muscle of the femur." One head attaches with the other hamstring muscles, on the ischial tuberosity, while the second head attaches to the femur. The biceps femoris is a muscle that connects *laterally* on the proximal leg. The **semitendinosus** also attaches to the ischial tuberosity, and it can be distinguished by the long, pencil-like distal tendon.

As opposed to the biceps femoris, the semitendinosus attaches *medially.* The **semimembranosus** also attaches to the ischial tuberosity and has a broad, flat, membranous tendon on the proximal part of the muscle. The semimembranosus connects *medially* on the leg along with the semitendinosus (remember the two "semis" go together and the membranosus is medial). Examine these muscles in fig. 14.5.

Ⓐ Muscles of the Leg and Foot

Examine the models and charts in the lab as you study the following descriptions. If you have a cadaver, pay particular attention to the attachments of the muscles of the leg and foot. Table 14.2 provides the details of the muscles in this exercise, while the text descriptions point out major features of the muscles. When you are studying the attachment points of the muscles, it helps to have an articulated skeleton so you can locate the bony markings as you study them. The actions of the muscles will be easier to understand then.

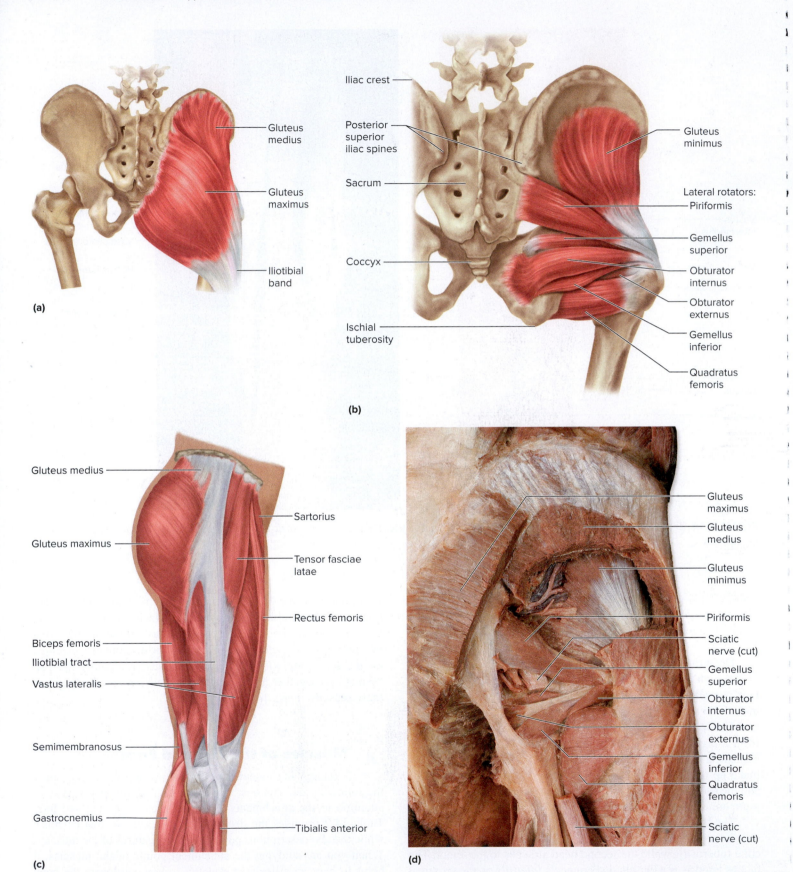

FIGURE 14.4 **Muscles of the Right Hip and Thigh.** Diagram (a) superficial muscles, posterior view; (b) deep muscles, posterior view; (c) lateral muscles. Photograph (d) lateral and posterior muscles.
(d) Christine Eckel/McGraw Hill

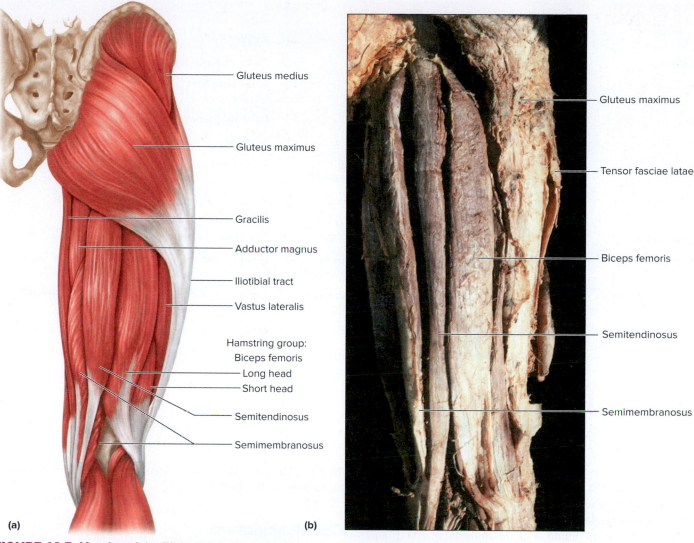

(a)   (b)

FIGURE 14.5 **Muscles of the Right Thigh, Posterior View.** (a) Diagram; (b) photograph.
(b) ©Eric Wise

Posterior Muscles

The **gastrocnemius** (GAS-trock-NEE-me-us) is a calf muscle with two heads that is the most superficial of the posterior leg group. The gastrocnemius crosses the knee joint and flexes the leg. It attaches to the calcaneus by way of the **calcaneal tendon** and plantar flexes the foot. The calcaneal tendon is also known as the Achilles tendon. Deep to the gastrocnemius is the **soleus** (SO-lee-us) muscle. Unlike the gastrocnemius, the soleus does not attach to the femur; therefore, it has no action on the leg. It attaches to the calcaneus, sharing the calcaneal tendon with the gastrocnemius, and it plantar flexes the foot.

(A) Activity

Palpate the lateral side of the gastrocnemius and feel the junction between it and the soleus.

The **plantaris** is a weak plantar flexor of the foot. The **popliteus** (pop-LIT-ee-us) is a small muscle that crosses the knee joint. It unlocks the knee joint. These muscles can be seen in fig. 14.6.

The **tibialis posterior** is a muscle deep to the soleus that plantar flexes and inverts the foot. The tendon of the muscle runs along the medial aspect of the ankle. Near the tibialis posterior is the **flexor digitorum longus,** which is a muscle that attaches to the distal phalanges of all of the digits of the foot except for the great toe. The flexor digitorum longus flexes (curls) the toes in addition to plantar flexing and inverting the foot. The **flexor hallucis longus** flexes the great toe (hallux = great toe) and aids the flexor digitorum longus in inverting the foot. Locate these muscles in the lab and compare them to fig. 14.6.

Anterior Muscles

The **tibialis anterior** is located just lateral to the crest of the tibia on the anterior side of the leg. Palpate the tibia. The tendon of the tibialis anterior crosses to the medial side of the foot and

TABLE 14.2	Muscles of the Leg and Foot			
Name	**Origin**	**Insertion**	**Action**	**Innervation**
Gastrocnemius	Condyles of femur	Calcaneus	Flexes leg, plantar flexes foot	Tibial nerve
Soleus	Posterior, proximal tibia and fibula	Calcaneus	Plantar flexes foot	Tibial nerve
Plantaris	Distal femur	Calcaneus	Flexes leg, plantar flexes foot	Tibial nerve
Popliteus	Lateral condyle of femur	Proximal tibia	Flexes leg	Tibial nerve
Tibialis posterior	Proximal portion of tibia and fibula	Second through fourth metatarsals and some tarsals	Plantar flexes and inverts foot	Tibial nerve
Flexor digitorum longus	Posterior tibia	Distal phalanges of second through fifth digits	Flexes toes, plantar flexes and inverts foot	Tibial nerve
Flexor hallucis longus	Inferior fibula	Distal phalanx of great toe	Flexes great toe, inverts foot	Tibial nerve
Tibialis anterior	Lateral condyle and proximal tibia	First metatarsal and medial cuneiform	Dorsiflexes and inverts foot	Deep fibular nerve
Extensor digitorum longus	Lateral condyle of tibia, shaft of fibula	Middle and distal phalanges of second through fifth digits	Extends toes, dorsiflexes foot	Deep fibular nerve
Extensor hallucis longus	Anterior shaft of fibula	Distal phalanx of great toe	Extends great toe, dorsiflexes foot	Deep fibular nerve
Fibularis longus	Head and shaft of fibula, lateral condyle of tibia	First metatarsal, medial cuneiform	Plantar flexes and everts foot	Superficial fibular nerve
Fibularis brevis	Shaft of fibula	Fifth metatarsal	Plantar flexes and everts foot	Superficial fibular nerve
Fibularis tertius	Medial, distal fibula	Fifth metatarsal	Dorsiflexes and everts foot	Deep fibular nerve

attaches to the first metatarsal and medial cuneiform. As the tibialis anterior contracts, it dorsiflexes the foot. Because the attachment of this muscle is medial, it also inverts the foot.

The **extensor digitorum longus** muscle is lateral to the tibialis anterior, and the tendons of this muscle splay out and attach to the middle and distal phalanges of all the digits of the foot except the great toe. Extend your toes and feel these tendons. The **extensor hallucis longus** extends the great toe and is a synergist to the extensor digitorum longus in dorsiflexion of the foot.

(A) Activity

Examine these muscles in the lab and compare them to fig. 14.7. These are major muscles of the foot. There are also smaller muscles such as the extensor digitorum brevis (fig. 14.7) and the extensor hallucis brevis.

The fibularis (peroneus) muscles are named for attaching to and running the length of the fibula. The **fibularis longus** has a tendon that travels from the lateral side of the foot underneath to the medial side, crossing under the arch of the foot. The **fibularis brevis** parallels the fibularis longus except that the tendon attaches to the lateral side of the foot at the fifth metatarsal. Both of these muscles have tendons that hook posterior to the lateral malleolus, and they both plantar flex and evert the foot. The **fibularis tertius** does not arch posterior to the lateral malleolus, and it attaches to the dorsum (top) of the fifth metatarsal. When it contracts, it dorsiflexes and everts the foot. Examine

these muscles on models or specimens in lab and compare them to the muscles in fig. 14.8.

Cat Musculature

The cat has many muscles of the hip and thigh that are good models for studying human muscles. Because cats are quadrupeds, there is a different size or placement of some of the thigh muscles. In the dissection, be careful that you *leave the blood vessels and nerves intact.* You will be looking at these structures in later exercises.

(A) Activity

If you have not already done so, remove the skin from the lower limb of the cat. Pay particular attention to the **great saphenous vein,** which runs under the skin and should be preserved. Be careful cutting the skin from the distal portions of the leg so you do not cut through the tendons of the foot. You may want to blunt dissect these areas by using a mall probe to separate structures. Remove any excess fat and fascia from the muscles as you dissect the material. If you need to cut a muscle, bisect it perpendicular to the fiber direction so half of it is attached to one attachment side and half of it is connected to the other.

Medial Muscles of the Thigh of the Cat

Two major muscles on the medial aspect of the thigh in the cat are the **sartorius** and the **gracilis** muscles. Locate these muscles

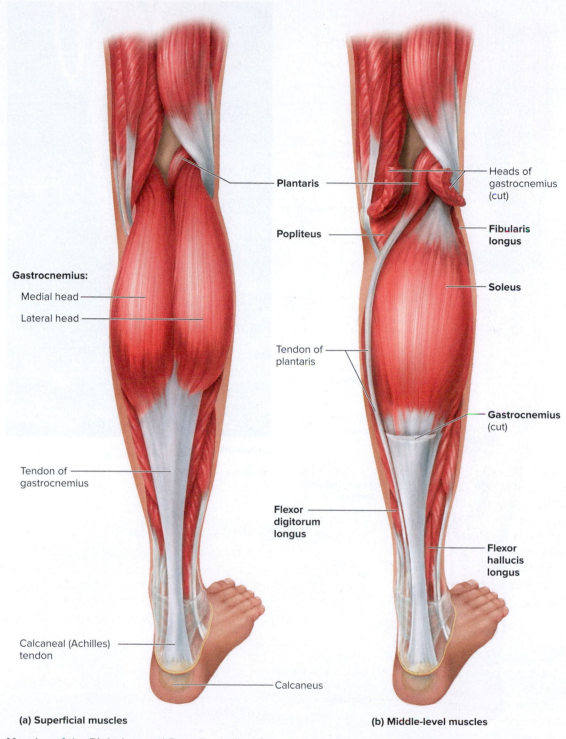

Plantaris

Heads of
gastrocnemius
(cut)

Popliteus

**Fibularis
longus**

Gastrocnemius:

Medial head

Lateral head

Soleus

Tendon of
plantaris

Gastrocnemius
(cut)

Tendon of
gastrocnemius

**Flexor
digitorum
longus**

**Flexor
hallucis
longus**

Calcaneal (Achilles)
tendon

Calcaneus

(a) Superficial muscles

(b) Middle-level muscles

FIGURE 14.6 **Muscles of the Right Leg and Foot, Posterior View.** (a–c) Diagram; (d) photograph.

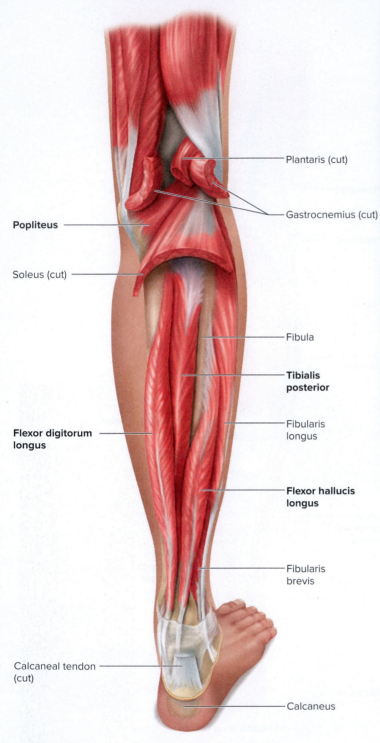

Plantaris (cut)

Gastrocnemius (cut)

Popliteus

Soleus (cut)

Fibula

Tibialis posterior

Flexor digitorum longus

Fibularis longus

Flexor hallucis longus

Fibularis brevis

Calcaneal tendon (cut)

Calcaneus

(c) Deep muscles

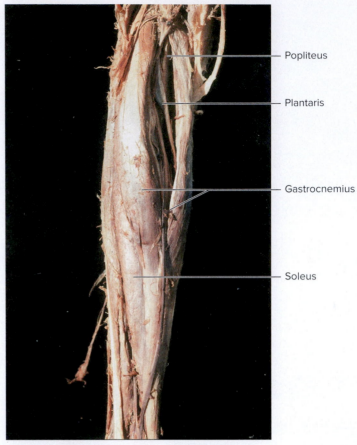

Popliteus

Plantaris

Gastrocnemius

Soleus

(d) Superficial muscles

FIGURE 14.6 *Continued.*
(d) ©Eric Wise

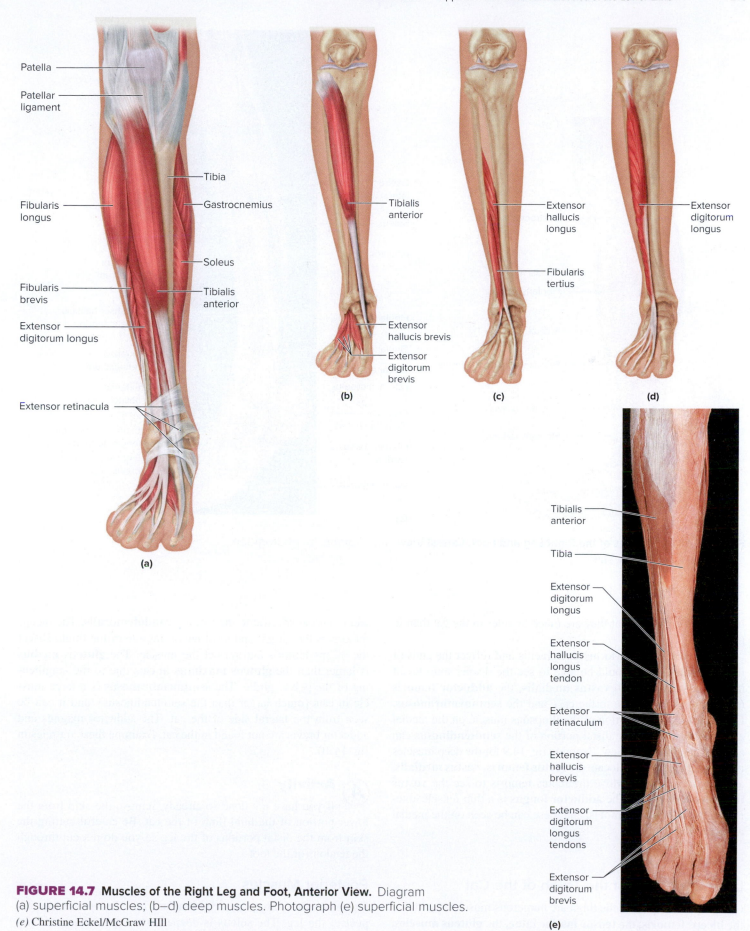

Patella

Patellar ligament

Tibia

Fibularis longus

Gastrocnemius

Soleus

Fibularis brevis

Tibialis anterior

Extensor digitorum longus

Extensor retinacula

(a)

Tibialis anterior

Extensor hallucis brevis

Extensor digitorum brevis

(b)

Extensor hallucis longus

Fibularis tertius

(c)

Extensor digitorum longus

(d)

Tibialis anterior

Tibia

Extensor digitorum longus

Extensor hallucis longus tendon

Extensor retinaculum

Extensor hallucis brevis

Extensor digitorum longus tendons

Extensor digitorum brevis

(e)

FIGURE 14.7 Muscles of the Right Leg and Foot, Anterior View. Diagram (a) superficial muscles; (b–d) deep muscles. Photograph (e) superficial muscles.
(e) Christine Eckel/McGraw Hlll

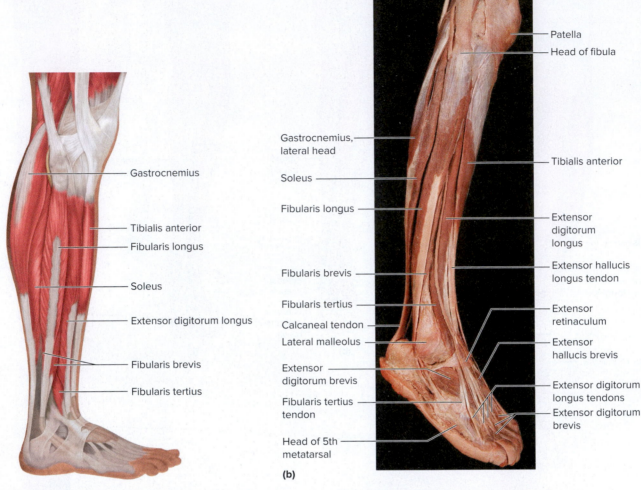

FIGURE 14.8 Muscles of the Right Leg and Foot, Lateral View. (a) Diagram; (b) photograph.
(b) Christine Eckel/McGraw Hill

in fig. 14.9 and note that they are much broader in the cat than in the human.

Cut through the sartorius and gracilis and reflect the ends of these muscles. You should be able to see the deeper muscles of the thigh, including the **vastus medialis,** the **adductor femoris** (a large muscle specific to the cat), and the **semimembranosus.** Locate the external portion of the **iliopsoas** muscle on the medial aspect of the thigh. The distal portion of the **semitendinosus** can also be seen from this view. Examine fig. 14.9 for the deep muscles of the medial thigh and locate the **rectus femoris, vastus medialis,** and **vastus lateralis.** Move the rectus femoris to see the **vastus intermedius** muscle. The **adductor longus** is a thin muscle anterior to the **adductor femoris,** and these can be seen on the medial aspect of the thigh.

Lateral Muscles of the Thigh of the Cat

On the lateral aspect of the thigh are numerous muscles, including the **biceps femoris,** the **tensor fasciae latae,** the **gluteus muscles,** and a muscle specific to the cat, the **caudofemoralis.** The biceps femoris is the largest and most lateral muscle of the thigh. Bisect the biceps femoris and reflect the muscle. The **gluteus medius** is larger than the **gluteus maximus** in cats due to the lengthening of the pelvic girdle. The **semimembranosus** is a large muscle in cats (much larger than the semitendinosus), and it can be seen from the lateral side of the cat. The adductor magnus and adductor brevis are not found in the cat. Examine these muscles in fig. 14.10.

(A) Activity

If you have not done so already, remove the skin from the lower portion of the hind limb of the cat. Be careful cutting the skin from the distal portions of the leg so you do not cut through the tendons of the foot.

Posterior Muscles

Examine the large **gastrocnemius** on the posterior aspect of the leg. The **soleus** is deeper to the gastrocnemius and

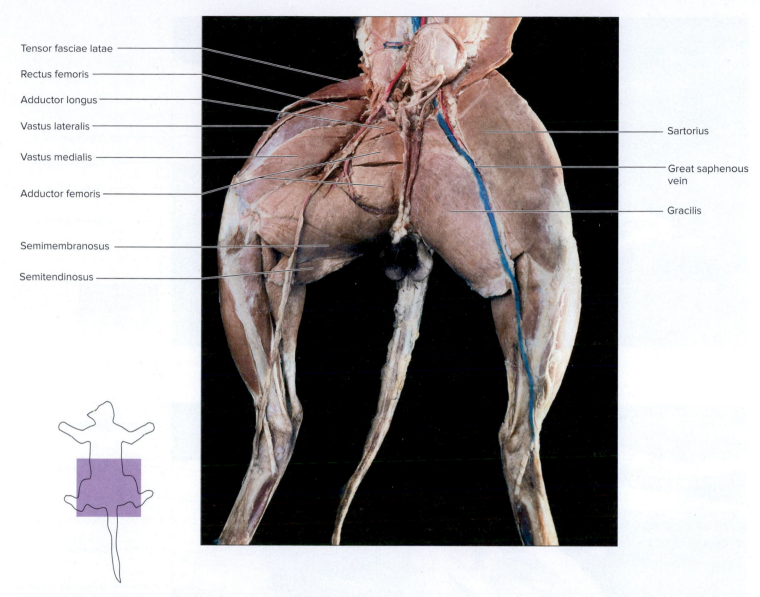

Tensor fasciae latae

Rectus femoris

Adductor longus

Vastus lateralis

Vastus medialis

Adductor femoris

Semimembranosus

Semitendinosus

Sartorius

Great saphenous vein

Gracilis

FIGURE 14.9 **Thigh Muscles of the Cat.** Superficial muscles (left side of cat); deep muscles (right side of cat).
©Eric Wise

inserts communally with the gastrocnemius on the calcaneus. In cats the soleus has only one attachment point, the fibula, while in humans it attaches to two bones. Examine fig. 14.11 with your dissection. Cut through the calcaneal tendon and lift the gastrocnemius and soleus so you can study the underlying muscles.

The **popliteus** is a small, triangular muscle that crosses the knee joint. Do not damage the nerves and blood vessels that pass over the popliteus because you will study them in the subsequent exercises.

The **flexor digitorum longus** is a muscle that runs along the medial side of the hind limb and flexes the digits of the cat. It joins with the **flexor hallucis longus,** which distally attaches to all the digits of the hind limb. The **tibialis posterior** is a narrow muscle that attaches to the tarsal bones of the foot. Locate these muscles in the cat and compare them to fig. 14.12.

Anterior Muscles

The **tibialis anterior** is a large muscle of the lower limb and runs to the dorsum of the foot. The **extensor digitorum longus** attaches to the femur in cats and goes to the distal phalanges in all of the digits in the cat. The extensor hallucis longus is not found in cats. These muscles can be seen in fig. 14.13. Examine these muscles and isolate them in your dissection.

The **fibularis longus** extends along the length of the fibula with the **fibularis brevis.** The fibularis longus ends at the base of the metatarsals, while the fibularis brevis goes to the fifth metatarsal. The **fibularis tertius** joins with the tendon of the extensor digitorum muscle. These muscles can be seen in fig. 14.14.

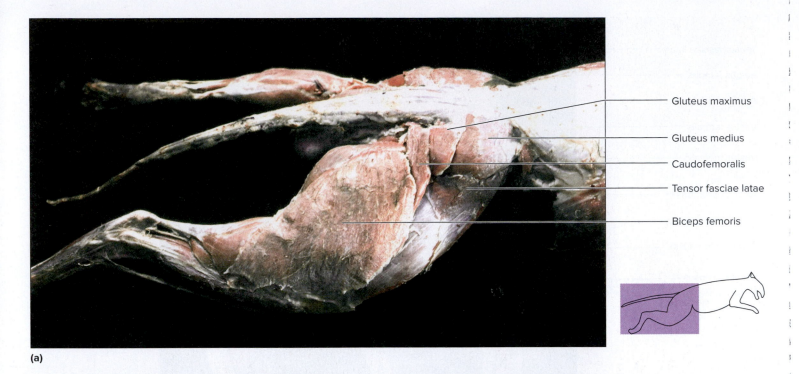

Gluteus maximus

Gluteus medius

Caudofemoralis

Tensor fasciae latae

Biceps femoris

(a)

Semitendinosus

Gluteus maximus

Gluteus medius

Caudofemoralis

Sartorius

Tensor fasciae latae

Semimembranosus

Sciatic nerve

Vastus lateralis

(b)

FIGURE 14.10 **Lateral Thigh Muscles of the Cat.** (a) Superficial muscles of the right side; (b) deep muscles of the right side.
(a), (b) ©Eric Wise

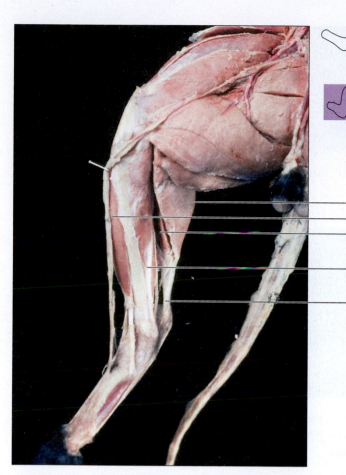

Gastrocnemius
Tibialis anterior
Soleus

Flexor digitorum longus

Calcaneal (Achilles) tendon

FIGURE 14.11 Superficial Muscles of the Right Leg of the Cat, Medial View.
©Eric Wise

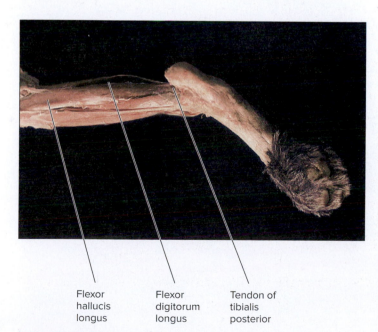

Flexor hallucis longus

Flexor digitorum longus

Tendon of tibialis posterior

FIGURE 14.12 Deep Muscles of the Left Leg of the Cat, Lateral View.
©Eric Wise

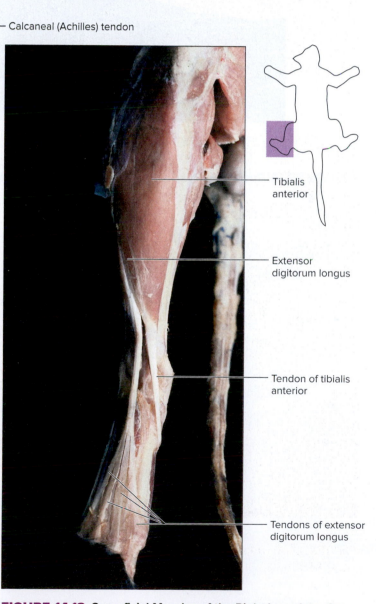

Tibialis anterior

Extensor digitorum longus

Tendon of tibialis anterior

Tendons of extensor digitorum longus

FIGURE 14.13 Superficial Muscles of the Right Leg of the Cat, Anterior View.
©Eric Wise

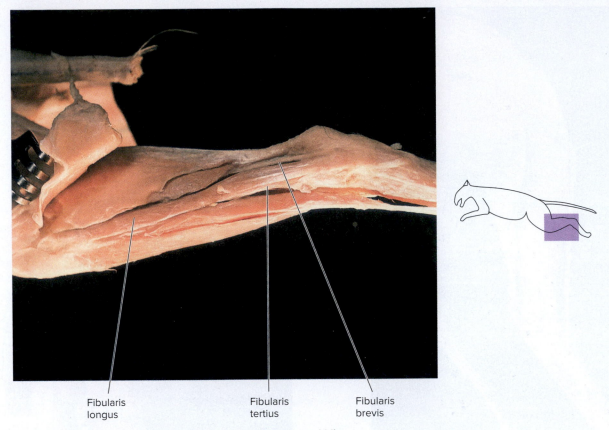

Fibularis
longus

Fibularis
tertius

Fibularis
brevis

FIGURE 14.14 Superficial Muscles of the Left Leg of the Cat, Lateral View.
©Eric Wise

REVIEW SECTION

Appendicular Muscles 2: Muscles of the Lower Limb

Name _____   *Date* _____

Lab Section _____   *Time* _____

Review Questions

1. If you were to ride a horse, what muscles would you use to keep your seat out of the saddle as you rode?
_____
_____

2. How do the gluteus medius and gluteus minimus prevent you from toppling over as you walk? _____
_____

3. What is a muscle that is an antagonist to the biceps femoris muscle? _____

4. What are two muscles that are synergists with the biceps femoris muscle? _____
_____

5. What is the common distal attachment of all the muscles of the quadriceps group? _____
_____

6. How does the action of the rectus femoris differ from those of the other quadriceps muscles? _____
_____

7. How many adductor muscles are there? _____

8. List two muscles in this exercise responsible for thigh flexion. _____
_____

9. Where do the hamstring muscles attach proximally as a group? _____
_____

10. What is the action of the vastus lateralis? _____
_____
_____

11. Which muscle group is found on the anterior part of the thigh? _____

12. Is abduction of the thigh movement away from or toward the midline? _____
_____

13. What muscle flexes the lumbar vertebrae as part of its action? _____
_____

14. Label the muscles in the following illustration. Use the terms provided.

adductor femoris   semimembranosus

gracilis   tensor fasciae latae

rectus femoris   vastus lateralis

sartorius   vastus medialis

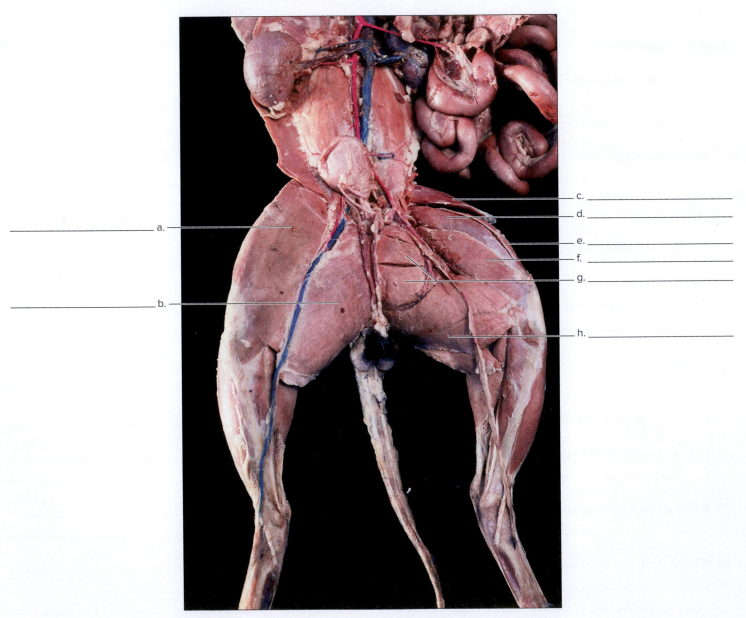

©Eric Wise

15. Label the muscles in the following illustration. Color the proximal attachment (origins) in red and the distal attachment (insertions) in blue where the attachments are visible.

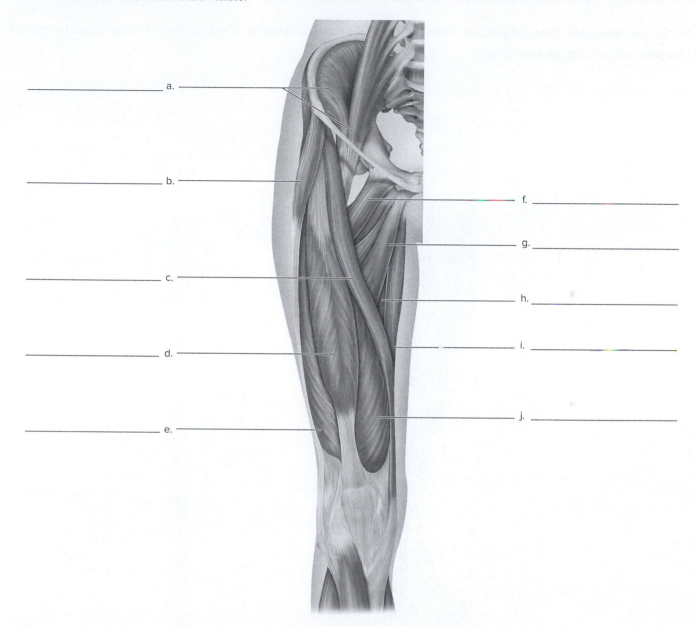

_____ a. _____

_____ b. _____

f. _____

g. _____

_____ c. _____

h. _____

i. _____

_____ d. _____

_____ e. _____

j. _____

16. What is the proximal attachment of the gastrocnemius? _____

17. What is the distal attachment of the tibialis anterior in humans? _____

18. How does the action of the fibularis longus in humans differ from that of the fibularis tertius? _____
_____
_____

19. What is the action of the extensor hallucis longus? _____
_____

20. The calf contains what two major muscles? _____

_____

21. Label the muscles in the following illustration. Color the proximal attachments (origins) in red and the distal attachments (insertions) in blue where the attachments are visible.

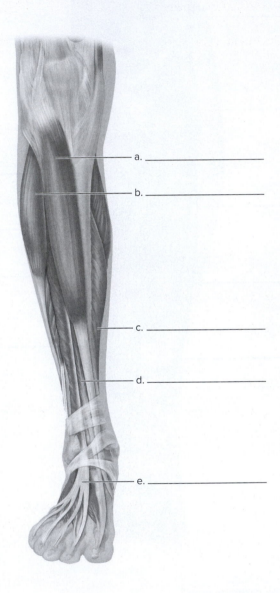

a. _____

b. _____

c. _____

d. _____

e. _____

22. Plantar flexion occurs by what muscles? _____

23. What muscle extends the toes? _____

24. Name a muscle in this exercise that dorsiflexes the foot. _____

25. Plantar flexion and eversion of the foot occurs by what muscle? _____

26. What is the distal attachment of the soleus? _____

27. What muscles do you use when you rise up on your toes? _____

28. What is the distal attachment of the fibularis tertius muscle? _____

29. What is the distal attachment of the flexor digitorum longus? _____

30. Label the muscles in the following illustration.

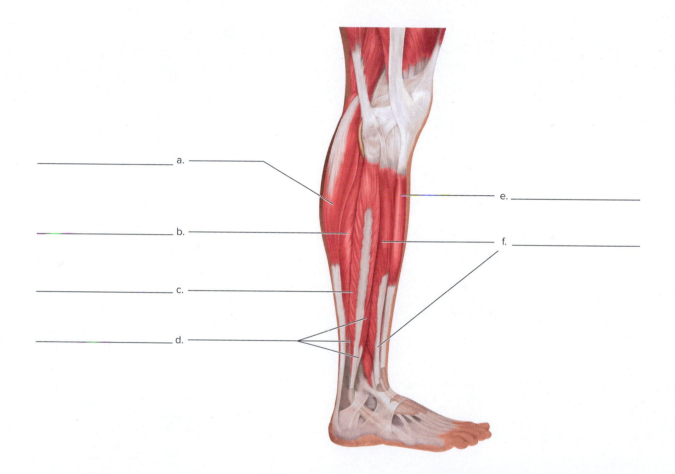

a. _____

b. _____

c. _____

d. _____

e. _____

f. _____

Notes

Muscle Physiology

INTRODUCTION

Skeletal muscles contract because of stimulation by nerves controlling them. In this exercise, you explore the nature of skeletal muscle contraction as initiated by external electrical stimulation and apply this information to the functions of the skeletal muscle in your body. These topics are covered in the Saladin text in chapter 11, "Muscular Tissue."

Two events are fundamental to an understanding of muscle physiology. One is an electrical event that occurs in muscle membranes, and the other is the physical contraction of the muscle. The normal contraction of skeletal muscle occurs when an electrochemical **nerve impulse** (action potential) travels down an axon and reaches the **synapse** or **neuromuscular junction** between the nerve and the muscle. **Acetylcholine (ACh)** is released by the neuron and stimulates an action potential that travels along the length of the muscle fiber.

Entire muscles in the body exhibit a **graded response,** where muscles progressively increase contraction strength. This is due to the presence of multiple fibers in a contracting muscle; the overall contractile strength of the entire muscle is determined by the number of muscle fibers in that muscle.

When a muscle is stimulated with a single, brief electrical impulse, the muscle undergoes a contraction known as a **twitch.** In a twitch, the muscle has three events: a latent period, a contraction phase, and a relaxation phase. After the initial **stimulus,** the muscle undergoes a **latent period (phase)** (fig. 15.1). This is the time when the action potential travels across the muscle cell membrane and calcium ions are released from the sarcoplasmic reticulum. After this short time, the muscle goes through the **contraction phase** as the muscle **filaments** or **myofilaments** slide across one another and the muscle shortens. Finally, there is a **relaxation phase,** characterized by the muscle returning to a resting state.

The impulse that causes a muscle to contract must exceed a **threshold** value before any contraction can occur. A subthreshold stimulus will not elicit a response in the muscle. A stimulus above threshold level that occurs too soon after a preliminary stimulus also does not cause the muscle to contract, and this is known as the **refractory period.** If a stimulus is applied shortly after the

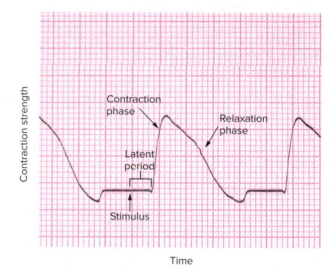

FIGURE 15.1 Three Phases of a Muscle Twitch.

refractory period, a muscle contracts and the contraction strength is more pronounced. This effect is called **temporal summation** or **wave summation** (fig. 15.2).

If rapid, repeated stimuli are sent to a muscle, then the muscle produces a series of contractions called **incomplete tetanus.** If the **frequency** (number of pulses per second) of the stimuli increases, the contractions fuse in a smooth contraction of the muscle known as **complete tetanus,** which occurs in lab experiments but not in living organisms.

Tetanus, as used in this exercise, is seen as muscles are stimulated electrically, and the word is used differently than the disease tetanus (lockjaw), described in your lecture text. Examine fig. 15.3 for recordings of incomplete tetanus and complete tetanus.

In this exercise, you examine the muscular contraction in a simulated frog or a real frog and what occurs as voltage is increased or the frequency is changed. Due to the decline in some frog populations (such as leopard frogs in North America) simulated activities may be preferable to the use of live frogs. Bullfrogs are not only plentiful but a pest species in many areas and, as

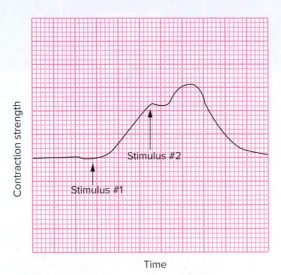

FIGURE 15.2 Wave Summation.

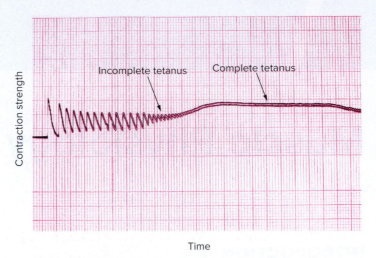

FIGURE 15.3 Incomplete and Complete Tetanus.

of today, may be used without significant impact to their population levels.

OBJECTIVES

At the end of this exercise, you should be able to

1. demonstrate the procedure to determine threshold stimulus;
2. describe incomplete tetanus and complete tetanus;
3. demonstrate maximum recruitment with lab equipment and a frog;
4. compare the lab experiments on frogs to muscular contractions in humans;
5. describe the process of muscle recruitment and fatigue in humans.

MATERIALS

Virtual Experiment Lab
Ph.I.L.S. Physiology Interactive Lab Simulations 4.0
Compatible computer

Live Frog Lab
Frog (one per experiment)
Frog Ringer's solution in dropper bottles; 3 bottles each of
    150 mL per table (iced, room temperature, and 37°C)
Nylon thread
Duograph, physiograph
Myograph transducer
Stimulator and cables
Scissors
Clean, live animal dissection pan
Sharp pithing probes
Glass hooks
Scalpel
BIOPAC setup

BSLSTM stimulator
Force transducer assembly—SS12LA
Pin electrodes ELSTM2
HDW100A tension adjuster
Data acquisition unit MP30 or MP150
BSL *PRO* template file: FrogMuscle.gtl

Ring stand
Anchoring board
Goggles
Protective gloves
Forceps
Dissection pins
Animal disposal container
Multiple weights – 2, 4, and 6 kg (5, 10, 15 lb)

PROCEDURE   PhILS

Simulated Frog Muscle Experiments— Physiology Interactive Lab Simulations

(A) Activity
Open Physiology Interactive Lab Simulations (Ph.I.L.S.) Version 4.0. For all of the virtual muscle experiments, you should follow the standard procedure outlined next.

A. Click on the number of the exercise you want to complete. These are under the **Skeletal Muscle Function** section and are listed as
   5. Stimulus-Dependent Force Generation
   6. Weight and Contraction
   7. Length–Tension Relationship
   8. Principles of Summation and Tetanus

After you have selected your choice, you should:

B. Read the objectives and introduction to the lab simulation. You can click on highlighted terms to see pictures of the material in question.

C. Take the pre-lab quiz. If you get a question wrong, the program will let you know and provide you with the correct answer.

D. Click on the Wet Lab tab and read the material.

E. Click on Continue, which opens up the Laboratory Exercise.

F. Turn the power on the computer screen and the Data Acquisition Unit. The green light lets you know it is on.

G. Connect all the plugs to their respective locations. You must drag the tip of the plug to the middle of the adapter in order for it to connect.

H. Adjusting the voltage, frequency, or tension is done by clicking on the up or down arrow repeatedly or clicking on the arrow and holding it until the desired value is obtained.

I. Take the post-lab quiz to determine your understanding of the lab.

The information for the specific lab exercises follows.

5. **PhILS**   Stimulus-Dependent Force Generation

In this laboratory exercise, you examine the impact that an increase in voltage stimulus has on the strength of contraction. You should see a picture that shows a virtual muscle, electrodes to stimulate the muscle, and an input that records the "pull" of the muscle after it has been stimulated. Click the "power switch" to turn on the computer screen. The switch will appear on the lower right of the simulated screen. The red and blue wires deliver electric impulses (shocks) to the frog muscle, and the black wire records how much force the muscle exerts. Follow the text directions in the program and attach the plugs to their proper connections. They are all color coded.

The directions ask you to select at least 1 volt to get a response. After you select 1 volt, click on the Shock button. You should see an upward curve.

To record the amplitude, you need to select a cross-hair cursor on the blue line to the right of the peak and drag it to the highest point of the curve on the graph. Click on the Journal entry icon. You should see the data points automatically plotted on the graph. Click on the X in the upper right corner of the graph table in the journal and you will return to the experiment.

Reduce the voltage to 0. Hit the Start button again. Enter the material in the journal as described above. Do this repeatedly until you fill in the entire journal from 0 to 1.6 volts. Notice that at low voltages the force line stays flat. This is because you are below the threshold of contraction. Record the threshold voltage (when you first see a contraction) for the muscle in the following space.

? Threshold voltage: _____ (1)

What happens to the muscle contraction strength as you increase the voltage?

Response: _____

Although it was not demonstrated in this exercise, there is a point at which the increase in voltage does not lead to an increase in contraction strength. This is known as **maximum recruitment,** and it is the voltage where all the muscle fibers are contracting.

6. **PhILS**   Weight and Contraction

In this part of the exercise, you will measure the latent period, the contraction phase, and the relaxation phase. Notice the flat line that precedes the actual contraction of the muscle. This is the **latent (latency) period,** the time between when the stimulus was delivered and when the muscle started to contract. The latent period is the time after the muscle is stimulated when calcium diffuses from the transverse tubules to the myofibrils. Follow the directions in the virtual lab, turn on the power switch on the computer monitor, connect the colored plugs, and set the voltage to 1.0 volts. Click on the cross-hair cursor and move it as far to the left as you can on the graph (at the beginning of when the muscle was stimulated). Move the other cross-hair cursor to the point where the blue line begins to rise up. This is the latent period. Enter the value in the following space.

? Latent period: _____ (2)

The left side of the curve is steeper than the right side of the curve. The left side of the curve is known as the **contraction phase.** How long does this phase take? It is measured by moving the left cross-hair cursor to the position of the graph where the line moves sharply up from the baseline and then moving the other cross-hair cursor to the peak of the blue line at the top of the curve. If the program will not allow you to enter the data, click Erase and redo the experiment. Select the appropriate areas and click on them.

Enter the time in the following space.

? Contraction phase: _____ (3)

Move the cursors so one is at the peak of the graph and the other is at the region of the blue line where it reaches the baseline and record the **relaxation phase.** Enter the time of the relaxation phase in the following space.

? Relaxation phase: _____ (4)

Follow the directions and complete the experiment by adding successively more weight to the muscle.

7. **PhILS**   Length–Tension Relationship

In this laboratory exercise, you examine the strength of contraction based on the initial length of the contracting muscle. Follow the directions as outlined above and in the directions in the program. You must click on the "zoom" button before you can set the voltage. Move the right cursor to the peak of the blue line and click the journal entry to get the tension. The muscle can be stretched prior to stimulation. Notice how the muscle is attached to a horizontal bar. On the left side of this bar are two arrows.

Click on the upper one to increase the tension on the muscle.

Move the arrow up, so you record the increase in length in 0.5 mm increments. This takes about two clicks of the upper arrow. What occurs to the strength of contraction as you stretch the muscle from 26 mm to 30 mm? _____

Where do you predict the greatest overlap of actin and myosin myofilaments? Why? _____

8. ⦙⦙⦙PhILS   Principles of Summation and Tetanus

In this exercise, you adjust the **frequency** (number of impulses per second) of the stimulus. Most muscles do not contract by a single twitch. Neurons typically send many impulses to many muscle fibers asynchronously, producing a smooth contraction. As in the previous experiment, you will need to read the material presented and answer the questions before doing the experiment. Once you get to the experimental stage, turn on the power button to the computer monitor and the data acquisition unit, drag the appropriate cables to their inputs, and set the voltage between 1.6 and 2 volts. If you click and hold the up arrow button, it will continue to increase the voltage. The voltage remains the same in all of the trials. Under low frequency (few stimulations per unit time), you will see that the muscle contractions all have the same height and return to the baseline. When you increase the frequency, by using the cursor to decrease the interval by holding the down arrow, the muscle contraction tracing leaves the baseline and you have **summation.** You can see the beginning of summation at an arrow on your graph by clicking on the highlighted term *summation*.

You will have to pay attention to when the blue line does not return to the baseline. Record the time interval for summation in milliseconds (ms) in the journal and in the following space.

❓ Summation: _____ (5)

As you decrease the interval between stimuli, notice how the peaks begin to merge. If you increase the frequency more, the contractions on the graph occur above the normal contraction peaks, but you still see individual contractions. This is **incomplete tetanus.** Record the time interval in milliseconds for incomplete tetanus in the following space.

❓ Incomplete tetanus: _____ (6)

As you increase the frequency, the peaks merge into a straight line. It represents continuous muscle contraction or **complete tetanus.** Record the time interval in milliseconds for complete tetanus in the following space.

❓ Complete tetanus: _____ (7)

Simulated Human Muscle Experiments

Click on number nine exercise, which is listed under the Skeletal Muscle Function section.

9. ⦙⦙⦙PhILS   EMG and Twitch Amplitude

The EMG (electromyography) records the electrical activity of contracting muscles. You will record two events. One is the electrical activity of contracting muscle, and the other is the force generated by the muscles of the arm and forearm squeezing on a rubber bulb. As opposed to the earlier exercises where you shock the muscle, in this exercise you are picking up recording information only. In this exercise, in addition to placing the plugs in their respective areas, you must also attach the recording electrodes to the proper location on the subject's right arm. After you press the "start" button, drag the cross-hair cursor to the peak of the muscle tension graph, which is the orange line. Then find the peak of the blue line (EMG) and move the cross-hair cursor there.

Close the journal box by clicking on the X in the right upper corner before you record the EMG data. Do this for several contractions by adjusting the arrows below the squeeze bulb in the subject's right hand.

What is the relationship between the force of contraction and the amount of electrical activity in the muscle? Record your answer in the following space. _____

Real Frog Experiment

Ⓐ Activity

You may do this experiment in small groups or your instructor may elect to do a demonstration for the class. If you are doing this experiment as a group, *read the entire exercise first* and then follow the directions.

Frog Preparation

If the frog has not been pithed, follow the directions under number 1. If the frog has been pithed, begin at number 2.

1. The most humane way to conduct frog muscle experiments is to pith the frog quickly by inserting a sharp probe into the braincase.
   a. To do this, firmly grasp the frog and bend the head under your index finger (fig. 15.4a).
   b. Insert the probe into the braincase and twirl it around in a conical manner, destroying the brain. This is known as a **single pith.** The frog will be killed at this point yet still have reflexes in the lower limbs.
   c. To stop the lower limb reflexes, insert the sharp probe into the vertebral canal and run it toward the caudal end of the frog. This is known as a **double pith.** Be careful not to thrust the probe into your hand as you try to locate the vertebral canal. The correct positioning is illustrated in fig. 15.4b.
2. Once the frog is pithed, carefully snip its skin above the hip joint and peel it back to the foot (fig. 15.5).
   a. Separate the posterior muscles of the thigh, and locate the **sciatic nerve,** which appears as a thin, white glossy thread that runs along the lateral aspect of the femur. Always keep the sciatic nerve moist with frog Ringer's solution during this experiment. You may have to use a scalpel and tease the muscles away from the sciatic nerve.
   b. Using a glass hook, carefully lift the sciatic nerve away from the thigh muscles (fig. 15.6).

(a)

(b)

FIGURE 15.4 **Pithing a Frog.** (a) Single pith; (b) double pith.
©Eric Wise

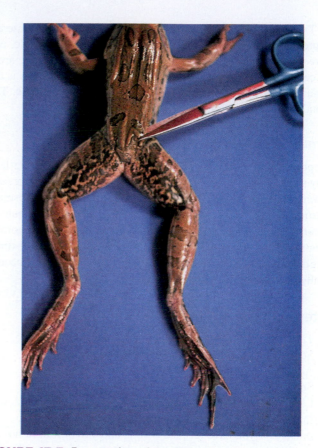

FIGURE 15.5 **Preparation of the Frog, Skin Removal.**
©Eric Wise

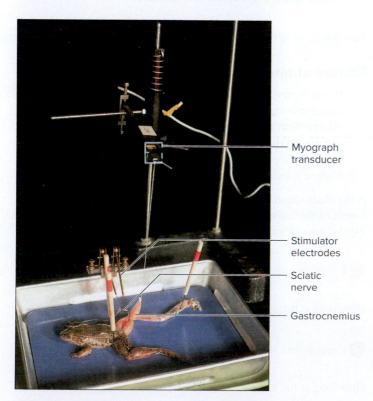

Myograph
transducer

Stimulator
electrodes

Sciatic
nerve

Gastrocnemius

FIGURE 15.6 **Frog Hookup to Recording Apparatus.**
©Eric Wise

c. **Ligate** the nerve by tying a thread around the proximal end of the nerve, near the sacrum, and cut the nerve above the ligature.

d. Separate the thigh muscles from the femur and remove them, leaving the femur exposed. Be careful not to cut or damage the sciatic nerve as you do this.

e. Locate the **gastrocnemius muscle** of the frog and tie the tendon of the muscle with thread.

f. Cut the calcaneal tendon distal to the ligature and lift the gastrocnemius away from the other muscles and from the tibiofibula (a fused bone in frogs).

g. Cut the muscles and the tibiofibula just distal to the knee so you have the femur, the sciatic nerve, the gastrocnemius, and the knee joint intact (fig. 15.6).

h. Anchor the femur to a board or femur clamp, or mount it on a tray and lay the sciatic nerve on the gastrocnemius muscle.

i. Do not tug on the nerve but gently lay it on the gastrocnemius muscle.

j. Attach the thread tied to the calcaneal tendon to the end of a myograph transducer leaf (fig. 15.6). This should be connected to a recording device, such as a physiograph, duograph, kymograph, or physiology computer. If you lightly tap on the leaf of the muscle transducer, you should see a response in your recording apparatus.

There are three critical areas of concern in this lab. These are the preparation of the muscle, the stimulation of the muscle, and the recording of the response. Failure in any of these three areas will cause poor results or no results. The first of these areas is the preparation of the muscle, which has already been outlined. The second and third areas are discussed next.

Stimulator Setup

The preparation of the stimulator first involves determining how many stimuli you deliver in a particular time. Most stimulators can deliver repeating stimuli or a single stimulus (pulse). These are measured as the **frequency** of the stimulus, and a good frequency to start out with is two pulses per second. Another important factor is the **duration** of the stimulus. This is how long each stimulus is delivered to the sciatic nerve. Durations of 2 to 10 milliseconds usually produce good results.

Determination of Threshold Stimulus

1. Turn the voltage to zero on the stimulator.
2. Set the stimulator frequency to two pulses per second.
3. Adjust the impulse duration to 2 milliseconds.
4. Set the switch to *repeat* on the stimulator.
5. Lay the sciatic nerve on the stimulator electrodes and slowly increase the voltage until you see the contraction of the muscle. If you have reached 8 volts and you still have no response, turn the voltage down, shut the stimulator off, and recheck your connections and settings. Once you see the muscle contract, then the lowest voltage that produces a response is the threshold stimulus.
6. Record this value in the space provided.

❓ Threshold stimulus: _____ (8)

Multiple Motor Unit (MMU) Summation

A nerve fiber and all the muscle fibers it attaches to are called a **motor unit.** The strength of contraction in muscle increases as more motor units are stimulated. The smoothness of muscle contractions is brought about by activating different motor units at different times so there is asynchronous firing of muscle fibers in a specific muscle. In this experiment, you examine the effect of increasing voltage and its impact on increasing the number of motor units stimulated in a muscle. There is a point where an increase in voltage will not stimulate any more motor units. This point is known as the **maximum recruitment,** which is the lowest voltage stimulus at which *all* of the muscle fibers are stimulated.

Recording Apparatus Setup

Your lab may be equipped with one or more different physiological recorders. Follow your instructor's directions to set up the apparatus if you are to do the experiment in groups or pay close attention if your instructor demonstrates the experiment. Pay particular attention to the settings of the apparatus.

1. To demonstrate this, set the duration of the pulse to 2 milliseconds, keep the frequency at one or two pulses per second, and set the stimulus on repeat.
2. Begin the recording and increase the voltage until you see a response (threshold) in the muscle. Continue to increase the voltage slowly, and you should see the contraction force increase, as indicated by an increase in the height of the tracing.
3. As you continue to increase the voltage slowly, the contraction tracings will not get any higher. The minimum voltage it takes to produce the maximum height is known as the maximum recruitment voltage.
4. Record the **maximum recruitment voltage** in the space provided and at the end of the exercise.

❓ Maximum recruitment voltage: _____ (9)

Temperature Effects on Muscle Contraction

1. Use the maximum recruitment voltage and stimulate the frog muscle at room temperature. Record the tracing.
2. Add 75 mL of warm frog Ringer's solution to the muscle and stimulate the muscle again. Record the tracing. Describe any change in the contraction strength below.

❓ Change in contraction strength: _____ (10)

3. Add 75 mL of iced frog Ringer's solution to the muscle and stimulate the muscle again. Record the tracing. Describe any change in the contraction strength below.

❓ Change in contraction strength: _____ (11)

Phases of Muscle Contraction

1. If you increase the chart speed to 50 mm per second, you can obtain tracings of the muscle where the **latent period, contraction phase,** and **relaxation phase** are seen. If you can record the time when you stimulate the muscle, then you can calculate the latent period of muscle contraction.
2. Make a tracing of the contraction and compare it to fig. 15.7.

If the chart speed is 50 mm per second in your tracing, what is the length of the latent period? This assumes you have a mark indicating the time of stimulation.

❓ Latent period: _____ (12)

How long is the contraction phase? Record the length of time.

❓ Contraction phase: _____ (13)

How long is the relaxation phase? Record the length of time.

❓ Relaxation phase: _____ (14)

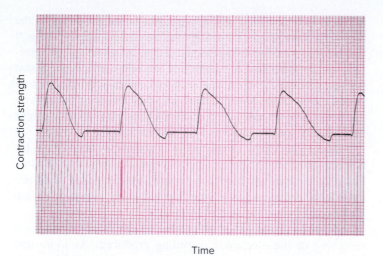

FIGURE 15.7 Phases of Muscle Contractions in a Single Twitch Contraction.

Fatigue in Frog Muscle

Fatigue in muscles can be demonstrated in the lab. You should set up the stimulator to the maximum recruitment voltage and increase the tension on the frog muscle. Stimulate the muscle at 20 pulses per second. The muscle should contract and reach a peak and then the tracing should decrease over time.

How long does it take for the muscle to fatigue? Record the time in the space provided.

❓ Length of time for muscle to fatigue: _____ (15)

Make sure that all members of your lab group see the frog demonstration and the recordings made from the experiment. Clean your dissection equipment and place the remains of the frog into the appropriate container designated by your instructor.

BIOPAC Setup for MP30 or MP150

1. Plug in data acquisition unit to wall outlet.
2. Connect data acquisition unit to computer.
3. Plug SS12LA into channel 2.
4. Turn on the data acquisition unit.

Preparation of Stimulator

1. Plug stimulator into wall outlet.
2. Attach Trigger to Analog Out of data acquisition unit.
3. Attach ELSTM2 Needle electrodes.
4. Set the voltage range to 10 volts but turn the voltage knob to zero.

Computer Setup

1. Start the BSL *PRO* software. You should see an untitled window.

2. Select the menu and choose "Show Stimulator."
3. Go to the File menu and choose "**Graph Template (*GTL) > File Name: FrogMuscle.gtl.**"
4. Do not close the stimulator window.
5. Select 0 volt–50 grams force range.
6. Select the Setup Channels from the menu and choose the wrench icon from the Channel menu and locate the scaling button. Make sure the stimulator is set to 0 volts (with the knob turned counterclockwise) and the pulse light NOT blinking. Click on the Cal 1 button.
7. Enter the Cal 2 value, which equals Cal 1 Input Value 50.
8. Click OK on the scaling window.
9. Go to the Channel 2 window and locate the wrench icon and the scaling window.
10. With only the S hook on the tension adjuster, click Cal 1.
11. Attach a 50-gram weight to the adjuster and click Cal 2.
12. Click OK and exit the scaling window.
13. Attach the tension adjuster to the ring stand with the holes facing down. The adjuster should be set so that, when the frog is attached, the string will be vertical. Attach a small S hook to the adjuster.

Frog Preparation

1. Loop and tie the thread to the calcaneal tendon. Cut the tendon from the calcaneus.
2. Pin the frog down to a board. Make sure that the thigh and leg are anchored. Moisten the muscle with frog Ringer's solution.
3. Cut gently along the length of the thigh muscles and expose the sciatic nerve.
4. Gently dissect the nerve using a glass hook. Do not damage the nerve or tug on it.
5. Loop the thread to the transducer hook and adjust the tension so the thread is not slack. The thread should be vertical as it attaches to the hook.
6. Remove the covers from the electrode tips (remember to replace them at the end of the experiment) and place them underneath the sciatic nerve. Make sure the electrode tips do not touch each other.

Running the Experiment

1. Click the Start button.
2. Adjust the level to 0.1 volt.
3. Click the on/off button in the stimulator window.
4. Increase the voltage by 0.2 volt until you see a response in the CH 2 window.
5. When you get a response, this is the threshold voltage. Record the value here: _____
6. Continue to increase the voltage until you get no further increase in contraction. This is the maximum recruitment.

   Record the value here: _____
7. Reduce the voltage and stop the recording.

Summation and Tetanus

1. Go to the stimulator window and select Continuous Pulses.
2. Adjust the pulses to 1 Hz with the scroll bar.
3. Adjust the voltage to what you obtained for maximum recruitment.
4. Start the recording and increase the frequency slowly by 1–2 Hz every second until the muscle shows a continuous contraction.
5. Turn off the stimulator and stop the recording.
6. Save your data by selecting the File **menu.** Choose Save As and select the file type: **BSL Pro files (*.ACQ) File name: (your name).** Choose Save.

Temperature Effects on Muscle Contraction

1. Open up a new file in FrogMuscle.gtl.
2. Use the maximum recruitment voltage and stimulate the frog muscle at room temperature. Record the tracing.
3. Add 75 mL of warm frog Ringer's solution to the muscle and stimulate the muscle again. Record the tracing. Describe any change in the contraction strength below.
   Change in contraction strength: _____
4. Add 75 mL of iced frog Ringer's solution to the muscle and stimulate the muscle again. Record the tracing. Describe any change in the contraction strength below.
   Change in contraction strength: _____

Fatigue

1. Go to the stimulator window and select Continuous Pulses.
2. Adjust the pulses to 1 Hz with the scroll bar.
3. Adjust the voltage to what you obtained for maximum recruitment.
4. Start the recording and increase the frequency slowly by 1–2 Hz every second until the muscle fatigues.

Human Muscle Physiology

In this part of the exercise, you study the effects of muscle recruitment and fatigue. These experiments are best done with the use of a device that can measure electrical activity, such as a physiograph or computer, and a hardware unit, such as the BIOPAC MP30 or MP150 electrode lead set, disposable electrodes, and the EMG1 student module.

Multiple Motor Unit Summation

Ⓐ **Activity**

   Clean the skin on your medial forearm and lateral epicondyle region with cotton and rubbing alcohol to remove sebaceous secretions on the surface of the skin. Attach the appropriate recording electrodes to the forearm (a positive, negative, and ground) and connect the electrodes to the physiograph or hardware unit. Place two electrodes of the same color on the medial side of the forearm

approximately one-third of the length of the forearm distal to the elbow. Place the third electrode by the lateral epicondyle. You will first need to establish a baseline recording by adjusting the sensitivity of the equipment to register the maximum grip strength that you can produce. To do this, you should clench your hand tightly around a squeezable hand exercise ball while recording the activity and make sure you have adequate displacement displayed. Ask your instructor for specific directions for the equipment in your lab.

   Once you have produced a reasonable recording for baseline data, begin the recording by slightly clenching your hand. Pause for a second and produce a slightly stronger grip. Pause once more for a second and close your hand as tightly as possible. Stop the recording.

   Examine the height of the displacement of the physiograph recording or the electronic recording produced. As more and more muscle fibers are recruited, the increase in the electrical activity produces a stronger signal.

   What do you predict to be the difference in your dominant arm (for example, the right arm in a right-handed person) versus your nondominant arm?

   Produce a recording of your *maximum* grip strength in your other arm and compare the two recordings.

Muscle Fatigue

You can examine the effects of muscle fatigue by using the same setup, but instead of using grip strength, examine the nature of muscles when they contract maximally for long periods. In this exercise, you should hold a 2- to 4-kilogram (5- or 10-pound) weight for a moment to establish your baseline recording. Once you have produced an adequate recording, begin the experiment as follows:

   Hold a weight (2, 4, or 7 kg) with your forearm held at a 90-degree angle to your body. Begin the recording and wait for the muscle to begin to fatigue. This is seen when the forearm begins to drop somewhat. Stronger individuals should hold heavier weights. When the weight can no longer be held and the forearm begins to fall significantly, stop the recording.

   How does the strength of the signal at the beginning of the recording compare to that at the end of the recording?

   _____

   _____

   Does the stability of the recording stay the same or is there variation in the electrical activity?

   _____

   How long does it take for the muscle to fatigue?

   _____

   If you measured fatigue in the frog, compare your results with those in the frog in terms of time for fatigue to begin.

REVIEW SECTION

Muscle Physiology

Name _____ Date _____

Lab Section _____ Time _____

❓ Chapter Summary Data

Use this section to record your results from questions within the exercise.

1. _____   9. _____
2. _____   10. _____
3. _____   11. _____
4. _____   12. _____
5. _____   13. _____
6. _____   14. _____
7. _____   15. _____
8. _____

Review Questions

1. Define subthreshold stimulus. _____
_____
_____

2. Describe complete tetanus. _____
_____
_____

3. What is maximum recruitment? _____
_____
_____

4. In the following illustration, place an "A" on the latent period, a "B" on the contraction phase, and a "C" on the relaxation phase.

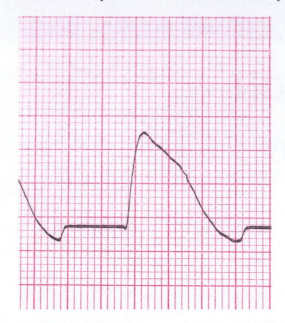

5. What was the threshold stimulus you obtained in lab? _____
_____

6. What was the voltage at which you first got maximum recruitment? _____

7. Explain a muscle spasm in terms of recruitment of muscle fibers. _____
_____

8. A skeletal muscle is stimulated to contract by what structure? _____

9. What chemical crosses the synapse, causing a muscle to contract? _____

10. Name the first event after a stimulus in a muscle contraction. _____
_____

11. What happens to the strength of contraction during wave summation? _____
_____

12. How does iced frog Ringer's solution change the contraction strength of frog muscle? _____
_____

Nervous Tissue, the Spinal Cord, and Spinal Nerves

INTRODUCTION

The function of the nervous system is communication between the various regions of the body, coordination of body functions (as in complicated processes such as digestion or walking), orientation to the environment, and assimilating (learning) information. The functional unit of the nervous system is the **neuron.** It is the cell that carries out the activity of nervous tissue. Neurons are located in the nerves of the body, ganglia, the spinal cord, and the brain. Neuroglia (glial cells) are also important cells of the nervous tissue. They aid the neurons in increasing the speed of neuron transmission, providing nutrients to the neurons and protecting the neurons.

The nervous system can be divided into three general divisions: the **central nervous system (CNS),** which consists of the brain and spinal cord, the **peripheral nervous system (PNS),** which consists of cranial nerves (for example, the facial nerve), spinal nerves that are adjacent to the spinal cord, ganglia (clusters of nerve cell bodies outside of the central nervous system), and peripheral nerves (such as the sciatic nerve or radial nerve), and the **enteric plexus** that has neurons in the digestive tract. The peripheral nervous system is functionally subdivided into a sensory division (one that transmits sensations to the CNS) and a motor division (one that carries information away from the CNS).

We cover the spinal cord and peripheral nerves together because of their anatomical proximity and their functional relationship. The spinal cord is part of the CNS and extends from the foramen magnum of the skull to the first or second lumbar vertebra. The spinal cord is shorter than the vertebral column because the vertebral bodies continue to grow after the spinal cord stops growing. The spinal cord receives sensory information from and transmits motor information to the spinal nerves, which radiate into the body as peripheral nerves. The peripheral nerves are part of the peripheral nervous system. Peripheral nerves **innervate** (IN-ur-vate) (functionally connect to) muscles and other parts of the body. In this exercise, you learn the major features of the spinal cord in longitudinal aspect and in cross section, as well as the major nerves and plexuses of the body. The spinal cord and somatic nerves are covered in the Saladin text in chapter 13, "The Spinal Cord, Spinal Nerves, and Somatic Reflexes."

OBJECTIVES

At the end of this exercise, you should be able to

1. describe the three parts of the neuron;
2. list the main divisions of the nervous system;
3. group the organs of the nervous system into the main divisions;
4. demonstrate the major regions in a cross section of the spinal cord;
5. describe the structure of the longitudinal aspect of the spinal cord;
6. list the major nerves that arise from each plexus;
7. name all the major nerves of the upper and lower limbs;
8. list the structures that carry impulses to and away from the spinal cord.

MATERIALS

Models or charts of the central and peripheral nervous systems

Charts or models of neurons

Microscopes

Prepared slides

    Spinal cord smear

    Longitudinal section of nerve

Prepared slide of a spinal cord in cross section

Model or chart of a spinal cord in cross section and longitudinal section

Cat

Materials for cat dissection

    Dissection tray

    Scalpels

    Protective gloves

Goggles
Blunt probe

PROCEDURE

Divisions

Ⓐ **Activity**

Examine the models or charts in the lab for the divisions of the nervous system. Write "PNS" for peripheral nervous system and "CNS" for the central nervous system for each of the organs listed below that are found in fig. 16.1. Compare the material in the lab to this figure.

Brain _____

Ganglia _____

Spinal Cord _____

Nerves _____

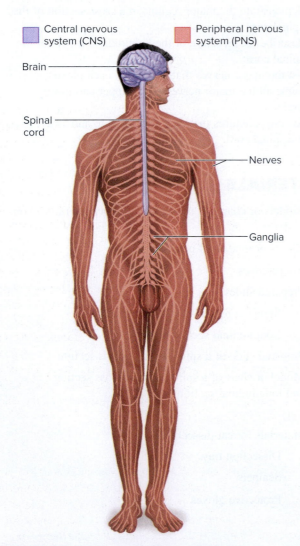

FIGURE 16.1 Divisions of the Nervous System.

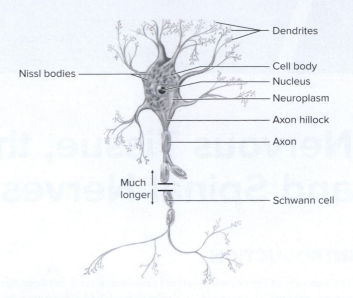

FIGURE 16.2 Parts of the Neuron.

Neuron

Ⓐ **Activity**

Examine fig. 16.2 and models or charts in the lab for the structure of neurons. The axon in fig. 16.2 is much shorter than what would normally be found in the body. The neuron is a remarkable cell. The nerves in your thigh and leg are composed of neuron fibers, and the neurons that pick up sensation in your toes continue as single cells up the leg and thigh to synapse (SIN-aps) (join) with other neurons in the lower back or brain stem. When you look at prepared slides of neurons in the microscope in this exercise, remember that some of these neurons are of great length.

Neurons consist of three main components: axons, dendrites, and the **cell body** (neurosoma or soma). When stimuli excite neurons, the impulses synapse with dendrites or the cell body. **Dendrites** are so named because they have branching structures that resemble a tree (*dendros* = tree). Because dendrites are numerous, they may have synapses with many other neurons. The cell body is the metabolic center of the neuron. It contains cytoplasm, **Nissl bodies** (rough endoplasmic reticulum of the neuron), and the **nucleus.** A triangular region of the cell body near the axon is the **axon hillock.** It is devoid of Nissl bodies, and it leads to the axon that exits the cell body.

Axons are long processes of neurons that transmit action potentials and are frequently wrapped in myelin sheaths. These sheaths are discussed later in the exercise.

Histology of the Neuron

Ⓐ **Activity**

Examine a prepared slide of a spinal cord smear under the microscope and locate a purple, star-shaped structure under low power. This is a cell body of a multipolar neuron. Switch to high power and locate the Nissl bodies, the nucleus, and the axon hillock. If you find the axon hillock (which may be difficult to do), you should be able to see the axon leading away from the hillock. The other processes attached to the cell body are dendrites. The small

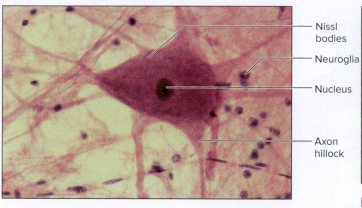

FIGURE 16.3 Spinal Cord Smear.
©Eric Wise

Labels: Nissl bodies, Neuroglia, Nucleus, Axon hillock

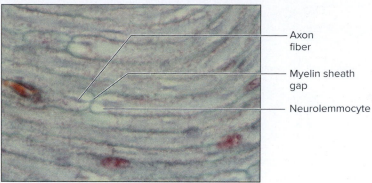

FIGURE 16.4 Nerve, Longitudinal Section (400x).
©Eric Wise

Labels: Axon fiber, Myelin sheath gap, Neurolemmocyte

nuclei scattered throughout the smear belong to neuroglia in the spinal cord. Compare your slide to fig. 16.3.

Neuroglia

Numerous cells aid the functioning of the neuron. These cells are called the **neuroglia** (noo-ROG-lee-uh), or **glial** (GLEE-ul) **cells**.

PNS Neuroglia   The common neuroglia of the peripheral nervous system, and the one examined in this exercise, is the **Schwann cell,** or **neurolemmocyte** (fig. 16.2). These neuroglia wrap around the axon, leaving small gaps between successive glial cells called the **myelin sheath gap (nodes of Ranvier)** (RON-vee-ay). Schwann cells wrap around the axon, much as a thin strip of paper can be wrapped around a pencil.

The neurolemmocyte consists of a significant amount of a lipoprotein material called myelin, and the series of neurolemmocytes produces a myelin sheath.

The nodes of Ranvier allow for the nerve transmission to jump from node to node, thus increasing the conduction speed of the neuron. This type of jumping transmission is called saltatory conduction.

Histology of the Neurolemmocyte

(A) Activity

Examine a prepared slide of a longitudinal section of nerve under high power. You should see the axon fibers as long, dark threads in the microscope. The clear areas on each side of the axon fibers are the myelin sheaths. Scan the slide closely and locate the junction of two neurolemmocytes. The gap between them is the myelin sheath gap. Compare your slide to fig. 16.4.

Longitudinal Aspect of the Spinal Cord

(A) Activity

Examine a model or chart in the lab of a longitudinal view of the spinal cord and locate the major features illustrated in fig. 16.5. The spinal cord terminates inferiorly as the **medullary** (MED-you-LAR-ee) **cone** at approximately vertebra L1 and is attached to the coccyx by a continuation of the pia mater known as the **terminal filum** (FYE-lum). The neural continuation of the spinal

cord exists as an extension of spinal nerves from L2 through S5. These parallel nerves resemble a horse's tail and are called the **cauda equina** (CAW-duh ee-KWY-nah). The spinal cord has a couple of expansions in two locations. The **cervical enlargement** occurs at C3 through T2 and represents a bulge in the spinal cord that has increased innervation of the upper limbs. The **lumbosacral enlargement** occurs at about T7 through T11, and this expanse is due to the increased innervation of the lower limbs.

Cross Section of Spinal Cord

(A) Activity

Examine a model or chart in the lab of a cross section of the spinal cord and locate the features in fig. 16.6. Locate the **gray matter,** which appears as an H pattern or a butterfly pattern in the middle of the spinal cord, with the **white matter** located on the periphery of the cord.

In a cross section of the spinal cord, you should see the gray matter divided into two narrow horns and two rounded horns. The narrow horns are known as the **posterior horns,** and these areas receive sensory information from the somatic nerves. The rounded horns are the **anterior horns,** and these send motor signals to the spinal nerves. In some parts of the spinal cord there are additional sections of gray matter known as **lateral horns.**

Each side of the gray matter is connected to the other by a crossbar known as the **gray commissure.** In the middle of the gray commissure is the **central canal,** which runs the length of the spinal cord. The white matter of the cord is divided into **columns,** or **funiculi,** which carry sensory information to the brain, in **ascending tracts,** or motor information from the brain, in **descending tracts.** You should also be able to see a depression in the posterior surface of the spinal cord. This is the **posterior median sulcus,** while the deeper depression on the anterior side is known as the **anterior median fissure.**

(A) Activity

Examine a prepared slide of the spinal cord under low power and locate the anterior and posterior horns, the gray commissure, the central canal, the posterior median sulcus, and the anterior median fissure. Examine the anterior horn of the spinal cord and look for the cell bodies of the **multipolar neurons** there. These are motor neurons. Compare your slide to fig. 16.7.

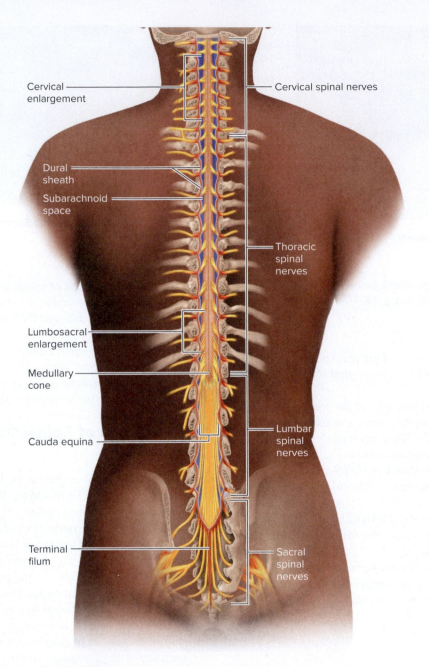

FIGURE 16.5 Spinal Cord, Longitudinal View.

Meninges

Examine a model of a cross section or three-dimensional representation of the spinal cord in the vertebral column and compare it to figs. 16.6 and 16.8. The spinal cord is covered by **meninges,** which continue into the brain. The outer one is the **dura mater,** which forms the dural sheath, which in turn encloses the spinal cord. Between the dura mater and the vertebra is the **epidural space,** a region where anaesthetics are given during some deliveries to relieve pain during childbirth. The next deeper membrane is the **arachnoid mater (membrane).** Deep to the arachnoid mater is the **pia mater,** the innermost of the meninges on the spinal cord proper. Between the pia mater and

the arachnoid is the **subarachnoid space,** which contains the **cerebrospinal fluid.**

Nerves Associated with the Spinal Cord

(A) **Activity**

Examine the material in the lab as you read the following text and compare them to fig. 16.9. The nerves that attach to the spinal cord are part of the peripheral nervous system. They are covered here because of their interplay with the spinal cord. The **posterior ramus** is a branch of sensory and motor nerve fibers that, with the **anterior ramus,** unite to form a **spinal nerve.**

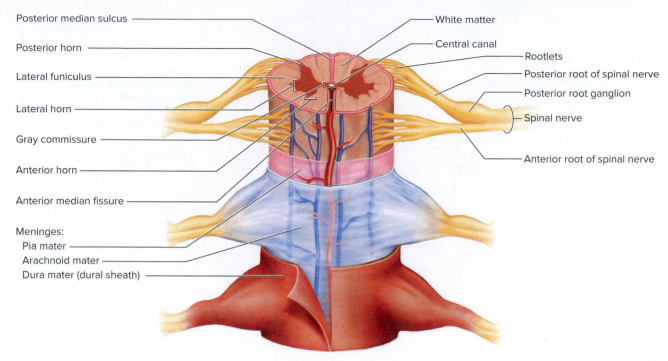

Posterior median sulcus
Posterior horn
Lateral funiculus
Lateral horn
Gray commissure
Anterior horn
Anterior median fissure
Meninges:
　Pia mater
　Arachnoid mater
　Dura mater (dural sheath)

White matter
Central canal
Rootlets
Posterior root of spinal nerve
Posterior root ganglion
Spinal nerve
Anterior root of spinal nerve

FIGURE 16.6 Spinal Cord, Cross Section.

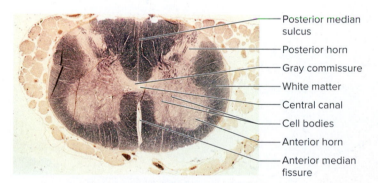

Posterior median sulcus
Posterior horn
Gray commissure
White matter
Central canal
Cell bodies
Anterior horn
Anterior median fissure

FIGURE 16.7 Histology of the Spinal Cord (10x). The white matter is darker due to the stained preparation of the specimen.
Biophoto Associates/Science Source

The spinal nerve enters the **intervertebral foramen** and divides into a **posterior (dorsal) root** and an **anterior (ventral) root.** The posterior root has a posterior root ganglion that contains the nerve cell bodies of the sensory nerves. The **posterior root** carries sensory information to the posterior horn of the spinal cord. The cell bodies of the motor nerves are located in the anterior horn of the spinal cord and exit via the anterior root.

Spinal Nerves and Plexuses

Nerve fibers (axons) are clustered in parallel arrangements called **nerves** if they are in the peripheral nervous system and **tracts** if they are in the central nervous system. There are 31 pairs of **spinal nerves** that exit from the spinal cord. The spinal nerves are **mixed nerves** carrying both sensory and motor information.

The spinal nerves pass through the intervertebral foramina and are named according to their region of origin. There are **8 pairs of cervical nerves, 12 pairs of thoracic nerves, 5 pairs of lumbar nerves, 5 pairs of sacral nerves,** and **1 pair of coccygeal nerves.** Some of these nerves exit the spinal cord, and branches of spinal nerves travel to parts of the body intact, while others exit the spinal cord and form a branching network with other spinal nerves. These networks are called **plexuses,** and there are four major plexuses. Some spinal nerves, such as C5 and L4, contribute to more than one plexus. The composition of the plexuses is outlined in table 16.1 and illustrated in fig. 16.10. **Communicating rami** are branches in parts of the spinal cord that innervate sympathetic ganglia of the autonomic nervous system.

Cervical and Brachial Plexus Nerves

The cervical plexus has many nerves that innervate the head and neck. An important nerve that exits from each side of the cervical plexus is the **phrenic (FREN-ic) nerve.** This nerve runs to the diaphragm and is responsible for its contraction in breathing.

(A) Activity

Examine the nerves of the **cervical plexus** and **brachial plexus** on models or charts in lab and compare them with figs. 16.11 and 16.12.

The nerves from the brachial plexus are those that primarily innervate the upper limbs. The innervation of the muscles of the upper limb is covered in the exercises on the specific muscles. The major nerves are divided into anterior and posterior divisions. In the anterior division is the **musculocutaneous**

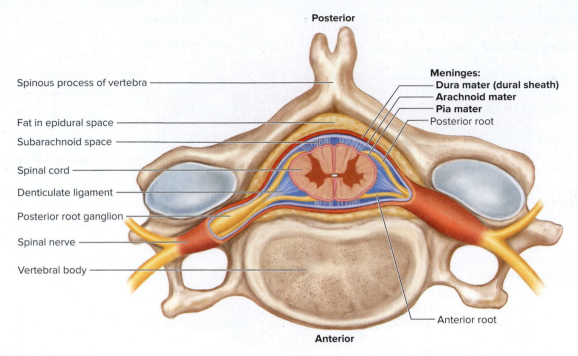

FIGURE 16.8 Spinal Meninges.

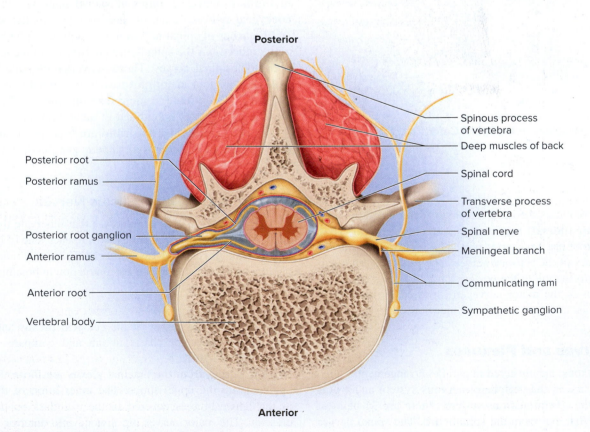

FIGURE 16.9 Nerves Associated with the Spinal Cord.

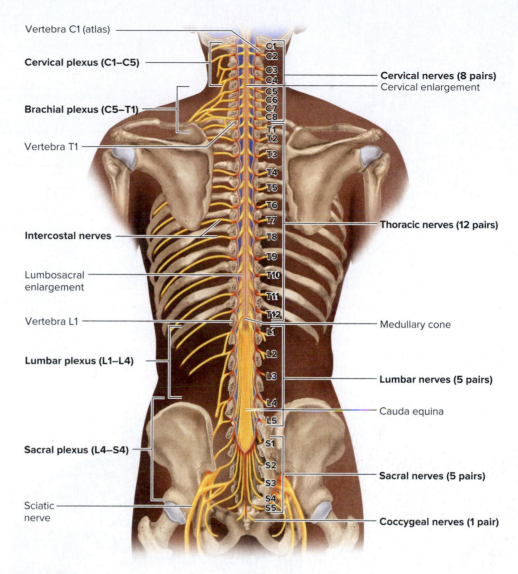

Vertebra C1 (atlas)
Cervical plexus (C1–C5)
Brachial plexus (C5–T1)
Vertebra T1
Intercostal nerves
Lumbosacral enlargement
Vertebra L1
Lumbar plexus (L1–L4)
Sacral plexus (L4–S4)
Sciatic nerve

Cervical nerves (8 pairs)
Cervical enlargement
Thoracic nerves (12 pairs)
Medullary cone
Lumbar nerves (5 pairs)
Cauda equina
Sacral nerves (5 pairs)
Coccygeal nerves (1 pair)

FIGURE 16.10 Nerve Plexuses.

TABLE 16.1	Plexuses	
Name	**Spinal Nerves Contributing to Plexus**	**Major Nerves of Plexus**
Cervical	C1–5	Phrenic
Brachial	C5–T1	Radial, median, ulnar, musculocutaneous, axillary
Lumbar	L1–4	Femoral, obturator
Sacral	L4–S4	Sciatic (tibial and common fibular)

(MUS-cue-lo-cyu-TANE-ee-us) **nerve,** innervating many of the muscles of the arm; the **median nerve,** running the length of each upper limb and serving important hand and forearm flexors; and the **ulnar nerve,** serving other forearm and hand flexors. The ulnar

nerve crosses behind the medial epicondyle of the humerus and is commonly known as the "funny bone." In the posterior division are the **axillary nerve,** innervating the upper shoulder, and the **radial nerve,** predominantly innervating the triceps brachii and the extensors of the hand.

Lumbar and Sacral Plexus Nerves

A Activity

Examine the models or charts of the lumbar and sacral plexuses in the lab and compare them with figs. 16.13 and 16.14. The **femoral nerve** is the most obvious structure arising from the **lumbar plexus.** This large nerve passes posteriorly to the inguinal ligament and mostly innervates the muscles of the anterior thigh. The **obturator nerve** also arises from the lumbar plexus and innervates

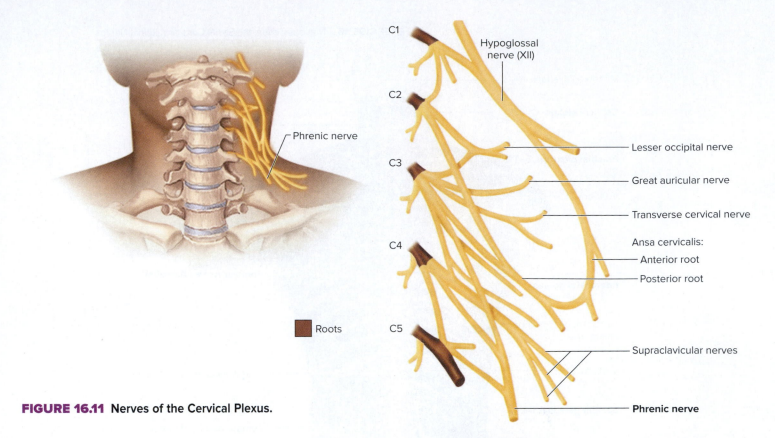

C1

Hypoglossal nerve (XII)

C2

Lesser occipital nerve

C3

Great auricular nerve

Transverse cervical nerve

Ansa cervicalis:

C4

Anterior root

Posterior root

Roots

C5

Supraclavicular nerves

Phrenic nerve

Phrenic nerve

FIGURE 16.11 Nerves of the Cervical Plexus.

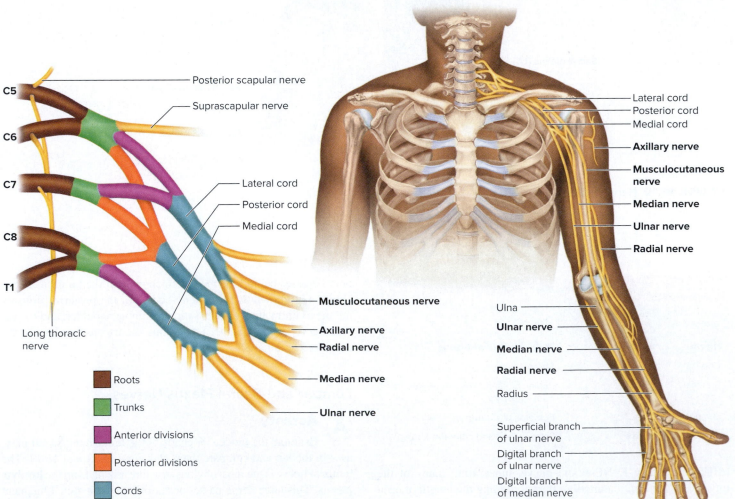

Posterior scapular nerve

Suprascapular nerve

C5

C6

C7

Lateral cord

Posterior cord

Medial cord

C8

T1

Musculocutaneous nerve

Long thoracic nerve

Axillary nerve

Radial nerve

Roots

Trunks

Median nerve

Anterior divisions

Posterior divisions

Ulnar nerve

Cords

Lateral cord
Posterior cord
Medial cord

Axillary nerve

Musculocutaneous nerve

Median nerve

Ulnar nerve

Radial nerve

Ulna

Ulnar nerve

Median nerve

Radial nerve

Radius

Superficial branch of ulnar nerve

Digital branch of ulnar nerve

Digital branch of median nerve

FIGURE 16.12 Nerves of the Brachial Plexus.

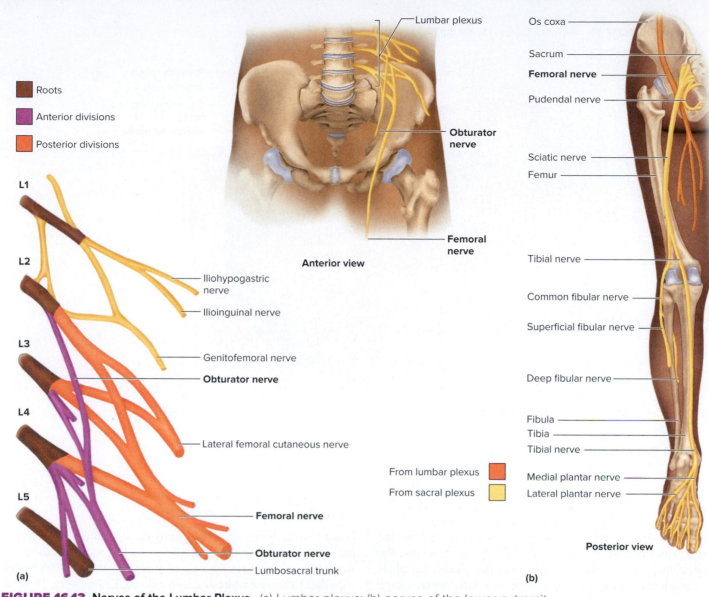

Legend:
- Roots
- Anterior divisions
- Posterior divisions

Lumbar plexus

Os coxa

Sacrum

Femoral nerve

Pudendal nerve

L1

L2

Iliohypogastric nerve

Ilioinguinal nerve

L3

Genitofemoral nerve

Obturator nerve

L4

Lateral femoral cutaneous nerve

L5

Femoral nerve

Obturator nerve

Lumbosacral trunk

(a)

Obturator nerve

Femoral nerve

Anterior view

Sciatic nerve

Femur

Tibial nerve

Common fibular nerve

Superficial fibular nerve

Deep fibular nerve

Fibula
Tibia
Tibial nerve

From lumbar plexus

From sacral plexus

Medial plantar nerve

Lateral plantar nerve

Posterior view

(b)

FIGURE 16.13 Nerves of the Lumbar Plexus. (a) Lumbar plexus; (b) nerves of the lower extremity.

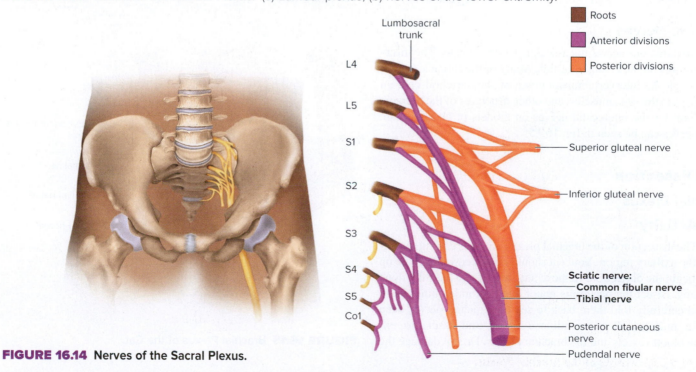

Lumbosacral trunk

Roots

Anterior divisions

Posterior divisions

L4

L5

S1

Superior gluteal nerve

S2

Inferior gluteal nerve

S3

S4

Sciatic nerve:
Common fibular nerve
Tibial nerve

S5

Co1

Posterior cutaneous nerve

Pudendal nerve

FIGURE 16.14 Nerves of the Sacral Plexus.

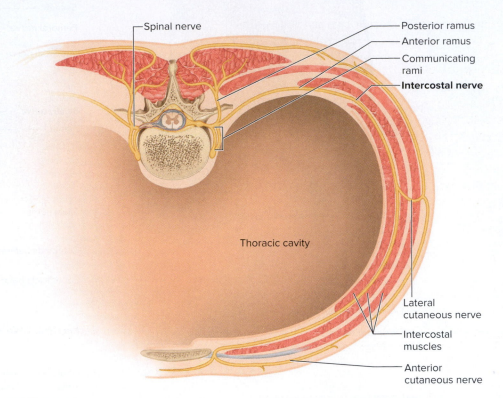

FIGURE 16.15 Intercostal Nerves.

the adductor muscles. There are many nerves that come from the **sacral plexus.** Many innervate the pelvis and muscles that move the hip, thigh, and leg. Two of the nerves from this plexus, the **tibial** and **common fibular nerve,** unite to form the **sciatic nerve,** a large nerve of the posterior thigh, which innervates the leg and foot.

Thoracic Nerves

There are numerous nerves not associated with a plexus. The **intercostal nerves** are an example of this. Many of the thoracic nerves exit through the intervertebral foramina of the vertebral column and innervate the ribs, muscles, and other structures of the thoracic wall. Look for the intercostal nerves on models or charts in lab. These nerves can be seen in fig. 16.15.

Cat Dissection

Brachial Plexus

Ⓐ **Activity**

The dissection of the brachial plexus involves careful dissection in the axillary region. Your cat should be placed ventral side up as you begin the dissection. Remove any skin, if you have not done so already. Bisect the pectoralis major and pectoralis minor muscles and carefully fold them back to see the brachial plexus deep to these muscles. Remove adipose tissue and fascia to expose the blood vessels and the brachial plexus. Do not damage the

blood vessels, because you will study these in future labs. Nerves are solid, commonly round or flattened, and will look different from arteries and veins because they are not injected with red or blue latex. They are white or cream colored. Examine fig. 16.16 as you study the cat.

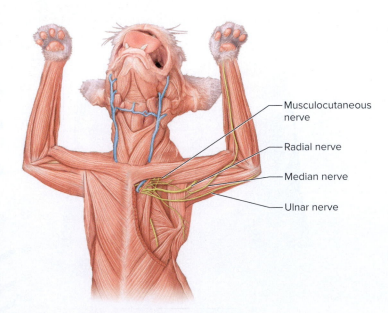

FIGURE 16.16 Brachial Plexus of the Cat.

The large and anterior nerve of the brachial plexus is the **musculocutaneous nerve.** It innervates the skin of the brachium and is the motor nerve of the biceps brachii muscle.

The **radial nerve** is the largest nerve of the plexus and is posterior to the musculocutaneous nerve. It is located on the posterior of the distal forelimb, where it innervates the muscles on the dorsal side of the forelimb, including many of the extensor muscles.

The **median nerve** is located alongside the brachial artery and is in the midline of the brachium and antebrachium. It innervates many of the flexor muscles.

The **ulnar nerve** travels posterior to the medial epicondyle of the humerus to innervate the muscles on the ulnar side of the antebrachium.

Sacral Plexus

(A) Activity

The sacral plexus is best seen from the dorsal view. Examine fig. 16.17 and locate the large **sciatic nerve** by dissecting the biceps femoris on the dorsal thigh. The sciatic nerve is composed of two nerves, the medial **tibial nerve** and the lateral **common fibular nerve.** These can be seen at the distal part of the thigh.

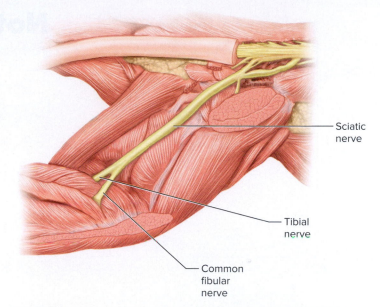

Sciatic nerve

Tibial nerve

Common fibular nerve

FIGURE 16.17 Sacral Plexus, Posterolateral View of Right Side.

Notes

REVIEW SECTION

Nervous Tissue, the Spinal Cord, and Spinal Nerves

Name _____   *Date* _____

Lab Section _____   *Time* _____

Review Questions

1. A spinal nerve belongs to what division of the nervous system? _____

2. What does CNS stand for? _____

3. What kind of cell performs the main function of the nervous system? _____

4. The nucleus is found in what specific part of the neuron? _____

5. A typical neuron has three main components. What are they? _____

6. What type of signal (sensory/motor) travels through the

    a. anterior horn? _____

    b. posterior horn? _____

    c. ascending spinal tracts? _____

    d. descending spinal tracts? _____

7. What major nerves arise from the following plexuses?

    a. cervical _____

    b. brachial _____

    c. lumbar _____

    d. sacral _____

8. In function, how does the posterior spinal root differ from the anterior spinal root? _____

    _____

    _____

9. What causes the cervical enlargement of the spinal cord? _____

    _____

    _____

10. Where is the terminal filum found? _____

_____

11. What is the medullary cone? _____

_____

12. What is the cauda equina? _____

_____

13. What is the area of gray matter found between the lateral halves of the spinal cord? _____

_____

_____

14. The subarachnoid space is filled with what fluid? _____

15. How do tracts differ from nerves? _____

_____

_____

16. What is a mixed nerve? _____

_____

17. Contractions of the diaphragm are regulated by what nerve? _____

18. The anterior muscles of the arm, such as the biceps brachii, have what innervation? _____

_____

19. The extensor muscles of the hand are controlled by what nerve? _____

20. The sciatic nerve is composed of two nerves. What are they? _____

_____

21. A person has feeling from the deltoid and biceps brachii region but no feeling from the wrist extensors. Where on the spinal cord

has injury occurred? _____

_____

22. Label the cross section of the spinal cord. _____

a. _____   g. _____

b. _____

c. _____   h. _____

                            i. _____

d. _____   j. _____

e. _____   k. _____

f. _____

The Brain and Cranial Nerves

INTRODUCTION

Two specific traits distinguish humans from other animals. One is our upright posture, and the other is the extensive development of our brain. In this exercise, you examine the anatomy of the human brain and the cranial nerves associated with it. You also compare the human brain to a sheep brain and note similarities and differences. There is a difference between humans and sheep not only in the overall size of the brains, but also in the relative size of various structures of the brain and the position of some of the anatomy of the brain. You may want to review the planes of sectioning in Exercise 2.

The brain, part of the central nervous system, is located in the cranial cavity of the skull and weighs approximately 1.4 kilograms (3 pounds). The brain can be subdivided into the **prosencephalon,** or **forebrain;** the **mesencephalon,** or **midbrain;** and the **rhomb-encephalon,** or **hindbrain.** Cranial nerves, though belonging to the peripheral nervous system, are studied along with the brain, as they are closely associated with it. Knowledge of the anatomy of the brain is important in locating the cranial nerves. These topics are covered in the Saladin text in chapter 14, "The Brain and Cranial Nerves."

OBJECTIVES

At the end of this exercise, you should be able to

1. name the three meninges of the brain and their location relative to one another;
2. locate the three major regions of the brain;
3. name the main structures in each of the three regions of the brain;
4. describe the function of the major structures in the brain;
5. trace the path of cerebrospinal fluid through the brain;
6. identify each of the 12 pairs of cranial nerves on an illustration, a model, or a real brain;
7. describe whether a cranial nerve is sensory, motor, or both; and
8. identify the structure that a particular cranial nerve innervates.

MATERIALS

Models and charts of the human brain

Preserved human brains (if available)

Cast of the ventricles of the brain

Chart, section, or illustration of the brain in coronal and transverse sections

Sheep brains

Dissection trays

Scalpels

Protective gloves

Blunt probes

PROCEDURE

Meninges

There are three membranes, called **meninges,** that surround the brain. The outermost of these is the **dura mater** (DOO-rah MAH-tur; Latin for "tough mother"), a tough, dense connective tissue membrane that encircles the brain. The next deeper layer is the **arachnoid** (uh-RAK-noyd; Greek for spider web–like) **mater,** a thin membrane. Deep to the arachnoid is the **subarachnoid space,** which contains **cerebrospinal fluid (CSF).** The deepest layer is the **pia** (PEE-uh; Latin for tender) **mater,** a membrane directly on the outer surface of the brain.

(A) Activity

Locate the meninges in preserved brains in the lab (if available). The dura mater resembles a swim cap in overall thickness and general appearance. The arachnoid is filmy and looks somewhat like a cobweb when pulled from the surface of the brain. The pia mater is attached directly to the outer surface of the brain.

Major Regions of the Brain

Examine a model or chart of a human brain and locate the major regions listed in table 17.1. These can be subdivided into smaller areas. The forebrain is the largest region of the brain and consists primarily of the cerebrum and the diencephalon (table 17.1 and fig. 17.1). The midbrain is the smallest region of the brain and is located between the forebrain and the hindbrain. The hindbrain is composed of the pons, the medulla oblongata (me-DULL-ah OB-long-GAH-ta), and the cerebellum. The hindbrain is the most inferior portion of the brain and connects to the spinal cord at the level of the foramen magnum.

Ventricles of the Brain

Ⓐ **Activity**

Examine models of ventricle casts of the brain. These were first obtained by pouring hot wax into the spaces inside the brain and waiting for them to solidify. Compare the models in lab to

TABLE 17.1	Regions of the Brain
Forebrain	*Midbrain*
Telencephalon	Peduncles
Cerebrum (cerebral hemispheres)	Tectum
Cerebral cortex (gray matter)	Superior colliculus
Basal nuclei (gray matter)	Inferior colliculus
Corpus callosum	
Diencephalon	*Hindbrain*
Pineal body	Metencephalon
Thalamus	Pons
Hypothalamus	Cerebellum
Pituitary gland	Myelencephalon
Mammillary bodies	Medulla oblongata

fig. 17.2. The hollow neural tube that develops in the first trimester of pregnancy becomes the ventricles of the brain in the adult. The two ventricles that occupy the center of each cerebral hemisphere are known as the **lateral ventricles.** These ventricles receive **cerebrospinal fluid** from tufts of blood capillaries called **choroid** (CO-royd) **plexuses.** In preserved brains the choroid plexuses are small, brown areas at the superior portions of the ventricles. Fluid from the lateral ventricles flows through the **interventricular foramina** and into the **third ventricle.** The third ventricle is medial to each thalamus and receives CSF from choroid plexuses in that area. The third ventricle drains into the **fourth ventricle** by way of the **cerebral aqueduct.** If this duct becomes occluded, then CSF accumulates in the lateral and third ventricles. This increases the size of the ventricles, producing a condition known as **hydrocephaly.** Normally, the cerebral aqueduct is open and passes through the region of the midbrain. Posterior to the cerebral aqueduct is the fourth ventricle, which occupies a space anterior to the cerebellum. The fourth ventricle also has a choroid plexus that secretes CSF. Cerebrospinal fluid flows from the fourth ventricle to the central canal of the spinal cord, and into the subarachnoid space of the spinal cord and brain. The CSF cushions and provides buoyancy to the brain. CSF flows through the arachnoid granulations under the dura mater of the skull to the venous sinuses, and the fluid returns to the rest of the cardiovascular system by the **internal jugular veins.** There is approximately 150 mL of CSF in the central nervous system, and it takes about 6 hours to circulate through the system.

Surface View of the Brain

Ⓐ **Activity**

Examine a model or chart of the brain and locate the **forebrain** and the **hindbrain.** In the forebrain, you should examine the large **cerebrum,** which can be seen with folds and ridges

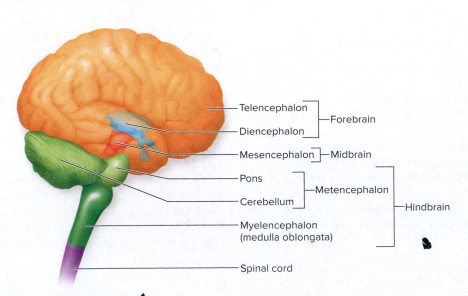

FIGURE 17.1 Major Regions of the Brain.

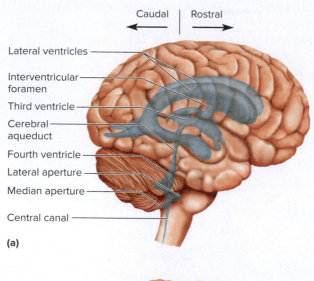

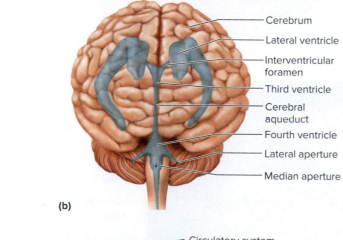

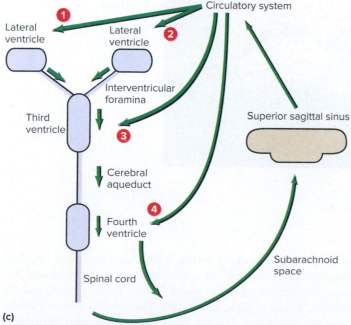

FIGURE 17.2 Ventricles of the Brain. (a) Lateral view; (b) anterior view; (c) schematic presentation of the flow of cerebrospinal fluid. Numbers 1–4 represent flow through choroid plexuses.

forming convolutions, which increase the surface area of the brain. The ridges of the convolutions are called **gyri** (JY-rye; singular, *gyrus*), and the depressions are either **sulci** (SUL-sye; singular, *sulcus*) or **fissures.** Fissures are deeper than sulci. The lateral view of the brain allows you to see the major lobes of each cerebral hemisphere. These major lobes are named for the bones of the skull under which they lie. They are the **frontal, parietal, occipital,** and **temporal lobes.** Deep to the lateral sulcus of the brain is a small lobe called the **insula.** The **lateral sulcus** separates the temporal lobe from the frontal and parietal lobes of the brain. Locate these features in fig. 17.3.

Frontal Lobe

The **frontal lobe** is responsible for many of the higher functions associated with being human. The frontal lobe is involved in intellect, abstract reasoning, creativity, social awareness, and language. An important area responsible for contributing to the formation of speech is called the **Broca area,** or the **motor language area.** This region is usually located in the left frontal lobe of the brain in most people, whether they are right-handed or left-handed. Locate the frontal lobe in fig. 17.3. The posterior border of the frontal lobe is defined by the **central sulcus.**

To find the central sulcus, look for two gyri that run from the superior portion of the cerebrum to the lateral sulcus, more or less continuously. The gyrus anterior to the central sulcus is part of the frontal lobe and is known as the **precentral gyrus,** or the **primary motor cortex.**

Parietal Lobe

The gyrus posterior to the central sulcus is known as the **postcentral gyrus,** or the **primary somatosensory cortex** as illustrated in fig. 17.4. This is part of the parietal lobe and is involved in receiving somatosensory information from the body. The primary somatosensory cortex receives information, yet the material is integrated just posterior to the somatosensory cortex in the **association areas.** The primary somatosensory cortex pinpoints the part of the body affected, and the association area interprets the type of sensation (pain, heat, cold, etc.). Locate the structures of the parietal lobe in fig. 17.3.

The **Wernicke** (WUR-ni-keh) **area (posterior speech area)** is involved in the formation of language, such as the recognition of written and spoken language and the forming of coherent sentences. The Wernicke area extends into the temporal lobe as well.

Occipital Lobe

Posterior to the parietal lobe is the **occipital lobe** of the brain. The occipital lobe is considered the primary visual area of the brain, and damage to this lobe may cause blindness. Shape, color, and distance of an object are perceived here. Recollection of past visual images occurs here as well. As you are reading these words, your occipital lobe is receiving the information and transferring it to other regions, which convert the words to thought. Between the occipital lobe and the cerebellum is a

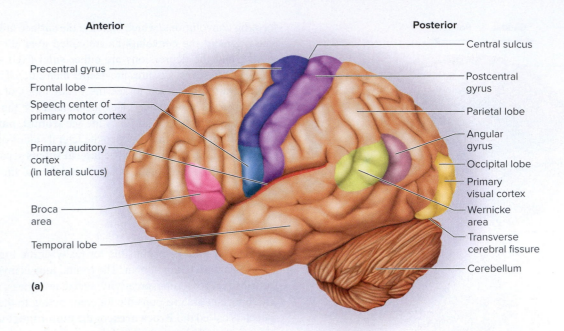

(a)

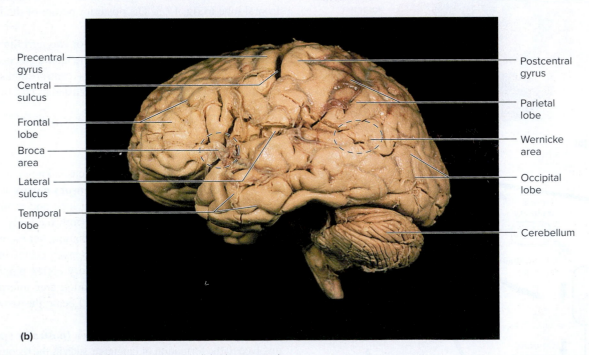

(b)

FIGURE 17.3 Brain, Lateral View. (a) Diagram; (b) photograph.
(b) ©Eric Wise

transverse cerebral fissure that separates these two regions of the brain. Locate the occipital lobe and the cerebellum in fig. 17.3.

Temporal Lobe and Insula

The **temporal lobe** is separated from the frontal and parietal lobes by the **lateral sulcus.** The temporal lobe contains an area known as the **primary auditory cortex,** which receives neural impulses sent from the inner ear. The nearby auditory association area distinguishes the nature of the sound (music, noise, speech) and the location, distance, pitch, and rhythm. The primary auditory cortex translates words into thought. The temporal lobe also has centers for the sense of smell (**olfactory centers**) and taste (**gustatory centers**). Deep to the temporal lobe is a small lobe of cortical material called the insula.

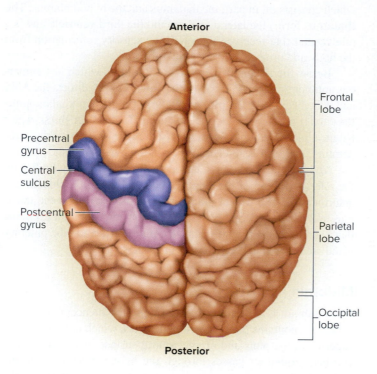

FIGURE 17.4 Primary Motor and Somatosensory Cortex. Motor cortex (precentral gyrus) and somatosensory cortex (postcentral gyrus).

Cerebral Hemispheres

Ⓐ **Activity**

Rotate the brain so you are looking at it from a superior view, and find the **longitudinal cerebral fissure** that separates the cerebrum into the left and right cerebral hemispheres. The **left cerebral hemisphere** in most people is involved in language and reasoning. For example, the Broca area is typically on the left side of the brain. The **right cerebral hemisphere** of the brain is involved in space and pattern perceptions, artistic awareness, imagination, and music comprehension. This specialization where one hemisphere of the cerebrum is involved in a particular task is known as **cerebral lateralization.** Examine the surface features of the brain in fig. 17.5.

Inferior Aspect of the Brain

Forebrain

Ⓐ **Activity**

Examine the frontal and temporal lobes of the cerebrum on models or on a real brain. You may be able to see the **pituitary gland** if it has not been removed. The **optic chiasm** (KY-AZ-um) (*chiasm* = cross) is anterior to the pituitary and transmits visual impulses from the eyes to the brain. Two small

processes posterior to the pituitary are the **mammillary bodies,** which relay information from the limbic system to the fornix.

Hindbrain

From an inferior view of the brain locate the **pons, medulla oblongata,** and **cerebellum** (SER-eh-BEL-um). The cerebellum has much finer folds of neural tissue called **folia.** The medulla

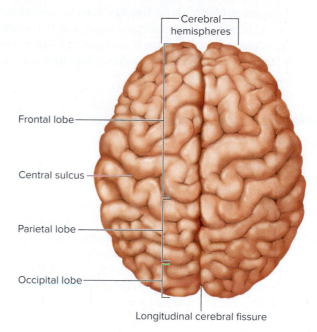

(a)

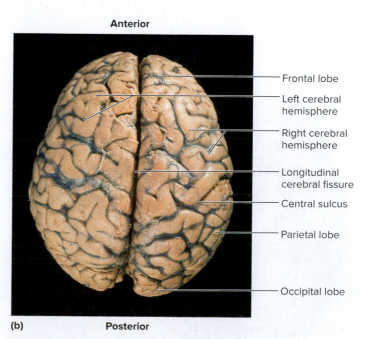

(b)

FIGURE 17.5 Brain, Superior View. (a) Diagram; (b) photograph.
(b) ©Eric Wise

oblongata is located inferior to the cerebellum. The medulla oblongata has centers for respiratory rate control and control of blood pressure, as well as other vital centers. Some information from the right side of the body crosses over to the left brain in the medulla oblongata. Information coming from the left side of the body crosses over to the right side of the brain. The area where motor tracts cross over in the medulla is known as the **pyramidal decussation.** A stroke on the right side of the brain might affect the muscle activity on the left (contralateral) side of the body. The medulla oblongata terminates at the foramen magnum, and the cervical region of the spinal cord continues inferior to the foramen magnum. The enlarged portion of the brain anterior to the medulla is the pons, which is a relay center for information. Examine these structures of the brain in fig. 17.6.

Median Section of the Brain

Forebrain

(A) **Activity**

Examine a median section of a brain, as illustrated in fig. 17.7, and locate the following structures. This section is seen by cutting the brain through the longitudinal cerebral fissure. Locate the C-shaped **corpus callosum,** which connects the two cerebral hemispheres. Just inferior to the corpus callosum is the **septum pellucidum,** which separates the lateral ventricles from one another. If the septum pellucidum is missing, you will be able to look into the lateral ventricle without obstruction.

Examine the material in the lab and locate the **diencephalon,** which consists of, in part, the thalamus and the hypothalamus. The **thalamus** forms the lateral wall around the third ventricle and is a relay center that receives almost all the sensory information from the body and projects it to the cerebral cortex.

Below the thalamus is the **hypothalamus,** which has numerous autonomic centers. The hypothalamus, in part, directs the ANS and is involved with the **pituitary gland** (in the hypothalamopituitary axis) in many endocrine functions. Centers for thirst, water balance, pleasure, rage, sexual desire, hunger, sleep patterns, temperature, and aggression are located in the hypothalamus.

You should also be able to locate the **mammillary bodies** on the inferior portion of the diencephalon and the **optic chiasm** just anterior to it. The **pineal gland** is located posterior to the thalamus and is an endocrine gland that secretes melatonin, a hormone that regulates daily rhythms. Both the pineal gland and the pituitary gland are covered in greater detail in Exercise 23.

Midbrain

The midbrain is a small area posterior to the diencephalon. This small area contains the **cerebral peduncles,** which occupy an area superior to the pons and on the ventral surface of the brain. The **cerebral aqueduct** passes through the mesencephalon, with the **peduncles** being inferior and the **tectum** as a roof superior to the aqueduct. Locate these features in the material in the lab and in fig. 17.7. The tectum has four hemispheric processes which consist of two **superior colliculi** (col-LIC-you-lye; areas of visual reflexes) and two **inferior colliculi** (areas of auditory reflexes).

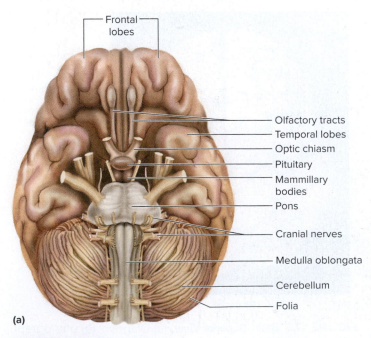

(a)

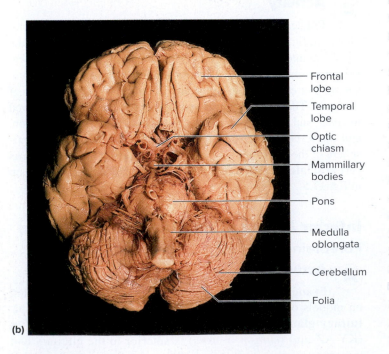

(b)

FIGURE 17.6 Brain, Inferior View. (a) Diagram; (b) photograph.
(b) ©Eric Wise

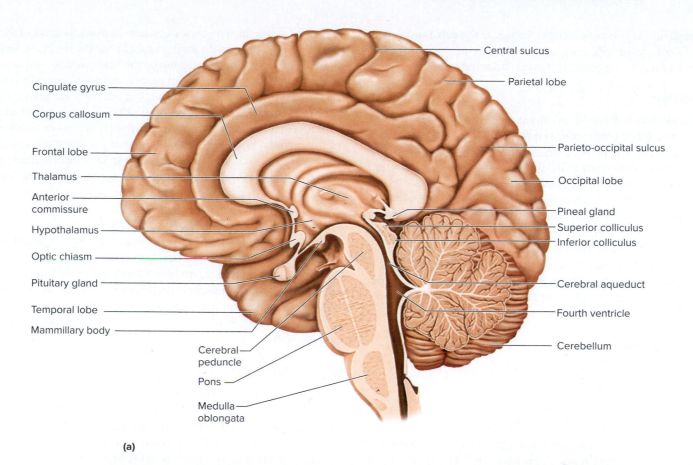

Central sulcus

Parietal lobe

Cingulate gyrus

Corpus callosum

Frontal lobe

Thalamus

Anterior commissure

Hypothalamus

Optic chiasm

Pituitary gland

Temporal lobe

Mammillary body

Cerebral peduncle

Pons

Medulla oblongata

Parieto-occipital sulcus

Occipital lobe

Pineal gland

Superior colliculus

Inferior colliculus

Cerebral aqueduct

Fourth ventricle

Cerebellum

(a)

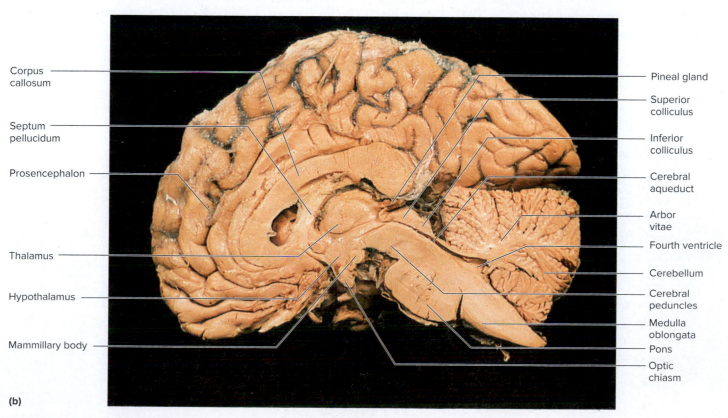

Corpus callosum

Septum pellucidum

Prosencephalon

Thalamus

Hypothalamus

Mammillary body

Pineal gland

Superior colliculus

Inferior colliculus

Cerebral aqueduct

Arbor vitae

Fourth ventricle

Cerebellum

Cerebral peduncles

Medulla oblongata

Pons

Optic chiasm

(b)

FIGURE 17.7 **Brain, Median Section.** (a) Diagram; (b) photograph.
(b) ©Eric Wise

The midbrain also houses a center known as the **substantia nigra** (sub-STAN-she-uh NY-gruh; not seen in median sections) that, when not functioning properly, causes Parkinson's disease.

Hindbrain

The **hindbrain** consists of an anterior bulge known as the **pons,** a terminal **medulla oblongata,** and the highly convoluted **cerebellum** (fig. 17.7). The pons carries sensory information from the inferior regions of the body to the thalamus. The pons has important respiratory centers involved in controlling breathing rate.

The cerebellum is a location noted for muscle coordination, the maintenance of posture, the conceptualization of the passage of time, and other cognitive functions. The cerebellum consists of an outer **cerebellar cortex** and an inner, extensively branched pattern of white matter known as the **arbor vitae.** The **folia** of the cerebellum can be seen in this section. The triangular space anterior to the cerebellum is the fourth ventricle and can be seen in fig. 17.7.

A   Activity

Examine the features of the hindbrain as described here and seen in material in the lab.

Coronal Section of the Brain

The brain consists of **unmyelinated** gray matter and **myelinated** white matter. The **gray matter** of the brain is extensive and forms the **cerebral cortex.** Most of the active, integrative processes of the brain occur in the cerebral cortex (fig. 17.8). The cerebral cortex is approximately 4 mm thick and occupies the superficial regions of the brain. The cortex is where humans do most of their "thinking" because it is the main metabolic area of the brain. Deep to the cortex is the **white matter** of the brain, which consists of **tracts** that take information from deeper regions of the brain to the cerebral cortex for processing. Tracts in the CNS are myelinated fibers and function as nerves do in the PNS. Sensory information coming from the spinal cord moves through the inferior regions of the brain and through the white matter for integration in the cerebral cortex. White matter can also take information from one region of the cerebral cortex to another for integration or from the cerebral cortex back to the spinal cord and to other parts of the body for action.

Gray matter is not restricted to the cerebral cortex, however. Deep islands of gray matter in the brain compose the **basal nuclei,** such as the **caudate nucleus, putamen,** and **globus pallidus.** Basal nuclei serve a number of functions in the brain, many of which involve subconscious processes, such as the swinging of arms while walking or the regulation of muscle tone. Basal nuclei are found not only in the cerebrum but in the thalamus and midbrain as well.

Limbic System

Part of the **limbic system** can be seen in a coronal section. The limbic system is a complex region of the brain involved in mood and emotion and has centers for feeding, sexual desire, fear, and satisfaction. The inferior portion of the limbic system has neural fibers that come from the olfactory regions of the brain. These are best seen in a model of the limbic system or in a transected brain. Examine fig. 17.9 for the major features of the limbic system.

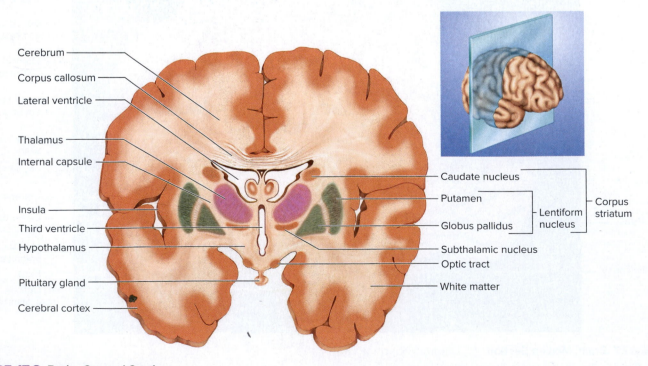

FIGURE 17.8 Brain, Coronal Section.

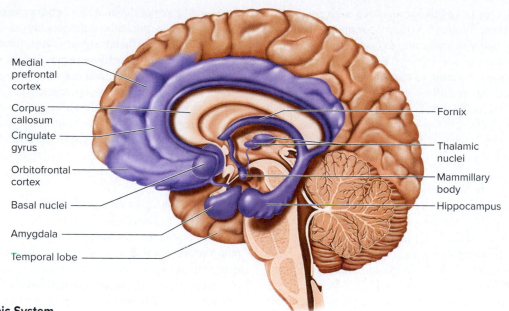

Medial prefrontal cortex

Corpus callosum

Cingulate gyrus

Orbitofrontal cortex

Basal nuclei

Amygdala

Temporal lobe

Fornix

Thalamic nuclei

Mammillary body

Hippocampus

FIGURE 17.9 **Limbic System.**

Brainstem

The brainstem consists of the brain, excluding the cerebrum and cerebellum. Look at a model or section of brain that has had the cerebrum removed. Locate the superior and inferior colliculi (fig. 17.10) along with the medulla oblongata and the pons.

Cranial Nerves

The cranial nerves are part of the PNS, but they are covered with the brain in lab because they are closely associated with it. There are 12 pairs of cranial nerves. The cranial nerves are listed by Roman numeral, and you should know the nerve by name and

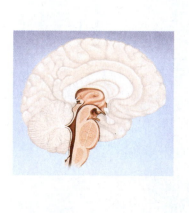

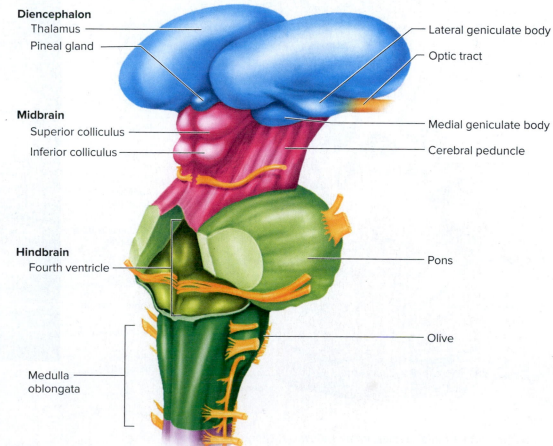

Diencephalon
Thalamus
Pineal gland

Midbrain
Superior colliculus
Inferior colliculus

Hindbrain
Fourth ventricle

Medulla oblongata

Lateral geniculate body

Optic tract

Medial geniculate body

Cerebral peduncle

Pons

Olive

FIGURE 17.10 **Brainstem, Posterolateral.**

by number. Examine a model of the brain along with fig. 17.11; the cranial nerves, except for nerve XII, are in sequence from anterior to posterior. When you study nerves, you should examine the anatomy of the brain to see where the nerve emerges from the brain. If you know, for example, that the abducens is found between the pons and the medulla oblongata, then you can use that nerve as a way to locate other nerves. If the predominant function of the nerve is to receive neural input, then the nerve is considered a **sensory nerve.** If the predominant function of the nerve is to innervate muscles, then the nerve is considered a **motor nerve.** Some nerves have significant sensory and motor functions, and these are **mixed nerves.**

The **olfactory nerves** enter the cranial cavity and terminate at the olfactory bulb. From here the olfactory tracts run along the anterior base of the brain at the inferior aspect of the frontal lobe. The **optic nerve** (*optic* refers to sight) is a sensory nerve from the eye to the base of the brain and ends at the optic chiasm. Some visual tracts remain on one side of the brain, while others cross to the other side. The **oculomotor nerve** is anterior to the pons, in the midline of the brain, while the **trochlear nerve** emerges at about a 45-degree angle from midline on the lateral aspect of the pons. The large **trigeminal nerve** is found at a 90-degree angle to the pons and is located on the lateral aspect of the pons. The **abducens nerve** is located close to the midline of the brain at the junction of the pons and medulla oblongata, while the **facial nerve** is more lateral. Lateral to the facial nerve is the **vestibulocochlear nerve,** and the nerve directly behind that is the **glossopharyngeal nerve.** The **vagus nerve** (*vagus* = wandering) is a large nerve or large cluster of fibers, and on the lateral aspect of the medulla oblongata is the **accessory nerve,** which arises from the superior spinal cord. More toward the midline is the **hypoglossal nerve,** the last of the cranial nerves.

Ⓐ Activity

Locate these nerves on material in lab and in fig. 17.11 and note their details in table 17.2.

The cranial nerves are listed in table 17.3 as sensory nerves, motor nerves, or both sensory and motor nerves. There have been many mnemonic devices constructed to remember the sequence of cranial nerves. If you use the first letter of the name of the cranial nerve, you can learn the sequence of these nerves. One such mnemonic is *Old Oliver Ogg Traveled Through Arches For Very Good Vacations And Holidays.* The first letters of the words in the mnemonic represent the first letters of the names of the nerves.

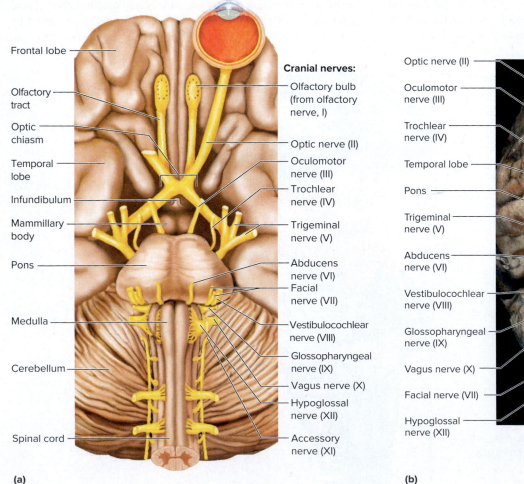

Cranial nerves:

- Frontal lobe
- Olfactory tract
- Optic chiasm
- Temporal lobe
- Infundibulum
- Mammillary body
- Pons
- Medulla
- Cerebellum
- Spinal cord
- Olfactory bulb (from olfactory nerve, I)
- Optic nerve (II)
- Oculomotor nerve (III)
- Trochlear nerve (IV)
- Trigeminal nerve (V)
- Abducens nerve (VI)
- Facial nerve (VII)
- Vestibulocochlear nerve (VIII)
- Glossopharyngeal nerve (IX)
- Vagus nerve (X)
- Hypoglossal nerve (XII)
- Accessory nerve (XI)

(a)

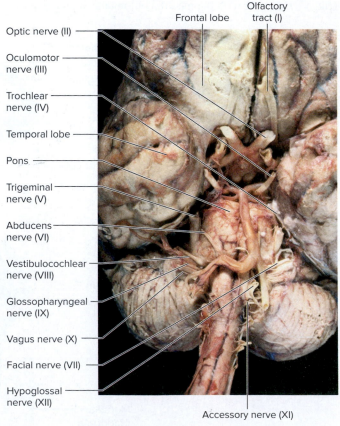

- Optic nerve (II)
- Oculomotor nerve (III)
- Trochlear nerve (IV)
- Temporal lobe
- Pons
- Trigeminal nerve (V)
- Abducens nerve (VI)
- Vestibulocochlear nerve (VIII)
- Glossopharyngeal nerve (IX)
- Vagus nerve (X)
- Facial nerve (VII)
- Hypoglossal nerve (XII)
- Frontal lobe
- Olfactory tract (I)
- Accessory nerve (XI)

(b)

FIGURE 17.11 Cranial Nerves, Inferior View. (a) Diagram; (b) photograph.
(b) ©Eric Wise

TABLE 17.2		Cranial Nerves
Number	**Name**	
I	Olfactory	**Function:** Receives sensory information from the nose
		Location: Begins in the upper nasal cavity and passes through the cribriform plate of the ethmoid bone. It synapses in the olfactory bulb on either side of the longitudinal cerebral fissure of the brain. The fibers take information on the sense of smell and pass via the olfactory tracts to be interpreted in the temporal lobe of the brain.
II	Optic	**Function:** Receives sensory information from the eye, conducting it to the brain
		Location: Takes sensory information from the retina at the back of the eye and transmits the impulses through the optic canal in the sphenoid bone. Some axons cross at the optic chiasm and pass via the optic tracts to the occipital lobe, where vision is interpreted. Other axons do not cross over the optic chiasm.
III	Oculomotor	**Function:** Conducts motor information to move the eye muscles, particularly to the medial, superior, and inferior rectus muscles and to the inferior oblique muscle
		Location: Emerges from the surface of the brain near the midline and just superior to the pons. It passes through the superior orbital fissure and innervates the inferior oblique muscle and the medial, superior, and inferior rectus muscles and carries parasympathetic fibers to the lens and iris.
IV	Trochlear	**Function:** Conducts motor information to move the eye muscles, particularly the superior oblique muscle
		Location: Seen at the sides of the pons at about a 45° angle from the midline of the brain. It passes through the superior orbital fissure to the superior oblique muscle.
V	Trigeminal	**Function:** A three-branched nerve; conducting both sensory information from, and motor information to, the face, also important for mastication (chewing)
		Location: Seen at a 90° angle from the midline at the lateral sides of the pons. The trigeminal has three branches: (1) The ophthalmic branch passes through the superior orbital fissure; (2) the maxillary branch passes through the foramen rotundum of the sphenoid bone; (3) the mandibular branch passes through the foramen ovale of the sphenoid bone and enters the mandible by the mandibular foramen and exits by the mental foramen.
VI	Abducens	**Function:** A motor nerve to move the eye muscles, particularly the lateral rectus muscle
		Location: Begins at the midline junction between the pons and the medulla oblongata and passes through the superior orbital fissure to carry motor information to the lateral rectus muscle of the eye.
VII	Facial	**Function:** A large nerve that receives sensory information from the anterior tongue and takes motor information to the head muscles. Innervates salivary and other glands.
		Location: Begins as the first of a cluster of nerves on the anterolateral part of the medulla oblongata. It passes through the internal auditory meatus and near the inner ear to the stylomastoid foramen of the temporal bone to innervate facial muscles and glands. It carries sensory information from the anterior tongue. Sensory information of the tongue is interpreted in the parietal and temporal lobes of the brain.
VIII	Vestibulocochlear	**Function:** Receives sensory information from the ear; the vestibular part transmits equilibrium information and the cochlear part transmits acoustic information
		Location: Comes from the inner ear and passes through the internal auditory meatus. The conduction passes to the pons, and hearing and balance are interpreted in the temporal lobe.
IX	Glossopharyngeal	**Function:** A mixed nerve of the tongue and throat including motor functions, that receives information on taste
		Location: Passes through the jugular foramen to innervate muscles of the throat (pharyngeal branches) and sensory receptors of the tongue. Motor portions of the nerve control a muscle of swallowing and a salivary gland, while sensory nerves carry information from the posterior tongue and from baroreceptors and chemoreceptors of the carotid artery.
X	Vagus	**Function:** Receives sensory information from the abdomen, thorax, neck, and root of the tongue; conducting motor information to the pharynx and larynx and controls autonomic functions of the heart, digestive organs, spleen, and kidneys
		Location: Passes through the jugular foramen and along the neck to the larynx, heart, lungs, and abdominal region. The sensory impulses travel in this nerve from the viscera in the abdomen, thorax, neck, and root of the tongue to the brain.
XI	Accessory	**Function:** A motor nerve to some muscles of the neck that move the head; also important for swallowing
		Location: Multiple fibers arise from the lateral sides of the superior spinal cord and pass through the jugular foramen to numerous muscles of the neck and back.
XII	Hypoglossal	**Function:** A motor nerve to the tongue
		Location: Begins at the anterior surface of the medulla and passes through the hypoglossal canal to innervate the muscles of the tongue.

TABLE 17.3	Mnemonics for Cranial Nerves			
Number	**Name**	**Name Mnemonic**	**Type**	**Mnemonic**
I	Olfactory	Old	Sensory	Sally
II	Optic	Oliver	Sensory	Sells
III	Oculomotor	Ogg	Motor	Many
IV	Trochlear	Traveled	Motor	Mangoes
V	Trigeminal	Through	Both	But
VI	Abducens	Arches	Motor	My
VII	Facial	For	Both	Brother
VIII	Vestibulocochlear	Very	Sensory	Sells
IX	Glossopharyngeal	Good	Both	Bigger
X	Vagus	Vacations	Both	Better
XI	Accessory	And	Motor	Mega
XII	Hypoglossal	Holidays	Motor	Mangoes

Dissection of the Sheep Brain

 Activity

1. Take a sheep brain back to your table along with a dissecting tray and appropriate dissection tools. If the brains still have the **dura mater,** examine this tough connective tissue coat that occurs on the outside of the brain.

2. Cut through the dura mater to examine the other meninges that occur underneath it. Deep to the **dura mater** is a filmy layer of tissue that contains blood vessels. This is known as the **arachnoid mater.** If you tease some of the membrane away from the brain, you will see that it has a cobweblike appearance in the **subarachnoid space.** In life, the subarachnoid space contains cerebrospinal fluid. Deep to this layer and adhering directly to the brain convolutions is the **pia mater,** the surface lining of the outer surface of the brain.

3. Work in pairs during the dissection of the sheep brain. Examine the features presented in the beginning of this exercise and locate the structures that appear in fig. 17.12.

4. Find the major lobes of the brain, the cerebellum, pons, and medulla oblongata. Sheep are quadrupeds, and the flexure of the brain does not occur in them as it does in humans. Sheep have a horizontal spinal cord, while humans have a vertical one. The average weight of the sheep brain is about one-tenth that of humans. The greatest difference between the two is the larger cerebrum in humans, which reflects our greater cognitive abilities.

The pons in human brains is larger than in the sheep. Sheep have an excellent sense of smell, and this is reflected in their larger olfactory bulbs.

5. Examine the inferior surface of the sheep brain. You should see the olfactory bulbs and tracts, the optic nerve, and the optic chiasm easily from this view. The pituitary gland will probably not be attached, but you should locate the infundibulum, caudad (posterior) to the optic chiasm. Locate these structures in fig. 17.13. If the sheep brain is intact, work with another student pair and decide which brain will be sectioned in the median plane (fig. 17.14) and which will be sectioned in the coronal plane (fig. 17.15).

6. For the median section, divide the brain along the length of the longitudinal cerebral fissure by cutting through the brain with your scalpel. Your cut should reflect a section illustrated in fig. 17.14.

7. From this median view, locate the corpus callosum, lateral ventricles, third ventricle, hypothalamus, pineal gland, superior and inferior colliculi, cerebellum, arbor vitae, pons, medulla oblongata, cerebral aqueduct, and fourth ventricle. Refer to models of human brains if you need to find specific structures.

8. The coronal section of a sheep brain is illustrated in fig. 17.15. Make a coronal section about midway through the cerebrum, and locate the cerebral cortex, cerebral white matter, lateral ventricles, corpus callosum, third ventricle, thalamus, and hypothalamus.

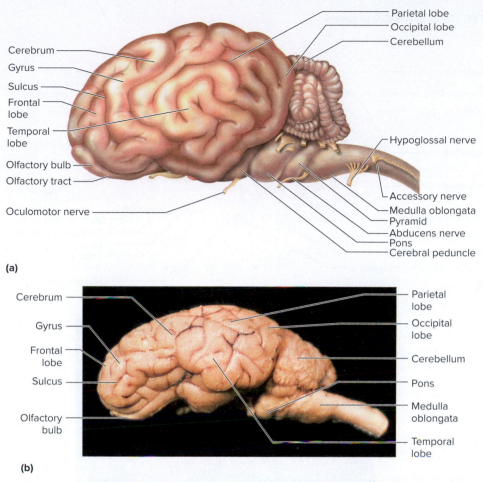

(a)

Cerebrum
Gyrus
Sulcus
Frontal lobe
Temporal lobe
Olfactory bulb
Olfactory tract
Oculomotor nerve

Parietal lobe
Occipital lobe
Cerebellum
Hypoglossal nerve
Accessory nerve
Medulla oblongata
Pyramid
Abducens nerve
Pons
Cerebral peduncle

(b)

Cerebrum
Gyrus
Frontal lobe
Sulcus
Olfactory bulb

Parietal lobe
Occipital lobe
Cerebellum
Pons
Medulla oblongata
Temporal lobe

FIGURE 17.12 Brain of the Sheep, Lateral View. (a) Diagram; (b) photograph.
(b) ©Eric Wise

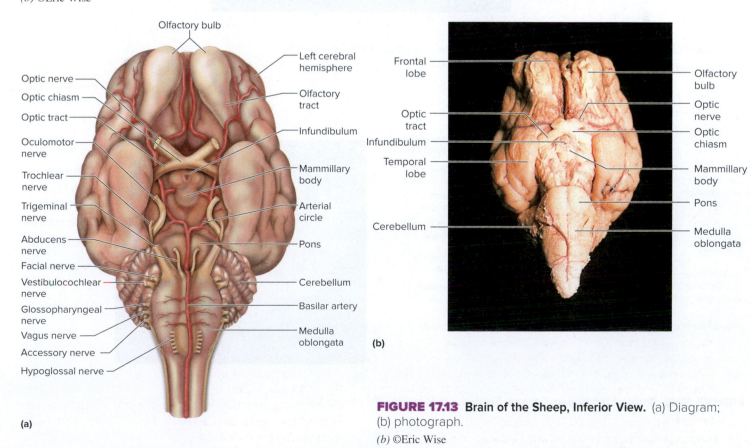

(a)

Olfactory bulb
Optic nerve
Optic chiasm
Optic tract
Oculomotor nerve
Trochlear nerve
Trigeminal nerve
Abducens nerve
Facial nerve
Vestibulocochlear nerve
Glossopharyngeal nerve
Vagus nerve
Accessory nerve
Hypoglossal nerve

Left cerebral hemisphere
Olfactory tract
Infundibulum
Mammillary body
Arterial circle
Pons
Cerebellum
Basilar artery
Medulla oblongata

(b)

Frontal lobe
Optic tract
Infundibulum
Temporal lobe
Cerebellum

Olfactory bulb
Optic nerve
Optic chiasm
Mammillary body
Pons
Medulla oblongata

FIGURE 17.13 Brain of the Sheep, Inferior View. (a) Diagram;
(b) photograph.
(b) ©Eric Wise

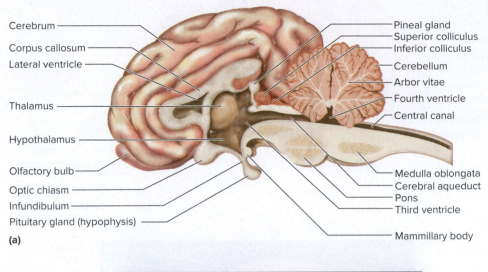

(a)

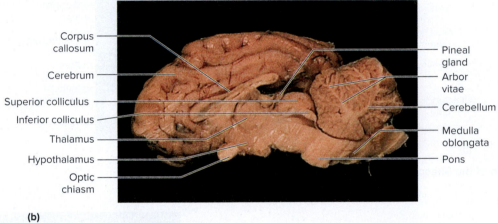

(b)

FIGURE 17.14 **Brain of the Sheep, Longitudinal Section.** (a) Diagram; (b) photograph.
(b) ©Eric Wise

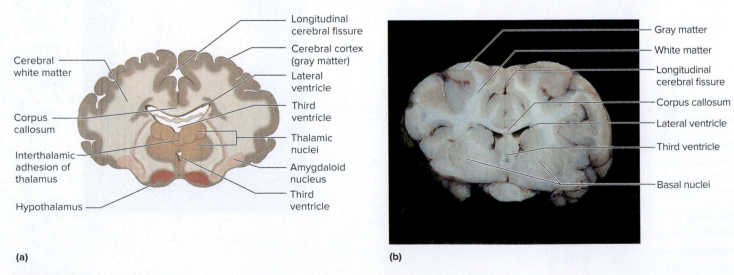

(a)

(b)

FIGURE 17.15 **Brain of the Sheep, Coronal Section.** (a) Diagram; (b) photograph.
(b) ©Eric Wise

REVIEW SECTION

The Brain and Cranial Nerves

Name _____ *Date* _____

Lab Section _____ *Time* _____

Review Questions

1. Which of the meninges is between the outer and inner meninges? _____
_____

2. What fluid is found in the ventricles of the brain? _____
_____

3. Where does fluid flow from the cerebral aqueduct? _____
_____

4. a. What is the difference between a gyrus and a sulcus? _____
_____

   b. What role do the convolutions play in the brain? _____
_____

5. Name the lobes of the cerebrum. _____
_____

6. What function does the precentral gyrus have? _____
_____

7. In addition to the sense of smell and taste, what other sense does the temporal lobe interpret? _____
_____

8. What depression separates the temporal lobe from the parietal lobe? _____
_____

9. What structure physically connects the cerebral hemispheres? _____

_____

10. Name the major regions of the midbrain. _____

_____

11. What function does the cerebellum have? _____

_____

12. What function does the optic nerve have? _____

_____

13. The trigeminal nerve is larger than the trochlear nerve. How does this correlate with the function of both nerves?

_____

_____

14. John pulled a "no-brainer" by hitting his forehead against the wall. What possible damage might he do to the function of his brain, particularly those functions associated with the frontal lobe?

_____

_____

15. One convenient excuse that people often make for their inability to do something is to describe themselves as left-brained or right-brained individuals. Describe the effect that the loss of an entire cerebral hemisphere would have on specific functions, such as

spatial awareness or the ability to speak. _____

_____

_____

16. Label the following illustration of the sheep brain using the terms provided.

arbor vitae   medulla oblongata

cerebral aqueduct   pineal gland

corpus callosum   pituitary gland

fourth ventricle   pons

hypothalamus   thalamus

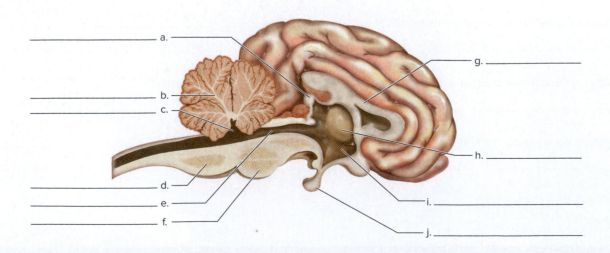

17. Label the following illustration using the terms provided.

arbor vitae   medulla oblongata

cerebral aqueduct   pineal gland

corpus callosum   pituitary gland

fourth ventricle   pons

hypothalamus   thalamus

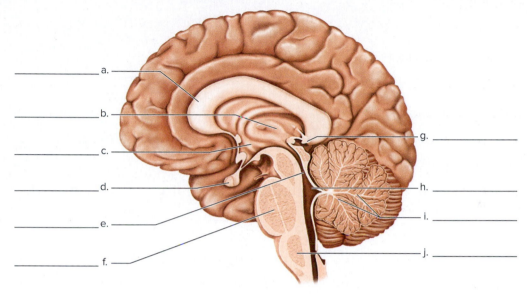

a. _____

b. _____

c. _____

d. _____

e. _____

f. _____

g. _____

h. _____

i. _____

j. _____

18. Label the following illustration using the terms provided.

cerebellum

hypoglossal nerves

medulla oblongata

oculomotor nerve

olfactory bulb

optic nerve

pons

trigeminal nerve

vagus nerve

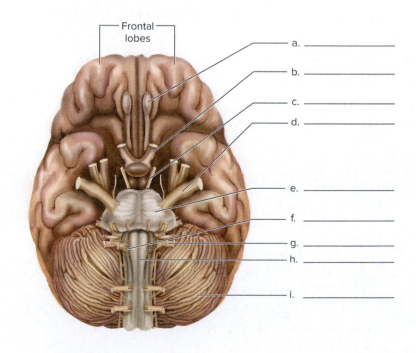

Frontal lobes

a. _____

b. _____

c. _____

d. _____

e. _____

f. _____

g. _____

h. _____

i. _____

Notes

Nervous System Physiology— Stimuli and Reflexes

INTRODUCTION

Neurons show two fundamental properties: excitability and conductivity. **Excitability** is the response of a nerve to some type of stimulus (chemical, mechanical, electrical), and **conductivity** is the movement of a nerve impulse along the length of the neuron. When nerves are stimulated by chemical, mechanical, or electrical means, voltage-regulated gates open and send electrochemical currents (action potentials) along the length of the neuron. Nerves receive information from a particular area (sense organ, regions in the CNS, etc.) and carry signals to either the brain for interpretation or an effector for some type of action. Reflexes are vital for the body in order to respond to external or internal stimuli. Visceral reflexes (such as the adjustment of the pupil to changing light conditions) occur in smooth muscle, cardiac muscle, and glands and are not covered in this exercise. Somatic reflexes involve skeletal muscles and are part of reflex arcs in the body. The nature of nerve conduction and reflexes are covered in the Saladin text in chapter 12, "Nervous Tissue," and chapter 13, "Spinal Cord and Spinal Nerves." In this exercise, you study the basic properties of neuronal conduction and their sensitivity to various stimuli and experiment with nerves as they form reflex arcs in several parts of the body.

OBJECTIVES

At the end of this exercise, you should be able to

1. describe the threshold nature of nerve responses;
2. list three things that cause a nerve to be stimulated;
3. name one substance that stimulates nerves and one that inhibits them;
4. describe reflex arcs;
5. list all the parts of a monosynaptic and polysynaptic reflex arc;
6. define *hyporeflexic* and *hyperreflexic*.

MATERIALS

Nerve Physiology Section

Ph.I.L.S. Version 4.0

Compatible computer

Frogs

Protective gloves

Glass rod with hook at one end

Hot pad or mitt

Beaker of boiling water on a hot plate

Frog Ringer's solution in dropper bottles (see Appendix B)

Microscope slide or small glass plate

Filter paper or paper towel

Dissection equipment for live animals

Scalpel or scissors

Cotton sewing thread

Stimulator apparatus with probe

Myograph transducer

Physiograph or physiology computer

Five small (50 mL) beakers

5% sodium chloride solution

0.1% hydrochloric acid solution (1 mL of concentrated HCl in 1 liter of water)

Procaine hydrochloride or other local anaesthetic solution

Cotton applicator sticks

Gauze squares

Ice bath

Reflex Section

Patellar reflex hammer

Rubber squeeze bulb

Models or charts of spinal cord and nerves

PROCEDURE   PhILS

Virtual Experiments in Neurophysiology

(A) **Activity**

Load the Physiology Interactive Lab Simulations (Ph.I.L.S.) Version 4.0.

For all of the virtual muscle experiments, you should follow the standard procedures that come up when you open the program.

1. Click on the number of the exercise you want to complete. These are under the **Resting Potentials** or **Action Potentials** headings.

After you have selected your choice:

2. Read the objectives and introduction to the lab simulation. You can click on highlighted terms to see pictures or animations of the material in question.
3. Take the pre-lab quiz. If you get a question wrong, the program will let you know and provide you with the correct answer.
4. Click on the **Wet Lab** tab and read the material. The highlighted terms open up videos of an actual procedure.
5. Click on **Continue,** which opens up the **Laboratory Exercise.**
6. Perform the exercises as outlined below. Print out any information your instructor directs after you have performed the experiment or download the information to a portable data storage device.
7. Take the post-lab quiz to determine your understanding of the lab.

The information for the specific lab exercises follows.

(A) **Activity**

Resting Potentials 10. Resting Potential and External [K⁺] Open the Resting Potential and External [K⁺] program. When the screen appears in the Ph.I.L.S. program, click on the **power** button on the virtual computer screen. Click on the **power** button on the Data Acquisition Unit and the Electrometer. You can then click and drag the blue cable to the Recording Inputs upper connection (#1). You should then click on the **Start** button, which is located in the upper right of the virtual computer screen. Follow the rest of the directions. In order to record the material in the journal, you must click on the journal and then click on the "X" to quit the journal. The micromanipulator control (up or down) is found on the circular button underneath the tube labeled "50" but at the level where the electrometer is found.

At the bottom of each tube is a stopcock indicated by a black rectangle. Click and hold the 5 mM tube until it empties. Move the microelectrode by clicking on the gray adjustment knobs. These are located to the right of where the orange wire attaches to the microelectrode (above and to the right of the micromanipulator control knob). Record the data in the journal and continue to move the microelectrode and take readings for 5, 10, 20, 50, and 100 mM KCl solutions by sequentially opening the stopcock for the respective fluids and taking readings.

What impact does increasing the potassium ion concentration have on membrane depolarization? Do you think that increasing potassium levels would speed up or slow down the transmission rate? Record your answer in the following space and in the review section at the end of the exercise.

❓ Effect of increasing potassium ion concentration:

_____ (1)

(A) **Activity**

Resting Potentials 11. Resting Potential and External [Na⁺] Follow the same procedure as in the Resting Potential and External [K⁺] for the **Resting Procedure and External Sodium Concentration** program. The setup is the same except that you will be starting with a more concentrated level of sodium and decreasing the sodium ion concentration rather than the potassium ion concentration.

Comparing the two solutions, which has a greater impact on membrane dynamics: sodium or potassium? Record your answer in the following space and in the review section at the end of the exercise.

❓ Greater impact on membrane dynamics: _____ (2)

(A) **Activity**

Action Potentials 12. The Compound Action Potential Open the program, read the introduction, and answer the pre-lab questions. Open the **Wet Lab** and click on the power switch to the virtual computer screen and the Data Acquisition Unit. Place all of the color-coded cables on the appropriate attachment points of the chamber holding the nerve. The drain tap is the small circular wheel located at the top of the chamber on the left-hand side. You must click and hold the knob in order for it to drain.

If you examine the graph, you will see the stimulus on the left side of the graph followed by a compound action potential on the right. As with the muscle physiology experiment, if you move the cursor to the graph you will see crosshairs as you move the cursor. Place the crosshairs on the top of the curve and click. Click again on the baseline of the curve. When you click on the **Journal** icon, it will enter the data into the journal.

You must click the "X" on the journal in order to proceed with obtaining more data. The threshold voltage is the voltage below which you get no response from the nerve. What was your threshold voltage? If there were no threshold for neuron conduction, how might this impact how nerves respond to environmental stimuli? Record your answer in the following space and in the review section at the end of the exercise.

❓ Threshold voltage: _____ (3)

(A) Activity

Action Potentials 13. Conduction Velocity and Temperature

In this exercise, you examine the speed of nerve conduction by altering the temperature of the nerve. The experiment runs at room temperature (22°C) and cold temperature (10°C). Follow the instructions as presented. Begin by clicking on the power switch to the Temperature Unit. Make sure you click on all power buttons to turn on the equipment. Insert all of the plugs into their color-coded receptacles. Place all of the color-coded cables on the attachment points on the chamber holding the nerve. The tap is the small circular wheel located at the top of the chamber on the left-hand side. You must click and hold the knob in order for it to drain. Follow the directions to decrease the length of the space between the electrodes. Move the cross hairs and note the conduction velocity. Record this velocity below.

? Conduction velocity: _____ (4)

After you run the first experiment at room temperature, make sure you enter your data into the journal by clicking on the **Journal** icon. Run the second experiment (make sure you drain the tap and adjust the electrodes) and compare the results of nerve conduction at different temperatures.

In which sequence (room temperature or cold) did you see a slowing down of the nerve conduction? Remember that a lower conduction velocity refers to a slower speed. Record your answer in the following space and in the review section at the end of the exercise.

? Slower nerve response: _____ (5)

How does this response correlate with decreasing enzyme function based on temperature? Record your answer in the following space and in the review section at the end of the exercise.

? Correlation: _____ (6)

Frog Nerve Conduction

Review the structure of the neuron in Exercise 16 for descriptions of the dendrites, neurosoma, and axon, and the anatomy of the nerve. Make sure you read through all of this exercise prior to beginning the experiments. Wear protective gloves as a general precaution when working with fresh specimens, such as frogs.

In the first part of this lab exercise, you will determine if nerves respond to only a specific stimulus or if they are more general and respond to many stimuli. You will also determine if certain materials or environmental conditions inhibit nerve response.

(A) Activity

In this exercise, you observe the process of nerve impulse conduction by experimenting on frogs or by watching a demonstration, depending on the wishes of your instructor. If you experiment on frogs, obtain a doubly pithed frog, dissection equipment, frog Ringer's solution, and various test solutions and take them to your table. If you are to pith a frog, refer to Exercise 15 in the *Real Frog Experiment* section. Keep the frog nerve preparation moist with

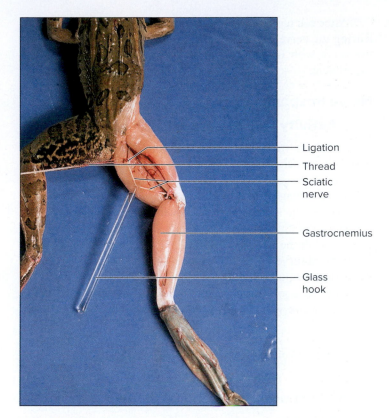

FIGURE 18.1 **Ligation of the Sciatic Nerve of the Frog.**
©Eric Wise

Ringer's solution during the entire experiment. Prepare the frog by cutting the skin away from the hip (fig. 18.1). Do not cut, pinch, or otherwise damage the sciatic nerve on the posterior side of the thigh.

Gently remove the nerve from between the muscles with a glass rod (fig. 18.1). Do not stretch the nerve; leave it intact alongside the muscles. You can attach the gastrocnemius to a myograph transducer, as you did in Exercise 15, or just examine the muscle for a response.

Nerve Response to Physical Stimuli

(A) Activity

Flush the nerve with frog Ringer's solution and make sure it remains moist. Nerve stimulation can be determined by the corresponding muscular contraction. You measure the effects of the nerve stimulation by the contraction of the gastrocnemius muscle. If the nerve is stimulated, then the gastrocnemius muscle should contract.

Cut a small (10 cm) section of cotton thread and gently slip it under the sciatic nerve with a pair of fine forceps. Gently move the thread up toward the hip. When you reach the proximal location where the nerve descends into the muscle, tie off the nerve with thread and examine the effects. Loop the thread and ligate (tie off) the nerve close to the sacrum (fig. 18.1) while watching the gastrocnemius muscle.

As the thread begins to tighten on the nerve, record the response in the space provided and in the Chapter Summary Data section.

? Response of the nerve to physical stimulation:

_____ (7)

After you have ligated the nerve, cut it from the anterior side, leaving the nerve attached to the gastrocnemius muscle. Moisten the nerve with frog Ringer's solution and prepare for the next experiment.

Nerve Response to Electrical Stimuli

Ⓐ Activity

Place a stimulator probe connected to a stimulator underneath the sciatic nerve, lifting the nerve away from the gastrocnemius muscle. Keep the nerve moist as you determine the minimum voltage **(threshold voltage)** required for nerve conduction. Set the stimulator to a frequency of two pulses per second and a duration of 10 milliseconds. The voltage should be set at zero (with the knob on 0.1 volt). Slowly increase the voltage until you observe the gastrocnemius twitch. As soon as you see the gastrocnemius twitch at the *lowest voltage,* record this as the threshold voltage in the space provided. Turn the voltage to zero, flush the nerve with frog Ringer's solution, and let it rest for a moment.

❓ Threshold voltage: _____ (8)

Nerve Response to Chemical Stimuli

Ⓐ Activity

In the following two experiments, you test the response of the nerve to different chemical agents. Make sure you observe the nerve as soon as you apply the solution and rinse it as soon as the observation is made.

Acid Solution   Apply a 0.1% hydrochloric acid solution to a cotton applicator stick and gently touch the applicator to the nerve. Record the nerve response.

❓ Response to hydrochloric acid: _____ (9)

Flush the nerve with frog Ringer's solution and let it rest for a moment.

Salt Solution   Gently apply a 5% sodium chloride solution to the nerve with a new cotton applicator stick. Record the response. Flush the nerve with frog Ringer's solution and let it rest for a moment.

❓ Response to sodium chloride solution: _____ (10)

Nerve Response to Anesthetics

Ⓐ Activity

Apply a solution of **procaine hydrochloride** (Novocain) or other local anaesthetic to the nerve by soaking a small square of gauze or cotton with anaesthetic solution and placing it on the nerve for a moment. As the gauze remains on the nerve, set up the stimulator apparatus and place the nerve over the stimulator probes. Remove the gauze and stimulate the nerve with a single pulse stimulus at the voltage that produced a twitch in the previous

experiment. If the nerve responds to the stimulus, leave the anaesthetic on longer. When the nerve does not respond, remove the gauze, flush with frog Ringer's solution, and stimulate the nerve once every 30 seconds until it recovers from the local anesthetic. Keep the nerve moist at all times with frog Ringer's solution. Record the recovery time.

❓ Recovery time: _____ (11)

Nerve Response to Changes in Temperature

Ⓐ Activity

Gently touch the nerve with a glass rod at room temperature. Record the response.

❓ Response to gentle touch: _____ (12)

Place the glass rod in an ice bath. As it equilibrates in the ice water, place a small chip of ice on the nerve and let it stay there for a moment. Gently touch the nerve with the cold rod and record the response. Flush the nerve with frog Ringer's solution and let it rest for a moment.

❓ Response to gentle touch with cold stimulation:

_____ (13)

Take another glass rod in a hot pad or mitts and place the rod in a beaker of boiling water for 30 seconds. Touch the nerve with the hot end of the glass rod. What is the response? Record your result.

❓ Response to gentle touch with hot stimulation:

_____ (14)

_____

Clean Up   Make sure you clean your station before continuing. Place the specimen in the appropriate container, and use care when cleaning sharp instruments, such as scalpels.

_____

Reflexes

Reflexes are involuntary, predictable responses to stimuli. A reflex involves a motor response to a stimulus produced without conscious thought. There are reflexes that pass through the spinal cord, known as **spinal reflexes,** and those that pass through the brain, known as **cranial reflexes.** The patellar reflex is a spinal reflex, while blinking in response to dust in the eye is a cranial reflex. Reflexes occur through **reflex arcs,** and these arcs have the following structure:

1. **Receptor** (structure that receives the stimulus and converts it to an action potential)

2. **Afferent** (sensory) **neuron** (the neuron taking the stimulus to the CNS)
3. **Integrating center** (brain or spinal cord)
4. **Efferent** (motor) **neuron** (the neuron taking the response from the CNS)
5. **Effector** (the structure causing an effect)

If the effector is skeletal muscle, it is a **somatic reflex.** If the effector is a gland, smooth muscle, or cardiac muscle, it is a **visceral,** or **autonomic, reflex.**

Most reflexes involve many neurons with many synapses and are called **polysynaptic reflex arcs.** A few reflexes involve only two neurons—a sensory neuron and a motor neuron with one synapse

between them. These are **monosynaptic reflex arcs.** They are illustrated in fig. 18.2.

Reflexes depend on a **stimulus,** or environmental cue; a **receptor** sensitive to the stimulus; an **afferent neuron** (or sensory neuron); an **efferent neuron** (or motor neuron); and an **effector.** The polysynaptic reflex has these structures, as well as an **integrating center** consisting of one or more **interneurons** located between the afferent and efferent neurons. Some interneurons take information to the brain via ascending tracts of the spinal cord.

Testing for reflexes is important for clinical evaluation of the condition of the nervous system. Decreased response or exaggerated response to a stimulus may indicate disease or damage to the nervous system. In this experiment, you test several reflexes

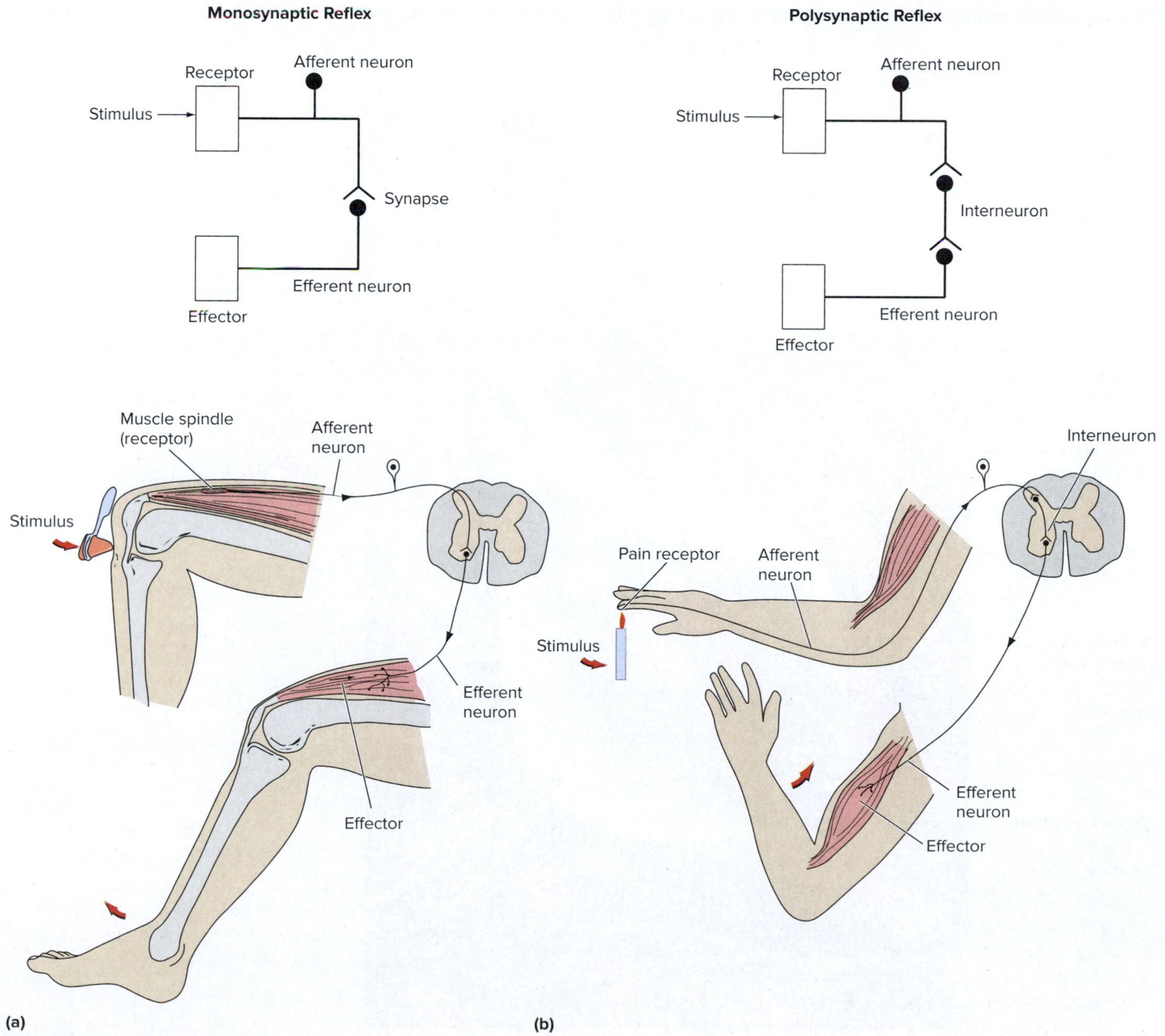

FIGURE 18.2 Reflex Arcs. (a) Monosynaptic; (b) polysynaptic.

and determine if the response is **normal** (movement of an inch or two), **hyporeflexic** (showing less than average response), or **hyperreflexic** (showing an exaggerated response). In clinical settings, hyperreflexia might indicate central nervous system damage. No reflex may indicate spinal cord damage, and hyporeflexia may be indicative of hypothyroidism.

Stretch Reflexes

Receptors for stretch reflexes are in the **muscle spindle** (fig. 18.3). There are many types of nerve fibers in a muscle spindle. When a muscle is stretched, many muscle spindles are stimulated, and **sensory nerve fibers** send strong signals to the spinal cord, resulting in contraction of the muscle that was stretched. This dampens the effect of stretching the muscle. Alpha motor neurons innervate the main muscle tissue and are primarily responsible for muscle contraction, while gamma motor neurons innervate the muscle

spindle and regulate the sensitivity of the spindle. These allow you to make changes in the contraction of muscle such as those required when trying to maintain your balance.

Stretching the muscle causes an increase in action potentials in the sensory neuron. The impulse travels to the motor neuron, causing contraction of the muscle that is stretched. A typical example of a stretch reflex is the patellar reflex, in which striking the patellar ligament stretches the quadriceps muscles. Sensory neurons transmit this information to the spinal column, where motor neurons stimulate the quadriceps muscle to contract, thus extending the leg.

Patellar Reflex

The **patellar reflex** tests the femoral nerve. Sensory neurons in the quadriceps muscle are stimulated by rapid lengthening of the muscle and conduct nerve impulses to the spinal cord.

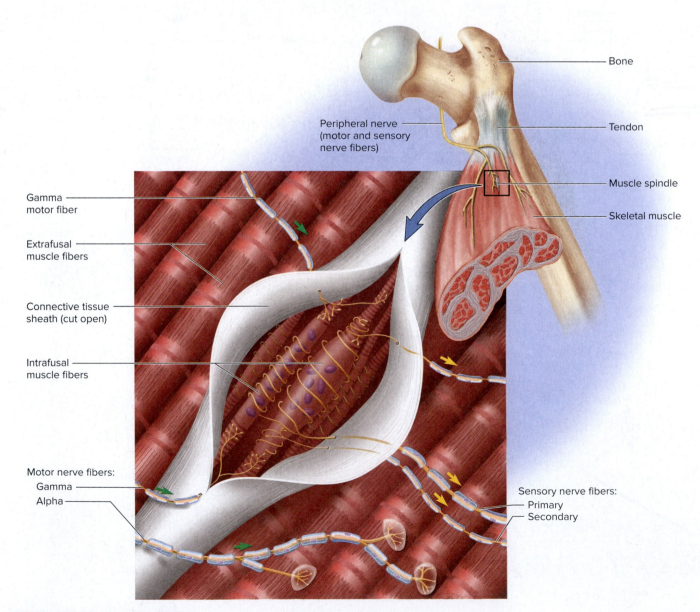

FIGURE 18.3 A Muscle Spindle.

Sensory neurons synapse with motor neurons that innervate muscles, stimulating them to resist the stretch. It is the most frequent reflex test performed in clinical settings.

(A) Activity

Sit on the lab table with your leg hanging over the edge and have your lab partner tap you on the patellar tendon with a patellar reflex hammer. The percussion should be placed about 3 to 4 cm below the inferior edge of the patella, and it should be firm but not hard enough to hurt. Look for extension of the leg as a response to the patellar reflex (fig. 18.4).

The tap stimulates the stretch receptors in the muscle and is representative of a monosynaptic reflex. Record the degree (hyperreflexic, normal, hyporeflexic) of the response.

? Patellar reflex: _____ (15)

In CNS damage, the dampening response to muscle contraction is reduced and often a hyperreflexic response is seen.

Triceps Brachii Reflex

(A) Activity

The **triceps brachii reflex** tests the seventh cervical nerve. Have your lab partner sit on a chair or lie down on his or her back on a clean lab table or cot resting his or her forearm on the abdomen. Tap the distal tendon of the triceps brachii muscle about 2 inches proximal to the olecranon. Look for the triceps muscle to twitch (fig. 18.5). Record your result.

? Triceps brachii reflex: _____ (16)

Biceps Reflex

(A) Activity

The **biceps reflex** tests the fifth cervical nerve. Sit comfortably and place your finger on the biceps tendon on the medial side of the cubital fossa (fig. 18.6).

Tap your finger with the pointed end of the reflex hammer, while they remain on the tendon, and look for the biceps brachii muscle contraction. Record your results.

? Biceps reflex: _____ (17)

Calcaneal (Achilles) Tendon Reflex

(A) Activity

To test the **calcaneal tendon reflex,** have your lab partner kneel on a chair with the foot dangling over the edge (fig. 18.7). Put a little pressure on the plantar surface (pad side) of the toes and tap the calcaneal tendon to test the first sacral nerve.

As you tap the calcaneal tendon, look for plantar flexion of the foot. You may see an initial movement of the foot due to the depression of the tendon by the reflex hammer, but there should

FIGURE 18.4 Patellar Reflex.
©Eric Wise

FIGURE 18.5 Triceps Reflex.
©Eric Wise

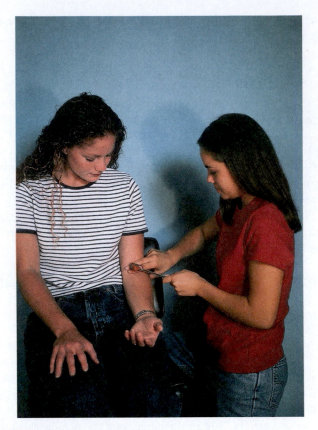

FIGURE 18.6 Biceps Reflex.
©Eric Wise

FIGURE 18.7 Calcaneal Tendon Reflex.
©Eric Wise

be a slight pause and then another quick movement of the foot. Record your results.

? Calcaneal tendon reflex: _____ (18)

Eye Reflexes

 Activity

The automatic blinking of the eye is a reflex and is important to keep material, such as dust, away from the outer layer of the eye, the cornea. In the first part of this experiment, have your lab partner try to make you blink by flicking his or her fingers near your eyes. They should not come close enough to touch your eyes. Can you prevent the blinking response? Record your answer.

? Control of blink reflex: _____ (19)

Now have your lab partner take a *clean* rubber squeeze bulb (a large pipette bulb works well) and squirt a sharp blast of air across the surface of the eye (fig. 18.8). This is a corneal reflex.

Caution! Use either new bulbs or ones free of debris to avoid damage to the eye.

CAUTION

Can you inhibit this response? Record your response.

? Control of corneal reflex: _____ (20)

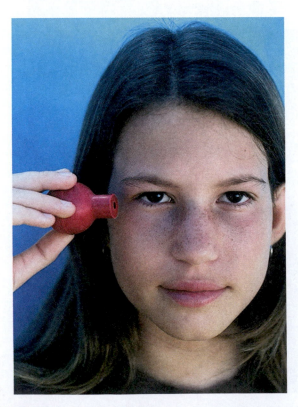

FIGURE 18.8 Corneal Reflex.
©Eric Wise

Plantar Response, or Babinski Reflex

Ⓐ Activity

Using the *metal end* of the patellar hammer, stroke the foot from the heel along the lateral, inferior surface and then toward the ball of the foot (fig. 18.9). The pressure should be firm but not uncomfortable.

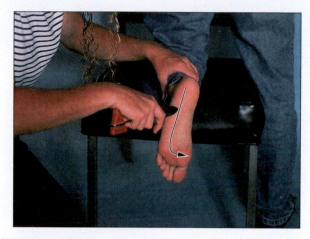

FIGURE 18.9 Babinski Reflex.
©Eric Wise

Stroking the plantar surface normally results in flexion of the toes in adults. Damage to pyramidal tracts causes extension of the big toe (known as the Babinski reflex). This is important for determining spinal damage. Adults normally do not show a Babinski reflex, and the normal response in adults is to flex the foot and toes. The plantar response, or Babinski reflex, occurs in newborns and is normal. It is seen when the extension of the big toe or the fanning of the toes occurs when the plantar surface of the foot is stroked. Once myelination of the nerves occurs, the Babinski reflex disappears and is not normally found in the adult. Babinski responses in the adult are indicative of CNS problems.

Notes

REVIEW SECTION

Nervous System Physiology— Stimuli and Reflexes

Name _____   Date _____

Lab Section _____   Time _____

? Chapter Summary Data

Use this section to record your results from questions within the exercise.

1. _____   11. _____

2. _____   12. _____

3. _____   13. _____

4. _____   14. _____

5. _____   15. _____

6. _____   16. _____

7. _____   17. _____

8. _____   18. _____

9. _____   19. _____

10. _____   20. _____

Review Questions

1. What structure receives stimuli from the external environment and relays those stimuli to the afferent neuron?

   _____

2. What is another name for an efferent neuron? _____

3. What is a reflex? _____

_____

_____

4. What kind of reflex has only two neurons? _____

5. Polysynaptic reflexes have a neuron specific to them. What is the name of that neuron?

_____

6. In numbers of synapses, what kind of reflex is a patellar reflex? _____

_____

7. After patients leave the operating room, they are transferred to an area called the "recovery room." Correlate the meaning of the word *recovery* in this context with what you have learned about the recovery of nerves in this exercise. _____

_____

8. Draw a monosynaptic reflex arc in the space provided. Label your illustration using the terms provided.

   effector

   motor neuron

   receptor

   sensory neuron

   synapse

9. What action occurs with a hyperreflexic response? What action happens with a hyporeflexic response?

_____

_____

10. List the positive responses obtained in the frog experiment, and correlate this with the specificity of neuronal sensitivity. _____

_____

11. What was the threshold voltage observed in the nerve response? _____

_____

12. Based on your experiments, are nerves generally sensitive to few stimuli or many? _____

_____

Cutaneous Senses

INTRODUCTION

The gateway to understanding our world comes from our ability to sense the environment within our bodies and the environment around us. There are two main classes of sense—**general (somesthetic) senses** and **special senses.** General senses occur in many locations of the body and are found in places such as the skin, muscle, joints, and viscera. The senses of touch, pressure, changes in temperature, pain, blood pressure, and stretching are general senses. Special senses occur in specific locations, such as the eye, ear, tongue, and nose, and include taste, smell, sight, hearing, and balance.

Sense receptors are not uniformly distributed throughout the body. In some areas, specific sense receptors are absent, or few in number, while in other areas they are densely clustered. This pattern of uneven distribution is called **punctate distribution.**

Our perception of the environment is dependent on environmental **stimuli.** These stimuli are classified by type, or **modalities,** such as light, heat, sound, pressure, and specific chemicals. **Receptors** are the receiving units in the body that respond to an adequate stimulus. They transform the stimulus to neural signals transmitted by sensory nerves and neural tracts to the brain, which interprets the message. If any link in this sensory chain is broken, or if the stimulus is not sufficient, the perception of stimuli cannot occur.

Receptors respond to specific modalities, and each receptor can be classified according to the type of stimulus it responds to. **Photoreceptors** detect light (for example, the retina in the eye); **thermoreceptors,** located in the skin and other areas, detect changes in temperature; **proprioceptors** detect changes in the position of the body and in tension, such as those in tendons when a muscle contracts; **pain receptors,** or **nociceptors,** are present as naked nerve endings throughout much of the body; **mechanoreceptors** are receptive to mechanical stimuli (for example, touch receptors or receptors in the ear that respond to sound or motion); **baroreceptors** respond to changes in blood pressure; and **chemoreceptors** respond to changes in

the chemical environment (for example, taste and smell). These receptors are discussed in the Saladin text in chapter 16, "Sense Organs."

The skin has several types of receptors and therefore makes a good starting point for understanding sense organs. There are receptors for pressure, pain, temperature, and light touch in your skin. You may want to review the major sensory receptors in the skin, such as **tactile (Meissner)** corpuscles, **lamellar (pacinian)** corpuscles, and pain receptors in Laboratory Exercise 6 before you begin this exercise.

OBJECTIVES

At the end of this exercise, you should be able to

1. define modality and receptor;
2. list the major receptor types in the body;
3. compare and contrast general senses with special senses;
4. distinguish between tonic and phasic receptors;
5. define adaptation in reference to a stimulus;
6. distinguish between relative and absolute determination of stimuli;
7. define *referred pain.*

MATERIALS

Blunt metal probes

Dishpan (or large finger bowls) of ice water (2 L)

Dishpan of room temperature water

Dishpan of warm water (45°C)

Towels

Small centimeter ruler

Black, washable, fine-tipped felt markers

Red, washable, fine-tipped felt markers

Blue, washable, fine-tipped felt markers

Two-point discriminators (or a mechanical compass)

Three lab thermometers

Hand lotion

Von Frey hairs (horsehair glued on a wooden stick)

Tweezers

PROCEDURE

Mapping Fine-Touch Receptors

Tactile corpuscles respond to light touch. You can map these receptors by testing the ability of your lab partner to determine if they have been touched by a hair.

Ⓐ **Activity**

1. Draw a square, 2 cm on a side, with a black, washable marker on the anterior surface of the forearm. If you apply a little hand lotion to the skin before doing this test, the ink comes off more easily after the experiment.
2. Use a Von Frey hair (a stiff bristle hair attached to a matchstick) to map the number of areas in the square that can be perceived by your lab partner. Your lab partner should close his or her eyes during the experiment. Press only until the hair bends slightly to stimulate the tactile corpuscles.
3. Record the location of each positive result with a small dot from the marker. How many positive responses did you get in the square on the forearm? Record your result in the following space and in the Chapter Summary Data section at the end of the exercise.

❓ Number of anterior forearm responses: _____ (1)

4. Repeat the experiment on the *posterior* surface of the *arm* using another square 2 cm on a side. How does the number of receptors here compare to those on the forearm? Record your results.

❓ Number of responses on posterior side of the arm:

_____ (2)

Two-Point Discrimination Test

The sensitivity of touch is dependent on the number of tactile corpuscles per unit area of the skin. You can map the relative density of the receptors in the skin by performing a two-point discrimination test. The idea behind the test is to determine the *minimum distance* that your lab partner is able to recognize as two points. This is illustrated in fig. 19.1.

Ⓐ **Activity**

1. Have your lab partner sit with eyes closed and his or her hand, palm up, resting on the lab counter.

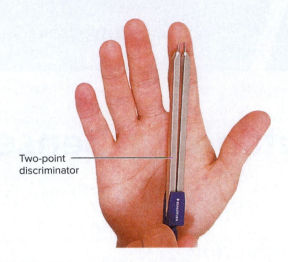

Two-point discriminator

FIGURE 19.1 **Two-Point Discrimination.**
©Eric Wise

2. Using the two-point discriminator (be careful as some two-point discriminators can have sharp points), or a mechanical compass, touch your lab partner's fingertip simultaneously with both points of the instrument and see if your lab partner can sense one or two points. In order to establish an accurate reading, make sure you gently touch both points of the discriminator at the same time. If you touch with one point of the discriminator and then another, your lab partner may perceive two points in time and not in space.
3. A good way to establish accuracy is to occasionally touch just one of the points on your lab partner's fingertip. Another way is to vary the spread of the discriminator. You might begin with 3 cm and then adjust it to 0.5 cm followed by a 2 cm spread. Record your results.

❓ Minimum distance perceived as two points on

fingertip: _____ (3)

4. Now move to the posterior surface of the arm. Establish the minimum distance that is perceived as two points by your lab partner. Record your results.

❓ Minimum distance perceived as two points on the

posterior arm: _____ (4)

❓ Is there a difference between the fingertip and the

posterior arm? _____ (5)

❓ If there is, how might the difference in distance perceived

be explained in terms of the number of nerve endings per

unit area? _____ (6)

_____

5. Now try the palm and then the back of the shoulder or neck. Record the results.

? Minimum distance perceived as two points on palm:

_____ (7)

? Minimum distance perceived as two points on back of

shoulder or neck: _____ (8)

Mapping Temperature Receptors

The skin has receptors that are sensitive to cool or warm temperatures. In this part of the exercise, you determine the relative numbers of these receptors.

Ⓐ Activity

1. Mark off a square that is 2 cm on a side on the anterior forearm of your lab partner using a washable marker.
2. Take two blunt metal probes and place the tip of one in an ice water bath and another in a warm water bath (45°C).
3. Let the probes reach the temperature of each bath, which should take a few minutes.
4. Lab partners should close their eyes and rest one of their arms on the lab counter.
5. Remove one of the probes and quickly wipe it on a clean towel. Test the ability of your lab partner to distinguish between cool and warm by using the tip of the blunt probe on your lab partner's forearm. Systematically test areas in the square. When your lab partner perceives cold (not just touch) in a location, mark it with a blue felt-tipped marker. Retest the area with the warm probe and make a dot with a red felt-tipped marker in the location where warm is perceived.

? Number of cool receptors in the square: _____ (9)

? Number of warm receptors in the square: _____ (10)

? What is the ratio of blue to red (cold/warm receptors)

in the square? _____ (11)

Adaptation to Touch

Receptors can be classified by the length of time it takes for them to adapt to a stimulus. Sensory adaptation occurs when neural firing decreases due to exposure to a stimulus, thus allowing you to adjust to the environmental stimulus. **Tonic receptors** continuously perceive stimuli (they adapt very slowly), while **phasic**

receptors perceive the stimulus initially and then adapt more quickly. In this experiment, you try to determine if the sense of fine touch is tonic or phasic.

Ⓐ Activity

1. Cut a small piece of paper, about 2 cm on a side, and crumple it into a small ball the size of a pea.
2. Have your lab partner close their eyes and place the hand, anterior side up, comfortably on the lab desk.
3. With a pair of tweezers, place the ball of paper on your lab partner's palm. Is the paper ball perceived after a few seconds?
4. Record your results in the following space and determine whether the sense of light touch is tonic or phasic.

? Results: _____ (12)

Locating Stimulus with Proprioception

In this exercise, you use washable markers of two colors. Location of the stimulus is dependent on both skin receptors and cerebellar function.

Ⓐ Activity

Have your lab partner close their eyes and rest a forearm on the lab counter. Touch your lab partner's forearm with a felt marker and have your lab partner try to locate the same spot with a felt marker of another color. Test at least five locations on various parts of the forearm, and repeat each location at least twice. Now try the fingertip and palm of the hand and record the result.

? Maximum distance error on the forearm: _____ (13)

? On the fingertip: _____ (14)

? On the palm: _____ (15)

Ⓐ Activity

Another method to test proprioreception is to close your eyes and *gently* try to touch the lateral corner of your eye with your fingertip. Have your lab partner watch you and determine the accuracy of your attempt. While your eyes are still closed, bring your hand far behind your head and then try to touch the bottom part of your earlobe or the exact tip of your chin. Record the error distance, if any, for each location.

? Corner of eye: _____ (16)

? Earlobe: _____ (17)

? Tip of chin: _____ (18)

Temperature Judgment

In this exercise, you examine the *adaptation* of thermoreceptors to temperature and whether the determination of temperature is absolute or relative. If the determination is absolute, you can distinguish the exact temperature, whether the surrounding temperature is warm or cold.

(A) **Activity**

On the lab counter, locate three dishpans or large finger bowls full of water. One bowl, located on the right, should be marked "Cold" (it should be about 10°C); another bowl, located on the left, should be marked "Warm" (it should be about 40–45° C); and the middle bowl should be marked "Room Temperature." Place one hand in the cold dish and the other in the warm dish and let them adjust to the temperature for a few minutes. If your hand begins to ache in the cold water, you may remove it for a short time, but try to keep it in the cold water for as long as possible during the adjustment time. After your hands have equilibrated, place them both in the room temperature water and describe to your lab partner the temperature (cold, warm, hot) of the water as sensed by each hand. How does the hand that was in cold water feel in the room temperature water, and how does the hand that was in warm water feel in the room temperature water? Record your results.

❓ Cold hand perception: _____ (19)

❓ Warm hand perception: _____ (20)

If the determination of temperature is absolute, the room temperature water should feel the same whether you are testing it with your cold hand or your warm hand. If it is relative, the room temperature water should feel warmer with your cold hand and colder with your warm hand.

❓ Is the determination of the temperature of water

absolute or relative? _____ (21)

❓ How did your experiment prove this? _____ (22)

_____

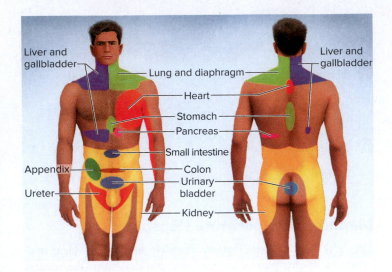

Liver and gallbladder — Lung and diaphragm — Heart — Stomach — Pancreas — Small intestine — Colon — Urinary bladder — Kidney — Appendix — Ureter — Liver and gallbladder

FIGURE 19.2 Referred Pain.

Referred Pain

Referred pain is the perception of pain in one area of the body when the pain is somewhere else. An example of referred pain is the pain felt in the left shoulder and arm when a person is suffering from a heart attack or chest pain (angina pectoris) (fig. 19.2).

Frequently, neural impulses have convergent pathways where nerves receiving stimuli from the skin or muscle in one area follow the same general ascending tract that sensory information from an organ (such as the kidney or gallbladder) do. If there are pain stimuli in that organ, an individual may feel the pain in the skin that shares the tract.

(A) **Activity**

Place your elbow into a dish of ice water and leave it there for 2 painful minutes. Describe the sensation you feel and the location of the sensation. Record your results.

❓ Description of sensation: _____ (23)

❓ Initial location of sensation: _____ (24)

❓ Sensation after 2-minute period: _____ (25)

REVIEW SECTION

Cutaneous Senses

Name _____ Date _____

Lab Section _____ Time _____

❓ Chapter Summary Data

Use this section to record your results from questions within the exercise.

1. _____

2. _____

3. _____

4. _____

5. _____

6. _____

7. _____

8. _____

9. _____

10. _____

11. _____

12. _____

13. _____

14. _____

15. _____

16. _____

17. _____

18. _____

19. _____

20. _____

21. _____

22. _____

23. _____

24. _____

25. _____

Review Questions

1. An area with a great number of fine touch receptors is the upper lip. What can you predict about the ability of the upper lip to distinguish two points? _____

   _____

2. Cool receptors in the skin are activated between 12° and 35°C. Warm receptors are activated between 25° and 45°C. When the temperature is below 12°C or above 45°C, pain receptors in the skin are activated. You or your lab partner may have had an experience with very cold conditions, such as when cleaning out a freezer or holding an ice cube. What perception is sensed?

   _____

3. Adaptation is important to sensory stimulation. We are bombarded with stimuli during most of the day, and much of what we sense is filtered from conscious thought. How is adaptation used by pickpockets?

   _____

   _____

4. In terms of receptor density, describe why it is difficult to find the same location on the forearm when your eyes are

   closed. _____

   _____

5. In regard to sense organs, what is punctate distribution? _____

   _____

6. In reference to the sense organs, what is a modality? _____

   _____

7. What types of receptors are sensitive to the following modalities?

   a. light _____   d. sound _____

   b. touch _____   e. smell _____

   c. temperature _____

8. What type of receptor is responsive to extremely hot sensations? _____

9. Tactile corpuscles respond to what type of sensation? _____

10. What type of receptor determines the weight of an object when you pick it up? _____

11. Which type of receptor (tonic/phasic) adapts to light in a darkened movie theater? _____

12. When you drink a burning hot liquid, the "chest pain" felt in the region of the sternum does not really occur there. What is this

    type of pain called? _____

Taste and Smell

INTRODUCTION

Like most of our senses, we take for granted our senses of taste and smell. They are not fully appreciated unless they are lost. People who have lost the sense of smell find food difficult to eat, since they derive little or no pleasure from the act of eating. In this exercise, you examine the structure and function of the organs involved in these two important senses.

Both taste and smell are examples of **chemoreception,** in which specific chemical compounds are detected by the sense organs and interpreted by various regions of the brain. The sense of taste and the sense of olfaction are covered in the Saladin text in chapter 16, "Sense Organs."

OBJECTIVES

At the end of this exercise, you should be able to

1. list the two major chemoreceptors located in the head;
2. diagram a taste bud;
3. trace the sense of smell from the nose to the integrative areas of the brain;
4. trace the sense of taste from the tongue to the integrative areas of the brain;
5. list the five tastes perceived by humans;
6. describe what happens in an olfactory reflex;
7. compare and contrast the senses of taste and smell.

MATERIALS

Sterile cotton-tipped applicators

Five dropper bottles containing one of the following:

Solution of salt water (3%) labeled "Salty"

Quinine solution (tonic water) labeled "Bitter"

Vinegar solution (household vinegar or 5% food grade acetic acid solution) labeled "Sour"

Sugar solution (3% sucrose) labeled "Sweet"

Umami solution (15 g MSG in 500 mL water) labeled "Umami"

Biohazard bag

Prepared slides of taste buds

Microscopes

Roll of household paper towels

Small bowl of salt crystals (household salt)

Small bowl of sugar crystals (household granulated sugar)

Flat toothpicks

Small bottle (100 mL) of household ammonia

Several small screw-cap bottles (10–20 mL) with essential oils labeled "Peppermint," "Almond," "Wintergreen," and "Camphor" (keep vials in separate, wide-mouthed jars to prevent cross-contamination of scents)

Four small vials colored red and labeled "Wild Cherry" filled with benzaldehyde solution

Four small vials of "Almond" essence

One vial of dilute perfume (one part perfume, five parts ethyl alcohol)

Marking pens

Noseclips and alcohol swabs

Selection of four or five fruit nectars (such as Kern's nectars), two cans each: apricot, coconut/pineapple, strawberry, mango, peach, apple

Small, 3 oz paper cups (89 mL), five per student (or student pair)

Napkins

PROCEDURE

Examination of Taste Buds

Ⓐ Activity

Examine the prepared slide of taste buds and compare them to fig. 20.1. The taste buds are located on the sides of papillae on the tongue. The taste buds appear lighter than the surrounding tissue (like microscopic onions cut in long sections). Taste buds consist of **supporting cells** and **taste cells,** specialized epithelial cells, with hairs that project into the **taste pores** near the surface of the tongue. The basal portion of the taste cell synapses with sensory neurons.

Projection Pathway of Gustation

The sense of taste, or **gustation,** is picked up by the receptors in taste buds primarily in the tongue, although there are also receptors in the soft palate and pharynx. The sense of taste travels through the facial nerves, glossopharyngeal nerves, and vagus nerves to the medulla oblongata. From there, some fibers travel to either the hypothalamus or the amygdala, where autonomic reflexes (such as swallowing) occur. Other fibers travel to the thalamus and then to higher brain centers, such as the postcentral gyrus, where the sense of taste is determined. From the postcentral gyrus, fibers take neural impulses to the orbitofrontal cortex, where sight and smell are integrated with taste. The sense of taste is influenced by a food's smell, appearance, temperature, and texture, even the mood of the individual.

Taste Determination of Solid Materials

Gustatory receptors are stimulated by specific chemicals in solution. Fluid runs down the sides of the tongue papillae, where the taste cells of the receptors are located.

Ⓐ Activity

Blot your tongue thoroughly with a paper towel. Make sure the surface is relatively dry. Have your lab partner select either the sugar crystals or the salt crystals and place a few crystals (with the flat end of a toothpick) on your tongue. Keep your mouth open and do not swirl saliva around. Can you determine what the sample is?

Now close your mouth and see if you can determine the nature of the sample.

Determining Specific Taste Receptors

In this section, you determine if receptors for taste are functioning among members of your lab section. There are five primary tastes: sweet, sour, salty, bitter, and umami. Umami gives meat and cheese their characteristic tastes, and it can be referred to as "savory." Humans have varying degrees of sensitivity to these five tastes. Some of us find bitter tastes to be especially objectionable, while others do not seem to mind them as much. This is due, at

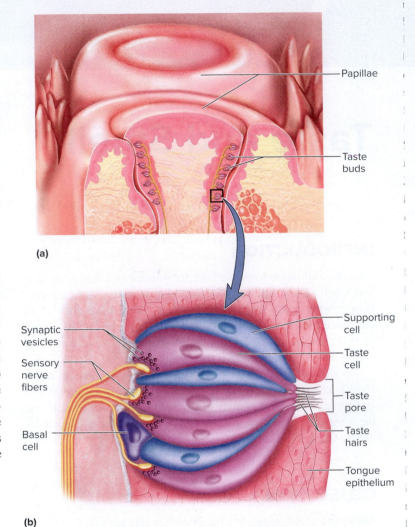

(a)

(b)

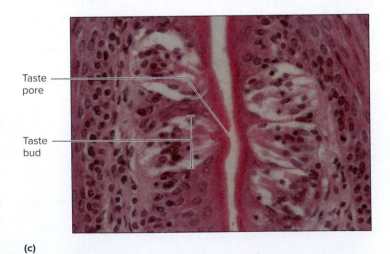

(c)

FIGURE 20.1 Taste Buds. (a) Taste buds on sides of a tongue papilla; (b) details of taste buds; (c) photomicrograph (100×).
(c) ©Eric Wise

least in part, to a genetic difference between people. As you perform the taste experiments, determine if members of your class are equally sensitive to the same tastes as you are.

 Caution! If you have food allergies, you may wish to omit this part of the lab. People with allergies, migraines, or heart problems should avoid tasting glutamate.

Ⓐ **Activity**

Select one of the dropper bottles labeled "Sweet," "Sour," "Salty," "Bitter," or "Umami." Using a sterile cotton-tipped applicator stick, apply one of the tasting solutions to the cotton tip. Saturate the cotton tip. Remember to keep track of the substance on your applicator stick. *Do not* reuse the applicator sticks. Dab the entire surface of your lab partner's tongue and determine by nods or hand signals when perception of the taste occurs. Certain regions of the tongue are more sensitive to a specific taste, and these are variable among individuals. Throw the applicator stick away. Repeat the test with a *new applicator stick* for each of the five solutions. *Do not contaminate* the solutions by reusing the applicator stick with the same or different solutions after it has been in your lab partner's mouth!

Projection Pathway of Olfaction

Ⓐ **Activity**

Examine a model, chart, or diagram of an inferior view of the brain and locate the olfactory nerves, olfactory bulb, and olfactory tract. The sense of smell begins in the nose, where odorant particles stimulate **olfactory cells** (specialized neurons). These neurons are clustered in bundles that make up the **olfactory nerve,** and they pass from the nasal mucosa to the **olfactory bulb** at the base of the frontal lobe (fig. 20.2). Some neurons travel to the temporal lobe, where the perception of smell occurs, while others travel to the hippocampus or amygdala, where the memory of smell is stored or the emotional response to smell occurs.

Olfactory Reflex

Ⓐ **Activity**

Take a small bottle of household ammonia and place it under the nose of your lab partner. Have your lab partner take a brief sniff from the bottle. If there is a visible movement of the head in a posterior direction, then your lab partner demonstrates an olfactory reflex to smell. Record the results of your experiment below.

Olfactory reflex:

❓ _____ (1 yes/no)

❓ What might be the adaptive benefit for people having

an olfactory reflex?_____ (2)

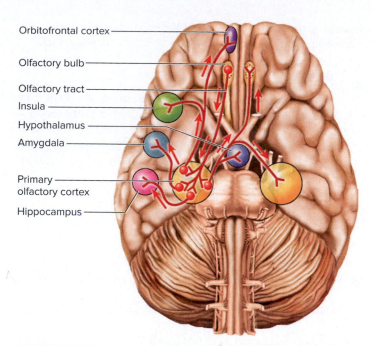

FIGURE 20.2 The Transmission of Olfaction from the Olfactory Bulbs to the Brain.

Labels: Orbitofrontal cortex, Olfactory bulb, Olfactory tract, Insula, Hypothalamus, Amygdala, Primary olfactory cortex, Hippocampus

Visual Cues in Smell Interpretation

In this section you examine the influence of visual cues on the interpretation of smell. You will seek to determine whether the color of a substance has any effect on what you perceive the smell to be.

Ⓐ **Activity**

Have your lab partner show you a small vial labeled "Almond" and then smell it. Now examine and smell the small red vial labeled "Wild Cherry." Do you perceive these as two separate smells? Close your eyes and have your lab partner select a vial for you. Can you tell which one it is?

❓ _____ (3 yes/no)

Olfactory Discrimination

Ⓐ **Activity**

Obtain four vials of different scents—peppermint, almond, wintergreen, and camphor. While keeping your eyes closed, try to determine the name of each essential oil as your lab partner presents it to you. Record how many of the smells you got correct out of the four. If you are hypersensitive to smells, have your lab partner do this section of the experiment.

❓ Number correct: _____ (4)

Adaptation to Smell

Adaptation to smell by the olfactory receptors occurs very rapidly, but the adaptation by the receptors is incomplete. Complete adaptation to smell probably occurs by additional CNS inhibition of the olfactory signals. In this section of the experiment, you are trying to determine approximately how long olfactory adaptation takes.

A Activity

Close your eyes and plug one nostril. Inhale the scent from a vial of dilute perfume or one of the scents from the olfactory discrimination test until the smell decreases significantly. Have your lab partner record the time when you begin the experiment and how long it takes for the perception of smell to decrease significantly. How long does this take?

? Length of time for significant reduction of the smell:

_____ (5)

? What might be the evolutionary advantage of adaptation

to smell? _____ (6)

Predict whether adaptation to one smell causes adaptation to another smell. Record your prediction that the smell of one material does or does not cause adaptation to another smell.

? Prediction (yes/no) _____ (7)

A Activity

Now smell the wintergreen or peppermint vial. Does the adaptation of one smell cause the olfactory receptors to adapt to other smells?

? Result _____ (8)

Taste and Olfaction Tests

A Activity

This experiment demonstrates the dependence of the sense of smell as a component of what we call *taste*.

A Activity

Obtain four or five small cups (3 oz) and, using a marking pen, label each with the name of the fruit juice or nectar it will contain. Have your lab partner select several types of fruit nectars and pour each into the proper cup.

Caution!

If you have a particular food allergy, you may want to have your lab partner do this test and you record the results.

Sit with your eyes and your nose closed (use noseclips or pinch off your nostrils with your finger and thumb) and try to determine the sample presented to you by your lab partner. Your lab partner should place the sample cup (sample unknown to you) in your hand. After you have "tasted" the sample, try to name it and record your results in the following chart. Test all the samples first with your nose closed. After you make your initial determination, release your nostrils but still keep your eyes closed and taste the samples again. Record the results with your nose open (unplugged). How accurate is your comparison?

Trial Number	Sample	Accuracy or Detection (Yes/No)	
		Nose Closed	Nose Open
1	Apricot		
2	Mango		
3	Coconut/pineapple		
4	Strawberry		
5	Peach		
6	Other		

REVIEW SECTION

Taste and Smell

Name _____ *Date* _____

Lab Section _____ *Time* _____

❓ Chapter Summary Data

Use this section to record your results from questions within the exercise.

1. _____   5. _____

2. _____   6. _____

3. _____   7. _____

4. _____   8. _____

Review Questions

1. Why does material have to be in solution for it to be sensed as taste? _____
_____

2. What are the primary classes of tastes? _____
_____

3. Did everyone in your lab have the same reaction to the taste tests in their like or dislike of sweet, sour, salty, umami, or bitter?
_____

4. Describe the projection pathway of smell from the olfactory receptors to the temporal lobes of the brain.
_____
_____

5. What structures are involved in taking the sense of taste from the taste buds to the brain?
_____
_____

6. Where are the taste buds located? _____

7. What is the exact region of the nasal cavity receptive to smell stimuli? _____

_____

8. Some individuals with severe sinus infections can lose their sense of smell. How can an infection that spreads from the frontal or maxillary sinus impair the sense of smell? What structure or structures might be affected?

_____

_____

9. Material must be in solution for it to be perceived by gustatory receptors. What process is used (olfaction, gustation) to perceive a lipid-based food, such as garlic or peppermint? _____

_____

10. Some smells that we perceive as two separate smells are actually identical. What are the other cues that we use to distinguish these two "smells" as being distinct? _____

_____

11. What physical mechanism is produced by a cold (rhinovirus) that influences our perception of the flavor of food? _____

_____

12. Does adaptation to one smell influence the adaptation to another smell?

_____

_____

Eye and Vision

INTRODUCTION

In most people, eyesight accounts for much of their accumulated knowledge. The importance of the eye can be inferred in that 4 of the 12 cranial nerves are dedicated, at least in part, to either receiving visual stimuli or coordinating the movement of the eyes. The anatomy and physiology of the eye are discussed in the Saladin text in chapter 16, "Sense Organs."

Anatomically, the eye consists of an anterior portion, visible as we look at the face of an individual, and a posterior portion, situated in the orbit of the skull. Light from the external environment travels through a number of transparent structures that bend and focus the light on the retina, the receptive layer of the eye that converts light energy to nerve impulses. Nerve impulses travel from the eyes to the optic nerves to the occipital lobes of the brain, where they are interpreted as sight. This exercise involves learning the structure of the eye and correlating those structures to the function of vision by performing basic physiology experiments.

OBJECTIVES

At the end of this exercise, you should be able to

1. identify the major structures of the mammalian eye;
2. describe the six extrinsic muscles of the eye and their effect on the movement of the eye;
3. distinguish between the pupil and the iris of the eye;
4. describe the position of the choroid in reference to the retina and the sclera;
5. describe the function of the rods and the cones of the eye;
6. define the near point of the eye;
7. determine the visual field for both eyes;
8. demonstrate the Snellen vision tests and those for accommodation and astigmatism.

MATERIALS

Models and charts of the eye

Snellen charts

Astigmatism charts

Vision Disk (Hubbard) or large protractor

Microscopes

Prepared microscope slides of eye in sagittal section

Preserved sheep or cow eyes

Dissection trays

Protective gloves

Scalpel

Animal waste disposal container

Card with fine print (8-point font)

3-by-5-inch cards

Ishihara color book or colored yarn

Ruler (approximately 35 cm)

Ophthalmoscope and batteries

Penlight

Paper card with a simple colored image (red circle, blue triangle) printed on it

Dark card with light image on it

PROCEDURE

External Features of the Eye

The external anatomy of the eye and the accessory structures are illustrated in fig. 21.1.

(A) Activity

Examine the eye of your lab partner and compare it to the figure. The **pupil** is an opening located in the center of the eye and is surrounded by the colored **iris**. The **sclera** is the white of the eye, and it is covered by a membrane, known as the **conjunctiva,** which continues underneath the eyelids. Numerous blood vessels traverse the conjunctiva, and if they become dilated anteriorly they give the eye the appearance of being "bloodshot." The eyelids join together at the

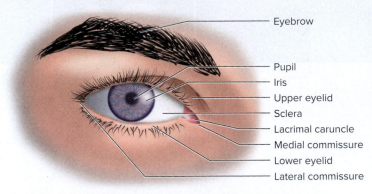

FIGURE 21.1 **External Anatomy of the Eye.**

lateral commissure and the **medial commissure.** There is a small piece of tissue near the medial commissure known as the **lacrimal caruncle.** Examine the **upper eyelid** and **eyelashes** and the **lower eyelid** and eyelashes, which prevent material from entering the eyes

and (in the case of the eyelids) reduce visual stimulation when we sleep. Look also at the **eyebrow** located on the supraorbital ridge.

The sclera is a protective portion of the eye that is an attachment point for the muscles of the eye and helps maintain the intraocular pressure (the pressure inside the eye). The pressure maintains the shape of the eye and helps keep the retina adhered to the back wall of the eye. The sclera is continuous with the transparent cornea in the front of the eye.

Attached to the sclera are the **extrinsic muscles** of the eye. There are six extrinsic muscles which coordinate to move the eye in quick and precise ways. Locate these muscles on a model and compare them to fig. 21.2. These muscles and their action on the eye are listed in table 21.1.

A structure important in the maintenance of the exterior of the eye is the **lacrimal apparatus.** This consists of the **lacrimal gland** located superior and lateral to the eye (fig. 21.3). Tears bathe and protect the eye and clean dust from its surface. The fluid drains through the **nasolacrimal duct** into the nasal cavity.

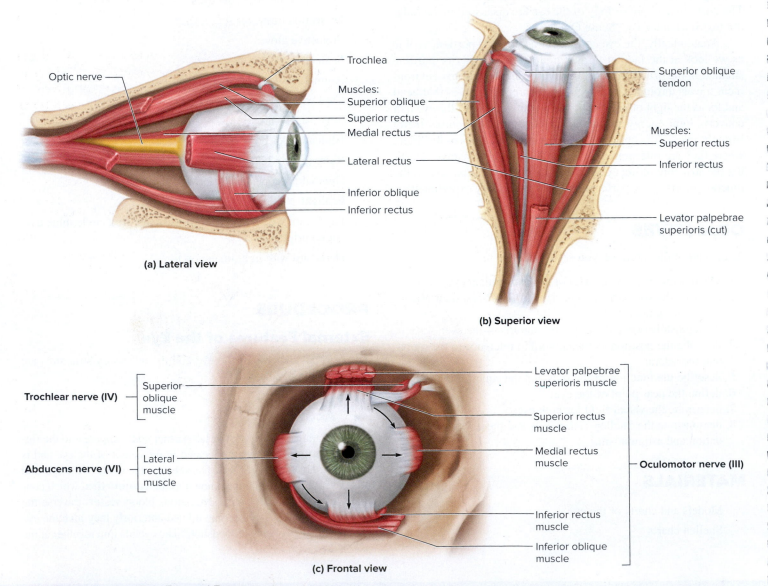

FIGURE 21.2 **Right Eye, External Features.**

TABLE 21.1	Extrinsic Muscles of the Eye	
Muscle Name	**Innervation**	**Direction Eye Turns**
Lateral rectus	VI (abducens)	Laterally
Medial rectus	III (oculomotor)	Medially
Superior rectus	III (oculomotor)	Superiorly
Inferior rectus	III (oculomotor)	Inferiorly
Inferior oblique	III (oculomotor)	Superiorly and laterally
Superior oblique	IV (trochlear)	Inferiorly and laterally

Interior of the Eye

From the anterior of the eye, the first layer covering the inside of the eyelid and extending across the exposed surface of the eye is the conjunctiva. The conjunctiva is composed of epithelial tissue and is an important indicator of a number of clinical conditions (for example, pink eye or conjunctivitis). In the center of the eye is the transparent **cornea** (fig. 21.4). The cornea is the structure of the eye most responsible for the bending of light rays that strike the eye. The lens adjusts the images in order to focus. It is composed of dense connective tissue and is avascular. Why would the presence of blood vessels in the cornea be a visual liability?

Directly posterior to the cornea is the **anterior cavity,** subdivided into the **anterior chamber,** between the cornea and the **iris,** and the **posterior chamber,** between the iris and the **lens.** The anterior cavity is filled with **aqueous humor,** produced by the **ciliary body.** Only a few milliliters of aqueous humor are produced each day, and this amount is absorbed by a **scleral venous sinus.**

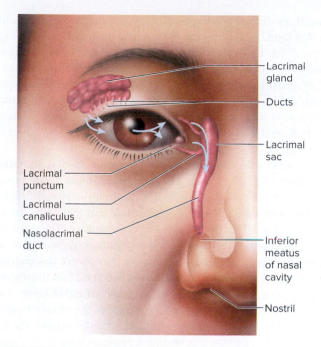

FIGURE 21.3 Lacrimal Apparatus.

The iris is what gives us a particular eye color. People with blue and gray eyes are more sensitive to bright light than those with brown eyes because of the greater percent of eumelanin found in brown eyes. The **pupillary constrictor** (circular) muscles of the iris constrict in bright light, reducing the diameter of the **pupil** (the space enclosed by the iris), and the

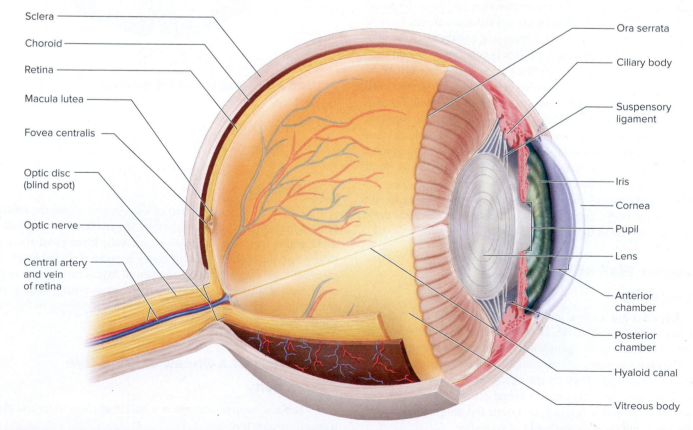

FIGURE 21.4 **The Left Eye.** Transverse (horizontal) section viewed from above.

pupillary dilator is composed of myoepithelial cells that contract in dim light, increasing the diameter of the pupil. Posterior to the pupil is the lens, made of a crystalline protein. The lens is more pliable in youth and stiffens as a person ages. Because of the loss of this elasticity, people in their forties usually begin to use reading glasses. The ciliary muscle in the ciliary body contracts, and the suspensory ligaments that attach to the lens loosen, decreasing the pull on the lens. The lens becomes rounder, allowing for close focusing.

Posterior to the lens is the **posterior cavity,** or **vitreous chamber** (fig. 21.4). This cavity occupies most of the posterior portion, or fundus, of the eye. The posterior cavity is filled with the **vitreous body,** a clear, jellylike fluid that maintains the shape of the eyeball. The eye has three layers or **tunics.** The most superficial layer is the **fibrous layer,** and it consists of the **sclera** (white of the eye) and the cornea (an anterior transparent continuation of the sclera). The cornea allows light to penetrate into the interior of the eye. The middle layer of the eye is the **vascular layer** or **uvea** (YOU-vee-uh) and consists of the **choroid,** the **ciliary body,** and the **iris.** The blood vessels found in the choroid nourish the retina, remove waste, and the pigmentation prevents light from scattering and blurring vision. The layer closest to the **vitreous body** is the **inner layer,** and it is composed of the **retina** and the **optic nerve.** The retina consists of an outer **pigmented layer,** which absorbs light passing through the eye and prevents light scattering, and an inner **neural layer.**

The retina converts visible light into action potentials. **Photoreceptor cells** (the **rods** and **cones**) are sensitive to light and generate an electrochemical signal. Rods and cones synapse with **bipolar cells**; thus, the visual stimulation that occurs in the posterior portion of the eye is transmitted toward the vitreous body to the bipolar cells. In the dark, photoreceptor cells release the neurotransmitter glutamate, which inhibits bipolar cells. When light strikes the retina, the inhibitory glutamate is not released from the photoreceptor cells, and the bipolar cells fire. The bipolar cells synapse with **ganglion cells,** and the axons of the ganglion cells form the **optic nerve.** From there the action potentials in the optic tracts travel to the lateral geniculate nucleus of the thalamus and then to the occipital region of the brain where they are interpreted in the visual association area of the occipital lobe (fig. 21.5). Some images from the left visual field cross over to the right side of the brain. The right side of the brain controls motor impulses for the left side of the body. Some images in the right visual field cross over to the left side of the brain.

Posterior Wall of the Eye

Ⓐ **Activity**

Obtain a microscope and a slide of the retina and compare what you see in the slide to fig. 21.6. The neural layer of the retina is composed of three layers: **ganglionic, bipolar,** and **photoreceptor cells.** The photoreceptor cells are composed of **rods** and **cones.** Rods function in dim light. They do not determine color vision. Cones are involved in color vision and in visual acuity (determining fine detail). They function in bright light. Locate the ganglionic layer, the bipolar layer, and the rods and cones in your slide.

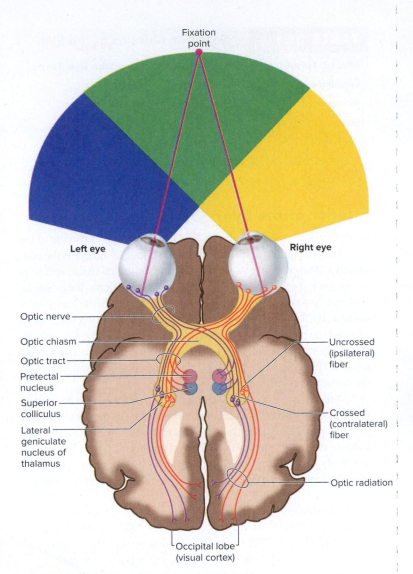

FIGURE 21.5 Visual Pathway to the Brain.

Ⓐ **Activity**

Examine a model or chart of the eye and locate the **macula lutea** at the posterior region of the eye. *Macula lutea* means "yellow spot," and in the center of this structure is the **fovea centralis,** a region where the concentration of cones is greatest. In the fovea, the cone cells are not covered by the neural layers as they are in the other parts of the retina. When you focus on an object intently, you are directing the image to the fovea. Locate the fovea in fig. 21.7 and in models available in the lab.

Dissection of a Sheep or Cow Eye

Ⓐ **Activity**

Rinse a sheep or cow eye in a bucket of clean water and place it on a dissection tray.

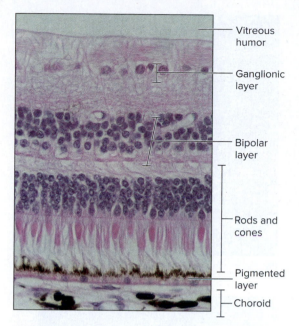

FIGURE 21.6 Retina (400×).
©Eric Wise

— Vitreous humor
— Ganglionic layer
— Bipolar layer
— Rods and cones
— Pigmented layer
— Choroid

1. Obtain a scalpel, scissors, and a blunt probe.

Caution!

Be careful with the sharp instruments and cut away from the hand holding the eye. Wear protective gloves while you perform the dissection.

---

2. Carefully remove the fat and muscles from the eyeball. Using a scalpel or scissors, make a coronal section of the eye behind the cornea (fig. 21.8). Do not squeeze the eye with force or thrust the blade sharply because you may squirt yourself with fluid.

3. Cut through the eye entirely and note the jellylike material in the posterior cavity. This is the vitreous body. Look at the posterior portion of the eye. Note the beige retina, which may have pulled away from the darkened choroid. The retina is only attached to the middle tunic at the optic nerve and the ora serrata. The vitreous body helps hold the retina to the back of the eye. Sharp blows to the head may dislodge the retina from the choroid causing a **detached retina.** The choroid in humans is very dark, but you may see an iridescent color in your specimen. This is the **tapetum lucidum,** which improves night vision in some animals. The tapetum lucidum produces the "eye shine" of nocturnal animals. Also examine the tough, white sclera, which envelops the choroid.

4. Now examine the anterior portion of the eye. Is the lens in place? The lens in your specimen probably will not be clear due to the preserving fluid denaturing the protein of the lens. Normally, the lens is transparent and allows for light penetration.

   The lens is held to the ciliary body by the suspensory ligaments (fig. 21.4). These ligaments pull on the lens and alter its shape for close or distant vision. Locate the ciliary body at the edge of the suspensory ligaments.

5. Turn the eye over. Is there any aqueous humor left in the anterior cavity? Make an incision through the conjunctiva and cornea into the anterior cavity. Can you determine the region of the anterior chamber and the posterior chamber that make up the anterior cavity? Locate the iris and the pupil.

6. Dispose of the specimen in a designated waste container and rinse your dissection tools.

Visual Tracking

(A) Activity

Eye movements are precisely controlled by the six extrinsic muscles of the eye. To determine their effectiveness, have your lab partner follow your finger as you move it in front of the eyes. Is the movement of the eye smooth (a normal function) or do the eyes move in a jerky fashion (an abnormal condition)? Record your results here.

Eye movement: (smooth/jerky) (select one).

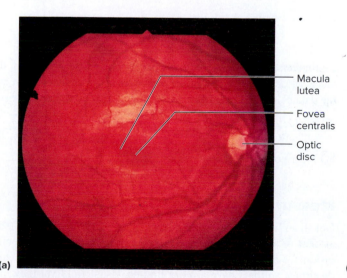

Macula lutea
Fovea centralis
Optic disc

(a)

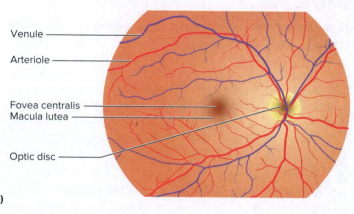

Venule
Arteriole
Fovea centralis
Macula lutea
Optic disc

(b)

FIGURE 21.7 Eye, Posterior Surface. (a) Photograph; (b) diagram.
(a) ©Eric Wise

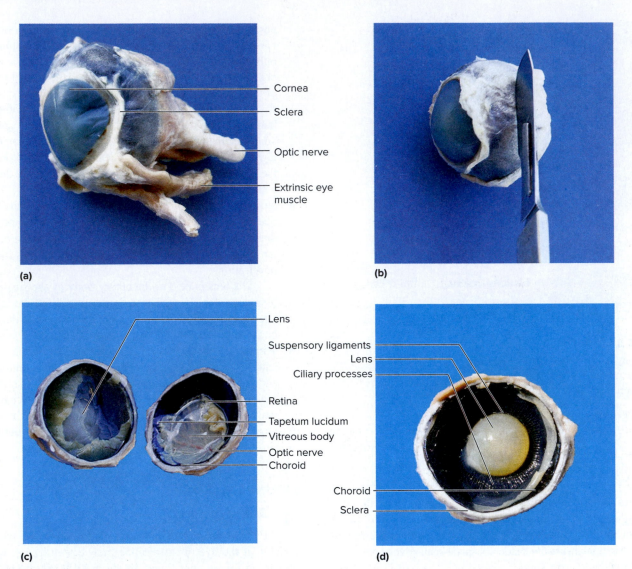

FIGURE 21.8 **Dissection of a Sheep Eye.** (a) External features; (b) coronal section; (c) eye with vitreous body; (d) anterior eye without vitreous body.

©Eric Wise

Determination of the Near Point

The minimum distance an object can comfortably be held in focus is called the **near point.** The eye's ability to focus is due to the elasticity of the lens. The elasticity of the lens decreases with age. A 10-year-old may be able to focus 8 to 10 cm away from the eye, yet a 65-year-old may not be able to focus closer than 80 to 100 cm. The decreased elasticity of the lens becomes noticeable around 40 to 50 years of age, when many people find reading glasses necessary because of their aging lenses.

Ⓐ **Activity**

You can measure your near point by holding a paper with fine print vertically at arm's length in front of you. Close one eye and slowly move the paper closer until either you see two objects or it becomes blurry. Have your lab partner measure the distance, in

centimeters, from your eye to the paper. This is the near point distance. Measure the near point for both eyes in centimeters and record the data below and in the review section at the end of the exercise.

❓ Near point of right eye: _____ (1)

❓ Near point of left eye: _____ (2)

Measurement of Binocular Visual Field

Not all animals have the same **visual field.** Some prey species (such as deer or sheep) have extensive visual fields with little binocular vision. On the other hand, many predators, birds, and arboreal (tree-dwelling) animals have a more limited visual field yet they have greater **binocular,** or **stereoscopic, vision.** Binocular vision allows arboreal animals (such as primates) to perceive depth—vital when

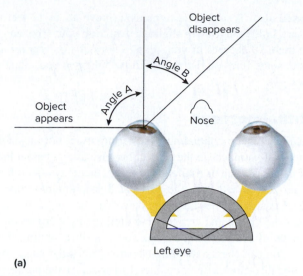

(a)

(b)

FIGURE 21.9 Measuring the Visual Field. (a) Left eye; (b) right eye.

judging how far away the next branch is! As humans are primates, this binocular vision of our arboreal ancestors has been retained.

A Activity

You can determine your visual field with the use of a Vision Disk or protractor. If you are using a Vision Disk, follow the instructions enclosed. If not, sit down and have your lab partner stand behind you. Close your right eye and look straight ahead with the left. While the right eye is closed, have your lab partner move an object (pen, paper disk, etc.) from behind your head from the left until you can just see the object. Measure the angle from the tip of the nose to where you see the object. This is illustrated in fig. 21.9 ("Angle A"). With the same eye directed ahead, continue moving the object until it is out of view (to the right of the nose somewhere). This is illustrated in fig. 21.9 ("Angle B"). Determine the angle from the nose to where the object disappears. Add these two values and record the result below and in the review section.

❓ Angle where object is perceived: _____ (3)

❓ Angle where object disappears: _____ (4)

Now close your left eye and repeat the exercise on the right side. This is illustrated in fig. 21.9 ("Angle C" and "Angle D"). Determine the total visual field for each eye by adding the sum of fig. 21.9, "Angle A" and "Angle B," for the left eye and the sum of "Angle C" and "Angle D" for the right eye. For both eyes, add the sum of "Angle A" and "Angle C," and for the degree of overlap between the eyes, "Angle B" and "Angle D" (fig. 21.10). Record these data in the following spaces and in the review section.

❓ Visual field of right eye: _____ (5)

❓ Visual field of left eye: _____ (6)

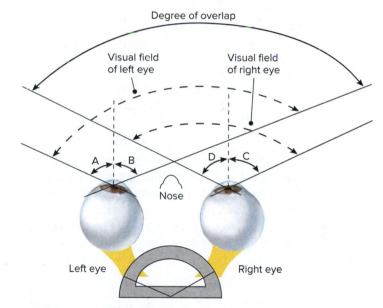

FIGURE 21.10 Visual Field.

❓ Complete visual field of both eyes: _____ (7)

❓ Degree of overlap: _____ (8)

Measurement of Visual Acuity (Snellen Test)

A Activity

Face the Snellen eye chart from 20 feet away and have your lab partner stand next to the chart. Cover one eye with a 3-by-5-inch card and have your lab partner point to the largest letter on the chart. Do *not* read the chart with both eyes open. Your lab partner should then progressively move down the chart and *note the line that has the smallest print in which you made no errors.* Your lab partner should record the numbers at the side of the line (such as 20/20). These numbers refer to your **visual acuity.** Switch the card to the other eye and repeat the test. Record your results below and in the review section.

❓ Left eye: _____ (9)

❓ Right eye: _____ (10)

A vision of 20/20 is considered normal. In 20/20 vision, you can read material at 20 feet that other people with normal vision read at the same distance. If your vision is 20/15, then you can read at 20 feet what people with normal vision read clearly at 15 feet. If your vision is 20/60, then you read at 20 feet what people with normal vision read at 60 feet.

Astigmatism Tests

If the cornea or lens of the human eye were perfectly smooth, the incoming image would strike the retina evenly, and there would be no blurry areas. The cornea is not a smooth spherical structure, and if it is irregularly shaped it can produce blurring of images. In a normal optical exam, the distance correction is made first (to determine visual acuity) and then, with the corrective lenses in place, an astigmatism test is performed.

Ⓐ **Activity**

If you do not normally wear corrective lenses, then cover one eye and examine the astigmatism chart in lab (as represented in fig. 21.11) from 20 feet away. The chart consists of a series of parallel lines radiating from the center. Stare at the center of the chart and determine which of the sets of parallel lines, if any, appears light or blurry. Have your lab partner note the corresponding number on the chart and record the number below and in the review section.

❓ Astigmatism numbers: _____ (11)

If you wear corrective lenses, then not only is the condition of nearsightedness or farsightedness corrected but astigmatism is corrected also. To test for astigmatism, move about 12 feet from the chart, hold your glasses slightly away from your face, and then rotate them 90 degrees. If your glasses normally correct for astigmatism, some lines on the chart will be blurry as you rotate your glasses.

Ophthalmoscope

Clinical use of the ophthalmoscope is important not only for the diagnosis of variances in the eyeball but also as a potential indicator of diseases, such as **diabetes mellitus,** that may affect the eye. Familiarize yourself with the parts of the ophthalmoscope (fig. 21.12). Locate the ring (rheostat control) at the top of the handle. Depress the button and turn the ring until the light comes on. The headpiece of the ophthalmoscope consists of a rotating disc of lenses. You can change the diopter (strength) of the lenses by rotating the disc clockwise or counterclockwise. As the numbers get progressively larger, you are looking through more convex lenses. If you rotate the disc in the other direction, the lenses become more concave. At the zero reading, there are no lenses in place.

Observation with the Ophthalmoscope

Caution! Examine the posterior region of the eye only for a short time. Extensive use of the ophthalmoscope can damage the eye.

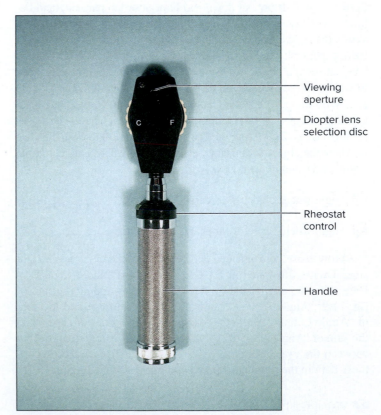

FIGURE 21.12 Ophthalmoscope.
©Eric Wise

- Viewing aperture
- Diopter lens selection disc
- Rheostat control
- Handle

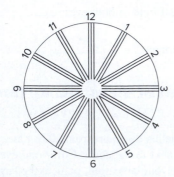

FIGURE 21.11 Astigmatism Chart.

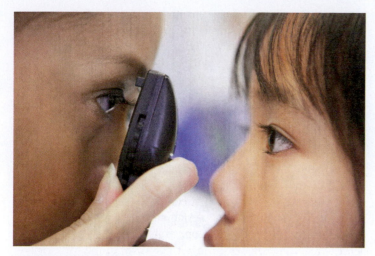

FIGURE 21.13 Use of Ophthalmoscope.
Fuse/Getty Images

(A) Activity

Sit facing your lab partner. Select the left eye to examine. Have the ophthalmoscope setting at zero and use your left hand to hold the ophthalmoscope. Use your left eye to look through the ophthalmoscope and move in close to examine the eye of your lab partner (fig. 21.13). You may rest your hand on the cheek of your lab partner if you need to steady your hand. Look into the pupil and examine the back of the eye. If the back appears fuzzy, rotate the disc clockwise (positive diopters) and see if it comes into focus. If a positive setting provides a clear view of the eye, then your lab partner has **hyperopia** (hypermetropic vision), or farsightedness. In farsightedness, the eyeball is too short and the image is focused posterior to the retina. Positive diopter lenses focus the image on the retina. If the image is indistinct, then adjust the disc counterclockwise to obtain a negative diopter reading. If the image is clear with a negative number, then your lab partner has **myopia** (myopic vision), or nearsightedness. In nearsightedness, the eyeball is too long and lenses with negative diopters focus the images farther back on the retina. This procedure is based on the assumption that your vision, as the examiner, is normal. Examine the retina of the eye and notice the blood vessels there.

Pupillary Reactions

(A) Activity

Sit in a dark room for a minute or two and examine your lab partner's eyes in dim light. Are the pupils dilated or constricted? Record the data below and in the review section.

? Pupil diameter in dim light: _____ (12)

Shine a penlight in the right eye of your lab partner while holding your hand vertically at the level of the nose to block light from going into the left eye. Record what happens to the pupil diameter.

? Right pupil diameter: _____ (13)

As you shine the light into the right eye, what occurs in the left pupil? This is called a **consensual reflex.** Do you think this reflex is controlled by the eye or in the brain? If you need to, review the structures of the midbrain in Exercise 17. Record the effect.

? Left pupil diameter: _____ (14)

Color Blindness

Color vision is dependent on three separate cone cell sensitivities. Cones may be red, green, or blue sensitive. Changes in the genes on the X chromosome are the most common cause of color blindness. Some individuals may be unable to see a particular color, while others may have a reduction in their ability to see a particular color. Color blindness is most common in males and relatively rare in females. This is due to the chromosomal makeup of the two sexes. The male chromosome makeup is XY. The Y chromosome does not carry the gene for color vision. If the X chromosome carries a gene for color blindness, then the male exhibits color blindness. On the other hand, if a female carries the gene for color blindness on the X chromosome, the chances are that she will have a normal gene on the other X chromosome. The normal gene is expressed, cone pigments are produced, and the female has normal color vision.

(A) Activity

Test for color blindness by using vision test kits with select color samples, or you can use Ishihara color charts, as represented in fig. 21.14. If you use the color charts, flip through the book with a lab partner and record which charts were accurately viewed and which plates, if any, were missed. You can calculate the type of color blindness and the degree by following your instructor's directions or those that come with the color chart. Record your data below and in the review section.

? Your color vision: _____ (15)

Determination of the Blind Spot

The **optic disc** is a region where the retinal nerve fibers exit from the back of the eye and form the **optic nerve.** The mass exit of the nerve fibers leaves a small circle at the back of the eye devoid of photoreceptors. This region, the optic disc, is also known as the **blind spot.** You can locate the blind spot by holding this lab manual at arm's length.

(A) Activity

Close your right eye and hold the lab manual about 30 cm from your eyes. Look at the **X** with your left eye and move the book slowly forward or back until the red dot disappears. Notice how the green bar still seems to be present. This is known as the **visual filling** phenomenon, which occurs as the brain fills in the missing information, instead of leaving a dark spot where the red dot used to be.

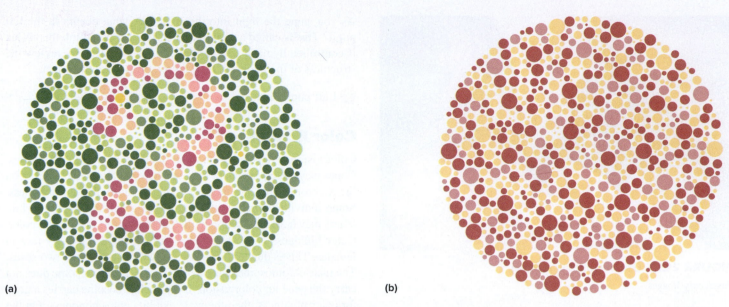

(a)   (b)

FIGURE 21.14 Ishihara Test for Color Blindness. (a) Plate as seen by a person with normal color vision; (b) plate as seen by a person who has red-green color blindness.

Strabismus

Normally the extrinsic muscles of your eyes adjust them so they are aligned, and you see a single image. When the eyes are not aligned, double vision occurs, which is known as **strabismus.** You can demonstrate this effect by the following activity.

Ⓐ Look at your text and gently press a clean fingertip at the lateral corner of the upper eyelid of one eye. What happens to your vision? Strabismus can occur due to the differences in the length of the rectus muscles in the eye or in the paralysis of cranial nerves III, IV, or VI. Trauma to the brain, nerves, or eye muscles can also cause strabismus.

Ⓐ If you look at a bright light or object against a dark background, then close your eyes, you can still see a light object in a dark background. This event is an **afterimage.** The afterimage can last from a few seconds to almost a minute, depending on the brightness of the light and the contrast between the object and the background. The afterimage is due to the photoreceptors in your eye continuing to fire. After this time, the reverse occurs, and you see a dark object against a light background. This is called a **negative afterimage.**

The photosensitive pigment of the eye is called **rhodopsin,** and it is composed of **retinal** and the protein **opsin.** Retinal is a metabolite of vitamin A. When light strikes the retina, the purple-colored rhodopsin splits into its two component parts and becomes pale, a process known as "bleaching."

You can test the time for separation and reassembly of the photopigments by staring at a light-colored image on a dark card under a bright light for a few moments. Stare long enough to get an image (about 10 to 20 seconds) and then shut your eyes. Have your lab partner record the time. You should see a colored image against a dark background. If the image is not visible, place your hands over your eyes to block out extraneous light. You should see a positive afterimage, which is due to the photoreceptors continuously firing. After a few moments, you should see the negative afterimage. A negative afterimage may reflect the impact of exhausting the rhodopsin. The dark parts of the normal image that were not bleached fire more than the light parts of the image, and the brain reverses the image based on the amount of action potentials it is receiving.

REVIEW SECTION

Eye and Vision

Name _____ *Date* _____

Lab Section _____ *Time* _____

❓ Chapter Summary Data

Use this section to record your results from questions within the exercise.

1. _____
2. _____
3. _____
4. _____
5. _____
6. _____
7. _____
8. _____

9. _____
10. _____
11. _____
12. _____
13. _____
14. _____
15. _____

Review Questions

1. Eye shine in nocturnal mammals is different from the "red eye" seen in some flash photographs. Eye shine is the reflection from the tapetum lucidum. What might produce "red eye"?

    _____

2. Fill in the following illustration using the terms provided.

anterior cavity	lens	sclera
choroid	optic nerve	suspensory ligaments
ciliary body	pupil	
cornea	retina	

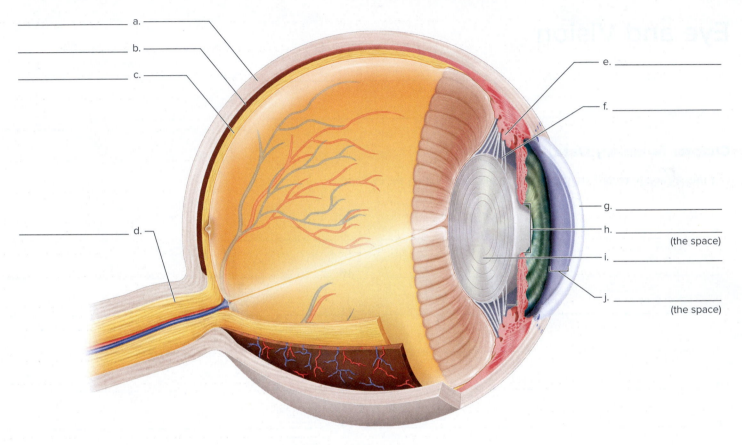

a. _____

b. _____

c. _____

d. _____

e. _____

f. _____

g. _____

h. _____
(the space)

i. _____

j. _____
(the space)

Sagittal section of the eye

3. Since the lens is made of protein, what effect might the preserving fluid used in lab have on the structure of the lens? How would this affect the clarity? _____

_____

4. What is the consensual reflex of the pupil? _____

_____

5. How does the vitreous body differ from the aqueous humor in location and viscosity?

_____

_____

6. What layer of the eye converts visible light into nerve impulses? _____

7. What nerve is composed of axons of the ganglion cells and conducts action potentials to the thalamus of the brain?

_____

8. What is the name of the outermost layer of the anterior eye? _____

9. How would you define an extrinsic muscle of the eye? _____
_____

10. What gland produces tears? _____

11. What is the name of the transparent layer of the eye in front of the anterior chamber?
_____

12. The iris of the eye has two main types of contractile structures. Name these two structures and how they affect the pupil.
_____
_____

13. Where is the vitreous body found? _____

14. What is the function of the choroid? _____

15. Is the lens anterior to or posterior to the iris? _____

16. Which retinal cells are responsible for vision in dim light? _____

17. How would you define the near point of the eye? _____
_____

18. What do the numbers 20/50 mean for visual acuity? _____
_____

19. What is an astigmatism? _____
_____

20. In what area of the eye is the blind spot located? Why is it called the blind spot?
_____
_____

21. Label the following illustration.

_____ a. _____
_____ b. _____
_____ c. _____
_____ d. _____
_____ e. _____
_____ f. _____
_____ g. _____
_____ h. _____

Notes

Ear, Hearing, and Equilibrium

INTRODUCTION

The ear is a complex sense organ that performs two major functions, hearing and equilibrium. It consists of three regions, an outer ear, a middle ear, and an inner ear. The structure and function of the ear are covered in the Saladin text in chapter 16, "Sense Organs." Hearing and balance are considered **mechanoreception,** because the ear receives mechanical vibrations (sound waves) and translates them into nerve impulses, and sensors in the ear detect gravity and motion. The hearing process begins with the vibrations reaching the outer ear and ends up being interpreted as sound in the temporal lobe of the brain. Sound is measured in both **wavelength** and **amplitude,** with distances between each wave's peak (wavelength) determining pitch and each wave's height (amplitude) producing louder or softer sounds. Equilibrium, on the other hand, involves receptors in the inner ear, along with other sensory perceptions, such as visual cues from the eye and **proprioception** in joints. There are two types of equilibrium sensed by the inner ear. These are **static equilibrium** and **dynamic equilibrium.** In static equilibrium, an individual is able to determine his or her nonmoving position (such as standing upright or lying down). In dynamic equilibrium, motion is detected. The body's perception of sudden acceleration, abrupt turning, and spinning is an example of dynamic equilibrium.

OBJECTIVES

At the end of this exercise, you should be able to

1. explain how mechanical sound vibrations are translated into nerve impulses;
2. describe the structures of the outer, middle, and inner ear;
3. perform hearing tests and indicate the significance of their results;
4. perform balance tests and indicate the significance of their results;
5. compare dynamic and static equilibrium and the structures involved in their perception;
6. describe the value of postural reflexes with regard to living in a dynamic environment.

MATERIALS

Models and charts of the ear

Microscope

Prepared slides of the cochlea

Tuning fork (middle C)

Rubber reflex hammer

Audiometer

Cotton balls

Model of ear ossicles

Meter stick

Ticking stopwatch or metronome application on smartphone

Bright desk lamp

Swivel chair

PROCEDURE

Anatomy of the Ear

The **outer (external) ear** consists of those auditory structures superficial to the **tympanic membrane (eardrum).** The **middle ear** contains the ear **ossicles (bones)** and **auditory (Eustachian) tube,** and the **inner ear** consists of the **cochlea, vestibule,** and **semicircular ducts.**

(A) Activity

Examine the charts and models in lab and compare them with fig. 22.1 as you read about the individual regions of the ear.

Structure of the Outer Ear

The outer ear consists of the **pinna (auricle),** which can further be subdivided into the **helix** and the **lobule (earlobe).** The helix is composed of stratified squamous epithelium overlying elastic cartilage. This cartilage allows the ears to bend significantly. The deepest part of the outer ear is the **auditory canal,** which penetrates into the temporal bone.

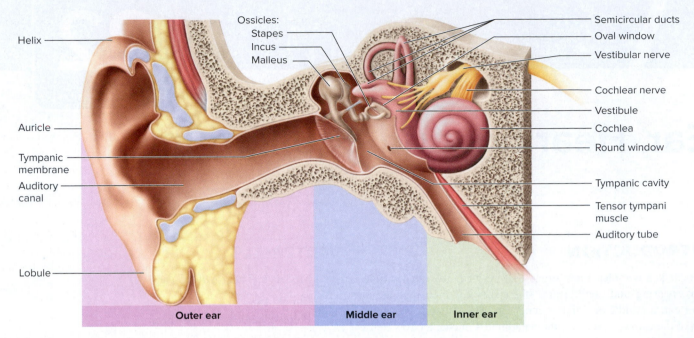

FIGURE 22.1 Anatomy of the Ear.

Otoscopy Procedure

Ⓐ **Activity**

Examination of the outer ear can be done with an otoscope if one is available in your lab.

Caution! Do not push the otoscope deep into the ear because you can rupture the tympanic membrane!

---

1. Wash your hands prior to examining the ear of your lab partner.
2. Use a new, disposable speculum (the funnel-shaped piece you insert into the ear), or if you use a reusable one, make sure you clean it with alcohol.
3. Choose the speculum with the largest diameter that will fit easily into your lab partner's ear.
4. Attach the speculum to the handle of the otoscope. Make sure there is a bright light when you turn the otoscope on. If not, replace the batteries or choose another otoscope.
5. If you examine the right ear, hold the otoscope with your right hand as you would a pencil and rest your fifth digit (little finger) on the cheek of your lab partner. This allows you to use the little finger as a fulcrum and move with your lab partner in the event of sudden movements of the head.
6. With your left hand, hold onto the superior, posterior part of the pinna of the ear, elevate the ear and move it posteriorly a little as you gently insert the otoscope into the ear. If this is painful to your lab partner, stop the procedure because the pain may indicate an ear infection. Insert the otoscope into the ear carefully in an inferior and anterior angle. Do not

bump into the ear canal, but "see your way in" as you insert the speculum into the ear.

7. Inspect the auditory canal for ear wax which may be yellow, gray, or white. Hair is also normal in the canal. Abnormalities include inflammation, pus discharge, or fungal infections causing conditions such as *swimmer's ear*. The tympanic membrane should be pink and translucent, and you should see the malleus attached in the superior part of the membrane. A bulging or yellow coloration of the membrane may indicate middle ear infection.
8. Clean the speculum completely with an alcohol swab or a cotton ball soaked in alcohol before putting the otoscope away.

Structure of the Middle Ear

The tympanic membrane is the border between the outer ear and the middle ear. It is composed of connective tissue covered by epithelial tissue. The membrane is sensitive to sound and vibrates as sound is funneled down the auditory canal. The middle ear consists of a main cavity known as the **tympanic cavity;** three small bones, or **ossicles;** and the auditory tube (fig. 22.1). The ossicle attached to the tympanic membrane is the **malleus** (*malleus* = hammer). The malleus is attached to the **incus** (*incus* = anvil), which is attached to the **stapes** (*stapes* = stirrup).

Sound consists of pressure waves. As these waves strike the tympanic membrane, it vibrates. This vibration is conducted by the ossicles to the oval window. The process of moving from a large-diameter structure (tympanic membrane) to a smaller-diameter structure (oval window) concentrates the sound energy about 17 times (fig. 22.2). Examine the ossicles in fig. 22.1 and on models in the lab. In addition to the ossicles, the **auditory tube** (Eustachian tube) occurs in the middle ear. This tube connects the middle ear to the nasopharynx (fig. 22.1) and equalizes pressure between the middle ear and the external environment when

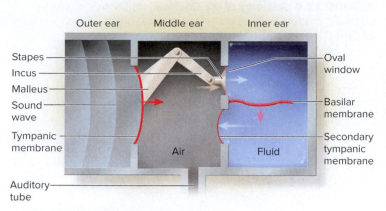

FIGURE 22.2 **Model of Hearing.** Pressure waves of sound vibrate the tympanic membrane, which transfers sound to the oval window and finally to the basilar membrane, where vibration is converted to neural impulses.

changes of pressure occur (such as during changes of elevation). The auditory tube can be a conduit for microorganisms that travel from the nasopharynx to the middle ear and lead to middle-ear infections, particularly in young children.

Structure of the Inner Ear

The inner ear is encased in two labyrinthine structures and filled with two separate fluids. The outermost structure is the **bony labyrinth.** Inside the bony labyrinth is **perilymph,** a clear fluid external to the **membranous labyrinth.** The fluid enclosed by the membranous labyrinth is the **endolymph,** important in both hearing and equilibrium. The membranous labyrinth is enclosed inside the bony labyrinth like a tube within a tube.

The inner ear is a complex structure composed of three regions, the cochlea, the vestibule, and the semicircular ducts (figs. 22.1 and 22.3). The cochlea is a spiral structure that resembles a snail

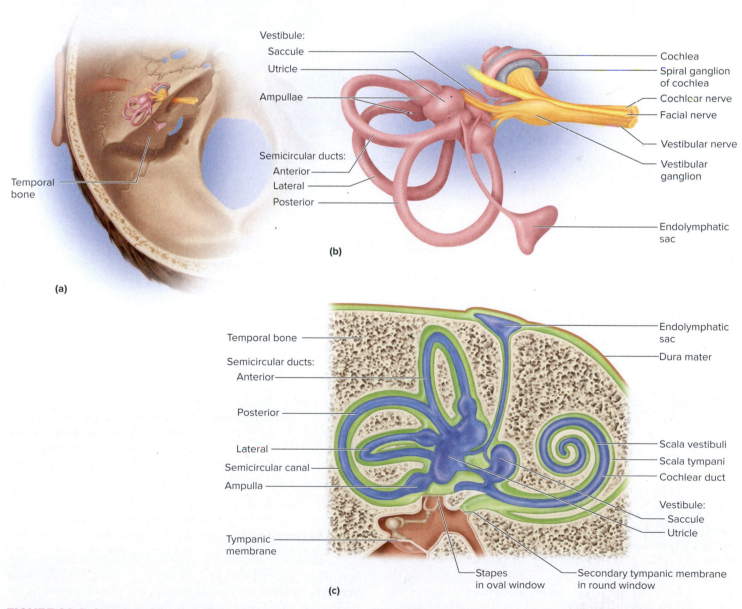

FIGURE 22.3 **Anatomy of the Inner Ear.** (a) Position of inner ear in skull; (b) major regions of the inner ear; (c) relationship of membranous labyrinth to bony labyrinth.

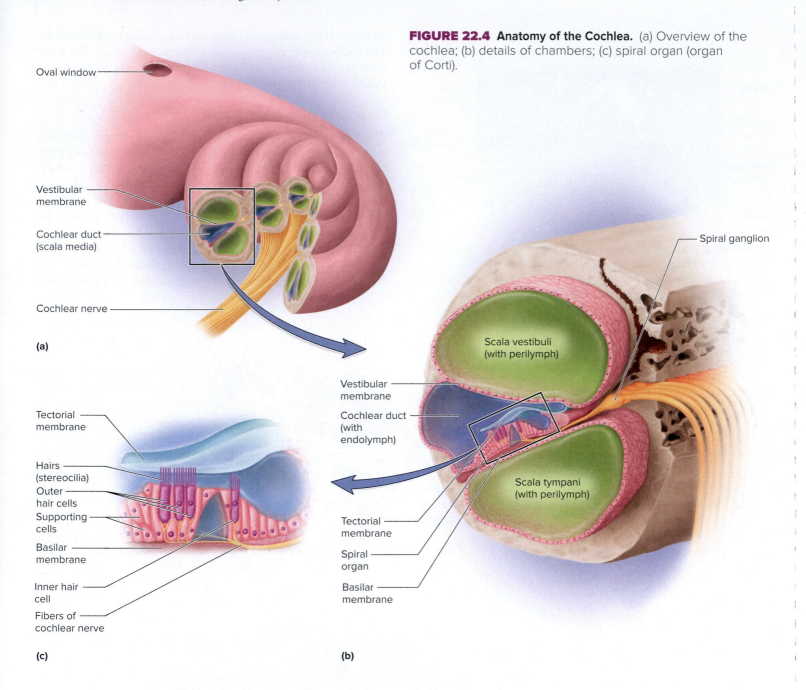

FIGURE 22.4 Anatomy of the Cochlea. (a) Overview of the cochlea; (b) details of chambers; (c) spiral organ (organ of Corti).

shell (*cochlea* = snail), while the semicircular ducts look like three loops. Between these two structures is the vestibule.

Cochlea   The cochlea (figs. 22.1, 22.3, and 22.4) is involved in hearing. As the sound waves travel down the auditory canal, they cause the tympanic membrane to vibrate. This vibration rocks the ear ossicles, which are connected to the inner ear. As the stapes vibrates, it moves back and forth in the **oval window,** and this causes fluid to move back and forth in the cochlea. The cochlea also has a **round window** with a secondary tympanic membrane, which allows the pressure wave from the ossicles to move fluid. Sound waves are measured by their amplitude (loudness) and frequency (pitch). The frequencies are measured in cycles per second, also known as hertz (Hz). The human ear is capable of hearing

high-pitched sounds up to 20,000 Hz and sounds as low as about 20 Hz. High-pitched sounds with vibrations of high frequency, up to 20,000 Hz in the human ear, stimulate the region of the cochlea closest to the middle ear. Low-pitched sounds with low frequency, down to 20 Hz, stimulate the region of the cochlea farther from the middle ear. In this way the cochlea can perceive sounds of varying wavelengths at the same time.

Ⓐ **Activity**

Examine a microscopic section of the cochlea in cross section. There are chambers in a coiled set of three. Find the **scala vestibuli, cochlear duct (scala media),** and **scala tympani** on the microscope slide. Compare these to figs. 22.4 and 22.5.

the **auditory cortex** of the **temporal lobe,** where they are interpreted as sound (fig. 22.6).

Vestibule Another part of the inner ear is the vestibule. The **vestibule** consists of the **utricle** and **saccule** (figs. 22.3 and 22.7). These two chambers are involved in the interpretation of static equilibrium and linear acceleration. The utricle and saccule have regions known as maculae, which consist of hair cells with stereocilia and a kinocilium grouped together in a gelatinous mass called the otolithic membrane and weighted with calcium carbonate stones called **otoliths.** These can be seen in fig. 22.7. As the head is accelerated or tipped by gravity, the otoliths cause the microvilli to bend, indicating that the position of the head has changed. Static equilibrium is perceived not only from the vestibule but from visual cues as well. When the visual cues and the vestibular cues are not synchronized, a sense of imbalance or nausea can occur.

Semicircular Ducts The third part of the inner ear consists of the **semicircular ducts,** each located inside a semicircular canal, involved in determining dynamic equilibrium. There are three semicircular ducts, each at 90 degrees to one another (in the horizontal, sagittal, and coronal planes) (fig. 22.8). Each semicircular duct is filled with endolymph and is expanded at its base into an **ampulla.** Inside each ampulla are clusters of hair cells and support cells (collectively called the **crista ampullaris**). The cells have stereocilia and a kinocilium enclosed in a gelatinous material called the **cupula.** The endolymph in the membranous labyrinth has inertia; that is, it tends to remain in the same place. As the head is rotated, the endolymph pushes against the **stereocilia.** In this way **angular** or **rotational** acceleration can be determined as shown in fig. 22.8c. The three semicircular ducts are at right angles to each other. If motion in the forward plane occurs (such as by doing

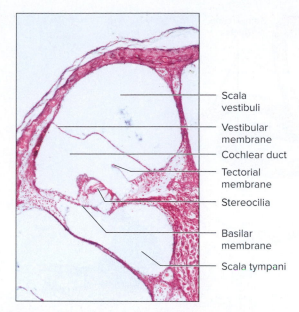

FIGURE 22.5 **Photomicrograph of Cross Section of the Cochlea (100×).**
©Eric Wise

Scala vestibuli
Vestibular membrane
Cochlear duct
Tectorial membrane
Stereocilia
Basilar membrane
Scala tympani

Note the **spiral organ (acoustic organ,** or **organ of Corti)** in the area between the **vestibular membrane** and the **basilar membrane.** The spiral organ in figs. 22.4 and 22.5 is seen in cross section, but remember that it runs the length of the cochlea. The spiral organ is sensitive to sound waves. As a particular region of the spiral organ is stimulated, the basilar membrane vibrates. This bends the cilia that connect the stereocilia to the tectorial membrane. The stereocilia send impulses to the cochlear branch of the **vestibulocochlear nerve.** These impulses travel to

FIGURE 22.6 **Auditory Projection Pathway.** Sound impulses from the cochlea travel via nerves and tracts to the primary auditory cortex.

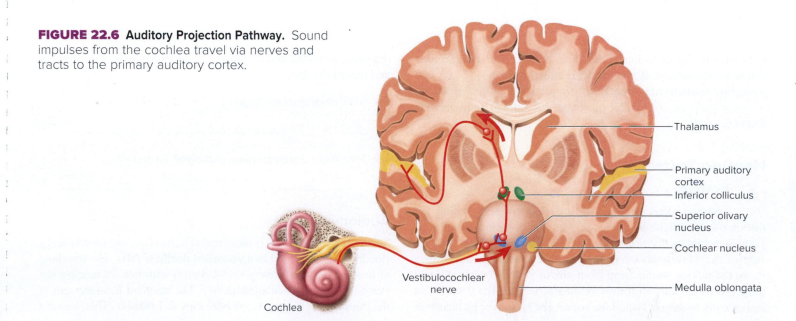

Thalamus
Primary auditory cortex
Inferior colliculus
Superior olivary nucleus
Cochlear nucleus
Medulla oblongata
Vestibulocochlear nerve
Cochlea

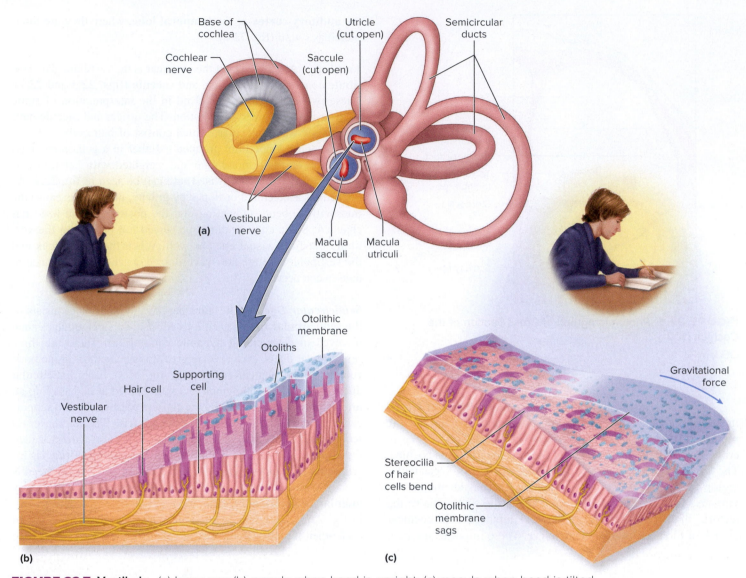

FIGURE 22.7 Vestibule. (a) Inner ear; (b) macula when head is upright; (c) macula when head is tilted.

back flips or flip turns in a pool), then the **anterior semicircular ducts** are stimulated. If you turn cartwheels or lean sideways, the **posterior semicircular ducts** are stimulated. If you spin around on your heels or do a ballet pirouette, the **lateral semicircular ducts** respond.

Hearing Tests

Ⓐ Activity

Obtain a meter stick and a ticking stopwatch. If your lab does not have a ticking stopwatch, you can download sounds on your smartphone by searching for "ticking clock" or "ticking stopwatch." Have your lab partner sit in a quiet room and slowly move the ticking sound away from one of your lab partner's ears until the sound can no longer be heard. Record the distance in centimeters or meters (when the sound can no longer be heard) in

the space provided and in the review section. Check the other ear and record the data.

❓ Maximum distance sound perceived for right ear:

_____ (1)

❓ Maximum distance sound perceived for left ear:

_____ (2)

Audiometer Test

Sound frequency (pitch) is measured in hertz (Hz), and sound energy (loudness) is measured in units called **decibels** (dB). The threshold of hearing is the quietest sound a person with normal hearing can perceive at any particular frequency. The standard measurement of the threshold of hearing is zero decibels at 1,000 Hz. This does not

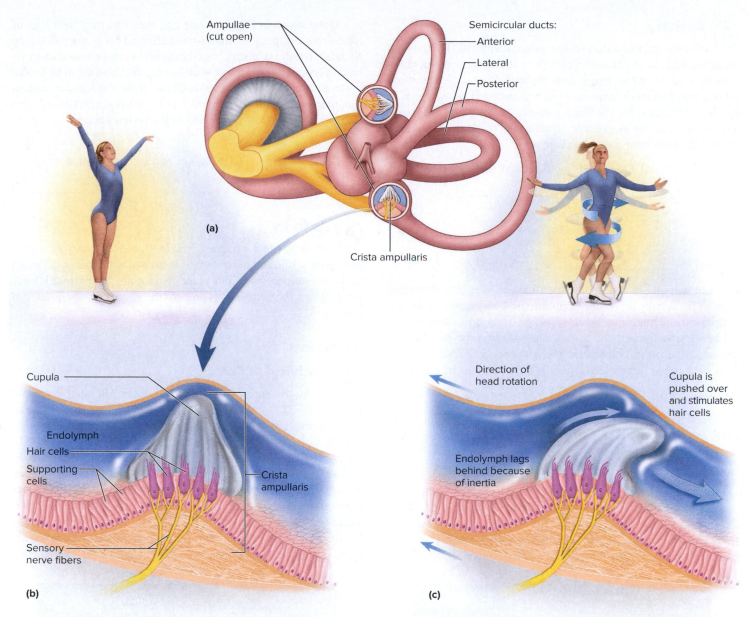

FIGURE 22.8 Semicircular Ducts. (a) Position of semicircular ducts; (b) position of semicircular ducts when still; (c) function of crista ampullaris and cupula during movement.

mean that there is no sound energy below zero decibels but rather that humans cannot perceive sound below that value. Decibels are measured on a logarithmic scale, so for each 10 dB increase in sound, 10 times more energy is produced. Therefore 30 dB is 10 times louder than 20 dB, but 40 dB is 100 times louder than 20 dB.

In clinical settings, a tuning fork test is frequently performed to assess a person's hearing, and if it appears that there is hearing loss, a more refined test using an audiometer may be suggested. Audiometry involves the use of a machine (an audiometer) that can produce sounds of single frequency at different decibels.

Ask your instructor to demonstrate how to use an audiometer to test hearing. **Hearing loss (presbycusis)** occurs after exposure to loud noises, or it can be an age-related event. Age-related hearing loss may be due to the loss of hair cells in the inner ear, stiffening of the tympanic membrane, loss of flexibility in the ossicles, or all of the above. The result is that voices may sound mumbled, or the

person has ringing in the ears (tinnitus), and there is a decrease in the perception of high-frequency sounds. Hearing loss is greatly accelerated by loud music, moderate to extensive headphone use, and exposure to machines that operate at high decibel levels. Another common source of hearing loss is the use of firearms without hearing protection. This cannot be reversed, so it is important to take care of your hearing as much as possible when you are young.

Measuring Hearing

A person with normal hearing can perceive sound at a minimum of 1 to 20 decibels over the range of 500 to 4,000 Hz. A person with mild hearing loss will not be able to hear some sounds below 20 dB in the 500–4,000 Hz range but will be able to hear sounds if the sound is increased between 20–40 dB. A person with moderate hearing loss will need to have the level increased to 40–60 dB, and a person with severe hearing loss will need the sound increased to above 60 dB.

A Activity

To measure this value, set the audiometer to 500 Hz and then lower the level in 10 dB increments from 50 dB to the point where your lab partner can no longer hear the tone. Record the threshold for hearing. Test other frequencies as well, as listed below. Hearing loss is significant when the hearing threshold is at 20 dB or more at any two frequencies in an ear or 30 dB at any particular frequency. Record the minimum sound levels in dB.

Right Ear (dB)   **Left Ear (dB)**

   500: _____   _____

1,000: _____   _____

2,000: _____   _____

4,000: _____   _____

The audiometer test is not specifically a gauge to determine whether the hearing loss is due to **conductive hearing loss,** where the tympanic membrane or ear ossicles may be damaged, or **sensorineural hearing loss,** where the damage may occur in the cochlea, vestibulocochlear nerve, or temporal lobe of the brain. Several tests have been used to determine which type of hearing loss may be involved.

Weber Test

Conductive deafness occurs when damage to the tympanic membrane or ear ossicles is present. You can test for this type of deafness with a tuning fork.

A Activity

Hold on to the handle of a middle C tuning fork. Strike the tines (the forked portion) on a soft but solid surface (such as the bottom of your shoe or the heel of your hand) and gently place the end of the handle on your lab partner's chin or forehead (fig. 22.9).

FIGURE 22.9 Weber Test.
©Eric Wise

If the sound is louder in one ear, then this may be a sign of conductive deafness. In this case the affected ear is not picking up sounds from the tympanic membrane and is more sensitive to vibrations coming through the skull bones. The sound will be louder in the ear that has conductive deafness. If the ear has nerve damage, then the sound will be louder in the normal ear. Record your results in the following space and in the review section.

? Sound perception (right/left or equal in both ears):

_____ (3)

A Activity

To simulate conductive hearing loss with the Weber test, have your lab partner close off one ear with the pad of a finger, strike the tuning fork, and place it on your lab partner's forehead. When the ear canal is closed, the sound will be louder in that ear.

Rinné Test

A Activity

Strike the tuning fork and place the handle on the mastoid process (fig. 22.10). When your lab partner indicates that the sound disappears, lift the tuning fork from the mastoid process and bring it to the outside of the pinna with the ends of the tines close to the auditory canal. Can the sound now be heard? If the sound cannot be heard after the tuning fork is placed near the outside of the ear, then there is damage to the tympanic membrane or ear ossicles (conductive deafness), a positive Rinné test. This may also be due to obstructions in the ear canal (such as earwax). The sound is normally louder in air because the auditory canal and tympanic membrane are more sensitive to sound traveling by air than by bone conduction.

FIGURE 22.10 Rinné Test.
©Eric Wise

Record your results in the following spaces and in the review section.

? Right ear:_____ (4)

? Left ear: _____ (5)

Bing Test

Another hearing test that examines hearing loss is the Bing test.

Ⓐ **Activity**
Strike the tuning fork and place it on the mastoid process as you did with the Rinné test. With your other hand, close off the auditory canal with the pad of a finger. A person with normal hearing or one with sensorineural hearing loss will hear the sound better when the ear canal is closed. This is a **positive Bing test.** A person with conductive hearing loss will not notice a change in sound. This is a **negative Bing test.** Record your results in the following space and in the section at the end of the exercise.

? Results from the Bing test: _____ (6)

Sound Location Test

Ⓐ **Activity**
Have your lab partner sit with eyes closed. Strike the tuning fork with a rubber reflex hammer above your lab partner's head. Have your lab partner describe to you where the sound is located. Strike the tuning fork behind the head, to each side, in front, and below the chin of your lab partner. Do not move the tuning fork after you strike it, but strike it at the location indicated in the chart below. Record the results in the following chart.

Location Where Sound Was Struck	Where Sound Was Perceived
Above head	
Behind head	
Right side	
Left side	
In front of head	
Below chin	

Postural Reflex Test

Unexpected changes that move the body away from a state of equilibrium cause postural reflexes that compensate for that change. If you are facing forward on a boat and it begins to tilt to the starboard (right), your body compensates and moves to the left.

Postural reflexes are important for maintaining the upright position of the body. These reflexes are negative-feedback mechanisms. If, for example, you lean slightly to the left, your left foot abducts to regain the center of balance.

Ⓐ **Activity**
1. Select an area free from any obstacles.
2. While you read this manual, stand on your tiptoes. Your lab partner should give you a little nudge to the left or right but (not enough to knock you off your feet) to push you off balance. Another lab partner should watch to see the movement of your feet. The postural reflex is reflected in the movement of the foot on the opposite side of your lab partner. If you are nudged to the left, then your left foot should move to the side to correct against the direction of the contact. Did your postural reflex work?
3. Record your result in the following space and in the review section.

? Postural reflex: _____ (7)

Barany's Test

Barany's test examines visual responses to changes in dynamic equilibrium. As the head turns, one of the reflexes that occurs is movement of the eyes in the opposite direction of the rotation. Nerve impulses from the semicircular ducts travel to the brain which subsequently sends neural signals that innervate the eye muscles and cause the eye movement. When rotating in one direction, the eyes move in the opposite direction so there is enough time for a visual image to be fixed. The volunteer for this test should not be subject to dizziness or nausea. Have one of the members of the lab record a video on the subject's smartphone so the subject can see the effect as well.

Ⓐ **Activity**
1. Place the subject in a swivel chair with four or five students close by, standing in a circle, in case the subject loses balance and begins to fall. The student chosen for this exercise should grab onto the chair firmly so as not to fall off. The subject should tilt their head forward about 30 degrees, which will place the lateral semicircular ducts horizontally for maximum stimulation.
2. One member of the group should spin the chair around about 10 revolutions while the subject keeps their eyes open.
3. Stop the chair and have the subject look forward. The twitching of the eyes is called **nystagmus** and is due to the stimulation of the endolymph flowing in the semicircular ducts. When the chair is stopped, the fluid in the endolymph will have overcome the inertia and will continue to flow in the ducts. Did the movement of the eyes occur in the direction of the rotation or in the opposite direction? Record your results in the following space and in the review section.

? Direction of chair rotation relative to subject: _____ (8)

? Direction of eye rotation: _____ (9)

Romberg Test

The Romberg test involves testing the static equilibrium function of the body. This should be done in pairs. An observer watches the degree of sway of the subject, who stands near a wall or blackboard.

 Activity

1. Your lab partner should stand with their back next to a chalkboard, whiteboard, or wall. Your lab partner should not be leaning on anything. If you cannot see the shadow of your lab partner, use a bright desk lamp to cast a shadow.
2. Make a mark at the edge of the shadow and record the amount of lateral sway that is produced by your lab partner over the time period of one minute.

3. Record your results in the following space and in the review section.

❓ Amount of lateral sway: _____ (10)

Have your lab partner close their eyes (to reduce visual cues) and repeat the test. Is the swaying motion increased or decreased? Record your results in the following space.

❓ Amount of lateral sway with eyes closed: _____ (11)

REVIEW SECTION

Ear, Hearing, and Equilibrium

Name _____   *Date* _____

Lab Section _____   *Time* _____

❓ Chapter Summary Data

Use this section to record your results from questions within the exercise.

1. _____   7. _____

2. _____   8. _____

3. _____   9. _____

4. _____   10. _____

5. _____   11. _____

6. _____

Review Questions

1. What are the three general regions of the ear? _____
   _____

2. The pinna of the ear consists of what two main parts? _____
   _____

3. To what sensory modality (or modalities) does the ear respond? _____
   _____

4. The ear performs two major sensory functions. What are they? _____
   _____

5. What structure separates the outer ear from the middle ear? _____

6. Trace the pathway of sound waves (pressure waves) from the outer ear to the inner ear.
   _____
   _____

7. From the following choices, select the function of the cochlea.

   a. static equilibrium

   b. taste

   c. hearing

   d. dynamic equilibrium

8. What area is found between the scala vestibuli and the scala tympani?

   _____

9. What is the name of the nerve that carries action potentials from the cochlea and vestibule to the brain?

   _____

10. What units are used to measure sound energy? _____

11. What part of the inner ear is involved in conducting signals of static equilibrium? _____

    _____

12. Name the parts of the ear that might be impaired if a person demonstrates conductive deafness.

    _____

13. What two diagnostic tests are used to determine conductive deafness? _____

14. What is the name of the canal that runs from the auricle to the tympanic membrane? _____

15. How would you differentiate conductive hearing loss from sensorineural hearing loss based on hearing tests?

    _____

    _____

16. What tube between the middle ear and the nasopharynx is responsible for the equalization of pressure when you change elevation?

    _____

17. What is the name of the space that encloses the ear ossicles? _____

18. Name the ear ossicles in sequence from the tympanic membrane to the oval window.

    _____

19. Name all the parts of the inner ear. _____

    _____

    _____

20. Background noise affects hearing tests. In the ticking watch test, what type of result, in terms of auditory sensitivity, would you have recorded if moderate background noise were present?

    _____

21. In the Weber test, the ear that perceives the sound as being louder is the deaf ear. Why is this the case?

    _____

    _____

22. Fill in the following illustration of the ear using the terms provided.

| auditory canal | cochlea | helix | middle ear | semicircular ducts | tympanic membrane |
| auditory tube | lobule | inner ear | outer ear | stapes | |

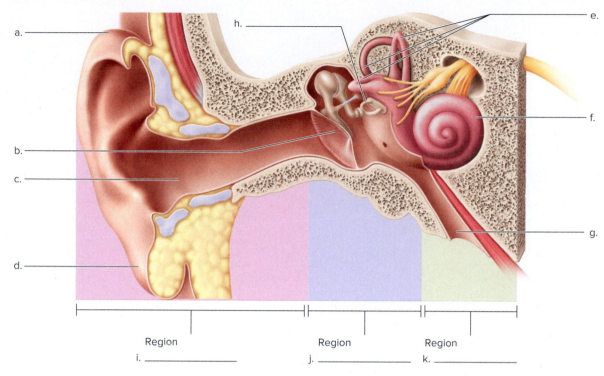

Region
i. _____

Region
j. _____   k. _____

23. Label the following illustration of the cross section of the cochlea using the terms provided.

hair cells

scala media (cochlear duct)

scala tympani

scala vestibuli

tectorial membrane

vestibular membrane

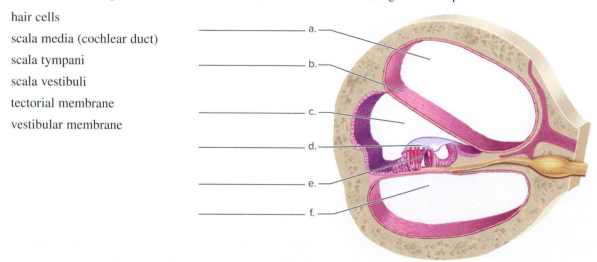

24. If you had an infection in the inner ear that affected the cochlea but not the vestibule, what would possibly be impaired?

Notes

Endocrine System

INTRODUCTION

The **endocrine system** consists of glands or organs that produce chemical messengers called **hormones,** which are picked up by blood capillaries and distributed throughout the body. This type of release is called a "ductless secretion," which is characteristic of the endocrine system. On the other hand, the glands that secrete material in tubules or ducts (such as sweat and salivary glands) or directly to a surface (ovary) are known as **exocrine glands.** The secretions of the endocrine glands enter the interstitial fluid and then travel by blood vessels, which act as highways to carry the hormones throughout the body. Areas receptive to hormones are called **target cells,** and these may be found in specific organs or tissues. Hormones produce many effects. Some of these are growth; changes and development, or maturation; metabolism; sexual development; regulation of the sexual cycle; and homeostasis. Every organ of the body, including the stomach, heart, and kidney, produce hormones and thus have endocrine functions. This exercise focuses on the glands that have a major endocrine component as their function. The endocrine system is presented in more detail in the Saladin text in chapter 17, "The Endocrine System."

OBJECTIVES

At the end of this exercise, you should be able to

1. list the major endocrine organs of the human body;
2. name the hormones produced by these endocrine organs;
3. list the hormones stored in the anterior and posterior pituitary glands;
4. discuss how the secretions of the endocrine glands differ from those of the exocrine glands;
5. identify endocrine organs in histological slides.

MATERIALS

Models and charts of endocrine glands

Microscopes

Microscope slides of thyroid, pituitary, adrenal glands; pancreas; testis; ovary

PROCEDURE

Anatomy of the Major Endocrine Organs

Ⓐ **Activity**

Examine models or charts in lab as you look at the endocrine glands described in this exercise. Locate the major endocrine glands in fig. 23.1, including the pineal gland, hypothalamus, pituitary gland, thyroid, parathyroids, thymus, pancreas, adrenals, and gonads (testes or ovaries). Once you have noted their location, you can proceed with a more detailed study.

Pineal Gland

Ⓐ **Activity**

The **pineal gland** develops from the diencephalon of the brain. It is named for its resemblance to a pine nut. Examine a model of the median section of the head or of the brain specifically. Locate the gland in the material in lab and in fig. 23.1. The pineal gland secretes the hormone **melatonin.** The role of melatonin is not completely understood in humans.

Hypothalamus and Pituitary Gland

Ⓐ **Activity**

Examine a model of the median section of the head or of the brain specifically. Locate the hypothalamus and pituitary in the material in lab and compare them with fig. 23.1. The hypothalamus and the pituitary gland are so closely interrelated that it is appropriate to discuss them together. The hypothalamus is located in the inferior portion of the forebrain and is connected to the anterior pituitary by the hypophyseal portal system. Primary capillaries in the hypothalamus lead to the secondary capillaries in the anterior pituitary. The hypothalamus has neurons that extend into the posterior pituitary by the hypothalamohypophyseal tract, so the posterior

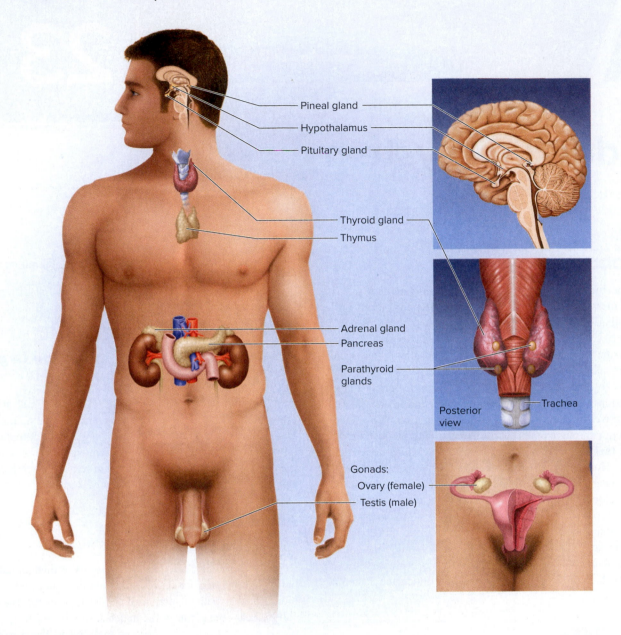

FIGURE 23.1 Major Endocrine Glands.

pituitary can actually be considered an extension of the hypothalamus. The hypothalamus secretes both stimulating and inhibitory hormones that either cause the release of hormones from their target areas or prevent the release of hormones from these areas.

The **pituitary gland,** or **hypophysis,** is seen in figs. 23.1 and 23.2. It is divided into an **anterior lobe,** or **adenohypophysis;** and a **posterior lobe,** or **neurohypophysis.** The pituitary is suspended from the hypothalamus by a stalklike extension of the posterior lobe called the **infundibulum.**

Adenohypophysis

The adenohypophysis originates from the roof of the oral cavity during embryonic development. The result of this development can be seen if you examine a prepared slide of the pituitary gland.

Ⓐ **Activity**

In the histological section, note that the adenohypophysis is composed of simple cuboidal epithelial cells (fig. 23.3). The cells of the anterior pituitary have the same embryonic origin, yet they have differentiated into several types of specialized cells. These cells produce a number of hormones that have broad effects throughout the body.

Some of the hormones produced in the anterior pituitary and their functions are as follows:

- **Thyroid-stimulating hormone (TSH),** or thyrotropin, stimulates the thyroid gland to produce thyroid hormones.
- **Growth hormone (GH),** or somatotropin, promotes growth of most of the cells and tissues of the body.

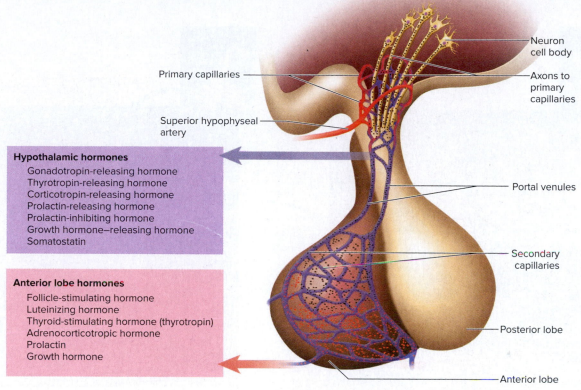

Hypothalamic hormones

Gonadotropin-releasing hormone
Thyrotropin-releasing hormone
Corticotropin-releasing hormone
Prolactin-releasing hormone
Prolactin-inhibiting hormone
Growth hormone–releasing hormone
Somatostatin

Anterior lobe hormones

Follicle-stimulating hormone
Luteinizing hormone
Thyroid-stimulating hormone (thyrotropin)
Adrenocorticotropic hormone
Prolactin
Growth hormone

Neuron cell body

Primary capillaries

Axons to primary capillaries

Superior hypophyseal artery

Portal venules

Secondary capillaries

Posterior lobe

Anterior lobe

(a)

Hypothalamus

TRH
GnRH
CRH

PRL

GH

Liver

IGF

Mammary gland

Hypothalamo–pituitary–gonadal axis

Fat, muscle, bone

Hypothalamo–pituitary–thyroid axis

Hypothalamo–pituitary–adrenal axis

TSH

ACTH

Thyroid

Adrenal cortex

LH
FSH

FIGURE 23.2 Pituitary Gland.
(a) Overview of the pituitary gland;
(b) target areas of the anterior
pituitary.

Testis

Ovary

(b)

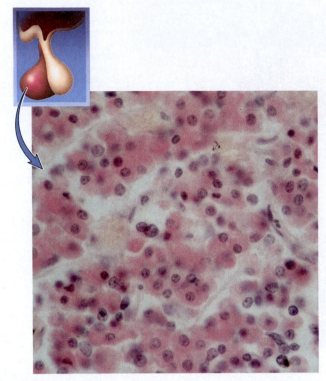

FIGURE 23.3 Histology of the Anterior Pituitary Gland (400×).
©Eric Wise

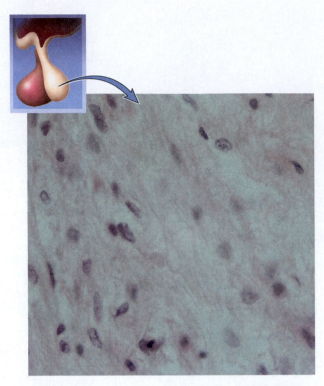

FIGURE 23.4 Histology of the Posterior Pituitary (400×).
©Eric Wise

- **Prolactin (PRL)** stimulates mammary glands to begin the production of milk.
- **Gonadotropins** stimulate ovaries and testes.
  —**Follicle-stimulating hormone (FSH)** stimulates the production of sex cells, causes ovarian follicles to mature (develop and produce egg cells), and stimulates testes to produce spermatozoa.
  —**Luteinizing hormone (LH)** stimulates ovaries to produce other hormones (for example, progesterone), causes the maturation of Graafian follicles and ovulation, and stimulates testosterone production in testes.
- **Adrenocorticotropic hormone (ACTH)** controls hormone production in the adrenal cortex.

Neurohypophysis

The tissue of the neurohypophysis, or posterior pituitary, is very different from that of the adenohypophysis. The neurohypophysis is composed of nervous tissue that originated from the base of the brain.

(A) **Activity**

Compare the tissue of the neurohypophysis in a prepared slide to fig. 23.4.

Hormones released from the posterior pituitary are synthesized by the **hypothalamus** and flow through axons to be stored in the posterior pituitary. These hormones and their actions include the following:

- **Antidiuretic hormone (ADH)** stimulates the reabsorption and retention of water by the kidneys. It is also known as

vasopressin because it causes arterioles to constrict, which elevates blood pressure.
- **Oxytocin** stimulates the contraction of the cells of the mammary glands, resulting in the release of milk, and causes uterine contractions. External influences on hormone actions can be seen in the release of oxytocin. Look at fig. 23.5 and note that the sucking action of an infant on the mother's breast sends impulses to the hypothalamus. The numbers indicate the sequence of events. The hypothalamus stimulates the posterior pituitary to release oxytocin, which causes milk ejection.

Thyroid Gland

Fig. 23.6 illustrates the **thyroid gland,** named for its location inferior to the thyroid cartilage of the larynx. The thyroid gland has two lateral **lobes** and a medial **isthmus** that connects them.

(A) **Activity**

The thyroid gland is easy to recognize histologically. Examine a slide of thyroid gland and compare it to fig. 23.7. Locate the **follicular cells** that surround the **colloid,** a storage region for thyroid hormones. Colloid is mostly composed of thyroglobulin, a large molecular weight compound. Thyroid hormones are formed inside the larger thyroglobulin molecule. Also locate the **parafollicular cells,** which occur in the spaces between the follicles.

The thyroid gland secretes three specific hormones. Two of these are T_3, or **triiodothyronine** (a molecule that includes three iodine atoms), and T_4, or **thyroxine** (containing four iodine atoms

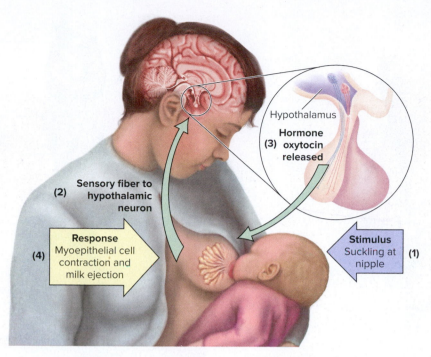

FIGURE 23.5 **External Influences on Hormonal Action.** (1) External stimulus of suckling; (2) sensory neurons take impulse to brain, where (3) hypothalamus stimulates posterior pituitary to release oxytocin and (4) milk is released from the breast.

per molecule). These hormones increase basal metabolic rates, are stored in the colloid, and will be studied in the physiology simulation experiment later in this exercise. Another hormone of the thyroid gland is **calcitonin.** Calcitonin decreases blood calcium levels by causing excretion of calcium by the kidneys and deposition of calcium in bone by decreasing osteoclast activity and formation. The role of calcitonin is important in children and may protect individuals from hypercalcemia. Calcitonin is produced by the parafollicular, or C, cells. Calcitonin is antagonistic to parathyroid hormone (discussed next).

Parathyroid Glands

Attached to the posterior portion of the thyroid are usually two pairs of organs called the **parathyroid glands. Chief cells** in the parathyroid gland secrete **parathyroid hormone (PTH),** or **parathormone,** responsible for increasing calcium levels in the blood. This occurs by increasing calcium uptake in the intestines, increasing kidney reabsorption of calcium, and releasing calcium from bone. Examine the location of the parathyroid glands embedded in the posterior surface of the thyroid gland (fig. 23.8).

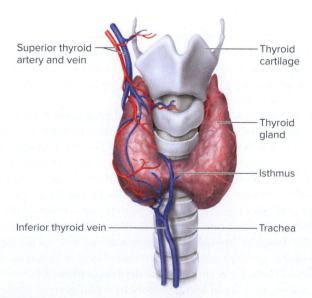

FIGURE 23.6 **Anatomy of the Thyroid Gland.**

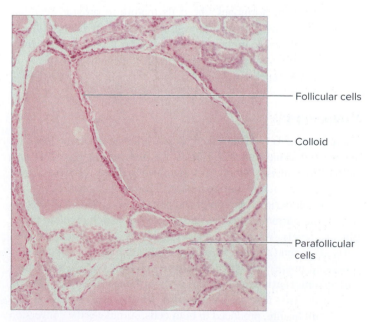

FIGURE 23.7 Histology of the Thyroid Gland (100×).
©Eric Wise

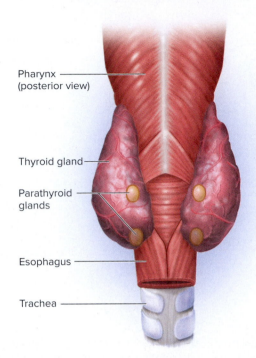

FIGURE 23.8 Parathyroid Glands.

Pharynx (posterior view)

Thyroid gland

Parathyroid glands

Esophagus

Trachea

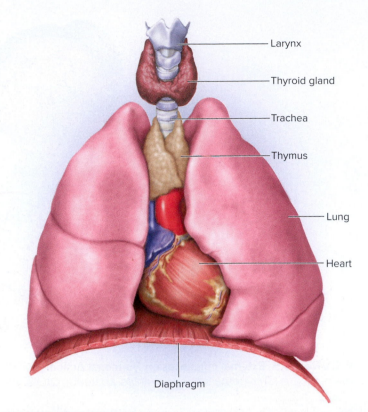

FIGURE 23.9 Thymus.

Larynx

Thyroid gland

Trachea

Thymus

Lung

Heart

Diaphragm

Thymus

The **thymus** is active in young individuals and plays an important part in immunocompetency. The thymus is located anterior and superior to the heart (fig. 23.9). It secretes **thymosin** and other hormones that causes the maturation of T cells. These cells originate as stem cells in the bone marrow and migrate to the thymus. Under the influence of hormones, the cells mature to provide "cellular immunity" against antigens (foreign material, such as bacteria and viruses, which cause immune reactions). The T cells migrate from the thymus predominantly to the lymph nodes and the spleen to carry out their functions.

Pancreas

The **pancreas** is a mixed gland in that it has an exocrine function and an endocrine function. Locate the pancreas in figs. 23.1 and 23.10. The exocrine function is digestive because pancreatic juice contains both sodium carbonate and digestive enzymes.

The endocrine function of the pancreas consists of the secretion of the hormones insulin and glucagon, which regulate blood glucose levels. When blood glucose levels drop, **glucagon** converts glycogen (a starch storage product) to glucose. Glucagon is produced in specialized cells **(alpha cells)** that occur in clusters called **pancreatic islets** (or the islets of Langerhans) (fig. 23.10). Pancreatic islets also produce **insulin,** which lowers the blood glucose level. Insulin, produced in **beta cells,** stimulates the conversion of glucose to glycogen. The pancreas also contains **delta cells,** which secrete **somatostatin.** Somatostatin is secreted after the ingestion of

a meal and inhibits the secretion of insulin and glucagons from the pancreas. This may play a role in increasing efficiency in digestion.

 Activity

Locate the pancreatic islets on a microscope slide and compare them to those in fig. 23.10.

Adrenal Glands

 Activity

The **adrenal glands** (**ad** = next to, **renal** = kidney) are superior in position to the kidneys. Each adrenal gland is composed of an outer **cortex** and an inner **medulla.** Examine a microscope slide of an adrenal gland and locate the cortex and the medulla. Compare these with figs. 23.11 and 23.12.

The hormones secreted from the adrenal cortex are called **corticosteroid hormones.** They are important in water and electrolyte balance (Na^+, K^+) in the body. They are also important for carbohydrate, protein, and fat metabolism, as well as stress management. The cortex can be divided into three regions. The outermost is the **zona glomerulosa** (figs. 23.11 and 23.12), which consists of clusters of cells that secrete **mineralocorticoids** (especially **aldosterone**). Mineralocorticoids affect salt and water balance. Deep to this layer (closer to the medulla) is the **zona fasciculata,** which consists of parallel bundles of cells that secrete **glucocorticoids,** especially **cortisol (hydrocortisone).** Glucocorticoids regulate protein and fat catabolism and gluconeogenesis (production of glucose from amino acids). They are also involved

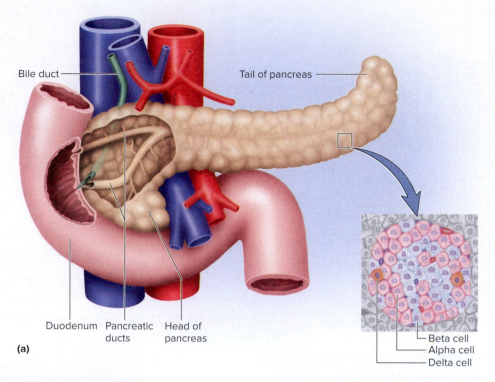

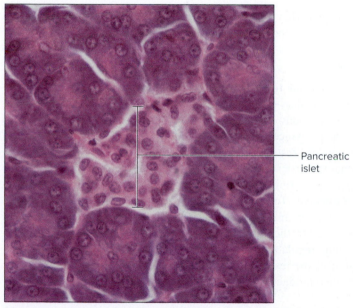

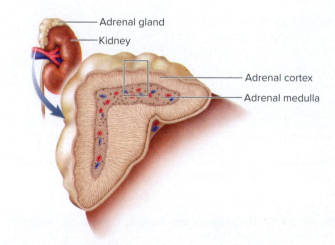

FIGURE 23.10 Pancreas. (a) Gross anatomy; (b) histology (400×).

(b) ©Eric Wise

FIGURE 23.11 Adrenal Cortex and Medulla. Boxed area is expanded in figure 23.12.

in reducing inflammation. The zona fasciculata also secretes **androgens** (male sex hormones). The deepest cortical layer is the **zona reticularis,** which consists of a branched pattern of cells that produce both **glucocorticoids** and **sex hormones** (androgens and estrogens). The hormone dehydroepiandrosterone (DHEA) is an androgen, a male sex hormone, and both males and females produce DHEA from the adrenal glands in small amounts. DHEA is secreted from the adrenal glands but converted to testosterone

or estrogens in other tissues. In males, the testes secrete significantly more androgens than the adrenal glands. Estrogen **(estradiol)** is secreted in small amounts in both sexes by the zona reticularis.

The hormones **epinephrine** and **norepinephrine** are produced in the adrenal medulla. Stimulation of the adrenal glands by the sympathetic nervous division causes the release of epinephrine and norepinephrine from the gland, which increases heart rate and

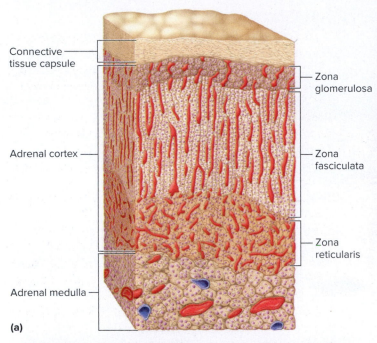

(a)

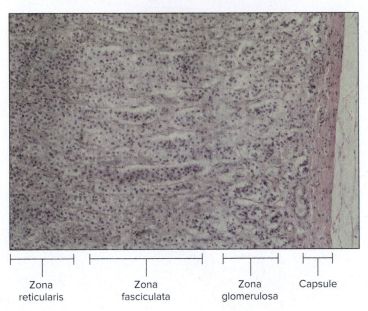

Zona reticularis   Zona fasciculata   Zona glomerulosa   Capsule

(b)

FIGURE 23.12 Histology of the Adrenal Cortex. (a) Diagram; (b) photomicrograph (100×).

(*b*) ©Eric Wise

prepares the body for "fight-or-flight" reactions by making the individual more alert and metabolizing fuel sources such as glucose and fatty acids.

Gonads

The gonads, testes and ovaries, produce not only sex cells but also hormones, so they are considered endocrine glands. Both are stimulated by follicle-stimulating hormone (FSH) from the anterior pituitary, which causes the production or maturation of the sex cells (spermatozoa or

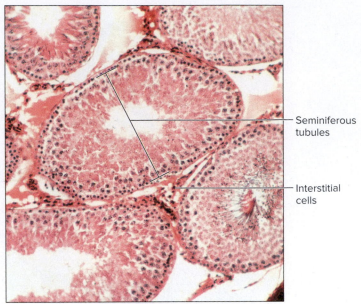

FIGURE 23.13 Histology of the Testis (100×).

©Eric Wise

oocytes). They are also under the influence of luteinizing hormone (LH), which increases the level of hormone production.

Testes

In males the **testes** produce **testosterone,** a hormone responsible for secondary sex characteristics, such as the development of facial and body hair, the expansion of the larynx (which produces a deeper voice), and the increased muscle and bone mass seen in males. Testosterone not only is responsible for secondary sex characteristics but also influences the development of the male genitalia (penis and scrotum) during embryological and fetal development. Testosterone, in conjunction with FSH, aids in the production of **spermatozoa. Inhibin** is a hormone produced and secreted by the testes, and it is involved in negative feedback, providing regulation for testosterone production. The exocrine function of the testis is explored in Laboratory Exercise 40, on the male reproductive system.

(A) Activity

Examine a microscope slide of a testis and find the **seminiferous tubules** and **interstitial cells.** Refer to fig. 23.13 for assistance. The interstitial cells are found between the tubules and produce testosterone.

Ovaries

In females the ovaries produce oocytes (eggs) and have an endocrine function in that they produce primarily the estrogens **estradiol** and **progesterone** (fig. 23.14). Female hormones are also responsible for secondary sex characteristics in women, such as the development of breasts, an additional subcutaneous adipose layer, and a higher voice. Estradiol and progesterone also influence the

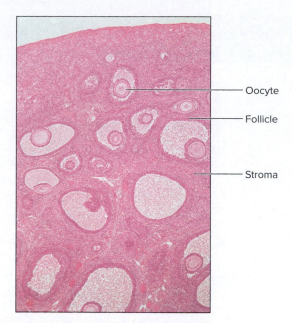

FIGURE 23.14 **Histology of the Ovary (40×).**
©Eric Wise

development of the endometrium, cause maturation of the oocytes, and regulate the menstrual cycle. *Estrogen* is a generic term for several hormones produced by females, including **estradiol.** Inhibin is also secreted by the ovary and regulates the levels of estrogen and progesterone. The reproductive functions of the ovary are covered in Laboratory Exercise 41.

Endocrine Physiology

Thyroid Hormone and Metabolic Rate   Most mammals maintain their body temperature within a fairly narrow range. Some of this is done by behavior, such as moving into the shade when it is hot. There are physiological changes as well, such as secretion of thyroid hormones that increase metabolic rate or shivering when the body temperature drops below what can be adjusted by hormones. In this part of the exercise, you will examine the response to a decrease in temperature by using two virtual mice. One subject is the control and the other is the experimental mouse that has been fed a diet including propylthiouracil (PTU at 0.15%), which lowers T_3 and T_4 production. In real mice, PTU shows a temporary and reversible decrease in thyroid hormone production.

(A) **Activity** **PhILS**

1. Open the Ph.I.L.S. 4.0 program and select Endocrine Function 19: Thyroid Gland and Metabolic Rate. There is an introduction and a segment where you need to answer five questions before clicking on the "continue" button to enter the Wet Lab part of the program.
2. Read the instructions, click the power switch and turn on the scale, and tare the scale (make sure that it reads zero).
3. Click the power switch on the thermostat.
4. Click and drag the normal mouse from the cage and place it on the top of the scale.
5. Click and drag the mouse to the wire cage on the left side of the screen.
6. Click on the orange bulb and move the eye dropper to the right of the 10 on the right side of the screen. You will release the bulb so that the bubble is at the end of the pipette.
7. At the upper left of the screen, there is a data table. Enter a 10 (the placement of the bubble) at time 0.00.
8. Start the procedure and every 15 seconds observe the pipette reading and enter the data in the data table.
9. Click the pause button at 2 minutes and click on the "Calc" button to enter your score.
10. You have recorded the data for the mouse at 24°C.
11. Start a new experiment by placing a new bubble to the right of the 10.
12. Adjust the thermostat to 22°C.
13. Enter a 10 in the data table and run the experiment again. Make sure you press the "Calc" button after the experiment.
14. Run the experiment at least one more time at a temperature below 15°C.
15. Return the mouse to where you got it originally and select the mouse fed with PTU.
16. Weigh the new mouse and run at least three trials, with one of them below 15°C.
17. Take the post lab quiz. You can screen grab the data and put it in a file on the desktop if you need to toggle between the post lab quiz and the data.

(A) **Activity**

Once you have studied the endocrine glands and material, complete table 23.1. You should fill in the blank spaces where needed.

TABLE 23.1

Organ	Hormones Produced	Effect of Hormones
_____	Antidiuretic hormone	_____
Thyroid	Thyroxine	_____
_____	Corticosteroid hormones	Regulate electrolyte balance
Ovary	_____	Regulates ovarian cycle
_____	Melatonin	May regulate sleep cycles, inhibits release of reproductive hormones
_____	Luteinizing hormone	_____
Neurohypophysis	Oxytocin	_____
Parathyroid	_____	Increases calcium in blood
_____	Glucagon	Increases blood glucose levels
Pancreas	_____	Decreases blood glucose levels
_____	Calcitonin	_____

REVIEW SECTION

Endocrine System

Name _____ *Date* _____

Lab Section _____ *Time* _____

Review Questions

1. What is the general name given for the types of organs that produce hormones? _____

2. What is the generic name for cells or tissues receptive to hormones? _____

3. Melatonin is secreted by what gland? _____

4. Where is ADH stored? _____

5. What is the effect of TSH, and where is it produced? _____
_____

6. What connects the two lobes of the thyroid gland? _____

7. Does parathormone increase or decrease calcium levels in the blood? _____

8. What does glucagon do as a hormone, and where is it produced? _____
_____

9. Which hormones in the adrenal gland control water and electrolyte balance? _____

10. What is the primary gland that secretes epinephrine? _____

11. Where is growth hormone produced? _____

12. What is another name for T_3? _____

13. Label the endocrine glands indicated in the following illustration.

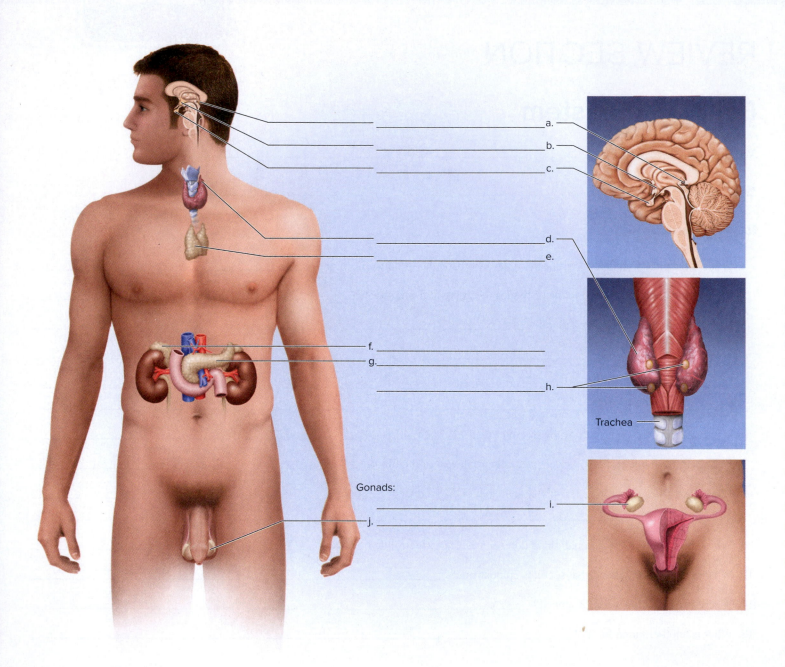

a. _____

b. _____

c. _____

d. _____

e. _____

f. _____

g. _____

h. _____

Trachea

Gonads:

i. _____

j. _____

14. Identify the parts of the pituitary as seen in the illustration, and list two hormones located in each.

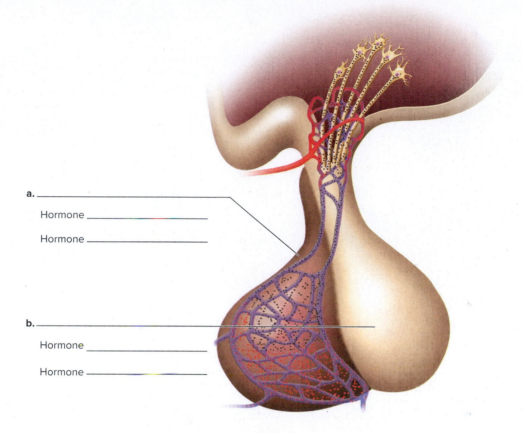

a. _____

   Hormone _____

   Hormone _____

b. _____

   Hormone _____

   Hormone _____

15. Identify the three layers of the adrenal cortex as illustrated, and list the hormones produced by each layer.

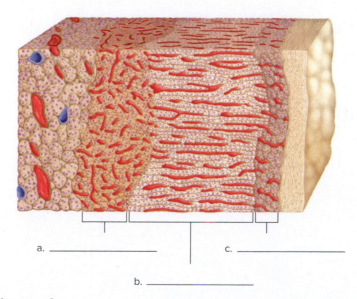

a. _____   c. _____

b. _____

16. Interstitial cells produce which hormone? _____

17. What endocrine organ produces estrogens? _____

Notes

Blood

INTRODUCTION

Blood is a connective tissue that consists of two parts: formed elements and plasma. **Formed elements** make up about 45% of the blood volume and are further subdivided into erythrocytes (red blood cells), leukocytes (white blood cells), and platelets (thrombocytes). Platelets are not cells, and this is why the fraction of blood is called formed elements and not blood cells.

Plasma constitutes approximately 55% of the blood volume and contains water, lipids, dissolved substances, colloidal proteins, and clotting factors. Plasma is the fluid portion of blood and is over 90% water by volume. The remainder mostly consists of proteins, such as albumins, globulins, and fibrinogen. Albumins are produced by the liver and make up the majority of the plasma proteins. They transport solutes and buffer the plasma, which has a pH of about 7.4. Some globulins are made by the plasma cells but most are made in the liver and these make up the next largest amount of proteins. Globulins have an important function for immunity (as antibodies) and they aid in solute transport and blood clotting. Fibrinogen is a clotting protein, and this and other clotting factors are produced by the liver. Plasma also contains electrolytes (such as Na^+, K^+, and Cl^+), nutrients, hormones, and wastes.

The study of blood cells is important because it is the fluid medium of the circulatory system and because it has significant clinical implications. Changes in the numbers and types of blood cells may be used as indicators of disease. In this exercise you examine the nature of blood cells, discussed in the Saladin text in chapter 18, "Blood."

Caution!

The risk of blood-borne diseases has been significantly minimized by the use of sterilized human blood and nonhuman mammal blood in this exercise. However, use the same precautions you would if you were handling fresh and potentially contaminated human blood. Your instructor will determine whether to use animal blood (from nonhuman mammals), sterilized blood, or synthetic blood. In any of these cases, follow strict procedures for handling potentially pathogenic material.

1. Wear protective gloves and goggles during the procedures.
2. Do not eat or drink in lab.
3. If you have an open wound, either make sure the wound is *securely* covered or do not participate in this exercise.
4. Your instructor may elect to have you use your blood. Because of the potential for disease transmission, such as AIDS or hepatitis in fresh blood samples, *make sure you keep away from other students' blood and keep them away from your blood.*
5. Do not participate in this part of the exercise if you have recently skipped a meal or if you are not well.
6. Place all disposable material in the biohazard bag.
7. Place all used lancets in the sharps container and all glassware to be reused in a 10% bleach solution.
8. After you have finished the exercise, clean and disinfect the countertops with a 10% bleach or other disinfecting solution.

OBJECTIVES

At the end of this exercise, you should be able to

1. discuss the composition of blood plasma;
2. distinguish among the various formed elements of blood;
3. determine the percent of each type of leukocyte in a differential white cell count;
4. describe the significance of an elevated level of a particular white blood cell and what it may indicate about a disease state or an allergic reaction.

MATERIALS

Prepared slides of human blood with Wright's or Giemsa stain

Colored pencils

Compound microscopes

Lab charts or illustrations showing the various blood cell types

Dropper bottle of Wright's stain

Large finger bowl or staining tray

Toothpicks

Clean microscope slides

Coverslips

Squeeze bottle of distilled water or phosphate buffer solution

Pasteur pipette and bulbs

Sterile cotton balls

Adhesive bandages

Alcohol swabs

Sterile, disposable lancets

Goggles

Protective gloves

Roll of paper towels

Hand counter

Biohazard bag or container

Container with 10% bleach or other disinfecting solution

Sharps container

PROCEDURE

Examination of Blood Cells

Your lab instructor will direct you to use a prepared slide of blood; fresh, nonhuman mammal blood; or your blood. If you are using a prepared slide of blood, you can move to the section entitled "Microscopic Examination of Blood Cells." If you are to make a smear, read the following directions.

Preparing a Fresh Blood Smear

Ⓐ **Activity**

If you are using nonhuman mammal blood, wear protective gloves and withdraw a small amount of blood with a clean Pasteur pipette. Place a drop of blood on a clean microscope slide; then proceed to step 8 in the following procedure.

Withdrawal of Your Blood

Ⓐ **Activity**

If you are using your blood, make sure you follow the safety precautions discussed at the beginning of the exercise. Obtain the following materials: a clean, sterile lancet; an alcohol swab; an adhesive bandage; protective gloves; sterile cotton balls; two clean microscope slides; and a paper towel.

1. Arrange the materials on the paper towel in front of you on the countertop.
2. Warm your hands with water or by rubbing them to increase blood flow to the fingers.
3. Select a finger on your non-dominant hand (the one you don't use for writing). Clean the end of the donor finger with the alcohol swab, let it dry momentarily, and let the hand from which you will withdraw blood hang by your side for a few moments to collect blood in the fingertips. You will be

puncturing the pad of your fingertip (where the fingerprints are located lateral to the center of the finger).

4. Peel back the covering of the lancet and hold on to the *blunt* end as you withdraw it from the package. Do not touch the sharp end or lay the lancet on the table before puncturing your finger.
5. Prick your finger quickly and wipe away the first drop of blood that forms with a sterile cotton ball. Throw the lancet in the sharps container. Never reuse the lancet or set it down on the table!
6. Place the second drop of blood from your finger on the microscope slide about 2 cm away from one of the ends of the slide (fig. 24.1a) and place another cotton ball on your finger.
7. Put an adhesive bandage on your finger.
8. Whether you are using nonhuman blood or your blood, use another clean slide to spread the blood by touching the drop with the edge of the slide and pulling the blood across the slide (fig. 24.1b & c). This should produce a smooth, thin smear of blood. Let the blood smear dry completely.
9. After the slide is dry, place it in a large finger bowl elevated on toothpicks or on a staining tray.
10. Cover the blood smear with several drops of Wright's stain from a dropper bottle. Let the stain remain on the slide for 1 to 2 minutes.

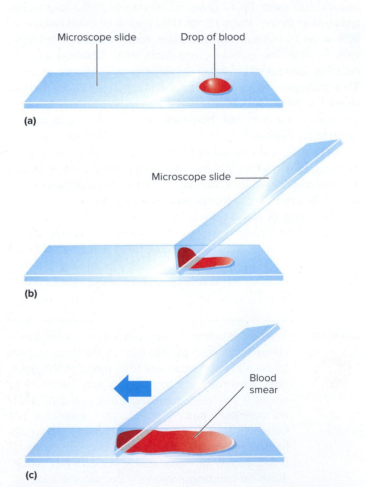

(a)

(b)

(c)

FIGURE 24.1 Making a Blood Smear. (a) Placement of blood on slides; (b) touching glass slide to front of blood drop; (c) spreading blood across slide.

11. After this time, add water or a prepared phosphate buffer solution to the slide. You can rock the slide gently with gloved hands or blow on it to stir the stain and water. A metallic green material should come to the surface of the slide. Let the slide remain covered with stain and water for 4 minutes.

12. Wash the slide gently with distilled water until the material is light pink and then stand it on edge to dry. You may also stain blood using an alternate stain (Giemsa stain), following your lab instructor's procedure.

Place all blood-contaminated disposable material in the biohazard container. Place the slide used to spread the blood in the 10% bleach or other disinfecting solution and place the lancets in the sharps container if you haven't done so already. Once the slide is completely dry it can be examined under the high-power or oil-immersion lens of your microscope.

Microscopic Examination of Blood Cells

Overview

As you examine the blood slide, you should refer to figs. 24.2–24.4. You will have to look at many blood cells to see all the cells included in this exercise.

Erythrocytes

Ⓐ **Activity**

Examine a slide of blood stained with either Wright's or Giemsa stain. Erythrocytes are the most common cells you will find

on the slide. There are about 5 million erythrocytes per microliter. They do not have a nucleus but appear as pink, biconcave disks (like a donut with the hole partially filled in). Erythrocytes contain the pigment **hemoglobin,** which carries oxygen.

Erythrocytes are about 7.5 μm in diameter, on average, and have a life span of about 120 days, after which time they are broken down by the spleen or liver. The iron portion of the hemoglobin is used by the bone marrow to make more hemoglobin, the proteins are broken down to amino acids and reabsorbed for general use by the body, and the heme is metabolized to bilirubin in the liver and secreted as part of the bile. Compare what you see under the microscope to fig. 24.2.

Platelets

There are about 150,000 to 350,000 platelets per microliter of blood. **Platelets,** or **thrombocytes,** are small fragments of **megakaryocytes** involved in clotting.

Ⓐ **Activity**

Examine the slide for small, purple fragments that may be single or clustered and compare them to the platelets in fig. 24.2.

Leukocytes

There are far fewer leukocytes in the blood than erythrocytes. The number of leukocytes in a healthy adult is about 4,300 to 10,800 cells per microliter of blood. Leukocytes are formed in bone marrow and lymphoid tissue. The life span of a leukocyte varies from a few hours to several years. Many are capable of ameboid

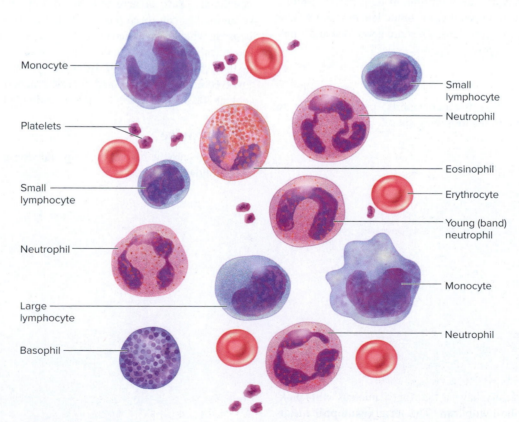

FIGURE 24.2 Formed Elements of Blood.

movement as they squeeze between cells of the vascular endothelium (a process termed *diapedesis*), and engulf foreign particles or cellular debris.

Leukocytes, or white blood cells, can be divided into two groups: **granulocytes** and **agranulocytes.** Examine the slide under high power or oil immersion and identify the different leukocytes as presented in the following sections. Leukocytes have two major types of responses to immune reactions. One of these is **nonspecific resistance,** which refers to a person passively being protected against microorganisms and foreign substances without prior exposure. Another kind of response is **adaptive immunity,** in which exposure to specific antigens initiates an immune reaction. Leukocytes have specific receptors and secrete specific antibodies for specific antigens.

Granulocytes   Granulocytes are so named because they have granules in their cytoplasm. The three types of granulocytes are neutrophils, eosinophils, and basophils.

Neutrophils are the most common of all leukocytes. They are also known as **polymorphonuclear (PMN) leukocytes** due to the variable shape of their nuclei (which are lobed, not round). Neutrophils typically have a half-life in the blood of about 6 hours and have ameboid capabilities. They squeeze between adjacent endothelial cells that make up postcapillary venules and move into the interstitial spaces of tissues, where they engulf foreign particles or cellular debris. They move into connective tissue, where they typically live for an additional 1 to 4 days. Neutrophils move toward infection sites and destroy foreign material. They are the major type of phagocytic leukocyte. The granules of neutrophils absorb very little stain so their granules are light pink to light purple. Mature neutrophils can be distinguished from other leukocytes by their typically two- to five-lobed nucleus. They are about one and a half times the size of erythrocytes.

(A) Activity

Examine your slide for neutrophils, compare them to fig. 24.3, and draw one in the following space.

Eosinophils typically have a two-lobed nucleus with pink-orange granules in the cytoplasm. The term **eosinophil** means "eosin-loving" (eosin is an orange-pink stain and is picked up by

the granules). They are about twice the size of erythrocytes. Eosinophils combat infections caused by multicellular pathogens, such as parasitic worms, and they are involved in allergic reactions.

(A) Activity

Compare fig. 24.3 to your slide as you locate the eosinophils and draw one in the following space.

Basophils are rare. The granules stain very dark (blue-purple), and sometimes the nucleus is obscured because of the dark-staining granules. The nucleus is S-shaped and the cell is about twice the size of an erythrocyte. Basophils contain the vasodilator histamine, which increases blood flow to the tissues. In times of infection it is beneficial to have an increase in blood flow so that phagocytic cells or antibodies can reach the infection site. Basophils also contain heparin, which is an anticoagulant. It keeps the blood from clotting too quickly. Clots decrease the effectiveness of combating infection as they can seal off the site from the immune reaction. They are involved in inflammatory and allergic reactions. You may have to look at 200 to 300 leukocytes before finding a basophil.

(A) Activity

Compare the basophil in your slide to fig. 24.3 and draw one in the following space.

Lymphocytes have a large, unlobed nucleus that usually has a flattened or dented area. The cytoplasm is clear and may appear as a blue halo around the purple nucleus. There are multiple types of lymphocytes. Two of them are the **B cells** and **T cells.** These cells cannot normally be distinguished from one another in standard histological preparations (for example, Wright's stain), and they will simply be regarded as lymphocytes in this exercise. Both B cells and T cells arise from fetal bone marrow, but they mature in different areas. B cells are thought to mature in the liver and spleen in the fetus, but they mature in bone marrow of adults. B cells become plasma cells when activated, which provides antibody-mediated immunity. That is, plasma cells secrete antibodies that travel in the blood plasma.

T cells mature in the thymus and provide **cell-mediated immunity.** In cell-mediated immunity, the cells (not antibodies in the blood plasma) move close to and destroy some types of bacteria or virus-infected cells. T cells also attack tumors and transplanted tissues. Most of the T cells are found in the lymph nodes, thymus, and spleen. They enter the bloodstream via the lymphatics. **Natural killer (NK) cells** are lymphocytes that attack bacteria, transplanted cells, and cancer cells.

Ⓐ **Activity**

Compare your slide to fig. 24.4 for lymphocytes and draw one in the following space.

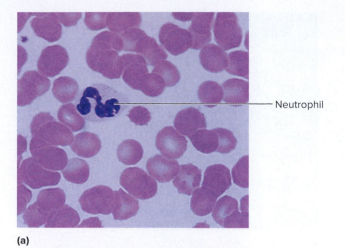

(a)

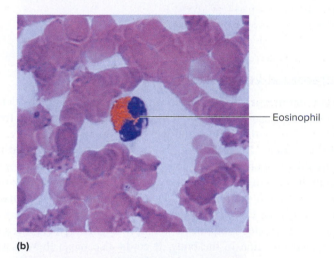

(b)

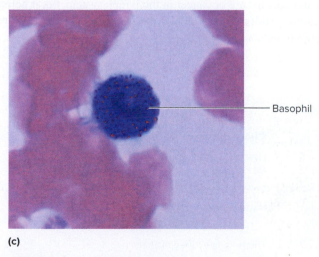

(c)

FIGURE 24.3 **Granulocytes.** (a) Neutrophil (1,000×); (b) eosinophil (1,000×); (c) basophil (1,500×). ©Eric Wise

Agranulocytes Agranulocytes are so named because, even though they have granules, they do not pick up significant amounts of stain. The nuclei are not lobed but may be dented or kidney bean–shaped. There are two types of agranulocytes: lymphocytes and monocytes.

Monocytes are large and often have a kidney-shaped or horseshoe-shaped nucleus. They are about three times the size of erythrocytes and are activated by T cells. Monocytes are important in that they can turn into **macrophages,** which are major phagocytic cells, and they are important in presenting foreign antigens to T lymphocytes. Monocytes move into tissues from the blood and become macrophages.

Neutrophil

Eosinophil

Basophil

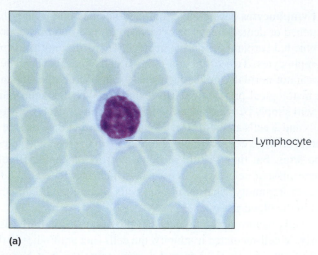

(a)   **(b)**

FIGURE 24.4 **Agranulocytes (1,000×).** (a) Lymphocyte; (b) monocyte.
©Eric Wise

(A) **Activity**

Locate the monocytes on your slide, compare them to fig. 24.4, and draw one in the following space. You can review the formed elements in table 24.1.

Differential Leukocyte Count

Rapid and inexpensive diagnosis of disease is a goal of modern health care. Narrowing the field of potential disease states is frequently done with a differential leukocyte count. For example, if a patient comes to the hospital with a fever, it could be due to many things, including viral infection, reactions to medications, metabolic disorders, and vascular disease. By taking a small sample of blood and examining the percentages of leukocytes, many diseases can be eliminated from the diagnosis. If, for example, there is an elevated lymphocyte count, it may mean there is a severe viral infection in the body. It could also mean there is an autoimmune disease or cancer in the lymph system, but it helps to direct the health care provider to look for answers in certain areas and not in others.

TABLE 24.1	Summary of Formed Elements in Blood		
	Size	**Number**	**Characteristics**
Erythrocytes	7.5 μm	5 million/μL	Live 120 days on average No nucleus
Platelets	2–4 μm	130,000–400,000/μL	Cell fragments
Leukocytes		7,000/μL	Nucleus present
Granulocytes			
Neutrophils	9–12 μm	60–70% of leukocytes	Three- to five-lobed nucleus Granules indistinct
Eosinophils	10–14 μm	2–4% of leukocytes	Two-lobed nucleus Orange granules
Basophils	8–10 μm	<0.5% of leukocytes	S-shaped nucleus Large, dark granules
Agranulocytes			
Lymphocytes	5–17 μm	25–33% of leukocytes	Nucleus appearing dented Thin rim of cytoplasm in some
Monocytes	12–15 μm	3–8% of leukocytes	Large, kidney-shaped nucleus

(A) **Activity**

In this part of the exercise you examine a blood smear and count leukocytes to determine their percentages. Once you have determined the percentages, compare your values with the standard values to see if they are within normal limits. Changes in the relative percentages of leukocytes may indicate a type of disease.

Neutrophils represent about 60–70% of leukocytes. They increase in number in appendicitis or other acute bacterial infections.

Eosinophils represent about 2–4% of leukocytes. Eosinophils increase in number during allergic reactions and parasitic infections (for example, trichinosis).

Basophils represent about 0.5–1% of leukocytes. They increase in number during allergies.

Lymphocytes make up 25–33% of leukocytes. These cells increase in times of viral infection, such as infectious mononucleosis and antibody-antigen reactions.

Monocytes make up about 3–8% of leukocytes. They increase in times of chronic infections, such as tuberculosis.

1. Work with a lab partner where one person calls out the names of different leukocytes while the other uses a hand counter and counts up to 100 as the cells are called out. Look into the microscope and methodically record the different types of leukocytes seen.
2. You should record how many of each type are found and keep track of the overall number with the use of a hand counter until 100 cells are counted.
3. Scan the slide in a systematic way by moving up one field of view and down another field of view, so that you don't count any cell twice. One method is illustrated in fig. 24.5. Tally your results in the following spaces.

Neutrophils: _____

Eosinophils: _____

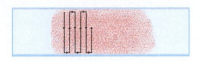

FIGURE 24.5 Counting Leukocytes.

Basophils: _____

Lymphocytes: _____

Monocytes: _____

Leukopenia is a decrease in the number of circulating leukocytes in the blood. Common causes are radiation therapy, chemotherapy, and some medications. **Leukemia** is an increase in the number of leukocytes due to cancer of the hematopoietic cells. Even though the leukocytes are abundant, these cells are often immature or deformed and they do not function normally.

(A) **Activity**

Fill in table 24.2 with what you know about the formed elements. Read all the information given before filling in the chart.

Clean Up Make sure the lab is clean after you finish. If you used immersion oil on the microscope, make sure the objective lenses are wiped clean (use clean lens paper only!). Place any slide with fresh blood on it or any material contaminated with bodily fluid in the bleach or other disinfectant solution. Place all gloves or contaminated paper towels in the biohazard container. All sharps material (broken slides or coverslips, lancets, etc.) should be placed in the sharps container. Clean the counters with a towel and a 10% bleach or other disinfecting solution.

TABLE 24.2	Characteristics of Formed Elements		
Formed Elements	**Granules (If Present)**	**Shape of Nucleus (If Present)**	**Cause for Increase**
Erythrocyte	No granules	No nucleus	_____
_____	Not obvious	_____	Mononucleosis
_____	Orange-staining	_____	Parasitic infections
_____	_____	Two- to five-lobed	_____
_____	Not obvious	Kidney	_____
Basophil	_____	_____	_____

Notes

REVIEW SECTION

Blood

Name _____   Date _____

Lab Section _____   Time _____

Review Questions

1. What are the three main groups of formed elements?

   _____

2. What is the most common plasma protein? _____

3. What is another name for a platelet? _____

4. Which is the most common blood cell? _____

5. What is another name for a leukocyte? _____

6. What leukocyte is most numerous in a normal blood smear? _____

7. How many erythrocytes are normally found per microliter of blood? _____

8. What is an average number of leukocytes found per microliter of blood? _____

9. B cells and T cells belong to what class of agranulocytes? _____

10. How does a differential leukocyte count aid in medical diagnosis? _____

   _____

11. In counting 100 leukocytes you are accurately able to distinguish 15 basophils. Is this a normal number for the white blood cell count, and what possible health implications can you draw from this?

   _____

   _____

12. What is the function of platelets? _____

13. Label the formed elements in the following illustration.

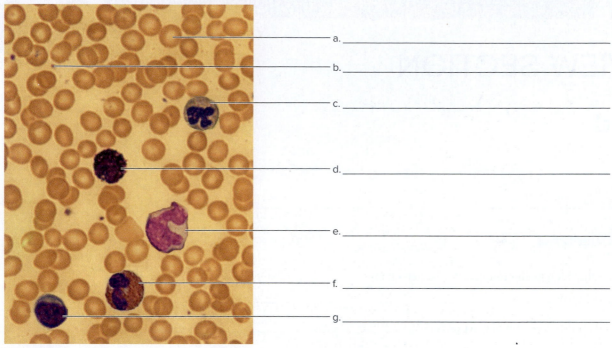

a. _____

b. _____

c. _____

d. _____

e. _____

f. _____

g. _____

©Eric Wise

14. In terms of volume, does blood normally contain more plasma or more formed elements?

_____

15. A patient has a lymphocyte count of 42% of the total white blood cell count. What might be the clinical reason for this?

_____

Blood Tests and Typing

INTRODUCTION

Blood is a connective tissue that is a complex medium with multiple functions in the body. Many clinical tests are performed on blood, including those to determine the number of erythrocytes or leukocytes in a known volume of blood, the blood type of an individual, and hemoglobin concentration. These topics are discussed in the Saladin text in chapter 18, "Blood." In this exercise, you evaluate different samples of blood. As in the previous exercise, handle the samples with care. This is important, so we've repeated the caution statement from that exercise here.

Caution! In this excercise, you may be using unsterilized human blood, sterilized human blood, nonhuman mammal blood, or synthetic blood. The risk of blood-borne diseases is significantly minimized by the use of sterilized human blood and nonhuman mammal blood; however, use the same precautions you would if you were handling fresh and potentially contaminated human blood. Your instructor will determine what type of blood to use. In all cases, follow strict procedures for handling potentially pathogenic material.

1. Wear protective gloves and goggles during the procedures.
2. Do not eat or drink in lab.
3. If you have an open wound, make sure the wound is securely covered or do not participate in this exercise.
4. Your instructor may elect to have you use your blood. Due to the potential for disease transmission, such as AIDS or hepatitis, in fresh blood samples, *make sure you keep away from other students' blood and keep them away from your blood.*
5. Do not participate in this part of the exercise if you have recently skipped a meal or if you are not well.
6. Place all disposable material in the biohazard bag.
7. Place all used lancets in the sharps container and all glassware to be reused in a 10% bleach or other disinfecting solution.

8. After you have finished the exercise, clean and disinfect the countertops with a 10% bleach or other disinfecting solution.

OBJECTIVES

At the end of this exercise, you should be able to

1. perform specific diagnostic tests, such as hematocrit and hemacytometer tests;
2. determine the antigens (agglutinogens) present in a particular ABO blood type;
3. list the antibodies (agglutinins) present in a particular ABO blood type;
4. relate Rh-positive or Rh-negative blood to antigens present;
5. correlate hematocrit with erythrocyte counts.

MATERIALS

5 mL of fresh, nonhuman mammal blood (dog, sheep, or cow)

Sterilized human blood, labeled 1 to 4

Blood typing antisera (anti-A, anti-B, anti-D), test cards, and toothpicks

Rh warming tray

Heparinized blood microcapillary tubes

Capillary tube centrifuge

Hematocrit reader (Criticorp Micro-hematocrit tube reader, Damon Micro-capillary reader)

Small ruler

Seal-ease

300 mL Erlenmeyer flask

1,000 μL micropipette or 1 mL pipette with pipette pump

0.9% saline solution

Hemoglobinometer

Hemacytometer, coverglass

Microscope

Protective gloves

Goggles

Tallquist paper

Tallquist chart

Cotton balls

Contamination bucket with autoclave bag

Container with 10% bleach or other disinfecting solution

Sharps container

PROCEDURE

Blood Typing

An understanding of blood type is critical in clinical work. Proper matching of blood between donor and recipient is a vital process, and you will learn the essentials of this process in this exercise. There are many different typing systems that match blood, but the ABO and Rh systems are the two most common in clinical settings. Blood types are genetically determined. There are other typing systems, such as the Kell, Lewis, MNS, and Duffy blood groups. These are antigen/antibody groups used to type blood, and they are frequently used in forensic studies.

The first blood typing system discovered was the ABO system. In this system a person has one of four blood types: type A, type B, type AB, or type O. These blood types are determined by the presence of **antigens.** Antigens are large molecules, such as glycoproteins, that occur on the outer membrane of a cell, with type A blood cells having the A antigen, type B blood having the B antigen, type AB blood having both the A antigens and B antigens, and type O blood having neither. Blood is "typed" by adding specific **antisera** to the blood sample. There are three clinical antisera used to type ABO blood. For example, antiserum-A causes type A blood to clump.

Antigens can react to **antibodies** that occur in the plasma of another person. Antibodies are secreted by plasma cells and, in the case of blood, attach to the antigens. Antibodies in the blood plasma attach to antigens on the surface of red blood cells, causing a reaction known as agglutination. These antigen-antibody complexes are broken down by immune cells. In the case of blood, the antigens are called **agglutinogens,** and the antibodies are called **agglutinins.** The agglutinins are found in the plasma portion of the blood.

In the ABO system, no prior exposure to the agglutinogen is needed for agglutination (granulation or clumping) to occur. For example, if a person has blood type A, that individual has agglutinins against blood type B. If the person with blood type A receives a transfusion of blood type B, then the anti-B agglutinins attack the agglutinogens in the blood introduced into the system. This causes a **transfusion reaction,** the agglutination and hemolysis of the transfused erythrocytes. If the reaction is severe enough, death may occur.

Table 25.1 gives details of the ABO blood system.

TABLE 25.1	ABO Blood System	
Blood Type	**Agglutinogens (Antigens)**	**Agglutinins (Antibodies)**
A	A	Anti-B
B	B	Anti-A
AB	A and B	None
O	None	Anti-A and anti-B

Procedure for Blood Typing

(A) Activity

1. Read the entire blood typing procedure before you begin this part of the exercise. Wear protective gloves and eyewear.
2. Obtain a sample of sterilized blood as directed by your instructor. The vial should be labeled 1, 2, 3, or 4.
3. Record the sample number you are using in the space provided.
4. Place two drops of the sample blood on a blood test card or on a clean glass microscope slide (fig. 25.1).
5. Place antiserum-A on one drop and antiserum-B on the other drop.
6. **Use separate toothpicks to stir each sample of blood and antiserum.**
7. Keep the antisera separate from one another. Examine the sample for agglutination within 2 minutes. Agglutination may appear as fine red dust or larger particles. If you are using a glass microscope slide, place the slide on a white piece of paper to see the granulations more clearly. Make sure you dispose of your toothpicks in the biohazard bag. If both blood samples agglutinate, you have a sample of type AB blood. If neither sample agglutinates, you have a sample of blood type O. If the blood with antiserum-A agglutinates, you have a sample of blood type A, and if blood with antiserum-B agglutinates, you have a sample of blood type B.

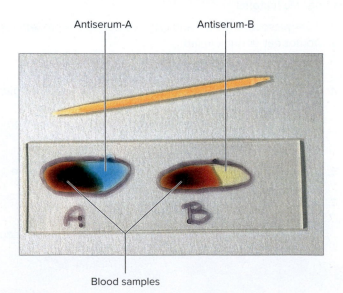

FIGURE 25.1 Blood Testing Procedures.
©Eric Wise

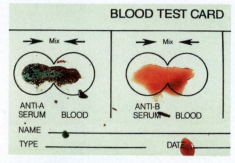

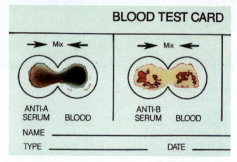

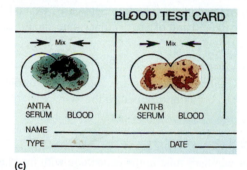

FIGURE 25.2 Blood Types. (a) Type A; (b) type B; (c) type AB.
©Eric Wise

8. Compare your sample to fig. 25.2.

Blood sample number: _____

Blood type: _____

Common ABO blood types for various groups in the United States are listed by percentages in table 25.2.

Determination of Rh Factor

The other significant blood group is the Rh system, which was named after rhesus monkeys, the animal model used in discovering this blood group. The Rh factor is determined by the blood genotype DD or Dd. People with this genetic makeup are said to be Rh+ (Rh-positive). About 85% of the people in the United States are Rh-positive. People who have the genotype dd are said to be Rh– (Rh-negative), and they represent about 15% of the U.S. population. In an Rh-positive person, the Rh antigen is present but the anti-Rh antibody is absent. In an Rh-negative person, the antigen is absent. Anti-Rh antibodies are normally not present in the blood. They develop in an Rh– person after exposure to the Rh antigen.

Determination of the Rh type is a much more subtle test than ABO typing. The Rh antiserum (anti-D) is more fragile in shipping and storage and should not be considered clinically relevant in this exercise. Use the same precautions as outlined at the beginning of the exercise regarding bodily fluids.

(A) Activity

1. Place one drop of blood on a slide and place a drop of antiserum-D (Rh antiserum) on the slide.
2. Use a new toothpick and stir the two drops together.
3. Place this mixture on a warming tray and gently rock the slide.
4. Examine the slide for agglutination after a minute or two. If agglutination occurs (which may appear as fine red granules), then the sample is Rh+ (the Rh antiserum reacted with the Rh antigen in the blood). If no agglutination occurs, then the sample is Rh–. This test is more difficult to determine, so look for slight agglutination. Rh antiserum should be used fresh.
5. Record your results.

Rh factor determination: _____

Rh determination is important if a pregnant woman is Rh–. If her unborn child is Rh+, she may develop antibodies against the Rh antigen during pregnancy or delivery (a time when fetal and maternal blood frequently mix). If her second child is Rh+, the antibodies she produced during her first pregnancy may cross the placenta and cause severe reactions in the fetus. This is called

TABLE 25.2	Blood Type Percentages			
Blood Type	**Caucasians**	**African Americans**	**Asian**	**Native Americans**
A	40	26	28	0
B	11	19	25	0
AB	4	4	7	0
O	45	51	40	100
Total	100	100	100	100

hemolytic disease of the newborn (HDN), or **erythroblastosis fetalis.** Rho-GAM (anti-Rh antibodies) prevents antibody formation in the mother against the Rh antigen and is frequently given to the mother during pregnancy and again after delivery. A woman who is Rh+ does not have the antibodies for the Rh factor and therefore will not produce antibodies against the developing fetus, regardless of the Rh factor of the fetus.

Blood Typing Problems

Extensive blood transfusions can lead to problems. Blood that is donated from one person carries antibodies, and these may react with the recipient's blood. There are other blood types in addition to the ABO and Rh systems, and sensitivities may develop due to antibodies present in the recipient's blood.

Hematocrit

The **hematocrit (packed cell volume** or **PCV)** is the percentage of erythrocytes in the total blood volume. The percentage of erythrocytes can be calculated after centrifuging a sample of blood. The cells end up as an extensive sediment in the bottom of the capillary tube, leaving the lighter leukocytes and plasma on top. The hematocrit varies in individuals, with values of 37% to 48% being normal in females and 45% to 52% being normal in males. An increase in the hematocrit above normal is known as **polycythemia** and can exceed 65%. Polycythemia may occur if the bone marrow produces more erythrocytes than normal or if a person is a heavy smoker or living at high altitude.

Anemia is a condition when blood is lost faster than it is replaced or when the production of erythrocytes is low. This may be due to a heavy menstrual flow, gastrointestinal ulcers, lack of iron in the diet, lack of vitamin B_{12}, chemotherapy, viral infections, or many other conditions. In severe anemia, the hematocrit may drop to 15% or less. Anemia may also be due to low levels of hemoglobin in the blood. Hemoglobin is a complex oxygen-carrying molecule composed of an iron-containing **heme group** and the protein **globin.** Another form of anemia is sickle-cell anemia. This genetic condition causes erythrocytes to form long, curved shapes when a person is under oxygen stress (such as when climbing a set of stairs quickly). The sickled erythrocytes are broken down, leaving the person anemic. You will next compare the hematocrit of sterilized human blood with that of another mammal.

Procedure for Determining the Hematocrit

Refer to the caution statement at the beginning of the exercise concerning working with blood.

 Activity

1. While wearing protective gloves, fill two capillary tubes with blood. One should have commercially prepared, sterilized human blood, and the other should have nonhuman mammal blood.
2. Fill each tube by touching the red end of the capillary tube to the blood sample. Let the blood flow up the tube by capillary action. Fill each tube to about three-quarters of the tube

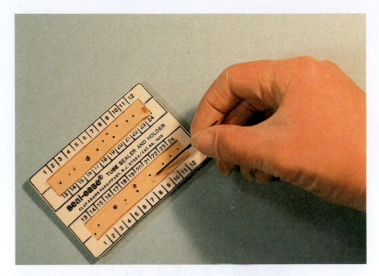

FIGURE 25.3 Preparation of a Hematocrit Tube.
©Eric Wise

length. Both tubes should have about the same volume of blood in them.
3. Push the red end of the tube into Seal-ease® or modeling clay. Only the red end of the capillary tube should be sealed (fig. 25.3).

 Caution! The capillary tubes are fragile and sharp! They can break and puncture the skin.

4. Place the capillary tube in the centrifuge with the clay plug against the outer rubber ring.
5. Make sure you keep track of the sample from each tube and don't mix them up!

Mammal blood tube #: _____

Sterilized blood tube #: _____

6. After many tubes have been placed into the centrifuge, make sure each has the clay seal facing the rubber gasket. If the seals are facing the center of the centrifuge, the blood will spray out of the tube when the centrifuge begins turning. Make sure the metal top is on the centrifuge and screwed securely in place. Close the cover and spin the capillary tubes for 3 to 5 minutes.
7. Once the centrifuge has come to a complete stop, remove each tube and calculate the percentage of red blood cells in the tube. This can be done with hematocrit readers. If you use a card reader, place the clay-blood interface at the zero line at the bottom of the card. The capillary tube should be vertical; move the tube with the clay-blood line still on the zero line on the card until the plasma-air interface reaches the 100% line. Read where the red cells intersect the chart on the sliding scale. This is the hematocrit.

8. Record the hematocrit in the following spaces and in the Chapter Summary Data section at the end of the exercise.

❓ Sterilized blood: _____ (1)

❓ Mammal blood: _____ (2)

❓ How do these values compare? _____ (3)

❓ What commercial procedure might lead to a difference in the hematocrit between these two samples? _____ (4)

❓ Plasma with red coloration indicates hemolysis of the erythrocytes. Normally, the plasma should be straw-colored or light yellow. Does either of your samples show hemolysis? _____ (5)

The buff-colored layer between the erythrocyte layer and the plasma layer is the leukocyte layer.

Hemoglobin Determination

Anemia is a condition whereby the oxygen-carrying capacity of the blood is diminished, which may be due to a reduced number of erythrocytes in the blood and thus may show up as a low hematocrit. In some cases, the hematocrit is normal but there is a decrease in hemoglobin levels in the individual red blood cells.

Hemoglobinometer Method

Levels of hemoglobin can be measured with the use of a hemoglobinometer. This instrument uses differences in the absorption of green light by hemoglobin to measure its concentration in the blood. Different hemoglobinometers are available.

Ⓐ **Activity**

The following test is for the Leica (AO) hemoglobinometer. You can determine the level of hemoglobin in the blood with the following procedure.

1. Make sure the hemoglobinometer has fresh batteries installed and that the light turns on when the light switch is depressed.
2. Place two glass plates together, so there is a chamber on the inside, and insert them into the metal clip.
3. Fill the space between the glass plates with blood and stir the blood with a hemolysis applicator which lyses the blood.
4. Insert the metal clip with the glass plates into the hemoglobinometer.
5. Move the lever on the side of the hemoglobinometer until the green colors are the same shade.
6. Read the scale to determine the hemoglobin concentration.
7. Carefully remove the glass plates and clean them with a cotton ball soaked in alcohol. Dispose of the cotton ball in a biohazard container and place the glass plates in a secure area where they can be used by other students or placed in a 10% bleach solution.

8. You may want to compare two different samples of blood, such as a sample of sterilized blood versus nonhuman mammal blood. Record your results.

Hemoglobin (Hb) levels are expressed as grams Hb/100 mL of blood. The average value for humans is 12 to 16 g/100 mL of blood. In males the normal range is 13 to 18 g/100 mL and for females it is 12 to 16 g/100 mL.

Tallquist Method

Another method to determine the level of hemoglobin in the blood is by comparing the color of the oxyhemoglobin in a drop of blood on Tallquist paper to a comparative chart.

Ⓐ **Activity**

1. Place a piece of Tallquist paper on a paper towel on your desk.
2. Place a drop or two of blood on the paper and wait 15 seconds.
3. Compare the color on your paper to the Tallquist chart.

❓ Hemoglobin from sterile blood sample: _____ (6)

❓ Hemoglobin from mammal blood sample: _____ (7)

In clinical settings, hemoglobin is usually measured with an optical cell counter. This instrument measures the amount of light that is transmitted through a solution or the amount of light absorbed by the solution. In terms of use with hemoglobin, the higher the hemoglobin level, the less light that passes through the blood.

Blood Cell Counts

Blood cell counts determine the number of blood cells in a known volume and subsequently determine the total number of cells per cubic millimeter (mm^3). Modern evaluation of erythrocyte and leukocyte counts is performed by injecting blood samples into an optical computer system, which automatically calculates the cell counts. Most hospitals use computer-driven optical systems to count cells.

The average number of erythrocytes in blood is approximately 5 million per cubic millimeter. This is too large a number to count during the lab period, so another technique will be used. To determine the number of cells in blood, you will dilute the blood and count the cells that occur in a small volume. The process used in lab will involve the use of a **hemacytometer,** a glass slide that has very fine lines etched on its surface. When a coverslip is placed on the hemacytometer and a sample of blood is introduced, the blood fills a precise volume. This volume is 1/50th of a cubic millimeter.

Even with this small volume, there are too many erythrocytes in the blood in a given volume to be counted accurately. The solution to this is to dilute the blood so that the cells can be counted individually. In the procedure you will perform, you will take 1 part of blood and dilute it to 200 parts diluent (dilution solution).

Procedure for Erythrocyte Counts

For males the average count is about 5.4 million erythrocytes per cubic millimeter of blood, with a range of 4.5 to 6.0 million cells per cubic millimeter. For females the average count is about 4.8 million erythrocytes per cubic millimeter of blood, with a range of 4.0 to 5.5 million cells per cubic millimeter.

(A) Activity

You will perform this experiment twice: once using commercially prepared, sterilized human blood and once using fresh, nonhuman mammal blood. Obtain the following:

Micropipette or pipette and pipette pump

300 mL Erlenmeyer flask

0.9% saline solution

Sample of nonhuman mammal blood

Sample of sterilized blood

Hemacytometer and coverslip

Protective gloves and eyewear

Microscope

Cotton balls

Because of the volume of solution made, you may want to use one dilution of blood for the entire class, depending on the wishes of your instructor. It is particularly important to accurately measure the blood withdrawn and the amount of diluent in order to make an accurate erythrocyte count. For withdrawing blood from your samples, you may use either a micropipette or a volumetric pipette. The following instructions provide directions for both. Make sure the blood sample you use is well mixed, so that blood cells are evenly distributed in the plasma. Wear protective gloves and eyewear during the entire procedure.

1. Add exactly 200 mL of 0.9% saline solution to an Erlenmeyer flask.
2. Use either a 1,000-microliter micropipette or a 1 mL pipette with a pipette pump. If you use a micropipette, depress the plunger until it stops and insert the tip into the blood sample. Slowly release the plunger to pull blood into the micropipette. If using the 1 mL pipette, use a pipette pump attached to the blunt end of the pipette and draw up exactly 1 mL of blood to the calibration line of the pipette. Never pipette anything by mouth! Use a pipette bulb or pump.
3. Transfer the blood to the Erlenmeyer flask. If you use a micropipette, depress the plunger until it stops and then push it a little farther. This expels the right amount of liquid into the flask. If you are using a volumetric pipette (one that has an expanded middle section), the blood should drain out of the pipette and you should NOT forcibly expel the remainder of the fluid in the pipette. Dispose of the pipette in a tray for that purpose or in a beaker of 10% bleach or other disinfecting solution.
4. Gently swirl the flask to distribute the blood cells. Withdraw the diluted sample with a Pasteur pipette.
5. Place the hemacytometer coverslip (not a regular coverslip) over the hemacytometer chamber and fill the chamber with the Pasteur pipette. Do not overfill the chamber.

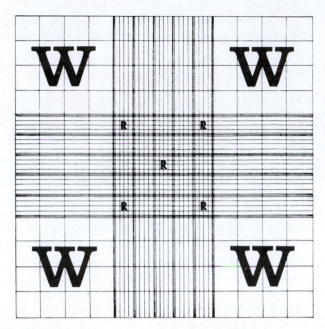

FIGURE 25.4 Erythrocyte Counts Using a Neubauer Hemacytometer.

6. Place the hemacytometer under the microscope and examine the cells under high power (400–450×).
7. Count all of the erythrocytes in each of the five areas marked with an "R" (fig. 25.4). Some cells will be on the border of the counting grid. Count only those cells touching the left line and the upper line as part of the sample. Do not count the cells touching the right line or the bottom line.

    The blood was diluted 1:200, and the sum of the volume in the five areas represents 1/50th of a cubic millimeter. If you multiply the dilution factor (200) and the volume that would occur to fill 1 mm³ (50), then you would multiply the number of erythrocytes counted by 10,000 to obtain the number of erythrocytes per cubic millimeter.

8. Record your results.

❓ Number of erythrocytes counted: _____ (8)

❓ Number of erythrocytes per cubic millimeter: _____ (9)

Use the sample of nonhuman mammal blood. Use the chamber on the opposite side of the hemacytometer and follow the same procedure you used for human blood. Record the results.

❓ Number of erythrocytes counted: _____ (10)

❓ Number of erythrocytes per cubic millimeter: _____ (11)

❓ How do the numbers of cells in sterilized blood and the number in nonhuman mammal blood compare?

_____ (12)

? Is there a correlation between the numbers of erythrocytes counted from these samples and the hematocrit taken from the same samples? _____ (13)

After you finish your counting, place the hemacytometer and coverslip *gently* into a 10% bleach solution or other disinfecting solution. Place all glassware that has come into contact with blood in a 10% bleach or other disinfecting solution.

Leukocyte Counts (Optional)

A **Activity**

Leukocytes are counted in a similar fashion except that the four large corner regions of the hemacytometer are used and the sum of the cell count from these four areas is multiplied by 50. Your instructor may provide you with additional instructions for a leukocyte count, using a specific solution for leukocytes.

? Number of leukocytes counted: _____ (14)

? Number of leukocytes per cubic millimeter: _____ (15)

Normal blood levels for leukocytes are about 7,000 cells per cubic millimeter, with a range of 4,300 to 10,800 cells per cubic millimeter. Leukocytosis is an elevated white blood cell count typically above 11,000 cells per cubic millimeter. Leukocytosis indicates inflammation. Leukopenia, on the other hand, is a decreased white blood cell count, with less than 4,000 cells per cubic millimeter. This may be due to chemotherapy, radiation therapy, HIV infection, infectious hepatitis, cirrhosis of the liver, typhoid fever, or chronic infections, such as tuberculosis.

Clean Up   Make sure the lab is clean before you leave. Wipe any blood from the microscope objective lenses with lens cleaner and lens paper. Make sure you do not leave glass or blood on the lab tables. Use a 10% bleach or other disinfecting solution and a towel to clean off your lab table before you leave.

Notes

REVIEW SECTION

Blood Tests and Typing

Name _____   *Date* _____

Lab Section _____   *Time* _____

❓ Chapter Summary Data

Use this section to record your results from questions within the exercise.

1. _____
2. _____
3. _____
4. _____
5. _____
6. _____
7. _____
8. _____

9. _____
10. _____
11. _____
12. _____
13. _____
14. _____
15. _____

Review Questions

1. What is the name of a surface membrane molecule that causes an immune reaction?

   _____

2. What is the average range of hematocrit for a normal female? _____

3. What is the average range of hematocrit for a normal male? _____

4. What percent of the blood volume consists of formed elements? _____

5. A person with blood type B has what kind of agglutinins (antibodies)? _____

6. A person has antibody A and antibody B in his or her blood, with no Rh antibody. What blood type does this person have? _____

7. A total of 240 erythrocytes are counted in the hemacytometer chamber. What is the red blood cell count of this individual in terms of erythrocytes per cubic milliliter? _____

8. A person with blood type B negative is injected with type A positive blood. From an immunological (antigen/antibody) standpoint, what will happen after the injection? _____
_____

9. How might changes in the pipette technique alter the final determined value of erythrocytes? What types of errors might you expect? _____
_____

10. Using the following illustration, calculate the hematocrit of the individual by using a ruler to measure the packed cell volume and the total volume of blood. Determine if it falls within normal limits.

Clay/red cell
interface

Red cell/plasma
interface

Plasma/air
interface

©Eric Wise

Hematocrit: _____

Within normal limits? _____(yes/no)

11. Define anemia. _____
_____

12. Explain the possible erroneous results you might get if you used just one toothpick to stir the various blood types in the ABO blood test. _____
_____

Anatomy of the Heart

INTRODUCTION

This exercise covers the anatomy of the heart, which is discussed in the Saladin text in chapter 19, "The Circulatory System: Heart." In this exercise, you examine models of the heart, as well as preserved sheep and human hearts, if available. As you look at the preserved material, try to see how the structure of the heart relates to its function.

The heart is located deep in the thorax between the lungs in a region known as the **mediastinum.** The mediastinum contains the heart, the membranes surrounding the heart (the **pericardium**), and other structures, such as the esophagus and descending aorta. The mediastinum is located between the sternum, lungs, and thoracic vertebrae. Vessels that return blood to the heart are called **veins,** and those that carry blood from the heart are called **arteries.** There are two main circulatory paths. One is the **pulmonary circuit,** where the heart pumps blood to the lungs and the return flow to the heart, and the other is the **systemic circuit,** where blood travels from the heart to the rest of the body and back.

OBJECTIVES

At the end of this exercise, you should be able to

1. list the three layers of the heart wall;
2. describe the position of the heart in the thoracic cavity;
3. describe the significant surface features of the heart;
4. describe the internal anatomy of the heart;
5. find and name the anatomical features on models of the heart and in the sheep heart;
6. describe the blood flow through the heart and the function of the internal parts of the heart;
7. discuss the functioning of the atrioventricular valves and the pulmonary and aortic valves and their role in circulating blood through the heart.

MATERIALS

Models and charts of the heart

Preserved sheep hearts

Preserved human hearts (if available)

Blunt probes (mall probes)

Dissection pans

Scalpels

Sharps container

Protective gloves

Waste container

Microscopes

Prepared slides of cardiac muscle

Fresh or thawed sheep heart

Clamp (hemostat) or string

PROCEDURE

Pericardium

If you were to open the chest cavity, the first structure you would see is the **pericardium.** The pericardium encloses the heart and has two layers: a tough, outer connective tissue sheath called the **fibrous pericardium** and an inner layer called the **serous pericardium.** The serous pericardium has two layers, one attached to the fibrous pericardium and the other attached to the heart wall called the **epicardium.** Between these two layers is the **pericardial cavity,** which contains a small amount of **serous fluid.** This fluid reduces the friction between the outer surface of the heart and the pericardial sac. Locate these pericardial layers in fig. 26.1.

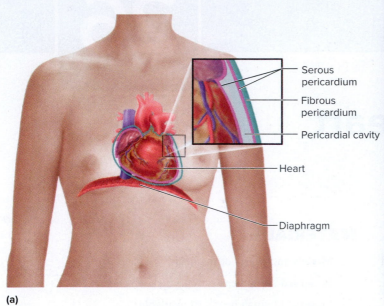

(a)

FIGURE 26.1 Heart in Thoracic Cavity and Heart Coverings.
(a) Coronal section; (b) photograph of cadaver.
(b) ©Eric Wise

(b)

Heart Wall

The heart wall is composed of three major layers. The outermost layer is the epicardium (the deep layer of the serous pericardium) composed of epithelial and connective tissue. The middle layer of the heart wall is the **myocardium** and is the thickest of the three layers. It is mostly made of **cardiac muscle.** The cardiac muscle is arranged spirally around the heart, and this arrangement provides a more efficient wringing motion to the heart. The inner layer of the heart wall is known as the **endocardium,** a membrane consisting of endothelium (simple squamous epithelium) and connective tissue.

(A) Activity

You may wish to review the slides of involuntary cardiac muscle and note the intercalated discs, branching fibers, and fine striations of cardiac tissue (as described in Laboratory Exercise 5).

It is best to examine heart models *before* dissecting a sheep heart unless your instructor directs you to do otherwise. Heart models are color coordinated and labeled to make the structures easier to locate.

Examination of the Heart Model

Overview

The heart is a four-chambered pump with two superior atria (AY-tree-uh) and two inferior ventricles. Blood enters the heart in the right atrium (fig. 26.2) and flows into the right ventricle. Once the blood is in the right ventricle, the contraction of the

ventricular wall sends the blood to the lungs. The blood is oxygenated in the lungs and returns to the heart by entering the left atrium. Blood moves from the left atrium to the left ventricle and is then pumped to the rest of the body.

Exterior of the Heart

(A) Activity

Examine the heart model and notice that the heart has a pointed end, the **apex,** and a blunt end, the **base.** The apex of the heart is inferior, and the great vessels leaving the heart are located at the base (therefore, in the case of the heart, the base is superior to the apex). Compare the model to fig. 26.2 and see how the **aorta** curves to the left in an anterior view of the heart and is posterior to the **pulmonary trunk.** The pulmonary trunk is colored blue on models and charts as vessels that carry deoxygenated blood are color coded blue. The vessels that are color coded red on models and charts represent those that carry oxygenated blood.

Locate the anterior features of the heart. The heart is composed of two large, inferior ventricles and two smaller and superior atria. The **left ventricle** extends to the apex of the heart and is delineated from the right ventricle by the **anterior interventricular sulcus.** In this anterior interventricular sulcus are some of the **coronary arteries** and **cardiac veins,** discussed later. The left ventricle looks larger than the right ventricle, but they both contain the same volume. Note the two earlike flaps on the anterior, superior atria. These are the **auricles.** At the junction of the right atrium and the right ventricle is the **coronary sulcus.**

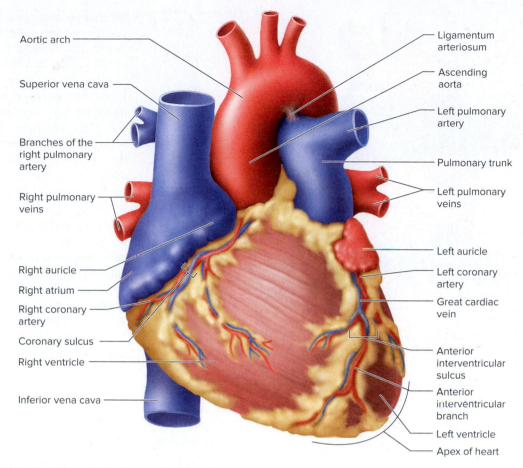

Aortic arch

Superior vena cava

Branches of the
right pulmonary
artery

Right pulmonary
veins

Right auricle

Right atrium

Right coronary
artery

Coronary sulcus

Right ventricle

Inferior vena cava

Ligamentum
arteriosum

Ascending
aorta

Left pulmonary
artery

Pulmonary trunk

Left pulmonary
veins

Left auricle

Left coronary
artery

Great cardiac
vein

Anterior
interventricular
sulcus

Anterior
interventricular
branch

Left ventricle

Apex of heart

FIGURE 26.2 Surface Anatomy of the Heart, Anterior View.

If you examine the heart from the posterior side you will see the atria more clearly. The **coronary sinus,** a large venous space that carries blood from the cardiac veins to the right atrium, is located in the coronary sulcus. Locate the **superior vena cava** and the **inferior vena cava,** two vessels that also return blood to the right atrium. Locate the **pulmonary veins,** which carry blood from the lungs to the left atrium. Compare the heart model to fig. 26.3.

The major vessels of the heart are illustrated in figs. 26.2–26.4. Locate the **pulmonary trunk, pulmonary arteries, ligamentum arteriosum** (a connective tissue band between the pulmonary trunk and aortic arch), **ascending aorta, pulmonary veins, superior vena cava, inferior vena cava, coronary arteries,** and **cardiac veins.**

The heart tissue is nourished by the coronary arteries. The **left coronary artery** arises from the ascending aorta (fig. 26.4) and then branches into the **anterior interventricular branch** and the **circumflex branch.** The **right coronary artery** also arises from the ascending aorta and branches to form the **posterior interventricular branch** and the **right marginal branch.** These major arteries of the heart supply blood to the myocardium. On the return flow from the heart muscle, the **great cardiac vein** follows the depression of the anterior interventricular sulcus and the **coronary sulcus** to the coronary sinus. On the right side of the heart, the **small cardiac vein** leads to the coronary sinus,

which empties into the right atrium. Locate these vessels on the external surface of the model of the heart and compare them to figs. 26.2–26.4.

Interior of the Heart

Ⓐ Activity

Examine a model of the interior of the heart and locate the right and left ventricles. The wall of the **right ventricle** is much thinner than the **left ventricle wall.** This is because blood from the right ventricle is pumped a short distance to the lungs, while blood in the left ventricle is pumped more extensively through the body. The ventricles are separated by the **interventricular septum,** which forms a wall between the two ventricular chambers. Compare the model to fig. 26.5.

Examine the **right atrium** and note how thin the wall is compared to the ventricles. The walls of the atria are thin because blood in the atria has to flow only a short distance to the ventricles. Examine the medial wall of the atrium, known as the **interatrial septum,** and locate a thin, oval depression in the atrial wall. This depression is the **fossa ovalis.** In fetal hearts this is the site of the **foramen ovale** (for-AYE-men o-VAL-eh), but closure of the foramen usually occurs just after birth. Note the extensive **pectinate** (PEK-tin-ate) **muscles** on the wall of the atrium. Blood in the

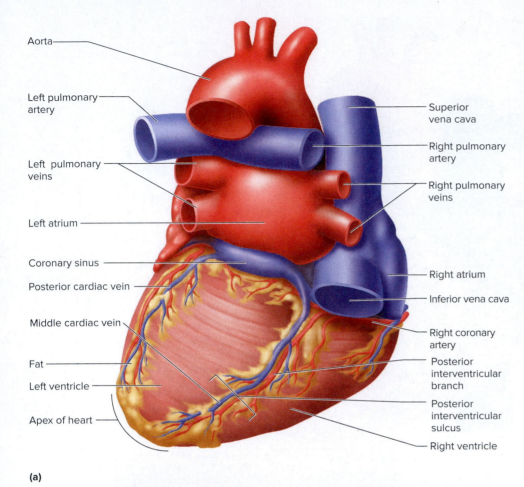

Aorta

Left pulmonary artery

Left pulmonary veins

Left atrium

Coronary sinus

Posterior cardiac vein

Middle cardiac vein

Fat

Left ventricle

Apex of heart

Superior vena cava

Right pulmonary artery

Right pulmonary veins

Right atrium

Inferior vena cava

Right coronary artery

Posterior interventricular branch

Posterior interventricular sulcus

Right ventricle

(a)

Aorta

Superior vena cava

Left atrium

Pulmonary veins

Right atrium

Inferior vena cava

Coronary sinus

Left ventricle

Posterior interventricular branch

Right ventricle

(b)

FIGURE 26.3 Surface Anatomy of the Heart, Posterior View. (a) Diagram; (b) photograph.
(b) ©Eric Wise

superior vena cava, the inferior vena cava, and the coronary sinus returns to the right atrium. Examine the features of the right atrium in figs. 26.5 and 26.6.

The blood from the right atrium flows into the right ventricle. Now examine the valve between the right atrium and right ventricle. This is the **right atrioventricular valve,** which prevents return of blood from the right ventricle into the right atrium during ventricular contraction. Examine the valve for flat sheets of tissue. These are the cusps of the **right atrioventricular (AV)** valve. The right AV valve has thin, threadlike attachments called **tendinous cords (chordae tendineae)** (COR-day TEN-din-ee). These tough cords are attached to larger **papillary muscles,** extensions from the wall of the ventricle. The right ventricle wall has small extensions called **trabeculae carneae** (trah-BEC-you-lee CAR-nee-ee). The blood from the right ventricle flows into the pulmonary trunk toward the lungs.

Locate the **pulmonary valve.** It appears as three small cusps between the right ventricle and the pulmonary trunk and keeps blood from flowing in reverse from the pulmonary trunk into the right ventricle during ventricular relaxation. Examine the details of the right ventricle in models in the lab and in fig. 26.5.

From the pulmonary trunk, blood flows into the pulmonary arteries then to the lungs, where it releases carbon dioxide and picks up oxygen. The **pulmonary veins** carry oxygenated blood from the lungs into the **left atrium.** These vessels are located in the superior, posterior portion of the left atrium. Blood from the left atrium flows into the left ventricle. Locate the cusps of the **left atrioventricular (AV)** valve between the left atrium and left ventricle. The left AV valve is also known as the **mitral valve.** Like the right AV valve, it has attached tendinous cords and papillary muscles, which you should locate in the models in the lab. Note the thickness of the left ventricle wall compared to the wall of the right ventricle. Compare the left side of the heart to fig. 26.5.

The **aortic valve** is located at the junction of the left ventricle and the ascending aorta. It has the same basic structure and general function as the pulmonary valve in that it prevents the

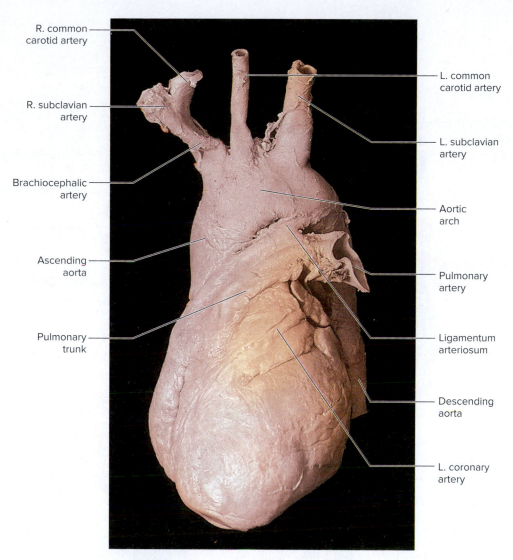

R. common carotid artery

R. subclavian artery

Brachiocephalic artery

Ascending aorta

Pulmonary trunk

L. common carotid artery

L. subclavian artery

Aortic arch

Pulmonary artery

Ligamentum arteriosum

Descending aorta

L. coronary artery

FIGURE 26.4 **Vessels of the Heart, Anterior View.**
©Eric Wise

flow of blood from the aorta into the left ventricle. Blood from the left ventricle moves into the aorta and subsequently to the rest of the body.

Caution! Be careful when handling preserved materials. Ask your instructor for the proper procedure for working with preserving fluid and for handling and disposal of the specimen. Do not dispose of animal material in the sinks. Place it in an appropriate waste container.

Dissection of the Sheep Heart

The sheep heart is similar to the human heart and is usually readily available as a dissection specimen. Dissection of anatomical material is valuable in that you can examine structures seen more accurately in preserved material than in models. Also, the preserved material has greater flexibility and is easily manipulated. There are some differences between sheep hearts and human hearts, especially in the position of the superior and inferior venae cavae. In sheep these are called the anterior and posterior venae cavae, but we refer to them using the human terminology.

If your sheep heart has not been dissected, you will need to open the heart. If your sheep or other mammalian heart has been previously dissected, you can skip the next two paragraphs.

Ⓐ Activity

Place the heart in a container of water for a few moments to rinse off the preserving fluid. Examine the external features of the heart, as seen in fig. 26.7. The sheep heart may still be in the pericardial sac. If this is the case, remove the sac before proceeding. Note the fat layer on the heart. The amount of fat on the human or sheep heart is variable. Locate the **left ventricle,** the **right ventricle,** the **anterior interventricular sulcus,** the **right atrium,** and the **left atrium.** Note the **auricles** that extend from

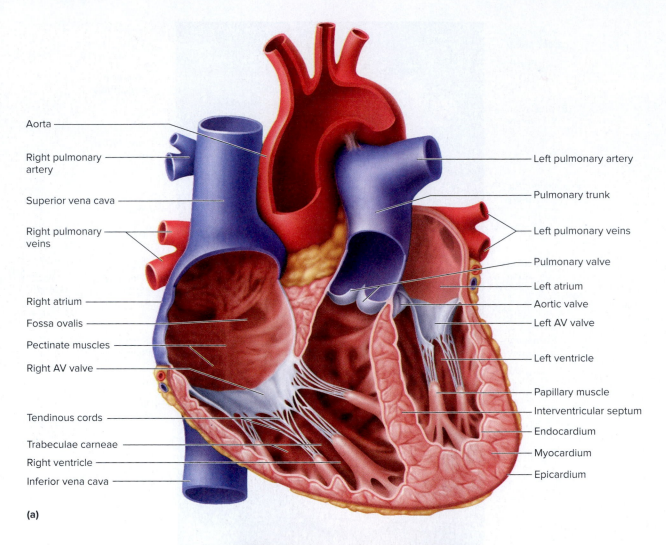

Aorta
Right pulmonary artery
Superior vena cava
Right pulmonary veins
Right atrium
Fossa ovalis
Pectinate muscles
Right AV valve
Tendinous cords
Trabeculae carneae
Right ventricle
Inferior vena cava

Left pulmonary artery
Pulmonary trunk
Left pulmonary veins
Pulmonary valve
Left atrium
Aortic valve
Left AV valve
Left ventricle
Papillary muscle
Interventricular septum
Endocardium
Myocardium
Epicardium

(a)

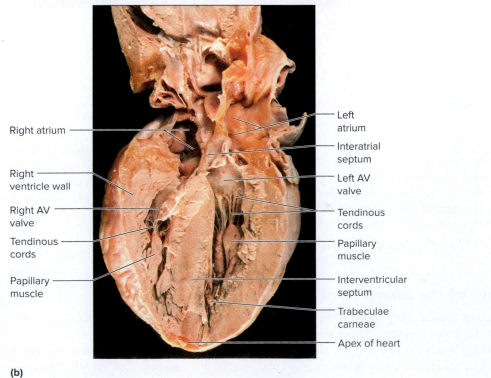

Right atrium
Right ventricle wall
Right AV valve
Tendinous cords
Papillary muscle

Left atrium
Interatrial septum
Left AV valve
Tendinous cords
Papillary muscle
Interventricular septum
Trabeculae carneae
Apex of heart

FIGURE 26.5 Heart, Frontal Section. (a) Diagram; (b) photograph.
(b) ©Eric Wise

(b)

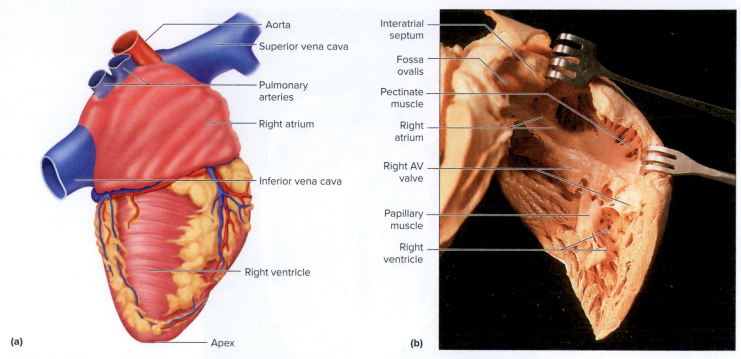

FIGURE 26.6 Right Side of the Heart. (a) Diagram of surface view; (b) photograph of interior view.
(b) ©Eric Wise

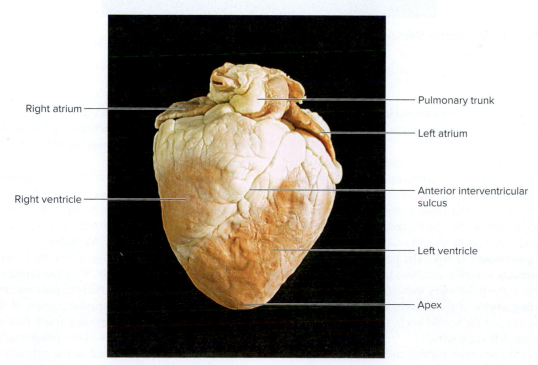

FIGURE 26.7 Sheep Heart, Anterior View.
©Eric Wise

the anterior surface of the atria. Carefully remove the adipose tissue from the major vessels of the heart.

Using a sharp scalpel, make an incision along the right lateral *side* of the heart from the apex of the heart to the lateral side of

the right atrium. If you are unsure about how to proceed during any part of the dissection, ask your instructor for directions. Make another long cut from the lateral side of the left atrium through the lateral side of the left ventricle. You will make a coronal section

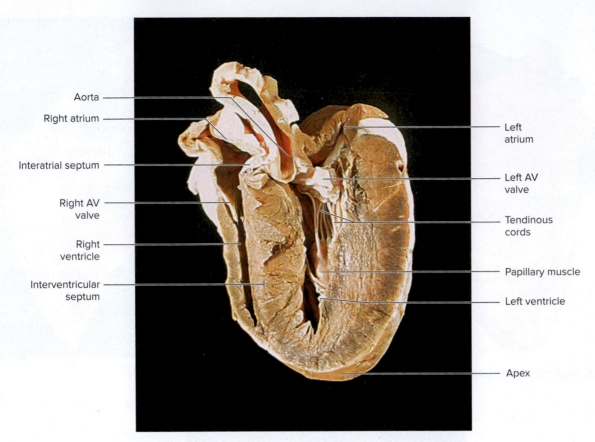

Aorta

Right atrium

Interatrial septum

Right AV valve

Right ventricle

Interventricular septum

Left atrium

Left AV valve

Tendinous cords

Papillary muscle

Left ventricle

Apex

FIGURE 26.8 Sheep Heart, Coronal Section.
©Eric Wise

of the heart if you cut through the **interventricular septum.** Once you have opened the heart, compare the structures of the sheep heart to fig. 26.8.

Locate the vessels of the heart by gently inserting a blunt metal probe into the vessels and determining which chamber the vessel goes to or comes from. Place the heart in anatomical position and insert the probe into the large, anterior vessel that exits toward the specimen's left side. Be careful and do not tear the heart valves. The blunt end of the probe should enter the right ventricle. This vessel is the **pulmonary trunk.** The pulmonary trunk may still have the **pulmonary arteries** attached. Locate the large vessel directly posterior to the pulmonary trunk (figs. 26.7 and 26.8). This is the **ascending aorta.** If the vessels are cut farther away from the heart, you can see the **aortic arch.** Insert the probe into this vessel and into the **left ventricle.**

Turn the heart to the posterior surface and locate the **superior (anterior) vena cava** and **inferior (posterior) vena cava.** Insert the probe into the superior and inferior venae cavae, pushing the probe into the **right atrium.** If you find only one large opening in the atrium, you may have cut through either the superior or the inferior vena cava during your initial dissection. The probe can be felt through the wall more easily here than in a ventricle because the atrial walls are thinner than those of the ventricles. On the left side, the **pulmonary veins** may appear as four separate veins, or

you may see a large hole on each side of the **left atrium** if the vessels were cut close to the atrial wall. Locate the same structures in the sheep heart that you found on the model and compare them to fig. 26.9.

Cut into the right atrium and use your blunt probe to locate the opening of the **coronary sinus** in the posterior, inferior portion of the atrium. It is small and can be difficult to find.

Examine the opening between the right atrium and the right ventricle to locate the **right AV valve.** You may have dissected through one of the cusps as you opened the heart. Locate the major features of the **right ventricle.** Find the **tendinous cords** and the **papillary muscles.** You can find the **pulmonary trunk** by inserting a blunt probe into the superior portion of the right ventricle. Make an incision in the pulmonary trunk near the right ventricle to expose the three thin cusps of the **pulmonary valve.** Note how the cusps press against the wall of the pulmonary trunk when the probe is pushed against them in a superior direction. These cusps close when blood begins to flow back into the right ventricle as the ventricle relaxes.

Locate the **left atrium, left AV valve,** and **left ventricle.** In the left ventricle, you should find the papillary muscles, tendinous cords, and **trabeculae carneae.** You can find the aorta and aortic valve by inserting a blunt probe toward the superior end of the left ventricle toward the middle of the heart.

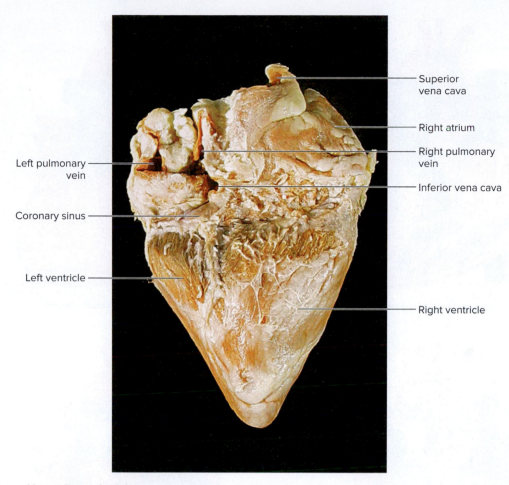

FIGURE 26.9 Sheep Heart, Posterior View.
©Eric Wise

Clean Up When you have finished your study of the sheep heart, make sure you clean your dissection equipment with soap and water and dry it. Be careful with sharp blades. Place the sheep heart either back in the preserving fluid or in the appropriate waste container as directed by your instructor.

Heart Valves

The pulmonary and aortic valves and atrioventricular valves prevent the backflow of blood. Your instructor may demonstrate the procedure, or you can do it yourself by using a new fresh or thawed sheep heart. These demonstrations may not work well with preserved sheep hearts.

Ⓐ Activity

Flush any remaining blood from the heart before you locate the valves.

Make an incision into the right atrium, exposing the **right AV valve.** Pour water into the right ventricle and notice how the water flows past the right AV valve and into the right ventricle. Clamp off the pulmonary trunk or tie it with string to prevent blood from flowing out of the pulmonary trunk. *Gently* squeeze the right ventricle and notice how the right atrioventricular valve closes and prevents blood from backing up into the right atrium.

What is the adaptive value for the closing of atrioventricular valves?

Cut the **pulmonary trunk** close to the right ventricle and slowly pour water into it as if you were trying to fill the right ventricle. The **pulmonary valve** should fill with water and close the entrance of the right ventricle, preventing backflow. This normally occurs as the right ventricle begins diastole (relaxation). Compare these valve closures to fig. 26.10.

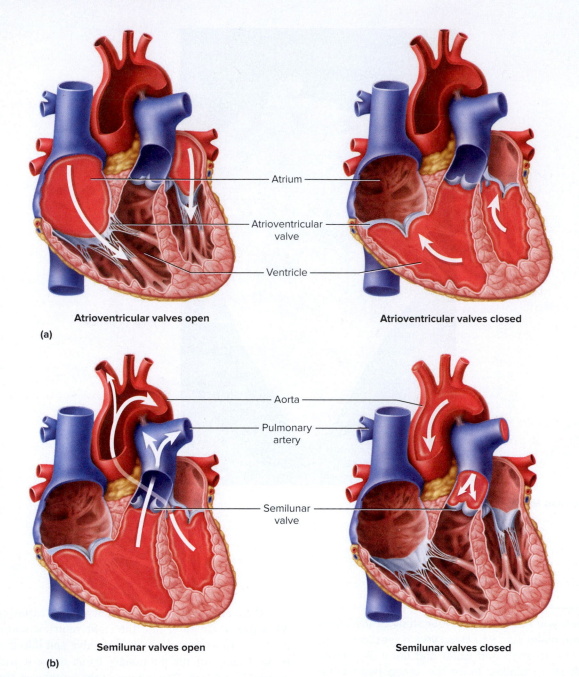

(a)

Atrioventricular valves open

Atrioventricular valves closed

Atrium

Atrioventricular valve

Ventricle

(b)

Semilunar valves open

Semilunar valves closed

Aorta

Pulmonary artery

Semilunar valve

FIGURE 26.10 Closure of the Heart Valves. (a) Atrioventricular valve; (b) pulmonary and aortic valves.

REVIEW SECTION

Anatomy of the Heart

Name _____ *Date* _____

Lab Section _____ *Time* _____

Review Questions

1. The heart is located between the lungs in an area known as the _____.

2. Name the outermost (superficial) layer of the pericardium. _____

3. What is the innermost layer of the heart wall called? _____

4. Name the depression between the two ventricles seen on the anterior surface of the heart.

   _____

5. Are auricles extensions of the atria or of the ventricles? _____

6. What three vessels take blood to the right atrium? _____

7. Where do the great cardiac vein and the small cardiac vein take blood?

   _____

8. Is the apex of the heart superior or inferior to other parts of the heart? _____

9. What blood vessels nourish the heart tissue? _____

   _____

10. What structure separates the left atrium from the right atrium? _____

11. The mitral valve is located between what two chambers of the heart?

    _____

12. What is the function of the aortic valve? _____

    _____

13. Name the structure found between the atrioventricular valve and the papillary muscle.

    _____

14. What is the cell type that makes up most of the myocardium? _____

15. The walls of the left ventricle are thicker than those of the right ventricle. What explanation can you give for this? _____

    _____

16. How does cardiac muscle resemble skeletal muscle? (Review Exercise 5, if necessary) _____
_____

17. In terms of function, how is cardiac muscle different from skeletal muscle? _____
_____

18. What is the function of the pulmonary valves? _____

19. Trace the flow of blood through the heart. _____
_____
_____
_____
_____

20. Label the following illustration using the terms provided.

aorta

apex

interventricular septum

left atrium

left AV valve

left ventricle (wall)

right atrium

right AV valve

right ventricle (wall)

tendinous cords

Electrical Conductivity of the Heart

INTRODUCTION

Before you begin the study of the function of the heart, review the anatomy of the heart in Laboratory Exercise 26. The topics for this exercise are covered in the Saladin text in chapter 19, "The Circulatory System: Heart."

The living heart is a phenomenal organ. It contracts an average of 72 times per minute for the lifetime of an individual. The heart has been described as a muscular pump or, more accurately, two pumps acting in unison. As a pump, the heart has a contraction mode and a relaxation mode. **Systole** is contraction of the heart muscle and can be described more specifically as **atrial systole** and **ventricular systole** (contraction of the atria and ventricles, respectively). **Diastole** is the relaxation of the heart muscle. **Atrial diastole** is relaxation of the atria, and **ventricular diastole** is relaxation of the ventricular muscles.

Electrical activity of the heart stimulates the heart muscle to contract. In this exercise, you observe the electrical activity of the heart and correlate it to the mechanical functioning of the heart. You learn the conductive structures of the heart, make a recording of the electrical activity of the heart, and correlate the recording with the activity of the heart.

The initiation of the electrical impulse in the heart begins at the **sinuatrial (SA) node,** which is commonly known as the **pacemaker.** The sinuatrial node is located in the superior portion of the right atrium. Cells of the sinuatrial node generate a difference in electrical voltage (about –60 mV) across the cells' membranes. The cells are said to be **polarized** because the inside of the cell membrane has more negative ions than the outside of the membrane. This separation of charged particles is reflected in the voltage difference across the membrane. The sinuatrial node spontaneously **depolarizes** at around –40 mV, which causes a change in the total amount of charge across the cell membrane. This change in membrane potential occurs approximately 72 times per minute (the average heart rate). Conduction from the sinuatrial node travels across the atria, causing the muscles of the atria to contract. The impulse that spreads out across the atria reaches the **atrioventricular (AV) node.** The impulse has a slight delay

(about 0.1 second) in the node before being conducted further. This delay allows the atrial cardiac muscle to contract prior to ventricular firing (fig. 27.1).

The electrical impulse then travels from the AV node to the **atrioventricular bundle (bundle of His),** to the **right** and **left bundle branches,** and finally to the **Subendocardial branches (Purkinje fibers).** The subendocardial branches stimulate the cardiac muscle of the ventricles to contract. The ventricles are thus stimulated from the apex toward the base, and the contraction proceeds from the inferior end of the ventricles toward the atria. Locate the structures of the heart's conduction system in fig. 27.1. Contraction of the heart muscle (a physical event) occurs just after depolarization of the heart muscle cells (an electrical event). Relaxation of the heart occurs just after repolarization of the heart muscle cells.

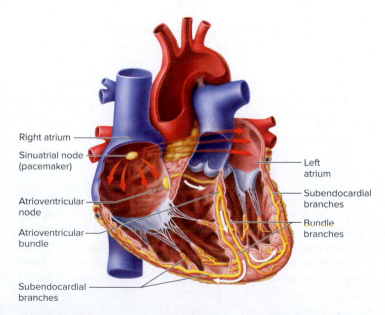

FIGURE 27.1 Conduction System of the Heart. Conduction follows the path indicated by arrows.

Two concepts are important in understanding the electrical activity of the heart. One is that the heart produces low-voltage **electrochemical impulses** in a way similar to the production of impulses in the nervous system. The "average" potential difference of **–90 millivolts** is a little less than one-tenth of a volt; therefore, the body is operating electrically at slightly less than 0.1 volt. The other important concept in understanding the electrical activity of the heart is that these impulses travel through the saline medium of the body and can be picked up by **sensors (electrode plates)** attached to the skin. The cells of the body are bathed in a saline solution of a little less than 1% salt, which is an excellent conducting medium for electrical impulses. The collective action potentials generated by the atria and ventricles depolarizing and then repolarizing can be recorded using an electrocardiograph machine attached to the electrode plates.

Early work done by Willem Einthoven established the modern techniques of recording the electrical activity of the heart by an **electrocardiograph (ECG)** machine. It is an instrument that measures slight changes in the voltage related to cardiac activity. The electrocardiograph produces a chart paper recording or a data file called an **electrocardiogram (ECG, or EKG).** Time is measured along the horizontal axis (or x-axis, with each millimeter equal to 0.04 second), and voltage difference is measured along the vertical axis (or y-axis, with 1 millimeter equal to 1 millivolt). The ECG has a **baseline** known as the **isoelectric line,** and deflections from that line record the electrical activity of the heart. Einthoven established three standard leads that record the heart's electrical events with the heart sitting in the middle of a theoretical shape known as Einthoven's triangle (fig. 27.2).

> **Lead I** connects the right arm and the left arm (RA-LA). It measures the potential voltage across the horizontal axis of the heart.
>
> **Lead II** connects the right arm and the left leg (RA-LL). This is the lead that records the potential voltage from the base to the apex of the heart.
>
> **Lead III** connects the left arm and the left leg (LA-LL). This is the lead that records the potential voltage along the left side of the heart.

In many college physiology labs, the ECG electrode plates are attached to four areas. These are the medial side of the **left** and **right ankles** and the anterior surface of the **left** and **right wrists.** The attachment of the electrode plate to the right ankle serves as an **electrical ground** and is not used for measurement. The ECG is measured as the potential voltage difference between selected electrode plates. There are numerous "V" electrodes, which are chest electrodes. You will not use the V electrodes in this exercise unless directed to do so by your instructor.

Before you record an ECG, examine the ECG in fig. 27.3 and read the accompanying description in the text. The ECG typically has three major events. The first event is the **P wave,** which is a small bump called a **deflection wave.** The P wave represents **atrial depolarization.** The **QRS complex** represents

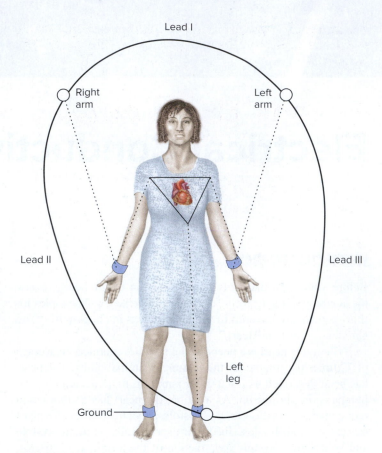

FIGURE 27.2 Leads of a Standard ECG.

the **ventricular depolarization.** Because the ventricles are more massive than the atria, the electrical events produced as the ventricles depolarize are much larger than the P wave generated by the atria. The **T wave** represents the **ventricular repolarization.**

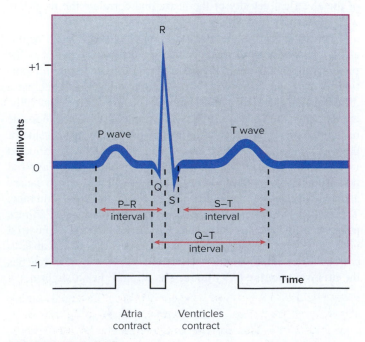

FIGURE 27.3 ECG.

The **atrial repolarization** causes an electrical deflection during ventricular depolarization but is masked by the larger QRS complex.

OBJECTIVES

At the end of this exercise, you should be able to

1. distinguish between systole and diastole;
2. list the structures in the conductive system of the heart;
3. describe how the electrocardiogram can be used to monitor electrical events in the heart;
4. correlate the electrical potentials of the heart to mechanical activity in a cardiac cycle;
5. describe the sequence of electrical conductivity of the heart;
6. associate the P wave, QRS complex, and T wave of an ECG with electrical events that occur in the heart;
7. relate how the electrical activity of the heart is transferred to the ECG;
8. calculate heart rate, PR interval, QRS interval, and QT interval from an ECG recording;
9. describe the variance in an ECG from normal if a person has a heart block;
10. describe the changes in an ECG immediately after exercise;
11. determine the mean electrical axis of the heart.

MATERIALS

Cot or covered lab table

Alcohol swabs

Electrode jelly, paste, or saline pads

BIOPAC Section

EL 503—general-purpose disposable electrode pads

SS2L—lead cables

CBLSERA—CBL serial for MP30 or MP150

MP30 or MP150—data acquisition unit and power supply (AC100A)

Compatible computer

ECG Machine Section

Electrocardiograph machine

PROCEDURE

Recording ECGs

There are a few ways to record ECGs. One is with a conventional ECG machine that puts out a paper chart strip, another is with a physiograph, and a third is with a computer program and hardware, such as BIOPAC. This exercise will describe the BIOPAC procedure and the ECG machine procedure.

BIOPAC Procedure

You will collect data for the electrical activity of your heart by attaching receiving electrodes to your skin. In this part of the exercise you will be using lead II. The resulting electrocardiogram is displayed on the computer monitor and saved as a file in your BIOPAC folder. If you are going to determine the mean electrical axis of the heart later in this exercise, you will need to use BIOPAC **Lesson 6.—L06 Electrocardiography 2.** The following instructions are for determining a simple ECG and are found in BIOPAC Lesson 5. These are the steps you will take:

(A) Activity

1. Setup

a. Make sure the computer is on.
b. Connect the data acquisition unit to an electrical outlet with the power supply.
c. With the data acquisition unit *off,* plug the SS2L lead cables into the CBLSERA cable, which should be connected to CH2 on the data acquisition unit.
d. Turn the data acquisition unit on.
e. Have your lab partner remove all jewelry and lie comfortably. Make sure you have three of the disposable electrode pads.
f. Clean the inner side of each ankle of your lab partner with an alcohol swab and attach one of the disposable EL 503 electrode pads to the inner left leg just proximal to the medial malleolus, as seen in fig. 27.2. You may want to place a small amount of electrode gel in the center of the pad in order to get proper electrical conduction. Attach another electrode pad to the right leg at the same height. Clean the distal, anterior, right forearm just proximal to the wrist with an alcohol swab and attach the third electrode pad there. You should have three electrode cables, a white, black, and red one.
g. Place the **white lead** on the right forearm. The clip has a squeeze lever and attaches to the metal button of the electrode pad. The clip connects on only one surface. If you cannot get it to attach one way, try flipping the connector over.
h. Place the **black lead** on the electrode pad of the right leg. This is a ground.
i. Place the **red lead** on the left leg.
j. Open the BIOPAC Student Lab (BSL) Software on the computer desktop and select BIOPAC **Lesson 5.—L05 Electrocardiography 1.**
k. Type in your (folder) name. When done, click OK. If you have a folder on this computer station, a window should appear with the message "A folder with this name already exists. Would you like to use it or create a new folder?" Choose **Use it.**

2. Calibration   Make sure your test subject is comfortable and relaxed. It is often beneficial to talk to the subject calmly as you prepare the test. Any extraneous movement may cause electrical interference. People who are nervous or jittery do not produce good ECGs.

Click on **Calibrate** in the upper left corner of the computer screen. The calibration will stop automatically after approximately 8 seconds. You should see a small ECG tracing on the computer screen with a flat baseline. If not, check your connections and click **Redo Calibration.**

3. Record Data   Have your lab partner relax, click on the **Record** button, and let several cycles occur. After about 15 seconds, click on **Suspend.** If the recording does not have a flat baseline, click on the **Redo** button. If your recording looks similar to fig. 27.3, you have made a successful recording.

4. Analyze Data   Find the **Review Saved Data** folder from the lessons menu and look for your file.

You can determine the length of various sections of your ECG by selecting the **Delta T** mode. This mode determines the time from one part of the ECG tracing you select to another part of the tracing you select. You can use the **I-beam** tool to select the appropriate area of your ECG. Choose the magnifying tool in the lower right side of the computer screen to give you a close-up view of the area you will choose.

Beats per minute—bpm Select the area from one R peak to another R peak using the I-beam tool. Bpm will show up in the channel measurement box. Record your bpm in table 27.1.

PR interval Use the I-beam and highlight a section of the ECG from the beginning of the P wave to the beginning of the QRS complex. Record this time (use the Delta T mode) in table 27.1.

QRS complex Use the same procedure as you did with the PR interval to determine the time for the QRS complex. The length of the QRS complex begins at the first deflection of the Q wave and ends when the S wave returns to the baseline. Record this time in table 27.1.

QT interval This interval is determined with the beginning of the Q wave to the end of the T wave. Record this time in table 27.1.

Procedure for Conventional ECG Machine

The ECG machine should be plugged in and turned on. Make sure the ECG machine is on "standby" when you attach the electrodes. Accidental grounding of the electrodes may cause damage to the equipment. Older ECG machines should warm up for a few minutes prior to running the experiment.

A Activity

1. Remove metal watches or any other jewelry that might interfere with the electrical signal between the heart and the extremities.
2. Have your lab partner scrub the inside of the ankle, about 2 cm above the medial malleolus, with an alcohol swab to remove dust and skin oil.
3. Clean the anterior sides of the wrist in the same way.
4. Saline pads or electrode jelly may be applied to these areas.
5. Attach the electrodes firmly, using the straps provided, but not so tightly that you cut off blood flow. The four connections are

    LA attaches to the left arm.

    RA attaches to the right arm.

    RL attaches to the right leg.

    LL attaches to the left leg.

    Your instructor will explain how to operate your electrocardiograph machine.

6. Have your lab partner rest comfortably on a cot or table for a few moments before monitoring the ECG. Some machines operate automatically and record all three leads. If you need to record leads manually, turn the knob on the ECG machine from "standby" to "run" and record lead I for about 10 cycles. Turn the knob to "standby" after recording each lead.
7. Record lead II and lead III for about 10 cycles each.
8. Tear the ECG from the machine and make sure you write your lab partner's name on the paper and write the appropriate lead on the paper.

TABLE 27.1	ECG Data			
Record your value for the various ECG data either from the BIOPAC ECG or the ECG machine and determine if your data falls within the normal values.				
bpm	Yours _____	Above 100 Tachycardia	Below 60 Bradycardia	Within norm? _____
PR interval	Yours _____	Normal 0.16–0.18 sec	Longer than 0.18 sec Partial AV heart block	Within norm? _____
QRS complex	Yours _____	Normal 0.08 sec	Longer than 0.08 sec Rt/lft bundle branch block	Within norm? _____
QT interval	Yours _____	Normal 0.3–0.4 sec	Shortens with increased heart rate	Within norm? _____

Interference in an ECG To illustrate the importance of being relaxed during an ECG test you can have your lab partner intentionally interfere with the recording.

Ⓐ Activity

You can demonstrate this electrical "noise" by the following activity. While your lab partner is still attached to the ECG machine, set the dial to lead II. Run three to four cardiac cycles and then have your lab partner clench their fists. Note the electrical interference that occurs as the skeletal muscles depolarize and repolarize. Emotional state also plays a part in electrical activity. If a patient is nervous or excitable, this will show in the ECG. Background electrical noise in the ECG varies from person to person, but you should be able to get a good ECG in the lab.

Prerecorded ECG Strip If, for some reason, you do not have access to ECG recordings or you did not obtain a good recording from your equipment, you can use the recordings in fig. 27.4a to fill in the data in table 27.1.

ECG and Exercise

Ⓐ Activity

Once you have compared your ECG to normal, you should exercise vigorously for a few minutes. You can do this by running outside of the lab room or doing jumping jacks for a minute or two. Once you feel your heart is pumping faster and harder, quickly lie down on the cot and re-measure your ECG.

How does your ECG compare to normal in terms of distance between the P waves?
How does the height of the QRS complex compare to the resting ECG?
The time between the T wave and the next P wave is the resting interval of the heart. As you exercise, this interval shortens.

ECG Evaluation

Ⓐ Activity

Select one of the leads (lead II usually works best) for the following evaluations.

Beats per minute The standard rate of paper travel in an ECG machine is 25 millimeters/second. Therefore, each millimeter is equal to 0.04 second. The small squares are millimeters and the darker lines indicate 5 mm or 0.5 cm. You can calculate the resting heart rate by counting how many millimeters ("Y" mm) occur between two peaks (such as R to R). This value is going to be multiplied by 0.04 second in the following equation:

$$\underline{\hspace{2cm}}\ \text{mm/beat} \times 0.04\ \text{second/mm} = \text{"Y" seconds/beat}$$

$$Y = \underline{\hspace{2cm}}$$

There are 60 seconds/minute and you want to calculate the beats/minute, so you can use the following equation to find out the beats per minute.

$$\frac{60\ \text{seconds/minute}}{\text{"Y" seconds/beat}} = \text{your number of beats/minute}$$

Calculated beats/minute: \underline{\hspace{3cm}}

Record your bpm in table 27.1.

PR interval Select a section of the ECG from the beginning of the P wave to the beginning of the QRS complex. Record this interval in table 27.1.
QRS complex The length of the QRS complex begins at the first deflection of the Q wave and ends when the S wave returns to the baseline. Record this time in table 27.1.
QT interval This interval starts at the beginning of the Q wave and terminates at the end of the T wave. Record this time in table 27.1.

Analysis of the ECG ECG recordings are important in assessing the health of the heart. **Conduction problems, myocardial infarcts (heart attacks),** and **heart blocks** are a few of the problems that may show up as variances on an ECG. The interpretation of ECGs in this exercise is for educational purposes only. Clinical evaluations of ECGs should be done only by trained health care workers. Locate the P wave, the QRS complex, and the T wave. The whole cardiac cycle should take, on average, 0.7 to 0.8 second.

Irregularities in Heart Rate An excessively high heart rate is termed **tachycardia.** In young adults a heart rate above 100 beats per minute is considered tachycardia. However, 100 beats per minute in small children may be perfectly normal. An excessively low heart rate is termed **bradycardia.** In young adults a rate below 60 beats per minute is considered bradycardia unless they are highly trained aerobic athletes.

PR Intervals PR intervals (also known as PQ intervals) are normally about 0.16 second. The PR interval is the time between the beginning of atrial depolarization and the beginning of ventricular depolarization. If the PR interval is longer than 0.2 second (5 mm on the chart paper), this might indicate a **heart block.** Heart blocks result from reduced conduction between the atria and the ventricles. This may be caused by damage to the AV node or decreased transmission in the AV bundle. In a **complete heart block,** the atria do not stimulate the ventricular depolarization at all; therefore, the atria fire independently from the ventricles. In a complete heart block the P waves may be spaced at 0.8 second apart, the SA node depolarization rate, but the ventricles fire at a much lower rate (1.5 to 2.0 seconds apart). Are the times measured on your recording within normal limits?

QRS Complex The **QRS complex** is 0.08 to 0.10 second, on average. If the QRS complex spans longer than 0.12 second, this may indicate a **right** or **left bundle branch block.** In this condition, the two ventricles contract at slightly different times, increasing the length of the QRS complex.

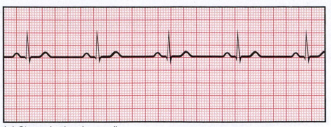

(a) Sinus rhythm (normal)

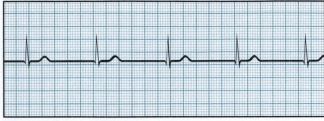

(b) Nodal rhythm – no SA node activity

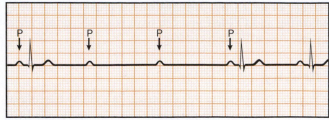

(c) Heart block

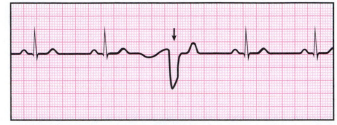

(d) Premature ventricular contraction

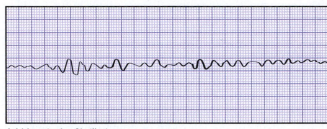

(e) Ventricular fibrillation

FIGURE 27.4 Normal and Pathological Electrocardiograms.
(a) Normal sinus rhythm; (b) nodal rhythm generated by the AV node in the absence of SA node activity—note the lack of P waves; (c) heart block, in which some P waves are not transmitted through the AV node and do not generate QRS complexes; (d) premature ventricular contraction (PVC), or extrasystole—note the inverted QRS complex, misshapen QRS and T, and absence of a P wave preceding this contraction; (e) ventricular fibrillation, with grossly irregular waves of depolarization.

QT Interval   The **QT interval** is 0.3 second, on average. The interval is shorter as the heart rate increases, and the interval becomes longer as the heart rate slows down.

Cardiac Arrhythmias   One of the major diagnostic uses of the ECG is to detect arrhythmias, abnormal rhythms, of the heart. Arrhythmias range from mild variations (due to emotions or stimulants) to severe abnormalities causing life-threatening conditions. Compare your ECG to fig. 27.4. The normal ECG is represented in the figure along with abnormal patterns. These are only a few examples of irregular ECGs.

Mean Electrical Axis of the Heart

If you have recorded an ECG from lead I and lead III, you can determine the electrical axis of the heart. If you are using BIOPAC you must use BIOPAC Lesson 6.—L06 Electrocardiography 2. If you are using a standard strip chart ECG, you will need to have lead I and lead III in hand. It is possible to calculate the relative position of the heart in the thoracic cavity by using the electrical data as was first analyzed by Einthoven. The mean electrical axis of the heart is essentially parallel to the interventricular septum. If you examine the three standard leads, they form a triangle known as Einthoven's triangle (fig. 27.2). Normally, the electrical axis of the heart runs from the superior right side to

the inferior left side, meaning that the heart is in the middle of the chest but the apex points to the inferior and left side. There are many reasons for deviations from the normal pattern, including loss of electrical activity in a portion of the heart (a bundle branch block), death of heart tissue, or an enlarged ventricle. Normal right axis deviations can occur in young and thin adults and normal left axis deviations can occur during pregnancy or in individuals who are obese. If you are using a standard ECG, you can calculate the mean electrical axis using the following information. You may be using a program that automatically calculates the mean electrical axis; however, the interpretation of the axis will be more meaningful if you also read the following description.

 Activity

1. To determine the mean electrical axis of the heart, you must acquire data from leads I and III.
2. Using your ECG, find the QRS complex in lead I. Count the number of millivolts (this is equal to the number of millimeters on chart paper) that the QRS complex projects above the isoelectric line (the baseline of the recording). If you are using BIOPAC L06, you can determine the amplitude of the wave by selecting the **delta mode.** Select the region of the graph using the I-beam tool. Subtract from that the sum

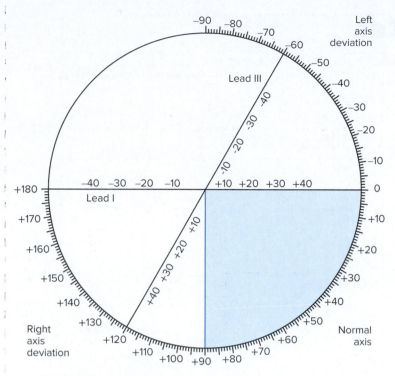

FIGURE 27.5 Electrical Axis of the Heart. Fill in this diagram with data you obtain from the ECG.

of the number of millivolts of both the Q and the S waves that project below the isoelectric line.

3. Mark this number on the scale for lead I and draw a vertical line perpendicular to the lead I axis on fig. 27.5. (If you obtained 15 mV, this would look like the dark yellow line in fig. 27.6.)

4. Repeat this procedure for lead III, but this time mark the area on the lead III scale and draw a line perpendicular to it on fig. 27.5. (If you obtained 5 mV, this would look like the blue line in fig. 27.6.) The lines for lead I and lead III should intersect.

5. Use a ruler and draw a line from the center mark of the circle through the point where the lead I and III lines intersect and record the number that represents the mean electrical axis on the edge of the circle that the line passes through. This is seen

as the black line in fig. 27.6 and shows a mean electrical axis of +49. As this is in the shaded area of 0 to +90, it is in the normal range.

6. Record your value.

Mean electrical axis: _____

Is yours within the normal range? _____

If the value is greater than +90, there is a right axis deviation. This may be due to right ventricular enlargement, displacement of the heart to the right side, or damage to the left side of the heart.

If the value is less than 0 (some physicians say less than −30), there is a left axis deviation. This may be due to left ventricular enlargement, left bundle branch block, displacement of the heart to the left side, or damage to the right side of the heart.

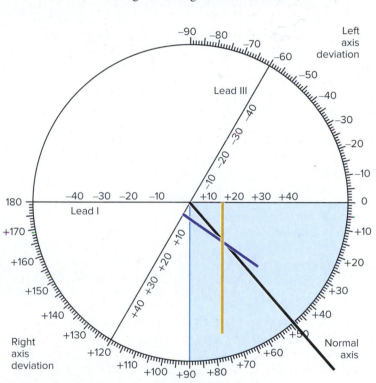

FIGURE 27.6 Electrical Axis of the Heart. Sample shown to determine mean electrical axis.

Notes

REVIEW SECTION

Electrical Conductivity of the Heart

Name _____   *Date* _____

Lab Section _____   *Time* _____

Review Questions

1. The sinuatrial node has a common name. What is it? _____

2. Which two chambers of the heart (atria or ventricles) contract last in a normal cardiac cycle?

   _____

3. What two chambers are stimulated immediately after the SA node depolarizes?

   _____

4. After the AV node depolarizes, what structures conduct the impulse to the myocardium of the ventricles?

   _____

5. What are the main events recorded by an ECG? _____

   _____

6. What electrical event in the heart does the QRS complex represent?

   _____

7. Ventricular repolarization is represented by what part of an ECG? _____

8. What ECG wave is represented by the atrial depolarization? _____

9. Why is the ECG event indicating atrial repolarization not seen in an ECG? _____

   _____

10. What does a heart block do to impulse transmission in the heart? _____

   _____

11. Fibrillation is uncoordinated cardiac muscle contraction. Predict what an ECG would look like if there were no uniform conduction of electrical activity in the heart. Draw what it might look like.

12. What consequence does fibrillation have for cardiac muscle contraction and for the pumping efficiency of the heart? Which is more serious—atrial or ventricular fibrillation? _____

_____

13. If a myocardial infarct (heart attack) destroyed a portion of the right or left bundle branches, what potential change might you see in an ECG? _____

_____

14. Tape or paste your ECG in the following space. Label the P wave, the QRS complex, and the T wave.

15. What was your value for the mean electrical axis of the heart? _____

Was this within normal limits? _____

Functions of the Heart

INTRODUCTION

One of the fundamental physiological activities of the body is to maintain adequate blood pressure. If blood pressure is too low, cells may not function correctly or they may die. If blood pressure is too high, damage may occur to the organs of the body or excess fluid may be expressed from the capillaries. One way to control blood pressure is through changes in **cardiac output.** When more oxygen is required, cardiac output increases with an elevated heart rate or an increase in the volume ejected per contraction, or both. The heart has an **intrinsic control,** which is the sinuatrial node (pacemaker) located in the heart wall that determines heart rate, and it has **extrinsic controls,** from organs other than the heart such as hormones or input from the nervous system that control the heart rate or contraction strength.

In this exercise, you explore some of the functions of the human heart, with particular emphasis on changes in heart rate and contraction strength. These topics are covered in the Saladin text in chapter 19, "The Circulatory System: Heart." You examine the resting rate of human hearts, the natural rate of heart contraction in a frog, and changes that occur when various solutions are added to the frog heart muscle. The heart of a frog is physically somewhat different from a human heart. The frog has two atria and only a single ventricle; however, the physiological response of the frog to various cardiac-influencing substances is similar to that of humans.

Due to the decline in some frog populations (the *leopard frogs* in North America), bullfrogs may be preferable as study specimens. Bullfrogs not only are plentiful but are frequently a pest species in many areas and, as of today, may be used without significant impact on their population levels.

Cardiac Muscle Characteristics

Cardiomyocytes, like neurons, undergo a normal **polarization** process. This produces a **resting membrane potential.** The mechanism by which electrochemical impulses occur in heart muscle depends on three ion channels, **fast sodium channels, slow calcium channels,** and **potassium channels.** In cardiomyocytes, polarization occurs by activation of the **sodium-potassium pump.**

The result is an increase in sodium and potassium ions outside the cell membrane. This produces a resting membrane potential.

When the cardiomyocyte is stimulated, the voltage-gated sodium channels open, causing depolarization to the cell's threshold. This causes other sodium channels to open, producing a peak at about +30 mV. This leads to a twitch of the muscle fiber. In skeletal muscle the twitch is quick (2 msec). In cardiac muscle the twitch is much longer (250 msec), caused by a **plateau** in the repolarization of the cell. This is due to the opening of calcium channels and the movement of calcium into the cytosol. The plateau can be seen in fig. 28.1. Depolarization travels across the cardiomyocyte. **Repolarization** occurs relatively quickly by reestablishing the resting membrane potential, and the heart muscle is ready for its next contraction.

Cardiac Pacemaker

Pacemaker cells in the **sinuatrial (SA) node** of the heart wall are specialized muscle cells that spontaneously and periodically depolarize, sending action potentials across the heart wall, causing the heart to contract. In neurons, by contrast, the resting membrane potential is stable. In the sinuatrial node, the depolarization is thought to occur due to the leaking of sodium ions into the pacemaker cells without the corresponding outflow of potassium ions. When threshold is reached, fast calcium-sodium channels open, and calcium and sodium ions flow into the cell. When 0 mV is reached, the **potassium channels** open, and potassium ions leave the cell.

Once this happens, a **field effect** occurs, depolarizing cardiomyocytes adjacent to the pacemaker. This depolarization spreads throughout the atria and eventually to the ventricles, stimulating the cardiac muscle to contract. Action potentials from the sinuatrial node open **voltage-gated slow calcium channels** in the cardiomyocytes. **Ligand-gated calcium channels** in the sarcoplasmic reticulum open and calcium moves through the cytosol, binding to troponin, as it does in skeletal muscle.

Because the heart muscle cells are linked together by intercalated discs with gap junctions between cells, they act as one electrical unit. An impulse generated in the nodal tissue spreads to

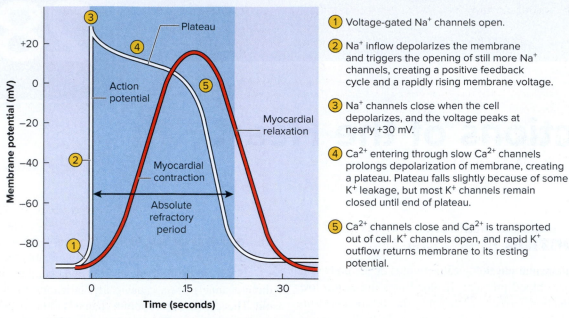

FIGURE 28.1 **Cardiac Muscle Action Potential and Muscle Tension.**

the surrounding muscle, causing it first to depolarize (an electrical event) and subsequently to contract (a mechanical event). Review the conduction pathway of the heart in Laboratory Exercise 27.

Normally, the resting heart rate of mammals is lower than the rate of excised hearts. In human hearts that have been removed for heart transplant surgery, the cardiac rate is about 100 beats per minute, which is the native sinuatrial depolarization rate.

This slowing of the heart rate from 100 beats per minute to an average of 72 beats per minute is due to the action of the **parasympathetic nervous system.** The **vagus nerve** conveys parasympathetic fibers that innervate the SA node. **Acetylcholine (ACh)** is a neurotransmitter released by this nerve, and it promotes potassium (K⁺) leakage from the nodal cells to the interstitial fluid, increasing the flow of potassium ions to the outside of the membrane, thus **hyperpolarizing** the cells. The greater the difference in voltage between the inside and the outside of the cell, the longer it takes for cells to reach threshold and depolarize. In this way the heart rate is slowed. Acetylcholine is a **parasympathomimetic substance,** one that mimics the effects of innervation by the parasympathetic nervous system.

The heart rate can be elevated by increasing the firing rate of the SA node. An increase in the heart rate from the normal resting rate to above 100 bpm is due to the progressive *inhibition* of the **parasympathetic nervous system** and subsequent *stimulation* of the **sympathetic nervous system.** As the vagus nerve secretes less acetylcholine at the node, the firing rate increases. Above 100 beats per minute, the heart rate is controlled by the sympathetic nervous system. Nerves from the cervical region of the spinal cord release **norepinephrine,** which binds to **beta-adrenergic receptors.** This causes calcium channels to open, *decreasing* the threshold and causing more rapid firing of the SA node. Like

norepinephrine, **epinephrine** increases the heart rate and strength of contraction by making the calcium channels more permeable. Epinephrine is a **sympathomimetic drug,** one that mimics the effect of sympathetic nervous innervation. Other chemicals function as stimulants. Caffeine is a central nervous system stimulant, and it has an effect on heart muscle. It increases heart rate and the strength of contraction. High concentrations of caffeine cause **arrhythmias,** or irregular heart rates.

The addition of calcium chloride to the heart tissue increases the heart rate and contraction strength, but it can also produce pacemakers in regions of the heart other than the SA node. These are called **ectopic foci.**

The electrical activity of the nodes of the heart occurs in specialized *muscle* cells, not in nervous tissue. The electrochemical transfer of impulses by cardiomyocytes is known as **myogenic conduction.** The muscle tissue may be influenced by the nervous system, but the activity is initiated and generated in specialized muscle fibers. In addition to the influence of the nervous system, certain hormones, some drugs, and ion concentration have an effect on the heart rate.

OBJECTIVES

At the end of this exercise, you should be able to

1. describe the basic contraction characteristics of the heart;
2. list the effects of epinephrine, calcium chloride, acetylcholine, and temperature on the heart rate;
3. outline the mechanisms by which the items in #2 change heart rate;
4. measure the resting pulse rate of the heart;

5. correlate the sounds the heart makes with the action of the heart;
6. explain how the valves of the heart increase the efficiency of the contraction.

MATERIALS

Pulse Rate and Heart Sounds

Clock or watch with second hand

Stethoscope

Alcohol wipes

Ph.I.L.S. Physiology Interactive Lab Simulations Version 4.0 (available through McGraw-Hill)

Compatible computer

Frog Heart Rate and Contraction Strength

Frog Preparation

Live frog

Clean dissection instruments

    Scissors

    Razor blades

    Dissection tray

    Forceps

    Dissection pins

Small heart hook (fishhook, copper wire, or Z wire)

Nylon thread

Frog anchoring board

Ring stand

Goggles

Animal waste container

BIOPAC Equipment

Force transducer assembly—SS12LA

HDW100A tension adjuster

Data acquisition unit MP30 or MP150

BIOPAC *BSL PRO* A04 Frog Heart (FrogHeart.gtl)

50-gram weight with hook

Physiograph Equipment

Myograph transducer

Physiograph or Duograph

Cable

Channel amplifier

Solutions

Frog Ringer's solution—room temperature, 37°C, and iced

2% calcium chloride solution

0.1% acetylcholine chloride solution

0.1% epinephrine solution

Saturated caffeine solution

PROCEDURE

Pulse Rate

Ⓐ **Activity**

Measure the pulse rate (heart rate) and listen to the heart sounds of your lab partner.

1. Place your fingers on the radial artery or the carotid artery of your lab partner and count the number of beats that occur in 1 minute and record the resting pulse rate.

❓ Resting pulse rate: _____ beats/minute (1)

2. You can calculate the average pulse rate for your class by having all the members of your class record their pulse rate in beats per minute on the chalkboard or whiteboard. Add all the pulse rates together and divide by the total number to determine the average. Compare the value you calculate from your class with the average heart rate.

❓ Average pulse rate for class: _____ bpm (2)

There are differences in pulse rate, depending on whether you are sitting, lying, or standing. The normal resting pulse rate is measured when you are sitting. One of the functions of changes in pulse rate is to alter blood pressure. The carotid sinus is one of the areas of the body that senses changes in blood pressure.

3. After you measure the pulse rate while sitting, lie down on a table or cot and measure the pulse rate. Record any change in the pulse rate in the following space.

❓ Pulse rate while lying down: _____ bpm (3)

4. Stand up and record the change in the pulse rate in the following space.

❓ Pulse rate while standing: _____ bpm (4)

Can you explain any changes in the rate from the one you recorded when you were sitting?

Heart Sounds

The S_1 and S_2 or **lubb/dupp** sounds of the heart reflect the closure of the heart valves. The first heart sound is the S_1 or lubb sound, and it occurs due to the closing of the atrioventricular valves. The second heart sound is the S_2 or dupp sound, and it is due to the closing of the pulmonary and aortic valves. You may hear an additional whooshing sound as you listen through the stethoscope. This is attributed to the imperfect closure of the valves, known as a **heart murmur.**

(A) Activity

1. Clean the earpieces of a stethoscope with alcohol swabs and examine the stethoscope. The earpieces should point in an anterior direction as you place them into your ears.
2. Listen for heart sounds generated by your lab partner by placing the diaphragm of the stethoscope inferior to the second rib on the left side of the body. The pulmonary valves are best heard in this location.
3. Place the diaphragm of the stethoscope on the right side of the body, in the intercostal space between the second and third ribs, for the location of the aortic valve.
4. Use fig. 28.2 as a guide and listen to heart sounds in these locations.

Examination of Frog Heart Rate and Contraction Strength in Frogs

Virtual Experiment **Ph.I.L.S.**
Frog Heart Functions 21. Thermal and Chemical Effects

(A) Activity

1. Open the program; click on and read the **Objectives and Introduction** tab.
2. Take the pre-lab quiz to make sure you understand the concepts presented in the virtual experiment. The **Wet Lab** portion describes the process as it is done on a live frog.
3. Follow the instructions carefully in the **laboratory exercise.**
4. Click the power button to turn on the computer screen and also the power button to the data acquisition unit. Connect

the blue plug to the Recording Input 1. Touch the tip of the plug to the unit and it should insert.
5. Click on the **Start** button on the computer screen. You should see the heart contractions.
6. Click the pipette once and drag it to the location where the simulated heart is attached to the myograph transducer.
7. Click the bulb of the pipette again to get it to dispense the regular frog Ringer's solution.
8. Let the heart beat for three or four cycles and then add the cold frog Ringer's solution. Once you get three or four more cycles, click on the **Stop** button. In order to record the data, you will have to click on the left arrow at the bottom of the chart to move to the first recording.
9. Move the crosshair cursor to the top of one of the curves and click it there.
10. Move the other crosshair cursor to the bottom of the curves and click it there. This will set the data points for the amplitude of the contraction.
11. Click on the **Journal** tab to get a record of the amplitude written in the journal.
12. Use the right cursor and move it to the peak of the next wave in order to determine the frequency of the contraction. Click on the **Journal** tab to enter the data.
13. Repeat the amplitude and frequency measurements for the cold Ringer's solution. You will probably need to move the chart by clicking on the right arrow at the bottom of the computer screen.
14. Make sure you push the **Start** button before adding new solutions. You will need to add regular frog Ringer's solution again and record the amplitude and frequency of regular frog Ringer's solution before you record the new solutions. If you make a mistake, you have to re-record the entire regular frog Ringer's solution and the experimental values before you can enter them into the **Journal.**
15. Wait for three or four cycles, then add the adrenaline solution via a dropper bottle. Record the amplitude of the contraction and the interval between contractions as you did in the first recording.

    Do the same with the acetylcholine solution as you did with the adrenaline.

    Record the amplitude of the various solutions below and at the end of the exercise.

❓ Regular frog Ringer's solution: _____ (5)

❓ Cold frog Ringer's solution: _____ (6)

❓ Frog Ringer's solution with adrenaline: _____ (7)

    _____

❓ Frog Ringer's solution with acetylcholine: _____ (8)

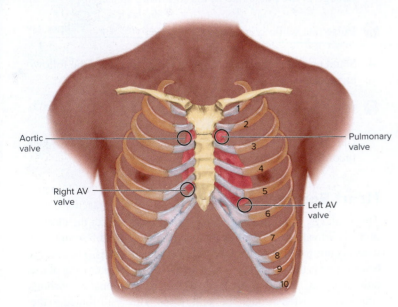

FIGURE 28.2 Auscultation Areas. Circles are locations where specific valvular sounds can best be heard.

Aortic valve
Pulmonary valve
Right AV valve
Left AV valve

Record the interval between contractions below and at the end of the exercise.

 Regular frog Ringer's solution: _____ (9)

 Cold frog Ringer's solution: _____ (10)

 Frog Ringer's solution with adrenaline:

_____ (11)

 Frog Ringer's solution with acetylcholine:

_____ (12)

Live Frog Experiment   Either you have a frog prepared for you, or your instructor may wish for you to double pith a frog for this exercise. If you need to pith a frog, follow the instructions outlined in Laboratory Exercise 15 and then follow the procedure described next. You may be using a BIOPAC setup or a physiograph setup. In either case, the frog will be prepared the same way.

Frog Preparation

Procedure

Ⓐ Activity

1. Once the frog is pithed, pin the frog down to a board.
2. Cut through the pectoral skin with scissors and through the sternum. The heart should be beating inside the pericardial sac. Moisten the heart with room temperature frog Ringer's solution.
3. Carefully cut the pericardial membrane of the heart away from the dorsal surface of the frog.
4. Cover the heart with a moist paper towel and prepare either the BIOPAC setup or the physiograph setup.

BIOPAC Section

BIOPAC Setup for MP30 or MP150 Data Acquisition Unit   You can access detailed instructions or customize the experiment at www.biopac.com.

1. Make sure the computer is on.
2. Connect the data acquisition unit to an electrical outlet with the power supply.
3. With the data acquisition unit *off,* connect it to the computer.
4. Plug SS12LA into Channel 1.
5. Turn the data acquisition unit on.

Computer Setup

1. Start the *BSL PRO* software. An untitled window should appear.
2. Go to the **File** menu and choose **Graph Template (*GTL) > File Name: FrogHeart.gtl.**
3. Physically attach the HDW100A tension adjuster to a ring stand.

4. Attach the force transducer assembly to the tension adjuster. Make sure that the place where the S-hook attaches is facing down. The adjuster should be set so that, when the frog heart is hooked up, the string will be vertical.
5. On the computer, select 0–50 grams force range.
6. Select the **Setup Channels** from the menu, choose the wrench icon from the **Channel 1** menu, and locate the wrench icon and scaling window.
7. With only the S-hook on the tension adjuster, click **Cal 1.**
8. Attach a 50-gram weight to the adjuster and click **Cal 2.**
9. Click **OK** and exit the scaling window.

Running the Experiment

1. Place the frog on the board underneath the force transducer assembly and remove the wet paper towel from the frog heart.
2. Attach a metal hook or wire to the membrane attached to the dorsal portion of the heart. You can pierce the apex of the heart with a wire or hook, but you must be very careful not to puncture the ventricle.
3. Attach the hook to the force transducer assembly (fig. 28.3).
4. Adjust the tension, so that the nylon thread is taut but not so much that you tear the hook from the heart.
5. Click the **Start** button and record 10 cycles of the heart.
6. Save your data by selecting the **File menu.** Choose **Save As** and select the file type: **BSL Pro files (*.ACQ) File name: (your name).** Choose **Save.**
7. Alter the temperature or add chemicals as described after "Physiograph Section."

Physiograph Section

Procedure

Ⓐ Activity

1. Attach a metal hook or wire to the membrane at the dorsal portion of the heart. You can pierce the apex of the heart with a wire or hook, but you must be very careful not to puncture the ventricle.
2. Attach the hook to a thread and tie the thread to the myograph transducer, adjusting the tension until you get a deflection in the myograph leaf. Your setup should look similar to fig. 28.3.
3. Set the chart speed to 0.5 cm/second and record the contraction rate and strength of contraction of the heart muscle. The force that is generated by the heart is measured by the height of the stylus on the chart paper. The rate can be determined by the distance between the peaks on the chart paper.

Record this value in the following spaces.

 Heart contraction strength: _____ (13)

 Heart contraction rate: _____ (14)

4. Alter the temperature or add chemicals as described in the following section.

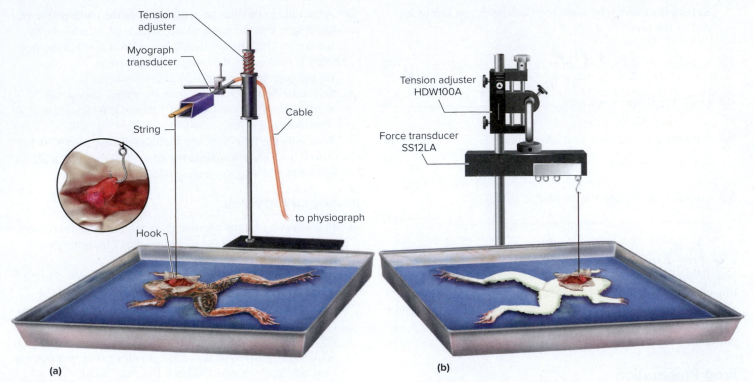

FIGURE 28.3 Frog Setup. (a) For physiograph; (b) for BIOPAC.

Data Acquisition

Keep the heart moistened with frog Ringer's solution. Unless directed to do otherwise, use the Ringer's solution that is at room temperature.

Changes in Heart Rate with Temperature   You have already recorded the heart rate at room temperature.

Ⓐ Activity

1. Run tests by flooding the heart with iced Ringer's solution. Note the change in heart rate.

❓ Heart rate with iced Ringer's solution: _____ bpm (15)

2. You can further study the effect of temperature on enzymes by flooding the frog heart with frog Ringer's solution at 37°C. Record the contraction rate.

❓ Heart rate at 37°C: _____ bpm (16)

The Q_{10} is an enzymatic relationship that states that, for every 10° rise in temperature (usually from 0°C to 50°C), there is a specific increase of enzyme activity relative to the species studied.

❓ Does this correlate with your findings on the temperature relationships with the heart? _____ (17)

Effect of Various Chemicals on Heart Rate and Contraction Strength   Various chemicals, including some potent cardiac drugs, can cause an increase in heart rate. These are called **positive chronotropic drugs.** Some chemicals cause a decrease in heart rate, and these are known as **negative chronotropic drugs.** Other chemicals cause an increase in the force of contraction, and these are the **positive inotropic drugs.** Those that lessen the force of contraction are known as **negative inotropic drugs.**

In the following tests, determine if any of the chemicals administered to the heart have these characteristics.

Calcium

Ⓐ Activity

Add several drops of 2% calcium chloride solution to the heart muscle. Wait until you get a change in the heart rate before you record the data. Turn on the recording device (BIOPAC or physiograph) and produce a tracing. Calculate the heart rate and record the number in the following space.

❓ Heart rate: _____ bpm (18)

What happens to the contraction strength?

_____

Rinse the heart with room temperature frog Ringer's solution.

Acetylcholine

Ⓐ Activity

Add several drops of 0.1% acetylcholine solution to the outer surface of the heart. You may have to wait a few minutes to observe any changes in heart rate. Record the heart rate for about 10 seconds after waiting 2 minutes. Wait 3 additional minutes and then record the rate for another 10 seconds. Record the rate in the following space.

❓ Heart rate after 2 minutes: _____ bpm (19)

❓ Heart rate after 3 additional minutes: _____ bpm (20)

When you have determined a change in the heart rate and recorded your results, flood the heart with frog Ringer's solution.

Epinephrine

Ⓐ Activity

Add several drops of 0.1% epinephrine solution to the outer surface of the heart. Examine the rate over the next 2 to 3 minutes, and when you notice a change, record the heart rate for about 10 seconds. Write your results in the space provided.

❓ Heart rate: _____ bpm (21)

What is the change in the contraction strength?

_____

After you record your results, flush the heart with room temperature frog Ringer's solution.

Caffeine

Ⓐ Activity

Add several drops of saturated caffeine solution to the outer surface of the heart and note the time when the rate changes. Once the rate changes, record the heart rate and strength of contraction of the heart muscle for about 10 seconds. Enter the rate in the following space.

❓ Heart rate with caffeine: _____ bpm (22)

Strength of contraction (increase/decrease):

_____

Clean Up When you are finished, dispose of the frog in the appropriate animal waste container. Clean your desk and equipment. Dispose of scalpel blades in the sharps container.

Notes

REVIEW SECTION

Functions of the Heart

Name _____   *Date* _____

Lab Section _____   *Time* _____

❓ Chapter Summary Data

Use this section to record your results from questions within the exercise.

1. _____   12. _____
2. _____   13. _____
3. _____   14. _____
4. _____   15. _____
5. _____   16. _____
6. _____   17. _____
7. _____   18. _____
8. _____   19. _____
9. _____   20. _____
10. _____   21. _____
11. _____   22. _____

Review Questions

1. Decreasing heart rate is under the control of what nervous division? _____

2. What is the resting heart rate of the average person? _____

3. Are there more sodium ions inside or outside a cardiac muscle cell during the resting membrane potential?
_____

4. What happens to sodium ions when a membrane depolarizes? _____
_____

5. What region in the heart depolarizes spontaneously? _____

6. What happens to the heart when an action potential is generated in the SA node? _____
_____

7. The movement of electrochemical impulses in the myocardium is called _____ conduction.

8. What effect do calcium slow channels have on shortening or lengthening the contraction time of the heart muscle? _____
_____

9. Beta-adrenergic blockers bind to norepinephrine sites, preventing these neurotransmitters from having an effect.

   What effect would the use of "beta blockers" have on heart rate? _____

10. How much of a change in the heart rate of the frog did you see after the addition of calcium chloride?
_____

   What was the change in the contraction strength? _____

   What process might account for this in terms of cardiac muscle interactions with calcium?
_____

11. What heart sound is produced by the closure of the atrioventricular valves in the heart? _____

12. A heart murmur is normally caused by what event? _____

13. When would a murmur occur in the S_1/S_2 cycle if the AV valves were not closing properly?
_____

14. Pilocarpine stimulates the release of acetylcholine from the vagus nerve, thereby increasing parasympathetic stimulation. What impact would this drug have on heart rate? _____

15. In the Ph.I.L.S. virtual heart experiment, what effect did the addition of the adrenaline solution have on heart rate and strength? _____

16. Did the addition of the adrenaline solution increase or decrease the threshold in the heart muscle tissue?
   increase/decrease (circle correct answer)

17. In the Ph.I.L.S. virtual heart experiment, what effect did the addition of the acetylcholine solution have on heart rate and strength? _____

18. Did the addition of acetylcholine solution hyperpolarize or depolarize the cardiac muscle cell membrane? hyperpolarize/depolarize (circle correct answer)

19. Which one of these is a parasympathomimetic substance? _____

Blood Vessels 1: Introduction to Blood Vessels and Blood Vessels of the Axial Region

INTRODUCTION

The blood vessels are the conduits of the cardiovascular system that carry oxygen, nutrients, and other materials to the cells, as well as remove wastes from the tissue fluid near the cells. These vascular tubes consist of numerous vessels, including arteries, arterioles, capillaries, venules, and veins. **Arteries** are defined as blood vessels that carry blood *away from* the heart. Most arteries carry oxygenated blood, but there are a few exceptions to this such as the pulmonary artery that takes deoxygenated blood to the lungs. Because the pulmonary trunk and pulmonary arteries carry deoxygenated blood, they are colored blue on models and charts. Arteries have higher blood pressure than veins, and therefore the walls of arteries are thicker. Arteries are frequently named for the region of the body they occur in, such as the *brachial* artery, which is found in the region of the arm, or the *femoral* artery located in the thigh region. Arteries are also named for the organs they serve, such as the *renal* artery that takes blood to the kidney or the *splenic* artery.

Arteries become progressively smaller and go from large arteries, to medium arteries, to small arteries, and then finally become **arterioles.** Arterioles become even smaller and take blood to **capillaries** where nutrients, water, and oxygen are provided to the cells of the body. The capillaries pick up carbon dioxide and waste from the interstitial fluid surrounding the cells. On the return flow, capillaries take blood to **venules,** then **veins,** and back to the heart under relatively low pressure. Veins are defined as blood vessels that carry blood toward the heart. Veins resemble tributaries of rivers because smaller veins flow into larger veins just as small creeks flow into rivers. Veins can be superficial (close to the skin) or deep. The deep veins of the body frequently travel alongside the major arteries and take on the arterial names (for example, the femoral vein, axillary vein, and subclavian vein), while superficial veins have names specific to themselves (for example, the great saphenous vein). Veins have thinner walls than arteries, and superficial veins contain valves for a one-way flow of blood to the heart. In this exercise, you learn the basic structure of blood vessels and study the arteries and veins of the axial region. These topics are covered in chapter 20, "The Circulatory System: Blood Vessels and Circulation" in the Saladin text.

OBJECTIVES

At the end of this exercise, you should be able to

1. draw a cross section of the wall of a generalized blood vessel showing the three layers;
2. compare the structure of the wall between arteries, veins, and capillaries;
3. differentiate between conducting arteries and distributing arteries;
4. describe the sequence of major arteries that branch from the aortic arch;
5. describe the sequence of major arteries that originate from the abdominal aorta or its derivatives;
6. identify major axial arteries or veins in models or a cadaver;
7. list the major blood vessels that take blood to or from the brain;
8. describe the major organs that receive blood from the axial arteries;
9. trace the blood flow from a selected organ back to the heart;
10. distinguish a portal system from normal venous return flow;
11. describe the major digestive organs that supply blood to the hepatic portal system; and
12. name the vessels that take blood to, or receive blood from, a particular axial artery or vein.

MATERIALS

Microscopes

Prepared slides of arteries and veins in cross section

Models of the blood vessels of the body

Charts and illustrations of the arterial and venous systems

Cadaver (if available)

Microscope

Prepared slide of arteriosclerosis

PROCEDURE

Overview of Blood Vessels

Obtain a prepared microscope slide of a cross section of an artery and vein. Both arteries and veins have walls that consist of three layers. The outer layer is known as the **tunica externa (adventia)** and consists of a connective tissue sheath. The middle layer, or **tunica media,** is composed of smooth muscle in both arteries and veins. The tunica media is thicker in arteries, and there may be pronounced elastic tissue in the wall of the tunica media in arteries. The innermost layer (near the blood) is the **tunica interna (tunica intima)** and consists of a thin layer of connective tissue and a thin layer of simple squamous epithelium, known as **endothelium.** Endothelium is the layer closest to the blood. In arteries there is an **internal elastic lamina,** a thin layer of elastic tissue. There is also an **external elastic lamina.** These layers are seen in fig. 29.1. The wall of a blood vessel requires oxygen, which is supplied by blood vessels that feed the tissue from the outside. Blood vessels that nourish other blood vessels are known as the **vasa vasorum.** Examine a prepared slide of an artery and vein and locate these layers.

Conducting (elastic) arteries are larger arteries close to the heart. Their appearance is made distinctive by the presence of significant amounts of elastic tissue in the tunica media. **Distributing (muscular) arteries** are found farther away from the heart and have more smooth muscle in the tunica media. There is a gradual transition between these two types of arteries.

A feature of veins not found in arteries is valves. You can easily distinguish veins from arteries in a prepared slide where the vessels are cut in cross section. Veins have a large lumen (though the vein is frequently collapsed in prepared sections) and a thin tunica media relative to its overall size. You may also find circular structures in these slides that are solid, not hollow. These are nerves. Compare your observations of your slide with fig. 29.1.

Major Arteries and Veins of the Body

Examine the models, charts, and illustrations in the lab and locate the major vessels of the body. Compare material in lab with figs. 29.2 and 29.3. Place a check mark next to the name of the vessel on the illustration when you locate it in lab material.

Pulmonary Circuit

Examine models or charts of blood flow to and from the lungs in lab and compare them to fig. 29.4. Locate the vessels

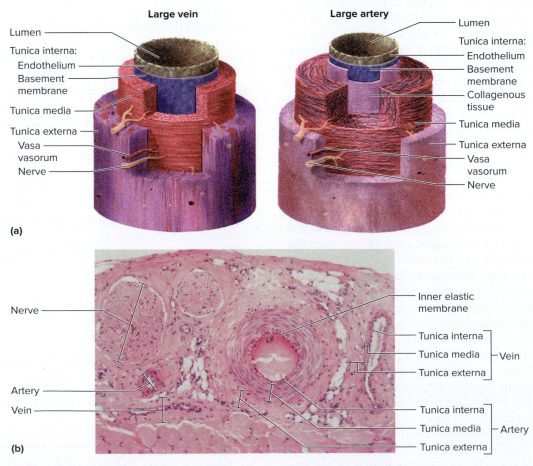

FIGURE 29.1 Artery and Vein Cross Section. (a) Diagram; (b) photomicrograph of an artery, a vein, and a nerve (40x).

(b) ©Eric Wise

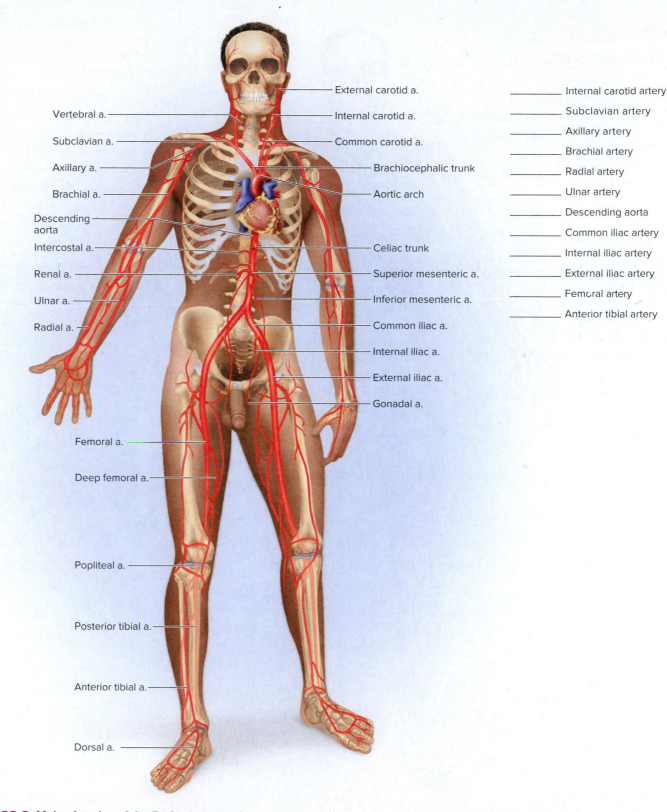

External carotid a.
Vertebral a.
Internal carotid a.
Subclavian a.
Common carotid a.
Axillary a.
Brachiocephalic trunk
Brachial a.
Aortic arch
Descending aorta
Intercostal a.
Celiac trunk
Renal a.
Superior mesenteric a.
Ulnar a.
Inferior mesenteric a.
Radial a.
Common iliac a.
Internal iliac a.
External iliac a.
Gonadal a.
Femoral a.
Deep femoral a.
Popliteal a.
Posterior tibial a.
Anterior tibial a.
Dorsal a.

_____ Internal carotid artery
_____ Subclavian artery
_____ Axillary artery
_____ Brachial artery
_____ Radial artery
_____ Ulnar artery
_____ Descending aorta
_____ Common iliac artery
_____ Internal iliac artery
_____ External iliac artery
_____ Femoral artery
_____ Anterior tibial artery

FIGURE 29.2 Major Arteries of the Body. (a = artery)

in this circuit as you read the following description. The pulmonary circulation involves the heart pumping blood to the lungs for oxygenation, removing carbon dioxide, and returning blood to the heart. Blood in the **right** **ventricle** exits the heart by the **pulmonary trunk.** The blood enters the left and right **pulmonary arteries.** These arteries are unusual in that they carry deoxygenated blood. Most arteries carry oxygenated blood.

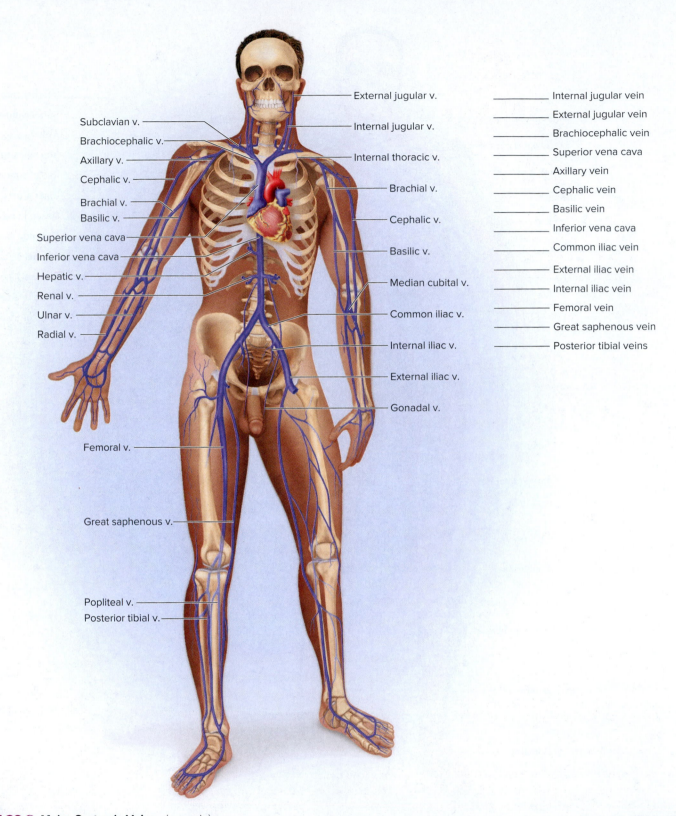

Subclavian v.
Brachiocephalic v.
Axillary v.
Cephalic v.
Brachial v.
Basilic v.
Superior vena cava
Inferior vena cava
Hepatic v.
Renal v.
Ulnar v.
Radial v.

External jugular v.
Internal jugular v.
Internal thoracic v.
Brachial v.
Cephalic v.
Basilic v.
Median cubital v.
Common iliac v.
Internal iliac v.
External iliac v.
Gonadal v.

Femoral v.

Great saphenous v.

Popliteal v.
Posterior tibial v.

_____ Internal jugular vein
_____ External jugular vein
_____ Brachiocephalic vein
_____ Superior vena cava
_____ Axillary vein
_____ Cephalic vein
_____ Basilic vein
_____ Inferior vena cava
_____ Common iliac vein
_____ External iliac vein
_____ Internal iliac vein
_____ Femoral vein
_____ Great saphenous vein
_____ Posterior tibial veins

FIGURE 29.3 Major Systemic Veins. (v = vein)

Anatomical models usually color these vessels blue to indicate that they carry deoxygenated blood. From the pulmonary arteries, blood enters the **lobar arteries** of the lungs and eventually flows to the capillaries, where the blood is oxygenated and carbon dioxide is released. The return flow from the lungs eventually enters the **pulmonary veins.** There are two pulmonary veins from each lung, and these enter the heart at the **left atrium.**

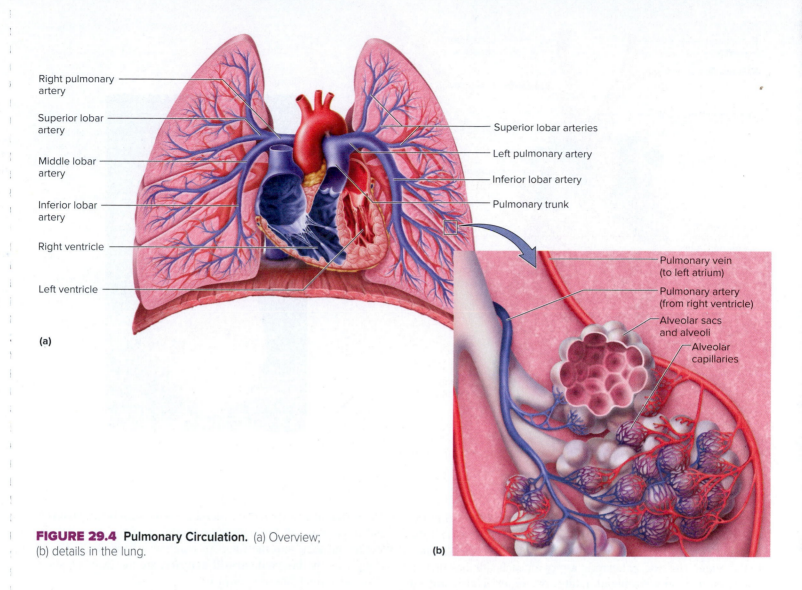

Right pulmonary artery

Superior lobar artery

Middle lobar artery

Inferior lobar artery

Right ventricle

Left ventricle

(a)

Superior lobar arteries

Left pulmonary artery

Inferior lobar artery

Pulmonary trunk

Pulmonary vein (to left atrium)

Pulmonary artery (from right ventricle)

Alveolar sacs and alveoli

Alveolar capillaries

(b)

FIGURE 29.4 Pulmonary Circulation. (a) Overview;
(b) details in the lung.

Systemic Vessels of the Axial Region

For the remainder of the lab exercise, you should find the arteries and veins of the regions that are described. Name the vessels that take blood to an artery or vein and those that receive blood from the artery or vein in question. Be able to identify organs that receive blood from an artery or provide blood to a vein.

Aortic Arteries

Locate the heart in models, charts, or a cadaver in lab and find the large **aorta** that exits from the left ventricle. This is the **ascending aorta,** and it is a large vessel about the size of a garden hose. The ascending aorta is relatively thick-walled due to the high pressure of the blood coming from the heart. The first two arteries that arise from the ascending aorta are the **coronary arteries,** which were covered in detail in Exercise 26. The ascending aorta curves to the left side of the body and forms the **aortic arch** (fig. 29.5). In humans, there are three main arteries that receive blood from the aortic arch. The first major artery to receive blood is on

the right side of the body and is called the **brachiocephalic trunk.** This artery shortly divides into arteries that feed the right side of the head (the **right common carotid artery**) and the right upper limb (the **right subclavian artery**), as seen in fig. 29.5. Two other arteries receive blood from the aortic arch. On the left side is the **left common carotid artery,** which takes blood to the left side of the head, and the **left subclavian artery,** which takes blood to the left upper limb.

As the aortic arch turns inferiorly behind the posterior part of the heart, it becomes the **descending aorta,** which is composed of two segments. Above the diaphragm the descending aorta is known as the **thoracic aorta,** and below the diaphragm it is known as the **abdominal aorta** (fig. 29.6). The thoracic aorta has numerous branches, called **intercostal arteries,** that run between the ribs (fig. 29.7).

Arteries of the Head and Neck   The two main sets of arteries that travel through the neck to the head are the **common carotid arteries** and the **vertebral arteries.** Each common carotid

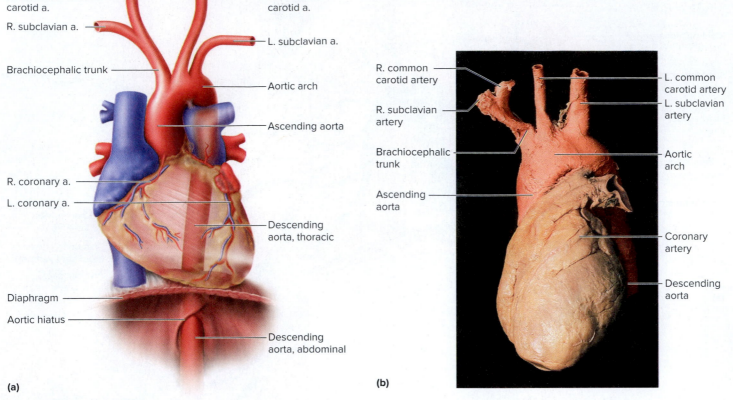

(a)

(b)

FIGURE 29.5 **Arteries of the Aortic Arch.** (a) Diagram; (b) photograph.
(b) ©Eric Wise

artery branches just below the angle of the mandible to form the **external carotid artery,** which takes blood to the face, and the **internal carotid artery,** which passes through the carotid canal to the brain. The external carotid artery on each side has several branches, including the **facial artery,** the **superficial temporal artery,** the **maxillary artery,** and the **occipital artery** (fig. 29.8).

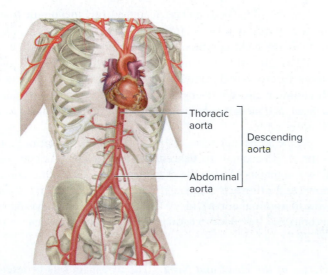

FIGURE 29.6 **Thoracic and Abdominal Aorta.**

The vertebral arteries receive blood from the **subclavian arteries** and travel through the transverse foramina of the cervical vertebrae and then into the foramen magnum of the skull. These, along with the **internal carotid arteries,** are the main suppliers of blood to the brain (fig. 29.2).

Blood Supply to the Brain

(A) Activity

Examine models or charts that demonstrate blood vessels associated with the brain. The blood vessels that supply nutrients and oxygen to the brain form a meshwork around the brain and penetrate into the ventricles. The main arteries that provide blood to the brain are the vertebral arteries and the internal carotid arteries. Locate the vertebral arteries shown in fig. 29.9. The basilar artery is at the base of the brain and branches to connect with the **arterial circle** (circle of Willis), which forms a loop around the pituitary gland. The basilar artery gives rise to the **cerebellar** (SER-eh-BEL-ur) **arteries,** which take blood to the cerebellum. The two internal carotid arteries join the arterial circle at the anterior end. Numerous **cerebral** (seh-REE-bral) arteries take blood superiorly to the surface of the cerebrum and into the interior of the brain. These lead to capillaries where an exchange of blood gases, nutrients, and waste material takes place between the cardiovascular system and the specialized brain cells. Note the various arteries in fig. 29.9 at the base of the brain.

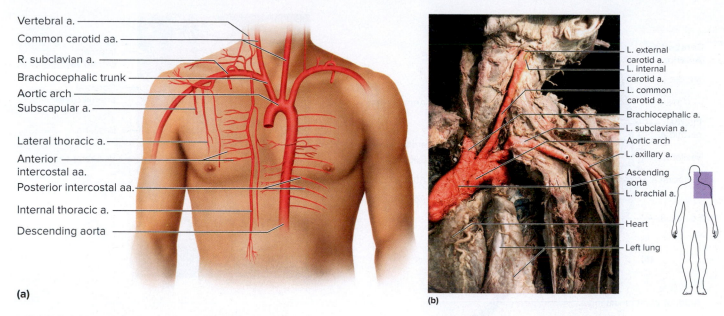

FIGURE 29.7 **Branches of the Subclavian Artery.** (a) Diagram; (b) photograph of cadaver.
(b) ©Eric Wise

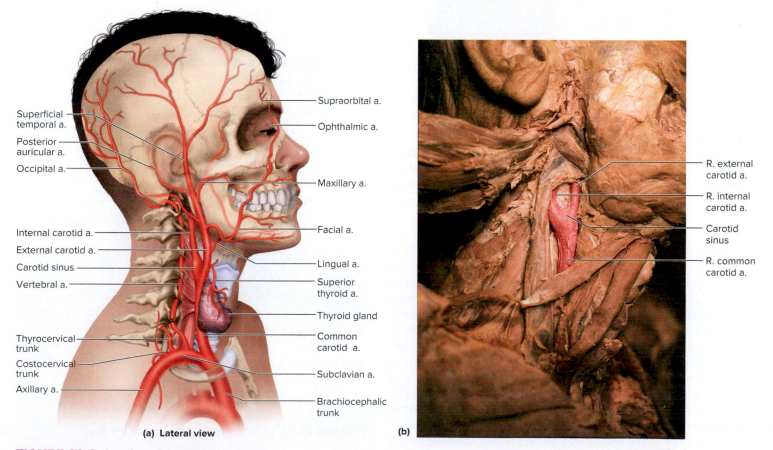

FIGURE 29.8 **Arteries of the Head.** (a) Diagram;
(b) photograph of cadaver.
(b) ©Eric Wise

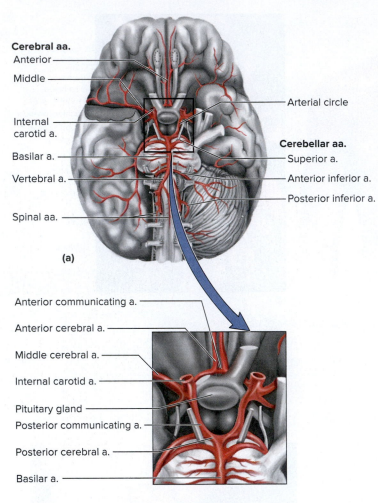

(a)

Cerebral aa.
Anterior
Middle
Internal carotid a.
Basilar a.
Vertebral a.
Spinal aa.
Arterial circle
Cerebellar aa.
Superior a.
Anterior inferior a.
Posterior inferior a.

(b)

Anterior communicating a.
Anterior cerebral a.
Middle cerebral a.
Internal carotid a.
Pituitary gland
Posterior communicating a.
Posterior cerebral a.
Basilar a.

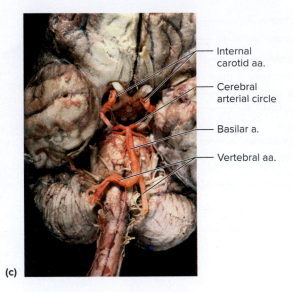

(c)

Internal carotid aa.
Cerebral arterial circle
Basilar a.
Vertebral aa.

FIGURE 29.9 Brain with Arteries, Inferior View. (a) Overview of arterial supply to brain; (b) close-up of arterial circle; (c) photograph.
(c) ©Eric Wise

The drainage of the brain occurs as veins take blood from the brain and pass through the subarachnoid space to the **venous sinuses.** Much of the drainage of blood from the brain flows into the **internal jugular veins** on the return trip to the heart. Examine fig. 29.10 for the venous drainage from the brain.

Veins of the Head and Neck   Read the following descriptions of the veins and find them as they are represented in lab. As you locate a specific vein, name the vessel that takes blood to the vein and those that receive blood from the vein. Examine the models and charts in the lab and locate the veins that drain blood from the head. These vessels are illustrated in fig. 29.11. Blood from the brain returns to the heart primarily by the **internal jugular vein.** This vein begins at the jugular foramen of the skull and passes along the lateral aspect of the neck as it moves toward the **brachiocephalic vein.** The brachiocephalic vein is formed by the union of the internal jugular vein and the **subclavian vein.** The superficial regions of the head (musculature and skin of the scalp and face) are mostly drained by the **external jugular vein.** The external jugular veins join with the subclavian veins prior to reaching the brachiocephalic veins. The left and right brachiocephalic veins take blood to the superior vena cava.

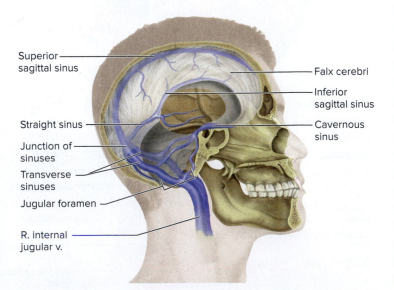

Superior sagittal sinus
Straight sinus
Junction of sinuses
Transverse sinuses
Jugular foramen
R. internal jugular v.
Falx cerebri
Inferior sagittal sinus
Cavernous sinus

FIGURE 29.10 Major Venous Drainage of the Brain.

The **vertebral veins** (fig. 29.11a) receive blood from the muscles and bones of the neck and, like the vertebral arteries, travel through the transverse foramina of the cervical vertebrae.

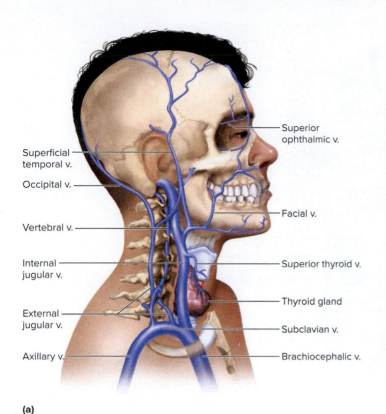

(a)

FIGURE 29.11 **Veins of the Head and Neck.** (a) Diagram of right side; (b) photograph of left veins.
(b) ©Eric Wise

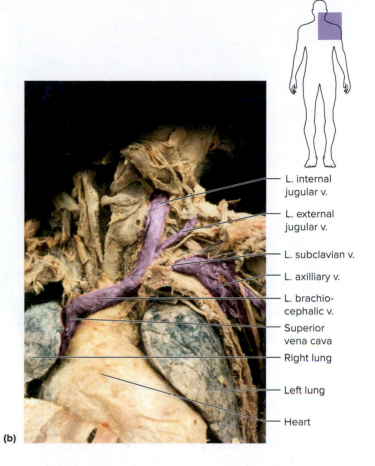

(b)

The vertebral veins take blood to the subclavian veins, which, in turn, flow to the brachiocephalic veins.

Arteries of the Thorax The arteries of the thorax supply blood to the region of the body superior to the diaphragm. Among these are the **anterior** and **posterior intercostal arteries** which supply blood to not only the intercostal muscles but also some thoracic muscles, such as the pectoralis muscles and the serratus anterior, as well as supplying blood to the vertebrae. The **internal thoracic artery** supplies blood to the breast tissue, the anterior thoracic wall, the pericardium, and diaphragm. Another significant artery of the thorax is the **subscapular artery,** which branches from the axillary artery and takes blood to the scapula and nearby muscles. Locate these arteries in lab and in fig. 29.7.

Veins of the Thorax If you look at models in lab for the veins of the thorax, you will have to remove the heart and lungs from the models at some time during your study. The primary veins of the thorax are the subclavian veins, brachiocephalic veins, and the **superior vena cava.** Note that there is one brachiocephalic artery but two brachiocephalic veins. The primary vein that drains the thoracic organs is the **azygos** (AZ-ih-goss) **vein,** which is found on the right, posterior

thoracic wall. The **hemiazygos vein** is smaller and is found on the left, posterior thoracic wall. It drains into the azygos vein which takes blood to the superior vena cava. Examine fig. 29.12 for these veins.

Arteries of the Abdomen and Pelvis The **abdominal aorta** is the portion of the **descending aorta** inferior to the diaphragm. It passes through a hole in the diaphragm known as the **aortic hiatus.** The first major branch of the abdominal aorta is the **celiac trunk** (also known as the **celiac artery**). The celiac trunk splits into three separate arteries: the **splenic artery,** taking blood to the spleen, pancreas, and part of the stomach; the **left gastric artery,** taking blood to the stomach and esophagus; and the **common hepatic artery,** taking blood to the liver, stomach, duodenum, and pancreas. Locate the celiac trunk and the branches of the celiac in fig. 29.13.

Just below the celiac trunk is the **superior mesenteric artery.** This vessel takes blood from the abdominal aorta and continues through the mesentery until it reaches the small intestine and proximal portions of the large intestine, including the cecum, the ascending colon, and part of the transverse colon. The major branches of the superior mesenteric artery are the **ileal and jejunal arteries,** the **ileocolic artery,** and the right and middle **colic arteries.** These are illustrated in fig. 29.14.

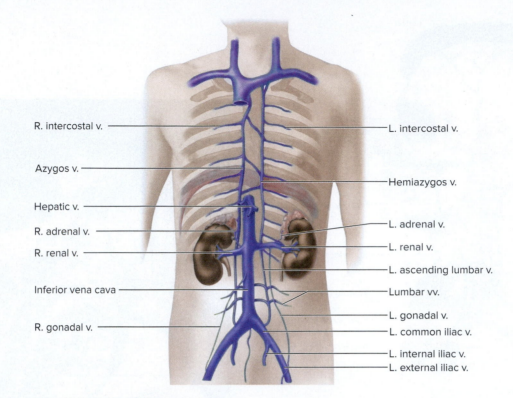

R. intercostal v.

Azygos v.

Hepatic v.

R. adrenal v.

R. renal v.

Inferior vena cava

R. gonadal v.

L. intercostal v.

Hemiazygos v.

L. adrenal v.

L. renal v.

L. ascending lumbar v.

Lumbar vv.

L. gonadal v.

L. common iliac v.

L. internal iliac v.

L. external iliac v.

FIGURE 29.12 Veins of the Abdomen and Thorax.

The next vessels to branch from the aorta are the paired **suprarenal arteries,** which take blood to the adrenal glands. Inferior to the suprarenal arteries are the **left** and **right renal arteries.**

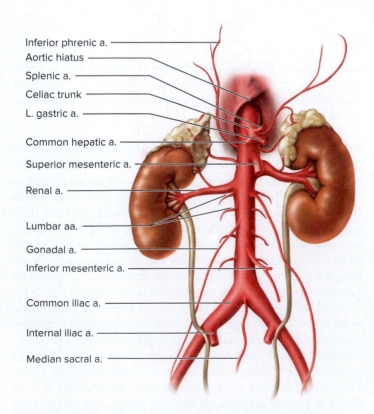

Inferior phrenic a.

Aortic hiatus

Splenic a.

Celiac trunk

L. gastric a.

Common hepatic a.

Superior mesenteric a.

Renal a.

Lumbar aa.

Gonadal a.

Inferior mesenteric a.

Common iliac a.

Internal iliac a.

Median sacral a.

FIGURE 29.13 Arteries of the Anterior Abdominal Region.

These arteries take blood to the kidneys. Locate these vessels in the lab and in fig. 29.15.

The **gonadal arteries** branch inferior to the renal arteries and descend to either the testes or the ovaries. The **inferior mesenteric artery** is the next vessel to branch from the aorta, and it takes blood to the lower portion of the large intestine, including part of the transverse colon, the descending colon, the sigmoid colon, and the rectum. These can be located in figs. 29.14 and 29.15. The abdominal aorta terminates by dividing into the two **common iliac arteries.** The common iliac arteries take blood to the internal and **external iliac arteries.** The **internal iliac artery** takes blood to the pelvic region, including branches that feed the rectum, pelvic floor, external genitalia, groin muscles, hip muscles, uterus, ovary, and vagina. These arteries are illustrated in fig. 29.15.

Veins of the Abdomen and Pelvis   Veins of the abdominal region drain the major organs of the digestive tract and other abdominal organs, such as the spleen. Frequently, these veins are named for the organs from which they receive blood (e.g., the splenic vein takes blood from the spleen). The veins that flow into the liver before returning to the heart are part of the **hepatic portal system.** Most veins take blood from capillaries and venules and return the blood to the heart. The portal system pattern is different from the normal venous blood flow. In a **portal system,** a series of vessels takes blood from the *capillary beds* of an organ (or organs) through a series of veins and then to another *capillary bed.* In the case of the **hepatic portal system,** the blood flows from the capillary beds of the abdominal organs through numerous veins to the capillary bed of the liver. The **inferior mesenteric vein** drains the distal part of the large intestine, receives blood from the **right gastro-omental vein,** and empties into the **splenic vein.** The splenic vein joins with the **superior mesenteric vein,** which drains the small intestine and the proximal portion of the large intestine. The splenic vein and the superior mesenteric vein unite to form the **hepatic portal vein,** which takes blood to the capillaries of the liver, where the blood is cleaned by macrophages, and nutrients are either stored in the liver or passed into the bloodstream. The liver receives blood from the digestive organs and processes it before sending it through the **hepatic veins** to the inferior vena cava and toward the heart. Examine fig. 29.16 for the main vessels of the hepatic portal system, identify the following veins, and check them off when you find them on models or in charts in lab.

_____ Inferior mesenteric vein

_____ Superior mesenteric vein

_____ Splenic vein

_____ Right gastro-omental vein

_____ Hepatic portal vein

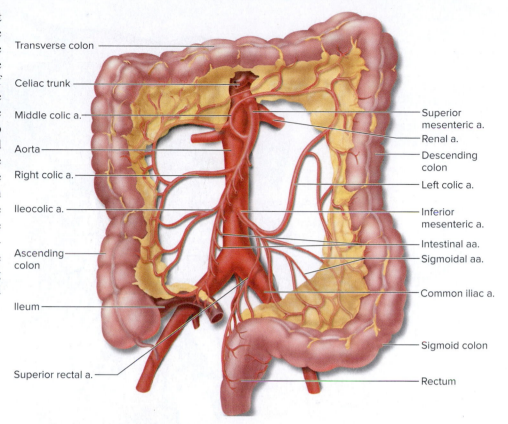

FIGURE 29.14 Middle Abdominal Arteries.

Labels: Transverse colon, Celiac trunk, Middle colic a., Aorta, Right colic a., Ileocolic a., Ascending colon, Ileum, Superior rectal a., Superior mesenteric a., Renal a., Descending colon, Left colic a., Inferior mesenteric a., Intestinal aa., Sigmoidal aa., Common iliac a., Sigmoid colon, Rectum

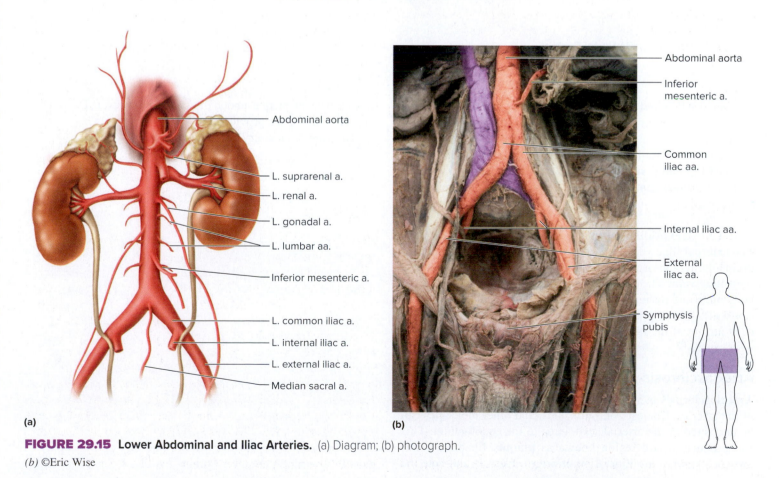

FIGURE 29.15 Lower Abdominal and Iliac Arteries. (a) Diagram; (b) photograph.

(a) Labels: Abdominal aorta, L. suprarenal a., L. renal a., L. gonadal a., L. lumbar aa., Inferior mesenteric a., L. common iliac a., L. internal iliac a., L. external iliac a., Median sacral a.

(b) Labels: Abdominal aorta, Inferior mesenteric a., Common iliac aa., Internal iliac aa., External iliac aa., Symphysis pubis

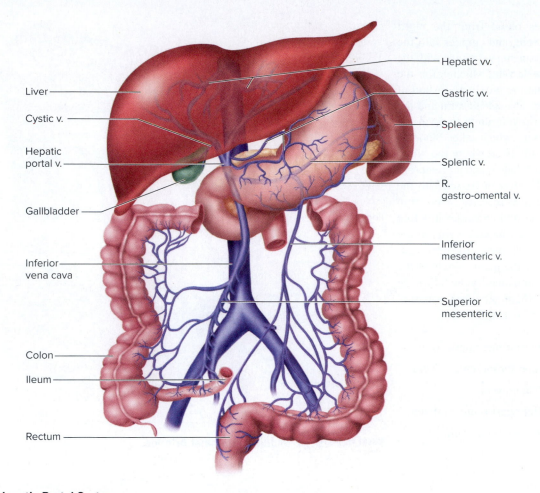

Liver

Cystic v.

Hepatic
portal v.

Gallbladder

Inferior
vena cava

Colon

Ileum

Rectum

Hepatic vv.

Gastric vv.

Spleen

Splenic v.

R.
gastro-omental v.

Inferior
mesenteric v.

Superior
mesenteric v.

FIGURE 29.16 **Hepatic Portal System.**

Some abdominal veins take blood directly to the inferior vena cava and some pass through the liver before reaching the inferior vena cava. Those that take blood directly to the inferior vena cava are the **renal veins,** the **suprarenal veins,** and the **lumbar veins.** The **right gonadal vein** takes blood directly to the inferior vena cava, but the **left gonadal vein** takes blood to the renal vein prior to flowing into the inferior vena cava. These vessels can be seen in fig. 29.17.

The **external iliac vein** takes blood from the femoral vein. The external iliac vein joins with the **internal iliac vein,** which drains the region of the pelvis, and their union forms the **common iliac vein.** The common iliac veins from both sides of the body unite and form the **inferior vena cava,** which travels superiorly along the right side of the vertebrae, taking blood to the right atrium of the heart. Locate these veins and compare them with fig. 29.17.

Arteriosclerosis

Arteriosclerosis is a condition known commonly as hardening of the arteries. Cholesterol, fatty acids, and other molecules can be deposited in the arterial wall, deep to the endothelium. This can produce a hard region known as **plaque.** Examine a microscope slide or an illustration of arteriosclerosis and note the

development of plaque under the endothelial layer. Draw what you see in the space provided.

Drawing of an artery with arteriosclerosis:

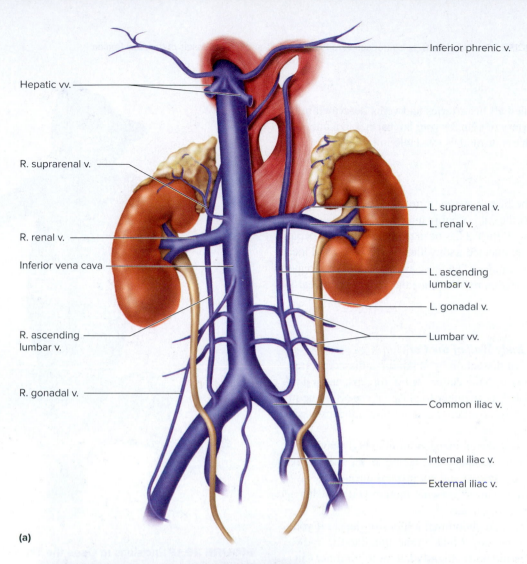

Inferior phrenic v.

Hepatic vv.

R. suprarenal v.

R. renal v.

Inferior vena cava

R. ascending lumbar v.

R. gonadal v.

L. suprarenal v.

L. renal v.

L. ascending lumbar v.

L. gonadal v.

Lumbar vv.

Common iliac v.

Internal iliac v.

External iliac v.

(a)

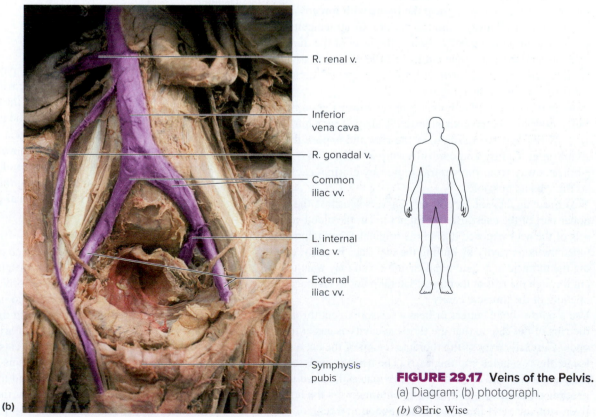

R. renal v.

Inferior vena cava

R. gonadal v.

Common iliac vv.

L. internal iliac v.

External iliac vv.

Symphysis pubis

(b)

FIGURE 29.17 **Veins of the Pelvis.**
(a) Diagram; (b) photograph.
(b) ©Eric Wise

Study Hint

Once you have studied all the arteries and veins described in this exercise, select an artery or vein for your lab partner to name. Quiz each other on the charts or models available in lab.

Cat Anatomy

Cat Dissection   The blood vessels in the cat have been injected with colored latex. If the cat is doubly injected, the arteries are red and the veins are blue. If the cat has been triply injected, the hepatic portal vein is probably injected with yellow latex. Be careful locating the arteries in the cat. You can tease away some of the connective tissue with a dissection needle to see the vessel more clearly.

(A) Activity

Opening the Body Wall of the Cat

1. Prepare for the cat dissection by obtaining a dissection tray, a scalpel, scissors or bone cutter, string, forceps, and a plastic bag with a label. Remember to place all excess tissue in the appropriate waste container and not in a standard wastebasket or down the sink!

2. Be careful as you make an incision into the body cavity of the cat. Cut with the scalpel blade facing to one side or, even better, use a lifting motion and cut upward through the tissue. If you use a downward, sawing motion you may damage the internal organs of the cat.

3. Open the thoracic and abdominal regions of the cat if you have not done so already. Make an incision into the body wall of the cat in the belly, slightly lateral to midline. Cut into the muscle and then grasp the tissue with forceps and lift the body wall away from the viscera. Continue cutting anteriorly through the belly. Stop at the level of the diaphragm, and then continue cutting a little off center as you cut through the costal cartilages. Use a scalpel or scissors to cut through the thoracic region.

4. Now make a transverse incision in the lower region of the belly, so that you have an inverted T-shaped incision (fig. 29.18). Lift the tissue near the ribs and expose the diaphragm. Gently and carefully snip or cut the diaphragm away from the ventral region by cutting as close to the ribs as possible.

5. Next make an incision by cutting transversely across the upper part of the chest. Be careful not to cut the blood vessels of the neck exposed during the original skinning process. Open the body cavity by pulling the two flaps laterally, exposing the thoracic and abdominal cavities. You may want to cut through the ribs at their most lateral point to facilitate the opening of the thoracic cavity.

6. Use a pair of bone cutters or heavy scissors to cut through the ribs of the cat, so that the thoracic cavity is easier to open. Carefully expose the thoracic region of the cat and locate the two lungs and the heart. The heart is enclosed in the pericardial sac. Look for the fatty material, called the **greater omentum.** The greater omentum covers the large liver, stomach, and intestines in the abdominal region.

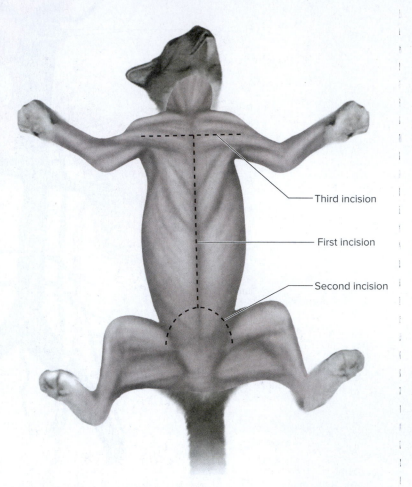

FIGURE 29.18 Incisions to Open the Thoracic and Abdominal Cavities.

7. As you dissect the arteries of the cat, pay careful attention to the veins and nerves that travel along with the arteries. *Do not dissect other organs.* You will study these systems later (for example, the respiratory, digestive, and urogenital systems).

8. As you open the abdominal cavity, you will see the space that contains the internal organs, known as the coelom, or body cavity. The lining on the inside of the body wall is the **parietal peritoneum,** and the lining that wraps around the outside of the intestines is the **visceral peritoneum.**

Arteries of the Head and Neck

1. Begin your study of the arteries of the cat by examining the **ascending aorta.** The ascending aorta bends and forms the **aortic arch** and then moves posteriorly to form the **descending aorta.** The descending aorta is composed of the **thoracic aorta** above the diaphragm and the **abdominal aorta** below the diaphragm. The pattern of arteries that leave the aortic arch is somewhat different in cats than in humans. In the cat, typically only two large vessels leave the aortic arch: the **large brachiocephalic trunk** and the **left subclavian artery.** The brachiocephalic trunk further divides into the **left common carotid artery,** the **right common carotid**

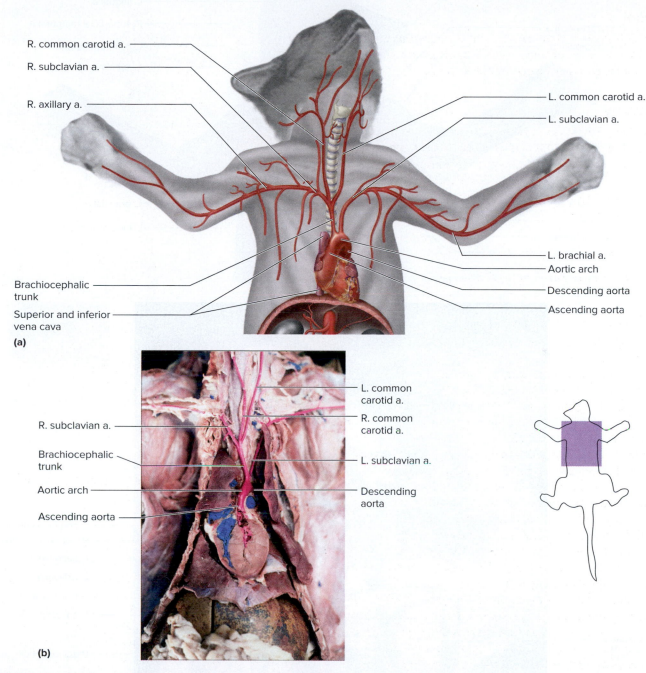

R. common carotid a.

R. subclavian a.

R. axillary a.

L. common carotid a.

L. subclavian a.

L. brachial a.

Aortic arch

Descending aorta

Ascending aorta

Brachiocephalic trunk

Superior and inferior vena cava

(a)

R. subclavian a.

L. common carotid a.

R. common carotid a.

Brachiocephalic trunk

L. subclavian a.

Aortic arch

Descending aorta

Ascending aorta

(b)

FIGURE 29.19   Aortic Arch Arteries of the Cat. (a) Diagram; (b) photograph.
(b) ©Eric Wise

artery, and the **right subclavian artery.** Examine fig. 29.19 for the pattern in the cat and compare it with your specimen. How is the pattern in the cat different from that in the human? Look for the **coronary arteries,** which take blood from the ascending aorta to the heart tissue.

2. Examine the branches of the subclavian arteries. Several arteries arise from the subclavian artery. The **vertebral artery** takes blood to the head, and the **subscapular artery** takes blood to the pectoral girdle muscles, along with the **ventral thoracic artery** and **long thoracic artery.** The ventral thoracic and long thoracic arteries supply the latissimus

dorsi and pectoralis muscles with blood. Additional arteries that branch from the subclavian artery are the **thyrocervical arteries,** which take blood to the region of the shoulder and neck (fig. 29.20).

3. The common carotid artery takes blood along the ventral surface of the neck to feed the small **internal carotid artery,** which supplies the brain. The common carotid artery also empties into the **external carotid artery,** which takes blood to the external portions of the head. The **occipital artery** receives blood from the common carotid artery and supplies the muscles of the neck.

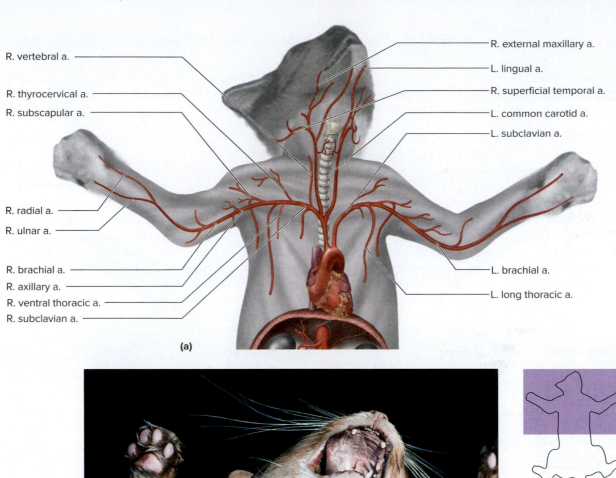

R. vertebral a.

R. thyrocervical a.

R. subscapular a.

R. radial a.

R. ulnar a.

R. brachial a.

R. axillary a.

R. ventral thoracic a.

R. subclavian a.

R. external maxillary a.

L. lingual a.

R. superficial temporal a.

L. common carotid a.

L. subclavian a.

L. brachial a.

L. long thoracic a.

(a)

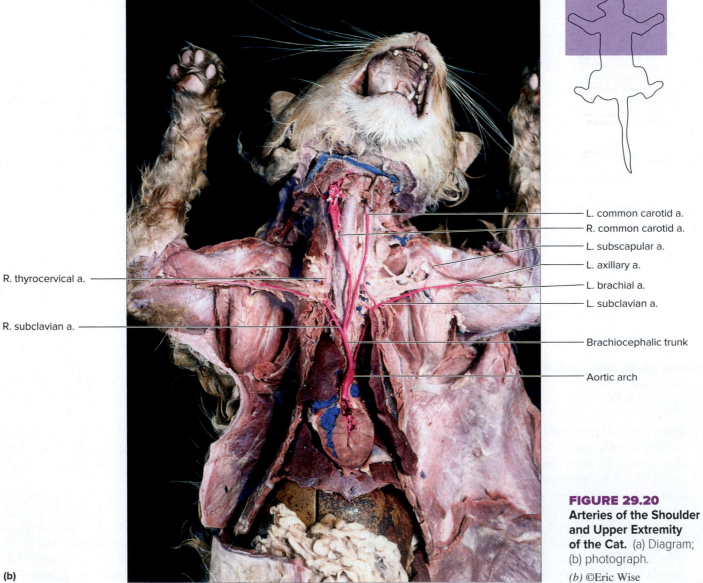

R. thyrocervical a.

R. subclavian a.

L. common carotid a.

R. common carotid a.

L. subscapular a.

L. axillary a.

L. brachial a.

L. subclavian a.

Brachiocephalic trunk

Aortic arch

FIGURE 29.20
Arteries of the Shoulder and Upper Extremity of the Cat. (a) Diagram; (b) photograph.

(b) ©Eric Wise

(b)

Veins of the Head and Neck

1. Look for the veins in the cat, which are generally injected with blue latex. The veins in this region of the cat
have a few variations from those in the human. In the
cat the **external jugular vein** is larger than the **internal
jugular vein.** The reverse is true in humans. What anatomical difference between the cat and human might
explain the difference in the volume of blood carried by
these two vessels?

2. The **transverse jugular vein,** present in cats but absent
in humans, connects the two external jugular veins.
The jugular veins, along with the **costocervical veins**
and the **subclavian veins,** unite to form the brachiocephalic veins (fig. 29.21). The vertebral veins take blood
from the neck muscles of the cat and empty into the
subclavian veins.

Arteries of the Thorax

Locate the arteries that emerge from the thoracic aorta in the cat.
Be careful not to damage other structures, such as organs of the
respiratory system, which you will study later. The thoracic aorta
not only takes blood to the abdominal aorta but also supplies the
rib cage with blood from the intercostal arteries. Esophageal arteries also branch from the thoracic aorta, supplying the esophagus
with blood. The thoracic aorta continues as the abdominal aorta as
it passes through the diaphragm.

Veins of the Thorax

Anterior to the diaphragm are the veins of the thorax. The **internal mammary vein** (internal thoracic vein) may be difficult to
locate (fig. 29.21) because it is frequently cut while opening up the
thoracic cavity. The internal mammary vein receives blood from
the ventral body wall. Another vein of the thoracic region is the
azygos vein. This vessel is found on the right side of the body and
takes blood from the esophageal, intercostal, and bronchial veins
(not illustrated).

Abdominal Arteries

1. Locate the abdominal arteries that come from the abdominal
aorta. The determination of the abdominal arteries is best
done by looking for both the exit of these arteries from the
aorta and their entrance into the organs supplied by these
vessels. The abdominal cavity of the cat should already be
opened. If not, make an incision in the lower abdominal wall
and carefully cut toward the sternum. Do not cut deeply into
the body cavity or you will puncture or cut into the viscera.
Locate the major organs of the abdominal region, such as
the stomach, spleen, liver, and small and large intestines.

2. Examine the specimen from the *left side,* gently lifting the
stomach, spleen, and intestines toward the ventral right side
as you look for the abdominal arteries. The descending
aorta is located on the left side of the body, while the inferior vena cava is located on the right side. Locate the short
celiac trunk, which quickly divides into three major vessels,
the **splenic artery,** the **left gastric artery,** and the **hepatic
artery.** The splenic artery takes blood to the spleen, the left

gastric artery supplies the stomach, and the hepatic artery
reaches the liver. Compare your dissection to fig. 29.22.

3. The **superior mesenteric artery** is the next major vessel to
take blood from the abdominal aorta. The superior mesenteric artery travels to the small intestine and the proximal
portion of the large intestine. It branches into the **jejunal
arteries,** the **ileal arteries,** and the **colic arteries.** Compare
the superior mesenteric artery illustrated in figs. 29.22 and
29.23 with your specimen.

4. The paired **adrenolumbar arteries** branch on each side
of the abdominal aorta and take blood to the adrenal
glands, parts of the body wall, and the diaphragm.
Just caudal to the adrenolumbar arteries are the paired
renal arteries, which take blood to the kidneys. The
gonadal arteries are the next set of paired arteries, and
these take blood to the testes in male cats, or ovaries in
female cats (fig. 29.22).

5. A single **inferior mesenteric artery** takes blood from the
abdominal aorta to the terminal portion of the large intestine. The inferior mesenteric artery can be found caudal to
the gonadal arteries.

6. The pelvic arteries are somewhat different in cats than
in humans. The abdominal aorta splits caudally into the
external iliac arteries, and a short section of the aorta
continues on and then divides to form the two **internal
iliac arteries** and the **caudal artery.** These are illustrated
in fig. 29.24. There is no common iliac artery in cats as
there is in humans. In cats the caudal artery takes blood to
the tail. The internal iliac artery takes blood to the urinary
bladder, the rectum, the external genital organs, the uterus,
and some thigh muscles.

Abdominal Veins

1. Make sure you identify the major organs in the digestive
tract before you proceed with the study of the veins of this
region. Pay particular attention to the stomach, liver, spleen,
kidneys, adrenal glands, small intestine, and large intestine.
The **inferior vena cava** is a large vessel that travels up the
right side of the body in the cat.

2. You may have to gently move some of the digestive organs
to the side to find the inferior vena cava. Examine where
the **common iliac veins** join to form the inferior vena cava.
The **caudal vein** joins the inferior vena cava at this point.
It takes blood from the tail of the cat. This vein is absent in
humans.

3. The **gonadal veins** receive blood from the testes or ovaries.
The right gonadal vein takes blood to the inferior vena
cava, and the left gonadal vein takes blood to the left **renal
vein,** which subsequently leads to the inferior vena cava.
The paired renal veins receive blood from each kidney and
empty into the inferior vena cava. The **iliolumbar veins**
take blood from the body wall and return it via the inferior
vena cava. The body wall is also drained by the **adrenolumbar veins,** which receive blood from the adrenal glands
in addition to the body wall. Compare the veins in your cat
to fig. 29.25.

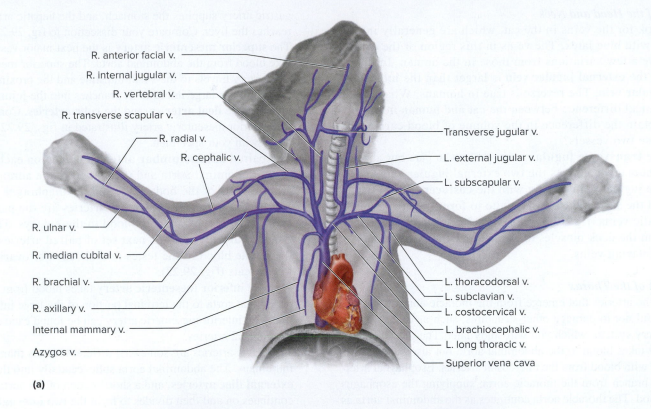

R. anterior facial v.
R. internal jugular v.
R. vertebral v.
R. transverse scapular v.
R. radial v.
R. cephalic v.
R. ulnar v.
R. median cubital v.
R. brachial v.
R. axillary v.
Internal mammary v.
Azygos v.

Transverse jugular v.
L. external jugular v.
L. subscapular v.
L. thoracodorsal v.
L. subclavian v.
L. costocervical v.
L. brachiocephalic v.
L. long thoracic v.
Superior vena cava

(a)

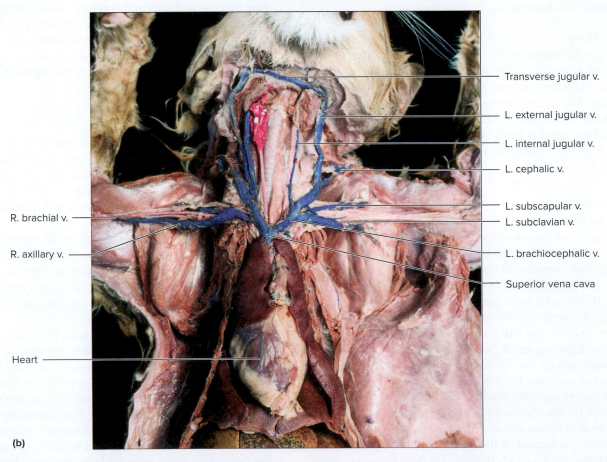

Transverse jugular v.
L. external jugular v.
L. internal jugular v.
L. cephalic v.
L. subscapular v.
L. subclavian v.
L. brachiocephalic v.
Superior vena cava

R. brachial v.
R. axillary v.

Heart

(b)

FIGURE 29.21 Anterior Veins of the Cat. (a) Diagram; (b) photograph.

(b) ©Eric Wise

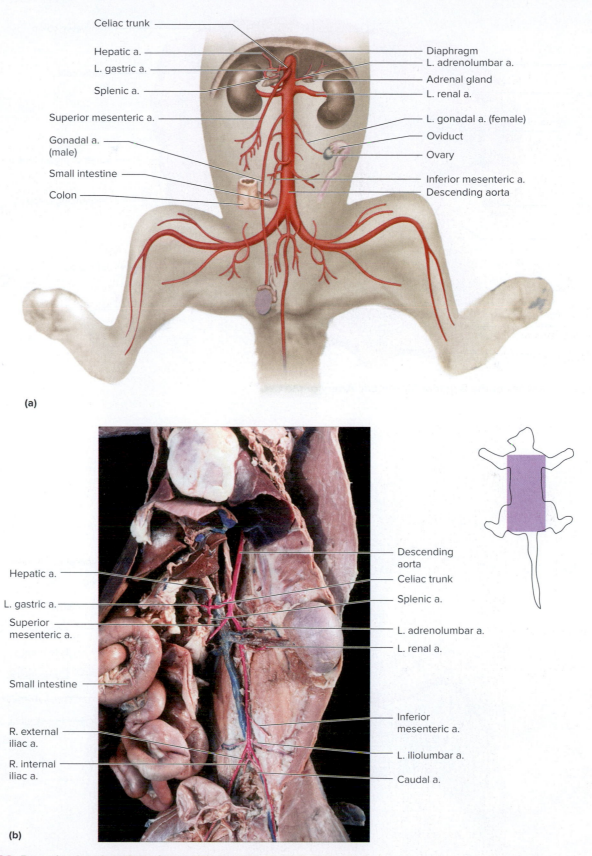

FIGURE 29.22 Posterior Arteries of the Cat. (a) Diagram; (b) photograph.

(b) ©Eric Wise

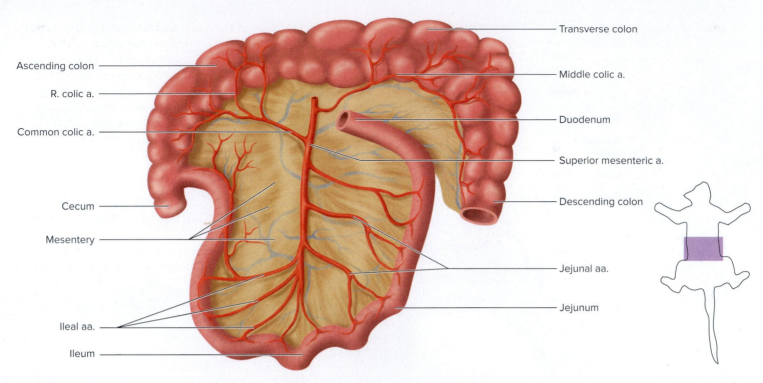

Ascending colon

R. colic a.

Common colic a.

Cecum

Mesentery

Ileal aa.

Ileum

Transverse colon

Middle colic a.

Duodenum

Superior mesenteric a.

Descending colon

Jejunal aa.

Jejunum

FIGURE 29.23 Branches of the Superior Mesenteric Artery of the Cat.

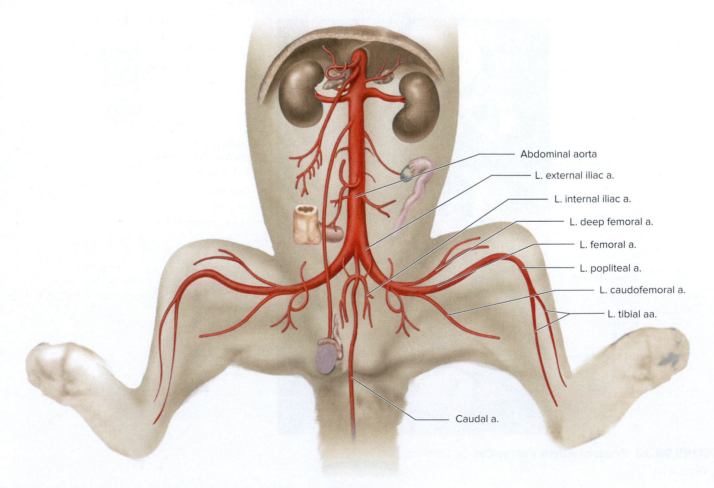

Abdominal aorta

L. external iliac a.

L. internal iliac a.

L. deep femoral a.

L. femoral a.

L. popliteal a.

L. caudofemoral a.

L. tibial aa.

Caudal a.

FIGURE 29.24 Arteries of the Pelvis and Lower Extremity of the Cat.

Hepatic Portal System

1. Examine the hepatic portal system of the cat while trying to maintain the integrity of the digestive organs. Compare your dissection to fig. 29.26. The vessels of the hepatic portal system take blood from a number of abdominal organs and transfer it to the liver, an organ where significant metabolic processing occurs.

2. Examine the lower portion of the digestive tract (transverse colon and descending colon), and locate the **inferior mesenteric vein.** The **superior mesenteric vein** receives blood from the duodenum, the pancreas, the jejunum, the ileum, and part of the large intestines. It also receives blood from the inferior mesenteric vein.

3. The **gastrosplenic vein** joins the superior mesenteric vein and takes blood to the **hepatic portal vein,** along with other digestive veins. The gastrosplenic vein, as its name implies,

receives blood from the spleen and the stomach. The hepatic portal vein thus receives blood from the spleen and the digestive organs and transports that blood to the liver. After the liver receives the blood, it is transferred to the inferior vena cava by the short **hepatic veins.** These veins are usually difficult to dissect because they are at the dorsal aspect of the liver and join the inferior vena cava at about the junction of the diaphragm.

Veins of the Pelvis

1. Examine the veins of the pelvis as seen in fig. 29.25. The **external iliac vein** and the **internal iliac vein** join to form the common iliac vein.

2. The two common iliac veins lead to the inferior vena cava along with the caudal vein, which takes blood from the tail.

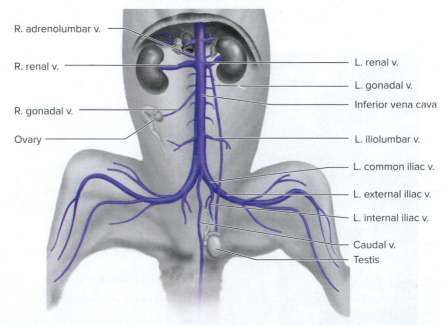

FIGURE 29.25 Veins of the Pelvis and Abdomen of the Cat. The right side of the cat represents a female, and the left side of the cat represents a male.

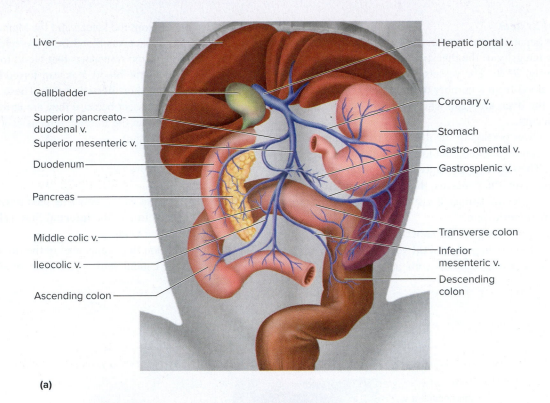

(a)

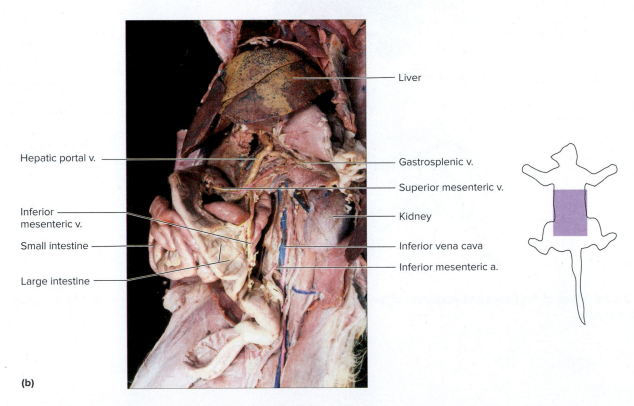

(b)

FIGURE 29.26 **Hepatic Portal System of the Cat.** (a) Diagram; (b) photograph (injected with yellow latex).
(b) ©Eric Wise

REVIEW SECTION

Blood Vessels 1: Introduction to Blood Vessels and Blood Vessels of the Axial Region

Name _____   Date _____

Lab Section _____   Time _____

Review Questions

1. Label the following illustration with the major arteries of the body. Try to complete the illustration first and then review the material in this exercise to determine your accuracy.

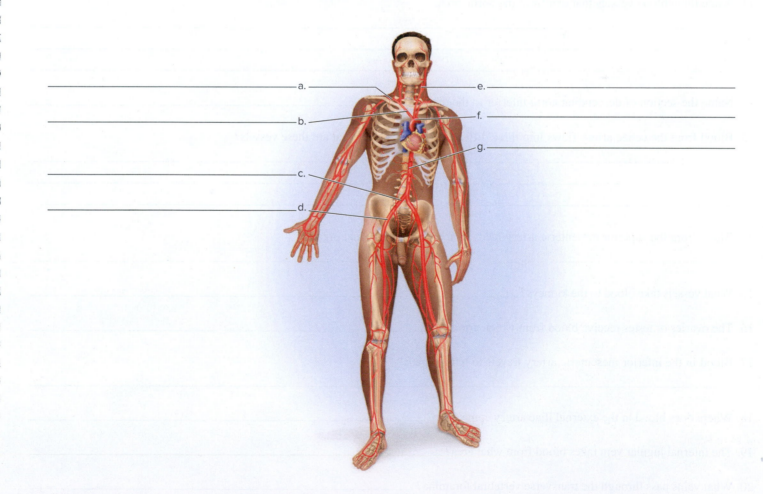

a. _____   e. _____

b. _____   f. _____

g. _____

c. _____

d. _____

2. Which veins (superficial/deep) have names that do not correlate with arteries? _____

3. What is the name of the outermost layer of a blood vessel? _____

4. What kind of blood vessels have valves? _____

5. An aneurysm is a weakened, expanded portion of an artery. Ruptured aneurysms can lead to rapid blood loss. Describe the significance of an aortic aneurysm versus a digital artery aneurysm.

_____

6. The pulmonary arteries carry deoxygenated blood from the heart to the lungs. Why are these blood vessels called arteries?

_____

7. Blood from the common carotid artery next travels to what two vessels? _____

8. The internal carotid artery takes blood to what organ? _____

9. The descending aorta receives blood from what vessel? _____

10. The right common carotid artery receives blood from what vessel? _____

11. Name three blood vessels that exit from the aortic arch.

_____

_____

_____

12. Name the section of descending aorta inferior to the diaphragm. _____

13. Blood from the celiac artery flows into three different blood vessels. What are these vessels?

_____

_____

_____

14. Blood from the superior mesenteric artery feeds which major abdominal organs?

_____

15. What vessels take blood to the kidneys? _____

16. The ovaries or testes receive blood from which arteries? _____

17. Blood in the inferior mesenteric artery travels to what organs?

_____

18. Where does blood in the external iliac artery come from? _____

19. The internal jugular vein takes blood from what area? _____

20. What veins pass through the transverse vertebral foramina? _____

21. The external jugular veins drain what area? _____

22. What is the functional nature of a portal system, and how is it different from normal venous return flow?

_____

23. What major vessels take blood to the hepatic portal vein?

_____

24. What is arteriosclerosis? _____

25. In what part of the arterial wall does cholesterol plaque develop? _____

26. Name the major veins that take blood from the brain.

_____

27. The basilar artery in the brain receives blood from what two arteries?

_____

Notes

Blood Vessels 2: Vessels of the Appendicular Region

INTRODUCTION

This exercise is a continuation of the study of blood vessels with an examination of the blood vessels of the upper and lower limbs as well as a study of fetal circulation. Many of the blood vessels of the appendicular region are named for their location, such as the *brachial* artery or the *femoral* vein. Superficial veins generally have names unique to them, such as the *basilic vein* and *great saphenous vein*. In this exercise, you locate the arteries and veins of the limbs and determine where the blood in each vessel comes from and where it goes. These topics are covered in chapter 20, "The Circulatory System: Blood Vessels and Circulation" in the Saladin text.

OBJECTIVES

At the end of this exercise, you should be able to

1. identify the arteries and veins of the upper limbs;
2. identify the arteries and veins of the lower limbs; and
3. name the vessels that take blood to or from a particular artery or vein.

MATERIALS

Models of the blood vessels of the body

Charts and illustrations of arterial and venous systems

Cadaver (if available)

Cats

Dissection equipment

PROCEDURE

Arteries of the Upper Limbs

If you review the position of the brachiocephalic artery, you will remember that it is a short tube that takes blood from the aortic arch and then branches into the **right subclavian artery** and the **right common carotid artery.** The right subclavian artery has a branch that takes blood to the **right axillary artery.** The **left subclavian artery** comes from the aortic arch and has a branch that takes blood to the **left axillary artery** (fig. 30.1a). The axillary artery on each side of the body takes blood to the **brachial artery.** The brachial artery has an obvious pulse that can be palpated by finding a groove on the medial, distal region of the arm between the biceps brachii and brachialis muscles. This location is the site for the placement of the diaphragm of a stethoscope during blood pressure measurement. The brachial artery supplies blood to the muscles of the arm and the humerus. The brachial artery bifurcates (splits in two) to form the **radial artery** on the lateral side of the antebrachium and the **ulnar artery** on the medial side. These two arteries supply blood to the forearm and hand. The radial artery can be palpated just lateral to the flexor carpi radialis muscle at the wrist, which is a common site for the clinical measurement of the pulse. The radial artery and ulnar artery become united in the palm as a **superficial palmar arch** and a **deep palmar arch.** They anastomose (join) and send out the small **digital arteries** that supply blood to the fingers (fig. 30.1).

Veins of the Upper Limbs

Examine the models and charts in the lab and locate the veins of the upper limbs. The fingers are drained by the small **digital veins,** which lead to the **venous palmar arches.** The major superficial veins of each upper limb are the **basilic vein,** on the anterior, medial side of the forearm and arm, and the **cephalic vein,** on the anterior, lateral side of the forearm and arm. The two vessels have many anastomosing branches (cross-connections) between them. One of

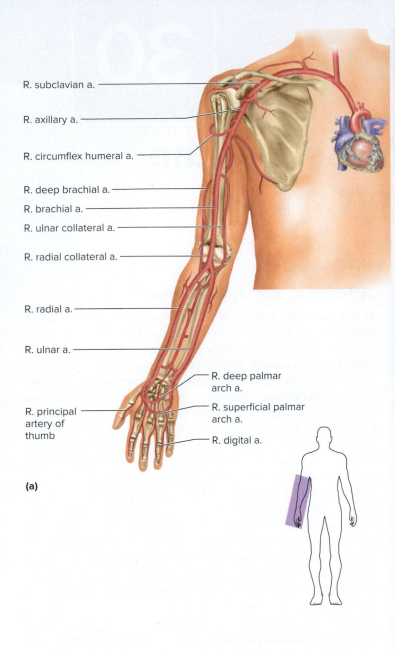

R. subclavian a.

R. axillary a.

R. circumflex humeral a.

R. deep brachial a.

R. brachial a.

R. ulnar collateral a.

R. radial collateral a.

R. radial a.

R. ulnar a.

R. deep palmar arch a.

R. principal artery of thumb

R. superficial palmar arch a.

R. digital a.

(a)

R. brachial a.

R. radial a.

R. ulnar a.

(b)

FIGURE 30.1 **Arteries of the Upper Limb.** (a) Diagram; (b) photograph.
(b) ©Eric Wise

the significant anastomosing veins is the **median cubital vein,** which crosses the anterior cubital fossa and is a common site for the withdrawal of blood. Locate these superficial veins in fig. 30.2.

The deep veins of the forearm are the **radial veins** and the **ulnar veins,** which can be found traveling near the arteries of the same name. The deep veins of the arm are the **brachial veins,** next to the brachial artery in the proximal portion of the arm. The brachial veins are formed by the union of the radial and ulnar veins and merge proximally with the **basilic vein** to form the short **axillary vein.** The axillary vein connects with the cephalic vein to form the **subclavian vein.** The blood of the upper limbs is carried to the heart by the left

and right subclavian veins, which flow to the **brachiocephalic veins,** then to the **superior vena cava,** and finally to the right atrium of the heart. Locate the deep veins of the upper limbs in fig. 30.2.

Arteries of the Lower Limb

Each external iliac artery branches from a common iliac artery and exits the body wall near the **inguinal canal.** Once the external iliac artery leaves the body cavity, it continues into the thigh as the **femoral artery.** The femoral artery has a superficial branch that feeds the thigh and a deeper branch, called the **deep femoral artery,**

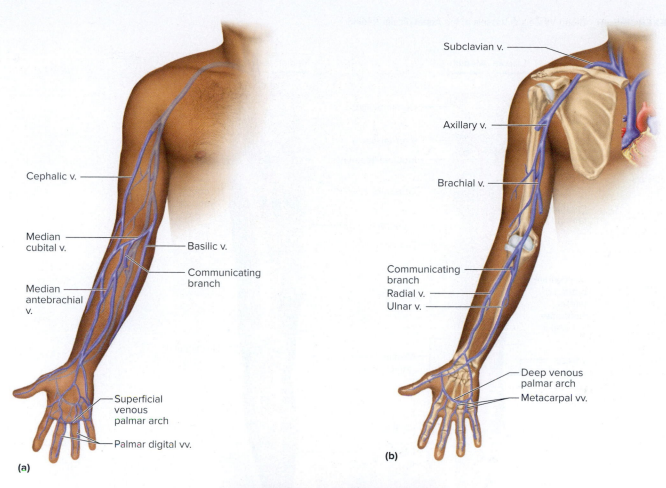

Cephalic v.

Median cubital v.

Basilic v.

Communicating branch

Median antebrachial v.

Superficial venous palmar arch

Palmar digital vv.

(a)

Subclavian v.

Axillary v.

Brachial v.

Communicating branch

Radial v.

Ulnar v.

Deep venous palmar arch

Metacarpal vv.

(b)

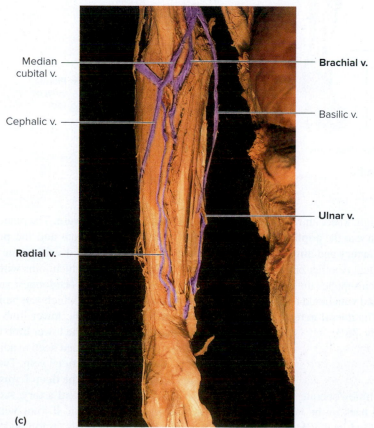

Median cubital v.

Cephalic v.

Radial v.

Brachial v.

Basilic v.

Ulnar v.

(c)

FIGURE 30.2 **Veins of the Upper Limb.** (a) Superficial veins, diagram; (b) deep veins, diagram and (c) photograph.
(c) ©Eric Wise

415

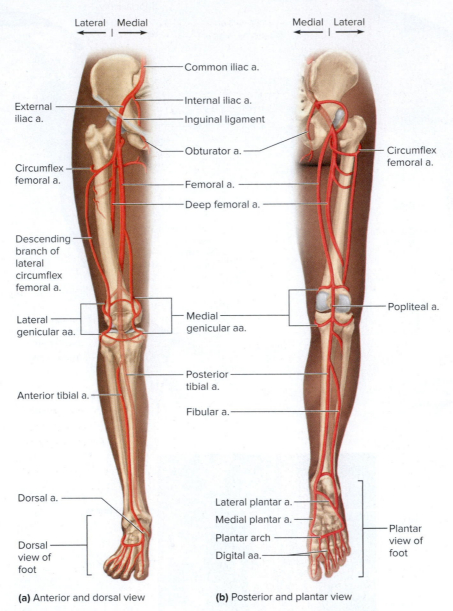

Lateral Medial   Medial Lateral

Common iliac a.

External iliac a.
Internal iliac a.
Inguinal ligament
Obturator a.
Circumflex femoral a.

Circumflex femoral a.
Femoral a.
Deep femoral a.

Descending branch of lateral circumflex femoral a.
Popliteal a.

Lateral genicular aa.
Medial genicular aa.

Posterior tibial a.

Anterior tibial a.
Fibular a.

Dorsal a.
Lateral plantar a.
Medial plantar a.
Plantar arch
Plantar view of foot
Digital aa.

Dorsal view of foot

(a) Anterior and dorsal view   **(b)** Posterior and plantar view

FIGURE 30.3 Arteries of the Lower Limb.

that takes blood to the muscles of the thigh, knee, and femur. The femoral artery continues posterior to the knee as the **popliteal artery** and then divides into the **posterior tibial artery** and **anterior tibial artery,** which take blood to the knee and leg. Another branch of the popliteal artery is the **fibular artery,** which supplies the muscles on the lateral side of the leg. The foot is supplied with blood from several arteries, including the **plantar arteries,** the **dorsal artery,** and the **digital arteries.** Locate these arteries in fig. 30.3.

Veins of the Lower Limb

The blood vessels that drain the lower limbs operate under relatively low pressure and must take blood back to the heart against gravity. Digital veins in the feet take blood to the **deep plantar venous arch** and **dorsal venous arch** veins. The dorsal venous arch takes blood to the **anterior tibial veins,** the **great saphenous vein,**

and the **small saphenous vein.** The plantar venous arch takes blood to the small saphenous vein and the **posterior tibial veins.** The anterior and posterior tibial veins unite posterior to the knee and form the **popliteal vein,** which joins with the small saphenous vein to form the femoral vein. The longest vessel in the human body is the great saphenous vein, which can be found just deep to the skin on the medial aspect of the lower limb at the level of the medial malleolus and traversing the lower limb to the proximal thigh. This vessel frequently is imbedded deep in adipose tissue below the skin, yet it is considered a superficial vein. Two other vessels in the thigh are the **femoral vein** and the **deep femoral vein.** The femoral vein travels alongside the femoral artery. As the great saphenous vein reaches the inguinal region, it joins with the femoral vein, which takes blood from the thigh region and passes under the inguinal ligament where the vessel becomes the **external iliac vein.** Examine these vessels in the lab and compare them with fig. 30.4.

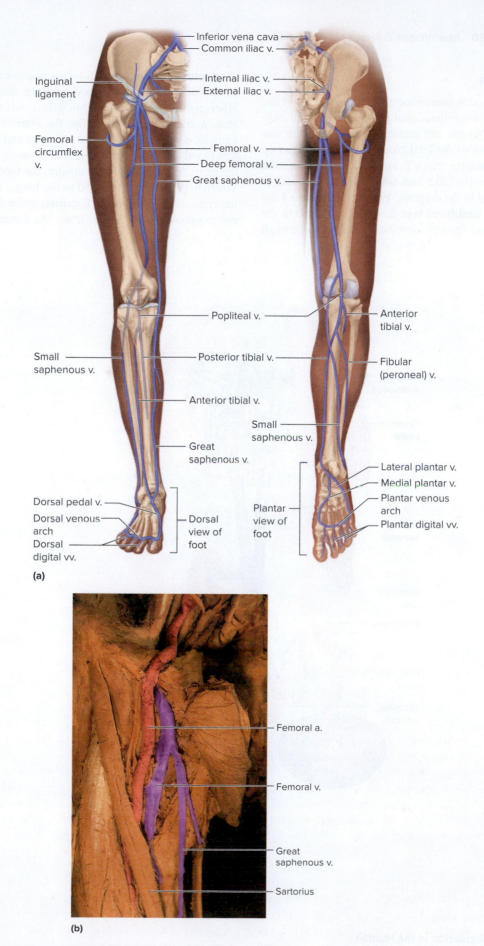

Inferior vena cava
Common iliac v.
Internal iliac v.
External iliac v.
Inguinal ligament
Femoral circumflex v.
Femoral v.
Deep femoral v.
Great saphenous v.
Popliteal v.
Anterior tibial v.
Small saphenous v.
Posterior tibial v.
Fibular (peroneal) v.
Anterior tibial v.
Small saphenous v.
Great saphenous v.
Lateral plantar v.
Medial plantar v.
Plantar venous arch
Plantar digital vv.
Dorsal pedal v.
Dorsal venous arch
Dorsal digital vv.
Dorsal view of foot
Plantar view of foot

(a)

Femoral a.
Femoral v.
Great saphenous v.
Sartorius

(b)

FIGURE 30.4 Veins of the Lower Limbs. (a) Diagram of entire lower limb; (b) photograph of proximal right thigh.
(b) ©Eric Wise

Fetal Circulation

The pathway of fetal blood is somewhat different from that of the adult in that the lungs are nonfunctional (in terms of blood oxygenation) in the fetus. Oxygen and nutrients move from the maternal side of the placenta to the fetal bloodstream, while carbon dioxide and metabolic wastes move from the fetal bloodstream to the placenta. Examine fig. 30.5 and note how the numbers in the description are placed in the diagram. From the **placenta 1** the blood flows through the **umbilical vein 2,** which is located in the **umbilical cord.** The blood from the umbilical vein travels through

the **ductus venosus 3,** which serves as a shunt to the **inferior vena cava 4** of the fetus. The blood from the umbilical vein, which is relatively high in oxygen and nutrients, mixes with the deoxygenated, nutrient-poor blood from the inferior vena cava; thus the fetus receives a mixture of oxygenated and deoxygenated blood. The blood from the inferior vena cava travels to the right atrium of the heart. While in the right atrium, the blood travels to the right ventricle (which pumps blood to the lungs) and through a hole in the right atrium called the **foramen ovale 5.** Since the lungs do not oxygenate blood in the fetus, the foramen ovale serves as a

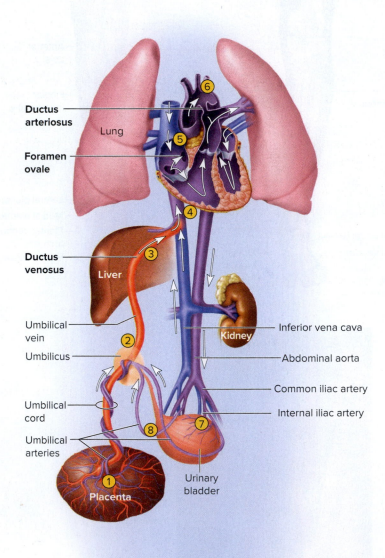

FIGURE 30.5 Fetal Circulation in the Human.

bypass route taking most of the blood away from the lungs and to the chambers of the heart that will pump blood to the body. Blood in the right ventricle is pumped to the pulmonary trunk, where another shunt vessel, the **ductus arteriosus 6,** carries blood to the aortic arch, bypassing the lungs. The lungs do receive some blood, but it is for the nourishment of the lung tissue, not for gas exchange. Blood from the heart exits the left ventricle and passes through the aorta to the systemic arteries. Blood travels down the **internal iliac arteries 7** to the pelvic organs and lower limbs, and some moves into the **umbilical arteries 8,** which carry blood to the placenta. At birth, the pressure changes in the newborn's lungs and heart commonly cause the closing of a flap of tissue over the foramen ovale, leaving a thin spot in the **interatrial septum** known as the **fossa ovalis.**

Cat Anatomy

Cat Dissection

(A) Activity

Prepare for the cat dissection by obtaining a cat and dissection equipment. Remember to place all excess tissue in the appropriate waste container and not in a standard classroom wastebasket or down the sink!

As stated in the previous exercise, the blood vessels in the cat have been injected with colored latex. If the cat has been doubly injected, the arteries are red and the veins are blue. If the cat has been triply injected, the hepatic portal vein is also injected, typically with yellow latex. The terms used in this lab manual will be human terms unless the vessels are specific to the cat.

1. Follow the dissection procedures discussed in the previous exercise. Make sure you do not cut the blood vessels from the organs that will be studied later.

Arteries of the Forelimb

2. Remove any remaining skin from the forelimb if you have not previously done so. Locate the **subclavian arteries.** The left and right subclavian arteries become the left and right **axillary arteries,** which in turn take blood to the **brachial arteries.** The brachial arteries run parallel to the humerus in the cat. These arteries take blood to the **radial** and **ulnar arteries** of the forelimb. Locate the arteries of the forelimb and compare them with fig. 30.6.

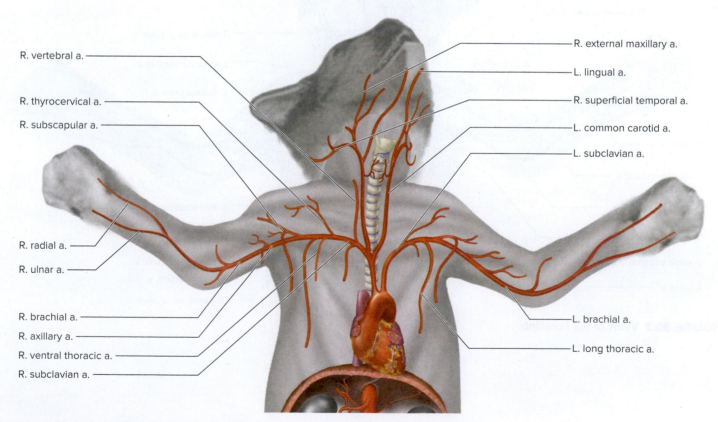

FIGURE 30.6 Arteries of the Forelimb of the Cat.

Veins of the Forelimb

3. Examine the veins of the forelimb. The basilic vein is not found in cats, but you should be able to find **the ulnar veins;** they join the **radial veins** taking blood to the **brachial veins.** The brachial veins are next to the brachial arteries. Locate the **median cubital vein** and the **cephalic vein.** The **axillary vein** and **subscapular vein** (from the shoulder) join to form the **subclavian vein,** which takes blood to the **brachiocephalic vein.** The cephalic vein continues into the **transverse scapular vein** that takes blood to the **external jugular vein.** Locate these veins in fig. 30.7.

Arteries of the Hindlimb

4. As the **external iliac artery** exits from the pelvic region and enters the thigh, it becomes the **femoral artery,** which supplies blood to the thigh, leg, and foot. The **deep femoral**

artery supplies blood to some of the thigh muscles. The femoral artery becomes the **popliteal artery** behind the knee, which further divides to form the **tibial arteries** (figs. 30.8 and 30.9).

Veins of the Hindlimb

5. Locate the superficial **great saphenous vein** in the cat on the medial side of the hindlimb. The great saphenous vein takes blood to the **femoral vein** just above the knee. Note the long **tibial veins** of the leg that join and form the **popliteal veins.** Find the femoral vein, found with the femoral artery. The vessels of the distal part of the hindlimb take blood to the femoral vein, which turns into the **external iliac vein** as it passes into the body cavity at the level of the inguinal ligament. Find these veins in the cat and compare them to fig. 30.10.

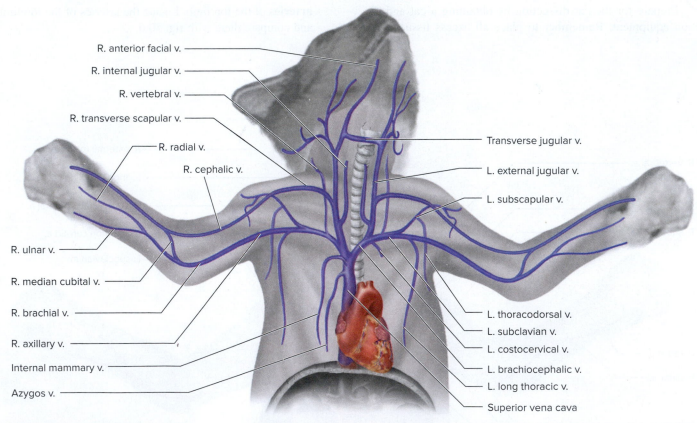

R. anterior facial v.
R. internal jugular v.
R. vertebral v.
R. transverse scapular v.
R. radial v.
R. cephalic v.
R. ulnar v.
R. median cubital v.
R. brachial v.
R. axillary v.
Internal mammary v.
Azygos v.

Transverse jugular v.
L. external jugular v.
L. subscapular v.
L. thoracodorsal v.
L. subclavian v.
L. costocervical v.
L. brachiocephalic v.
L. long thoracic v.
Superior vena cava

FIGURE 30.7 Veins of the Forelimb.

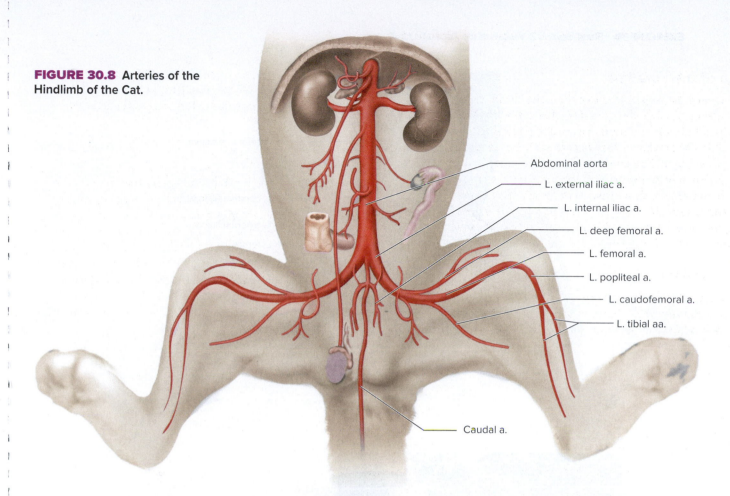

FIGURE 30.8 Arteries of the Hindlimb of the Cat.

Abdominal aorta

L. external iliac a.

L. internal iliac a.

L. deep femoral a.

L. femoral a.

L. popliteal a.

L. caudofemoral a.

L. tibial aa.

Caudal a.

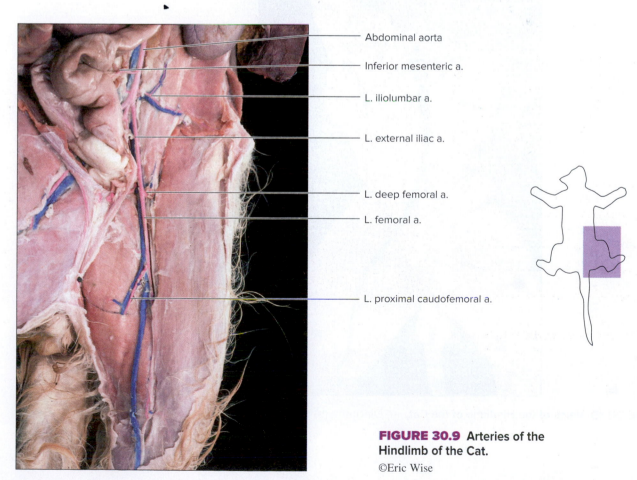

Abdominal aorta

Inferior mesenteric a.

L. iliolumbar a.

L. external iliac a.

L. deep femoral a.

L. femoral a.

L. proximal caudofemoral a.

FIGURE 30.9 Arteries of the Hindlimb of the Cat.
©Eric Wise

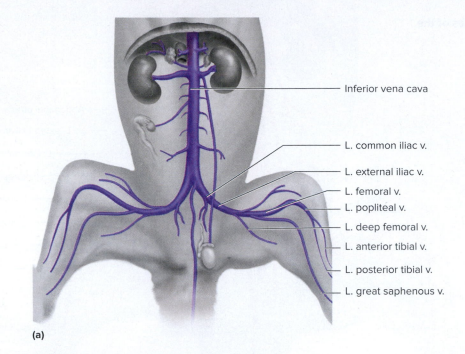

(a)

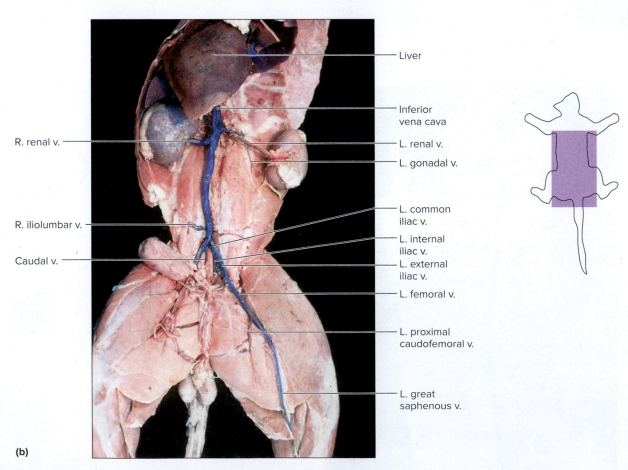

(b)

FIGURE 30.10 **Veins of the Hindlimb of the Cat.** (a) Diagram; (b) photograph.
(b) ©Eric Wise

REVIEW SECTION

Blood Vessels 2: Vessels of the Appendicular Region

Name _____ *Date* _____

Lab Section _____ *Time* _____

Review Questions

1. Where does blood in the right subclavian artery come from? _____

2. Blood from the left subclavian artery flows into what vessel as it moves toward the left arm? _____

3. Blood in the radial artery comes from what blood vessel? _____

4. Is the radial vein a superficial or deep vein? _____

5. Where is the median cubital vein found? _____

6. Where is the cephalic vein located? _____

7. Blood from the right axillary vein travels next to what vessel? _____

8. What vessel receives blood from the ulnar vein? _____

9. What artery takes blood directly to the femoral artery? _____

10. Blood from the popliteal artery comes directly from what artery? _____

11. What vessels take blood to the left femoral vein? _____

12. In what region of the body is the great saphenous vein? _____

13. Where does blood flow after it leaves the femoral vein? _____

14. In the fetal heart, what is the name of the shunt between the pulmonary trunk and the aortic arch? _____

15. Name the opening between the atria in the fetal heart. _____

16. Label the following illustration using the terms provided.

anterior tibial artery   femoral artery

axillary artery   radial artery

brachial artery   ulnar artery

dorsal artery

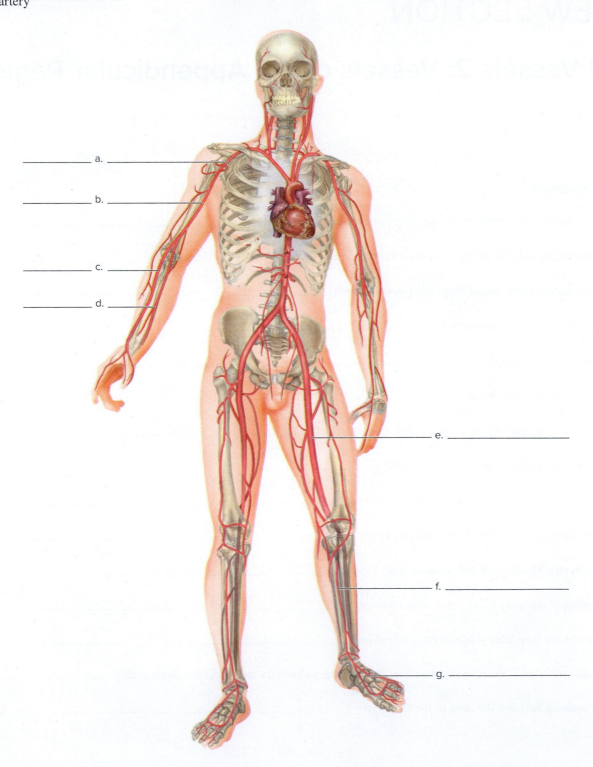

a. _____

b. _____

c. _____

d. _____

e. _____

f. _____

g. _____

17. Label the following illustration using the terms provided.

basilic vein   femoral vein

brachial vein   great saphenous vein

cephalic vein   median cubital vein

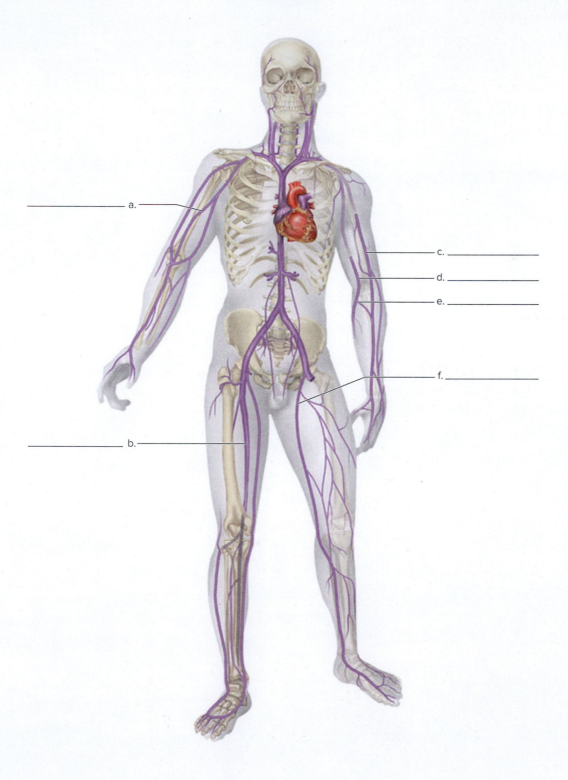

Notes

Lymphoid System

INTRODUCTION

Blood is pumped from the **heart** through **arteries** and distributed through the **arterioles** to the **capillaries,** where nutrients, water, and oxygen are exchanged with the cells of the body. The capillaries deliver oxygen to the cells and pick up carbon dioxide and waste from the interstitial fluid surrounding the cells. Blood cells and plasma proteins typically stay in the blood vessels, yet a significant amount of fluid from the plasma leaks from the capillaries and bathes the cells of the body. This fluid flows between the cells of the body and is known as **interstitial (tissue) fluid.** It provides nutrients to the cells and receives dissolved wastes from the cells along with cellular debris. Capillaries absorb most of the interstitial fluid, with the remainder picked up by **lymphatic capillaries,** which return the fluid, now known as lymph, to the lymphatic vessels. Lymphatic vessels return the lymph to lymphatic trunks and then to the venous system. Fluid leaked from the capillaries returns to the cardiovascular system. This process is illustrated in fig. 31.1.

The lymphoid system is also instrumental in the absorption of lipids from the digestive system. Lipids in the small intestine are converted to **chylomicrons** (phospholipids and other molecules) in the cells that line the digestive tract. These chylomicrons move to the lymphoid system, which then transports the material through the thoracic duct and into the subclavian vein where it enters the cardiovascular system.

As the fluid flows through the lymphatic vessels, macrophages in **lymph nodes** phagocytize the cellular debris and foreign material (bacteria, some cancer cells, and viruses) that may have entered into the lymphoid system. Lymph nodes also activate B cells and T cells, providing immunity. The spleen recycles old blood cells, the tonsils provide immunity, as does the lymphoid tissue, and the red bone marrow produces some lymphocytes. In this exercise you examine the nature of the lymphoid system. These topics are covered in greater detail in the Saladin text in chapter 21, "The Lymphoid and Immune Systems."

OBJECTIVES

At the end of this exercise, you should be able to

1. list the major functions of the lymphatic vessels;
2. identify the valves in lymphatic vessels;

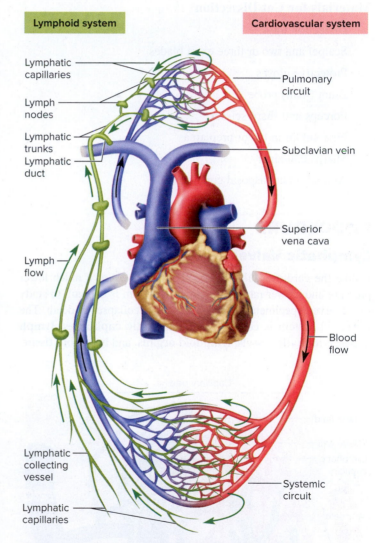

FIGURE 31.1 Flow of Interstitial Fluid and Lymph.

3. locate the major histological features of a lymph node;
4. demonstrate on a model the location of the tonsils and other lymphoid organs;

5. describe the function of the spleen and thymus;
6. discuss the function of a lymph node.

MATERIALS

Microscopes

Microscope slides of lymph nodes and lymphatics with valves

Torso models

Models of the lymphoid system

Cat

Materials for Cat Dissection

Dissection trays

Scalpel and two or three extra blades

Protective gloves

Blunt (mall) probe

Forceps and sharp scissors

First aid kit in lab or prep area

Sharps container

Animal waste disposal container

PROCEDURE

Lymphatic Valves

Unlike the cardiovascular system, lymph travels at a much lower pressure and slower rate. The lymphoid system is difficult to study in preserved specimens because much of it collapses at death. The lymphoid system is composed of **lymphatic capillaries, lymph nodes, lymphatic vessels, lymphoid organs,** and **lymphoid tissue.**

The lymphatic capillaries have blunt ends and flaplike valves that move lymph away from interstitial areas. Examine fig. 31.2 and describe how the valves might operate.

Valve operation: _____

_____

_____

Lymph Node

Ⓐ **Activity**

Look at a slide of a **lymph node.** The lymph node is enclosed by a sheath of tissue called the **capsule.** Locate the dark purple **lymphoid nodules** in the lymph node. The outer portion of the lymph node is called the **cortex,** and the inner part is the **medulla.** The cortex has numerous lymphoid nodules. In the medulla, **medullary cords** and **sinuses** cleanse the lymph of foreign particles and cellular debris. Look at models or charts in lab and find **afferent lymphatic vessels,** which take lymph to the node, and **efferent lymphatic vessels,** which take lymph away from the node. Compare your slide to fig. 31.3.

Lymphatic Vessels

Ⓐ **Activity**

Use fig. 31.4 to examine the one-way drainage pattern of lymph in the body. Compare this illustration to models or charts in lab. The **lymphatic vessels** lead to regions of the body where lymph nodes are found. Nodes are clustered in the groin (inguinal region), axilla, antecubital fossa, popliteal region, neck, thorax, and abdomen. The **thoracic duct** drains most of the body, taking lymph to the left subclavian vein, where the fluid is returned to the cardiovascular system. The **right lymphatic duct** drains the right side of the head and neck, the right thoracic region, and the

Capillary bed

Tissue fluid

Tissue cell

Lymphatic capillary

Arteriole

Venule

(a)

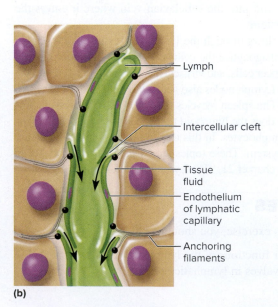

Lymph

Intercellular cleft

Tissue fluid

Endothelium of lymphatic capillary

Anchoring filaments

(b)

FIGURE 31.2 Lymphatic Capillary. (a) Overview; (b) detail.

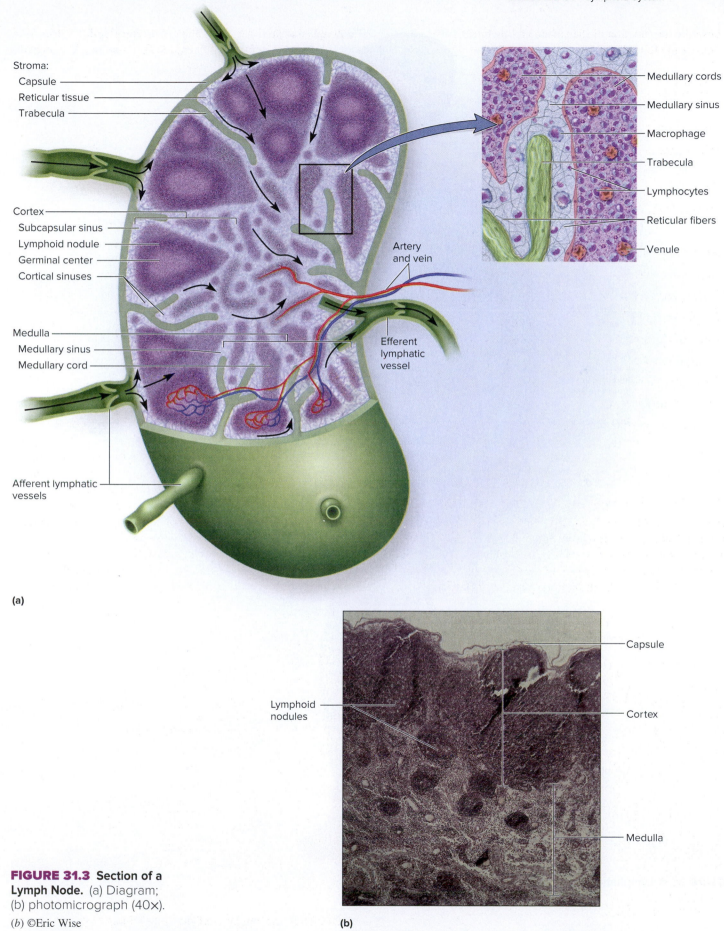

Stroma:
Capsule
Reticular tissue
Trabecula

Cortex
Subcapsular sinus
Lymphoid nodule
Germinal center
Cortical sinuses

Medulla
Medullary sinus
Medullary cord

Afferent lymphatic
vessels

Artery
and vein

Efferent
lymphatic
vessel

Medullary cords
Medullary sinus
Macrophage
Trabecula
Lymphocytes
Reticular fibers
Venule

(a)

Lymphoid
nodules

Capsule
Cortex
Medulla

(b)

FIGURE 31.3 **Section of a
Lymph Node.** (a) Diagram;
(b) photomicrograph (40×).
(*b*) ©Eric Wise

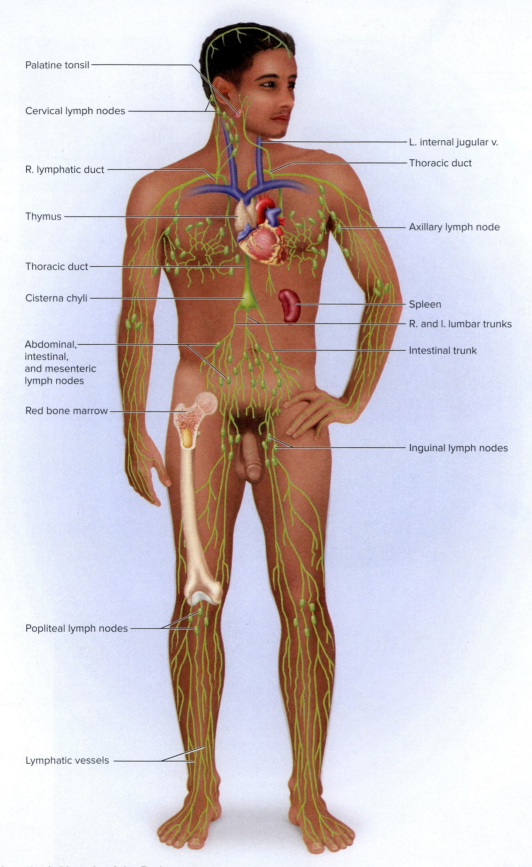

Palatine tonsil

Cervical lymph nodes

R. lymphatic duct

Thymus

Thoracic duct

Cisterna chyli

Abdominal, intestinal, and mesenteric lymph nodes

Red bone marrow

Popliteal lymph nodes

Lymphatic vessels

L. internal jugular v.

Thoracic duct

Axillary lymph node

Spleen

R. and l. lumbar trunks

Intestinal trunk

Inguinal lymph nodes

FIGURE 31.4 **Lymphatic Vessels of the Body.**

right upper extremity. The right lymphatic duct returns lymph to the right subclavian vein. Identify the **cisterna chyli,** an enlarged portion of the thoracic duct in the abdominal region.

(A) Activity

Examine a prepared slide of a lymphatic vessel and note the presence of the **valve.** Valves prevent the backflow of lymph away from the vascular system into which it drains. Compare your slide to fig. 31.5.

Lymphoid Organs

Tonsils

Tonsils are lymphoid organs in the oral cavity and nasopharynx.

(A) Activity

Look at illustrations in your text and models in the lab and locate the tonsils. These are the **palatine tonsils** along the sides of the oral cavity near the oropharynx, the **lingual tonsils** at the posterior portion of the tongue, and the singular **pharyngeal tonsil** in the nasopharynx. An adenoid is an enlarged pharyngeal tonsil. Palatine and lingual tonsils have crypts lined with **lymphoid nodules** that serve as a first line of defense against foreign matter that is inhaled or swallowed. Compare these to fig. 31.6.

Spleen

The **spleen** is an important lymphoid organ that removes aging erythrocytes and foreign particles from the blood. The spleen is located on the left side of the body adjacent to the stomach. It is a highly vascular organ that cleanses the blood and produces lymphocytes. The spleen contains **red pulp,** which filters blood, and **white pulp,** which contains lymphocytes from the thymus and bone marrow and sees an increase in these cells during an immune response.

(A) Activity

Locate the spleen in models and charts in the lab and compare it to fig. 31.7.

The appendix is another organ containing lymphoid tissue. It destroys intestinal bacteria and produces lymphocytes.

Thymus

The **thymus** is superficial to the vessels superior to the heart. It is an important site for immune competence. Lymphocytes travel from the bone marrow to the thymus, where they undergo development essential to immune responses. Once these cells become immunocompetent they are known as **T cells.**

(A) Activity

Examine the charts and models of the thymus in the lab and compare them to fig. 31.8.

Cat Dissection

(A) Activity

In the cat, locate the **spleen,** which is an elongated organ near the stomach on the left side of the abdominal cavity. It is a brown organ, and you should find the splenic artery and splenic vein associated with the organ.

You should also look for the **thymus,** which is anterior to the heart. It may be small in older cats or may have been removed in the study of the heart.

There are many locations where you can find **lymph nodes,** and these are illustrated in fig. 31.9. Use a forceps or a probe and tease away loose connective tissue to locate the nodes. They should appear as small lumps of beige tissue.

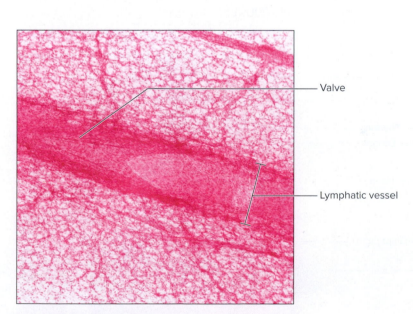

FIGURE 31.5 Lymphatic Vessel with Valve (100×).

©Eric Wise

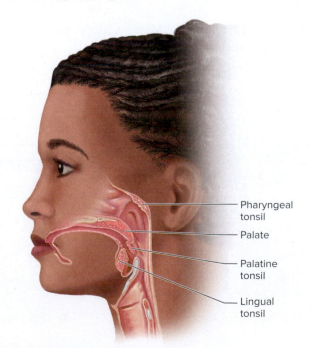

Valve

Lymphatic vessel

Pharyngeal tonsil

Palate

Palatine tonsil

Lingual tonsil

FIGURE 31.6 Tonsils.

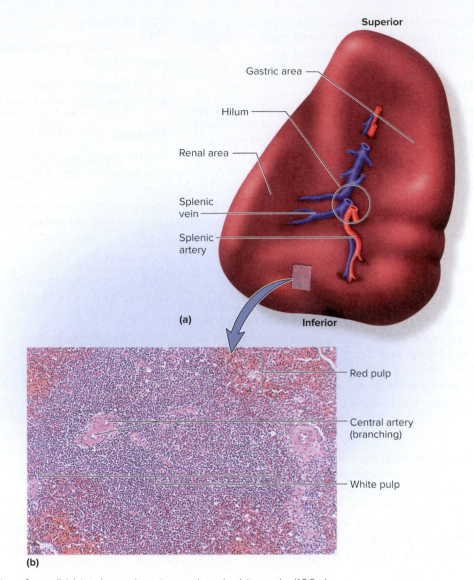

Superior

Gastric area

Hilum

Renal area

Splenic vein

Splenic artery

Inferior

(a)

Red pulp

Central artery (branching)

White pulp

(b)

FIGURE 31.7 Spleen. (a) Medial surface; (b) histology showing red and white pulp (100×).
Dennis Strete/McGraw Hill

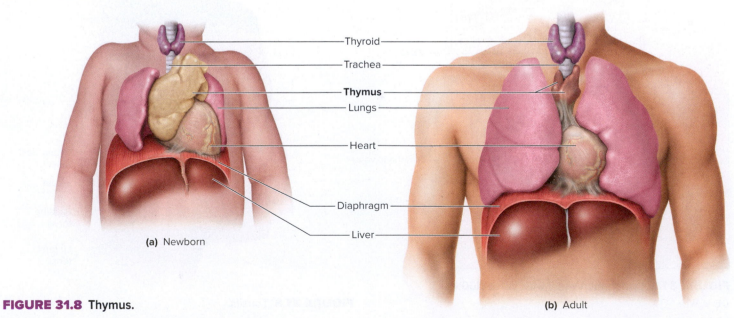

Thyroid

Trachea

Thymus

Lungs

Heart

Diaphragm

Liver

(a) Newborn

(b) Adult

FIGURE 31.8 Thymus.

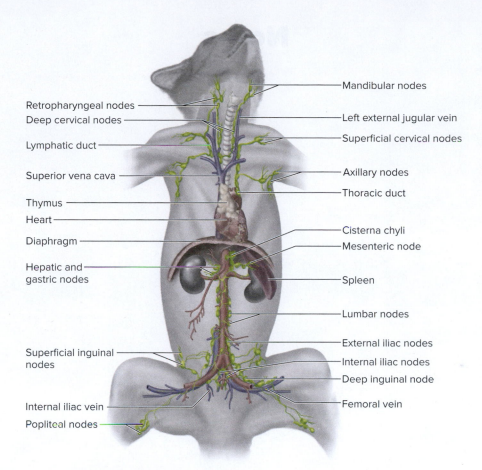

FIGURE 31.9 Lymphoid System of the Cat.

Notes

REVIEW SECTION

Lymphoid System

Name _____ *Date* _____

Lab Section _____ *Time* _____

Review Questions

1. Trace the flow of the lymphoid system from the region of the lymphatic capillaries to the subclavian veins.
   _____
   _____

2. What are the names of the inner region and the outer region of a lymph node? Describe their functions.
   _____
   _____

3. Once tissue fluid enters the lymphatic vessels, what is it called? What might be found in this fluid?
   _____
   _____

4. Name the vessel that takes lymph to a lymph node and the vessel that takes lymph away from a lymph node.
   _____

5. An adenoid is an enlarged _____ tonsil.

6. Which tonsils are found on the sides of the oral cavity? _____

7. What tonsils are located at the back of the tongue? _____

8. Blood is recycled by which lymphoid organ in the adult? _____

9. Where do T cells become immunocompetent? _____

10. From what you know of the functions of lymph nodes, make a prediction of the difference between lymph entering a node and lymph leaving a node. What materials may be missing from the lymph leaving the node?
    _____
    _____

11. Fill in the figure with names of the organs or vessels.

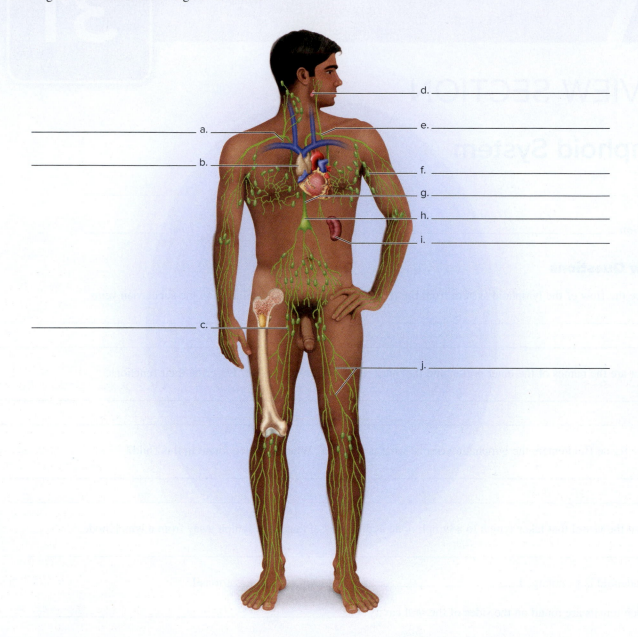

d. _____

a. _____   e. _____

b. _____   f. _____

g. _____

h. _____

i. _____

c. _____

j. _____

12. In the analysis of breast cancer, lymph nodes of the axillary region are removed and a biopsy is performed. The removal of the nodes is done to determine if cancer has spread from the breast to other regions of the body. What effect would the removal of lymph nodes have on the drainage of the pectoral region?

_____

_____

13. Lymphatic vessels have a one-way flow from the extremities to the heart. Damage to the lymphoid system can lead to edema, an increase in tissue fluid. From the standpoint of reducing edema, how does the use of medical leeches (segmented worms that drain tissue fluid) work for a region that has suffered trauma?

_____

_____

Blood Vessels and Blood Pressure

INTRODUCTION

Maintenance of blood pressure is important for the health of the heart and for proper functioning of various organs. High blood pressure (hypertension) increases the workload of the heart by increasing the force with which the heart must pump to provide blood to the body. Chronic (long-term) high blood pressure can harm the arteries, causing fatty deposits in the walls of arteries, a condition called **atherosclerosis** (ATH-er-oh-skler-OH-sis), which narrows the arteries and reduces blood flow to organs, such as the brain, heart, and kidneys. These deposits can harden, causing arteriosclerosis, or hardening of the arteries. Historically, blood pressure has been indirectly measured with a **sphygmomanometer** (blood pressure meter), which measures pressure in millimeters of mercury (mmHg) rising in a glass column. The greater the pressure, the higher the rise of mercury. Mercury-filled sphygmomanometers are still in use but aneroid sphygmomanometers (using air) and differential pressure transducers are also used.

Normal blood pressure is approximately 120/80 mmHg. Chronic hypertension is typically a blood pressure in excess of 130/80 mmHg for young adults, though the greatest concern is in an elevated diastolic reading. The long-term diastolic pressure should be less than 80 mmHg.

Caffeine and nicotine can temporarily cause an increase in blood pressure, along with exercise and emotional reactions. A short period of elevated blood pressure is called **acute hypertension. Chronic hypertension** lasts for much longer periods of time and is a significant health concern.

Many things influence chronic hypertension. As people age, their arteries typically become less elastic and blood pressure increases. Kidney disease and other illnesses can cause hypertension. Poor diets, such as those rich in fats or excess salt; lack of exercise; obesity; stress; some drugs; and alcohol intake can also increase blood pressure. Males generally have higher blood pressure than females until about the age of 55 and then females show an increase in hypertension after that age.

Hypotension can also be dangerous. Low blood pressure may occur when you stand up quickly. This is known as orthostatic hypotension. Severe dehydration, genetic makeup, and early pregnancy can all lead to low blood pressure. A decrease in pressure may lead to fainting or dizziness.

Blood pressure is not uniform throughout the body but is influenced by gravity; therefore, the pressure in the arteries of the head and neck is less than the blood pressure from the heart. The pressure in the arteries of the leg is greater than the blood pressure in the heart when a person is standing or sitting. When lying down, the blood pressure in the legs decreases significantly. Blood pressure is measured in the **brachial artery,** which is at the level of the heart and has the approximate pressure of the blood leaving the heart. Ausculatory measurement of blood pressure involves listening to sounds as blood passes through the brachial artery. Normally, no sound is heard through a stethoscope as blood passes through the brachial artery because the blood moves by laminar flow. The blood passes smoothly through the vessel, like water through a garden hose. When pressure is applied by pinching off a garden hose, the turbulence creates sound. When pressure is applied to the arm due to the constriction from the blood pressure cuff, the turbulence of the blood passing through the vessel creates sound. The significant difference between the flow of blood in the body and that of water in a hose is that the blood pulses rhythmically through the arteries as the heart contracts during ventricular systole. The sounds made by the flow of blood in a partially constricted artery are known as **Korotkoff sounds.** When the pressure from a blood pressure cuff (fig. 32.1) exceeds ventricular systole, no sound is heard. When the pressure from a blood pressure cuff is just lower than ventricular systole, turbulent blood causes the first Korotkoff sound. As the blood pressure from the cuff decreases further, laminar flow is established, and the turbulent sound disappears.

These topics are covered in the Saladin text in chapter 20, "The Circulatory System: Blood Vessels and Circulation." In this exercise, you learn how to determine blood pressure and some factors affecting blood pressure.

OBJECTIVES

At the end of this exercise, you should be able to

1. determine the pulse rate of an individual;
2. define hypertension in terms of millimeters of mercury pressure;

3. distinguish between systolic pressure and diastolic pressure;
4. properly take and record blood pressure;
5. list three factors that affect blood pressure;
6. describe the function of valves in veins.

MATERIALS

Blood Pressure Experiment

Stethoscope

Alcohol wipes or isopropyl alcohol and sterile cotton swabs

Washable felt pen

Sphygmomanometer

Watch with accuracy in seconds or timer function on smartphone

Live goldfish

Dissection scopes

Paper towels

Small squeeze bottle of water

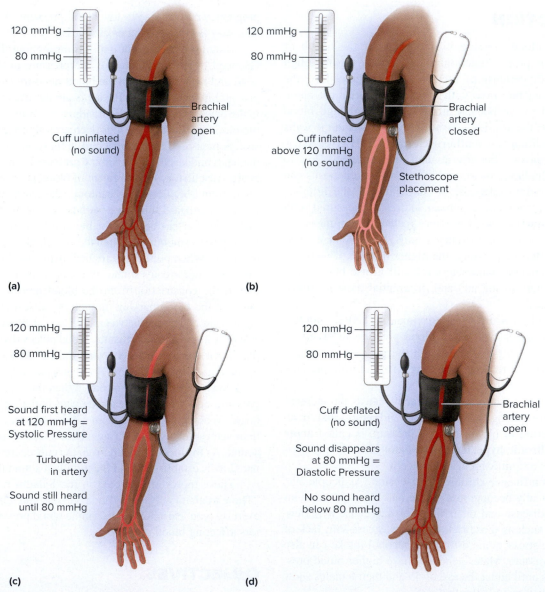

(a)

(b)

(c)

(d)

FIGURE 32.1 Blood Pressure Measurement. Assume a blood pressure of 120/80. (a) Normal arterial flow with no pressure on the brachial artery. No sound is heard in the stethoscope. (b) Cuff inflated to beyond systolic pressure. Brachial artery is closed and no sound is heard. (c) Release of pressure from the sphygmomanometer just less than blood pressure. Blood rushes into brachial artery and makes a rhythmic "whooshing" sound. Sound increases as cuff pressure is lessened. (d) Sound disappears at 80 mmHg because blood is no longer constricted. Pressure when sound disappears is known as diastolic pressure.

Flow Experiment

Corn syrup colored red

1 yard of ¼-inch plastic flexible tubing

1 yard of ½-inch plastic flexible tubing

Small funnel to fit small tubing

Large funnel to fit large tubing

Stopwatch

Two ring stands

Beakers

Clamps

Wash basin

Roll of paper towels

PROCEDURE

Measurement of Pulse Rate in Beats per Minute (bpm)

(A) Activity

Initially record a baseline pulse rate. The pulse rate is measured in beats per minute (bpm) by locating regions of the body where the pulse can be palpated—the radial artery or carotid artery. If you are recording your lab partner's pulse, make sure you use the tips of your fingers to measure the pulse rate, not your thumb, which has its own pulse and can give false results. Determine beats per minute by counting the pulse for 1 minute.

Pulse rate of lab partner: _____ bpm

Measurement of Arterial Blood Pressure

(A) Activity

1. To determine blood pressure, locate the pulse of the brachial artery on your lab partner. This can be done by placing two fingers on the medial side of the biceps brachii muscle near the antecubital fossa. You can place a small "X" with a felt pen on this location on the arm (fig. 32.2).
2. Place the blood pressure cuff around the arm of your lab partner at the level of the heart. Make sure the inflatable portion of the cuff is on the anterior medial side. Some cuffs have a metal bar that provides a loop through which a part of the cuff goes. This bar should *not* be located on the medial side of the arm because if the bar is located there, it may not constrict the brachial artery effectively as the cuff is inflated (fig. 32.3).
3. Clean the ear pieces of the stethoscope with alcohol wipes, place the diaphragm of the stethoscope on the "X" where you located the brachial pulse, and have your lab partner rest their forearm on the lab counter. You should *not* hear any sound at this time.

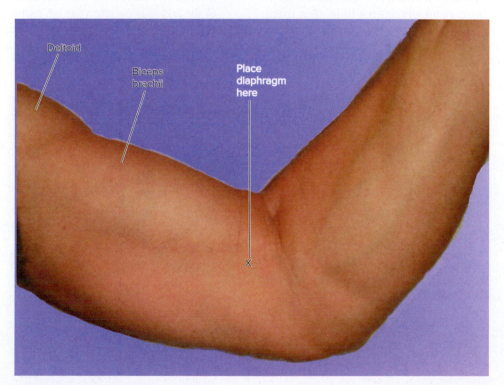

FIGURE 32.2 Location of Stethoscope Diaphragm, Left Arm.
©Eric Wise

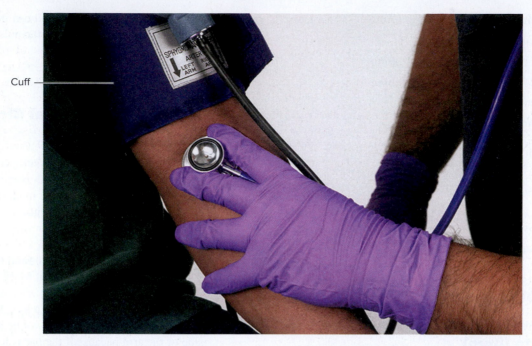

FIGURE 32.3 Proper Placement of Cuff.

Rick Brady/McGraw Hill

4. Hold on to the rubber squeeze bulb with the attached rubber tubing leading away from you. Turn the metal dial clockwise until it is closed. You can now begin to inflate the cuff (fig. 32.4).
5. Pump the cuff up to about 80 mmHg. Look at the mercury in the glass tube or the dial on an aneroid instrument. If the mercury or needle bounces up and down a little, then listen closely for the sounds in the stethoscope. (The ear pieces are inserted into the ears, facing anteriorly.) If you do not

hear any sound, then inflate the mercury or the dial on an aneroid instrument to 100 or 120 mmHg. If you see the mercury or needle pulsing, listen again for the sound. Make sure the diaphragm of the stethoscope is in the right place.
6. Once you are sure you have heard the pulse sound, remove the cuff and place it on the other arm. Excessive constriction of the arm may elevate your lab partner's blood pressure.
7. Inflate the cuff on the other arm, but make sure you exceed the level where you see motion in the needle or pulsing in

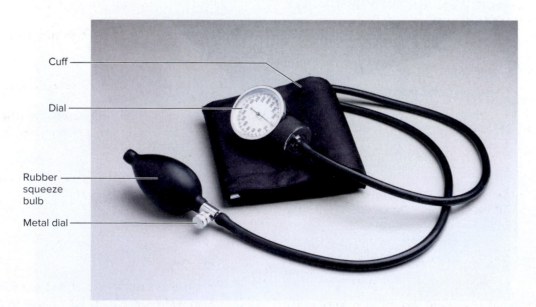

FIGURE 32.4 Parts of an Aneroid Sphygmomanometer.

Creatas/Getty Images

the mercury (usually around 150 mmHg). Do not leave the cuff inflated for a long period.

8. Release the knob slowly, so that the mercury or needle slowly begins to descend.

9. Record the number when the first sound is heard. This sound represents the **systolic pressure** of the ventricles, which is the pressure the heart generates that exceeds the pressure of the cuff. The sound may muffle a bit and then come in strong.

10. Continue to let air out of the cuff slowly until the sound starts to muffle. Listen very carefully until the sound completely disappears. The exact level at which the sound disappears is the **diastolic pressure.**

11. Record the level of blood pressure as the systolic pressure over the diastolic pressure:

$$\frac{\text{Systolic pressure}}{\text{Diastolic pressure}} = \underline{\hspace{3cm}}$$

If you are unsure of exactly where the pressure is, you can fine-tune your readings by increasing the cuff pressure at each approximate end to obtain an accurate reading. Do not subject your lab partner to more than two or three attempts. Ask your instructor for help if you have difficulty determining blood pressure.

Pulse Pressure

You can now determine the pulse pressure of your lab partner. The pulse pressure is determined by subtracting the diastolic pressure from the systolic pressure. It is normally around 50 mmHg.

Pulse pressure: _____

Mean Arterial Blood Pressure

Mean arterial blood pressure (MAP) is used clinically to measure the average blood pressure of a person. If the MAP is too low, then organs of the body are not receiving enough blood pressure and could become ischemic. Once you measure your systolic blood pressure (SP) and diastolic blood pressure (DP), you can get an approximation of the MAP by using the following formula.

$$\text{Approximated MAP} = \text{DP} + 1/3 \, (\text{SP} - \text{DP})$$

Enter your MAP in the following space:

MAP _____

Normally, it should fall between 70 and 110 mmHg. If it is below 60 mmHg, the organs of the body are not receiving adequate blood flow.

Influence of Blood Vessel Diameter on Resistance

In this part of the lab, you do a little investigative work to try to determine the influence that diameter has on the resistance to blood flow in blood vessels. The blood flow in a vessel increases with increasing pressure and decreases with increasing resistance. This can be expressed in the following formula: **Flow = pressure/resistance.** Therefore, if you increase the pressure of the system, the rate of flow increases. If you increase the resistance in the blood vessel, the flow decreases. There are a few factors that determine resistance of blood flow through vessels. These are the length of the vessel, the viscosity of the blood, and the diameter of the vessel in which the blood flows. Small changes in the diameter of the blood vessel cause significant changes in resistance, and this is the most critical physiological factor in blood flow.

(A) **Activity**

Using the items provided by your instructor, design an experiment that demonstrates how an increase or a decrease in diameter alters the resistance and subsequently the flow rate in blood vessels.

1. corn syrup colored red
2. 1 yard of ¼-inch plastic flexible tubing
3. 1 yard of ½-inch plastic flexible tubing
4. small funnel to fit small tubing
5. large funnel to fit large tubing
6. watch or timer function on smartphone
7. ring stand
8. clamps
9. wash basin
10. beakers

Flow is altered in humans by cardiac output (heart rate × stroke volume), peripheral resistance, precapillary sphincters, and general vasoconstriction.

Acute Factors Affecting Blood Pressure

Body Position

(A) **Activity**

Measure the blood pressure of your lab partner while lying down. Record this value.

Measurement lying down: _____

Have your lab partner stand suddenly, and quickly take a new measurement. *Be careful not to drop the sphygmomanometer!* Record the results.

Blood pressure on immediately standing: _____

How can you account for these two readings?

Exercise

(A) **Activity**

Have your lab partner run around the building for a while or do strenuous physical activity for a couple of minutes. Measure the blood pressure immediately after the cessation of exercise and record your results.

Caution! Do not do strenuous exercise if you are not feeling well, have a history of heart trouble, or have been advised by a physician to avoid exercise.

Blood pressure after exercise: _____

Wait for 1 minute and record your blood pressure in the following space.

Blood pressure after 1 minute: _____

Wait an additional minute and record your blood pressure 2 minutes after exercise.

Blood pressure after 2 minutes: _____

Is there a difference between the recovery of the heart rate in individuals in class who are in shape and the recovery in those who are out of shape?

Demonstration of Valves in Veins

(A) **Activity**

One way to examine the nature of valves in veins is to let your arm hang at your side until the veins become engorged with blood. As you hold your arm in this position, stroke the superficial veins from distal to proximal with the index finger of your free hand, applying uniform and constant pressure (fig. 32.5a). Maintain pressure on the vein at its proximal end and see if the vein fills with blood. Record your results, indicating if the veins fill with blood.

Results: _____

Try the experiment again, but keep pressure on the distal part of the vein with your index finger as you push the blood toward the heart with your thumb (fig. 32.5b). Release the vein from the proximal area and see if blood fills the vein. Record your results.

Results: _____

Do the valves prevent the blood from flowing in a superior or an inferior direction?

Movement Through Capillaries

Blood capillaries have a relatively simple structure in that they are composed of endothelium. The endothelium consists of a single layer of simple squamous epithelium that forms a tube slightly larger than the erythrocytes that pass through it. You can observe capillaries in living tissue by examining the blood flow through the fin of a live goldfish.

(A) **Activity**

Obtain a *living* goldfish and wrap its body in a wet paper towel. Leave the tail fin exposed. *Do not let the fish dry out while you are examining it and look at it for only a minute maximum.* Hold the fin under a dissecting microscope and fan it out. Make sure to hold the fish securely. Observe the flow of the blood through the capillaries, keeping the fish moist with water. Capillaries can be distinguished from arterioles because blood cells pass through the capillary one at a time due to the small size of the capillary. Describe the flow of the blood through the capillaries of the fin in terms of speed and size of the capillary relative to the diameter of an erythrocyte. Return the fish to the tank immediately after examination.

Your description: _____

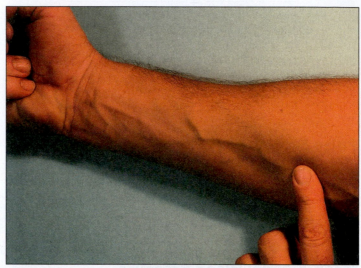

(a)

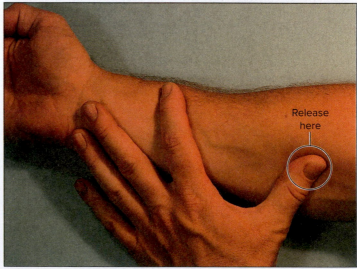

(b)

FIGURE 32.5 Testing Valves in Veins. (a) Holding proximal vein; (b) holding distal part of vein and releasing proximal section.
©Eric Wise

REVIEW SECTION

Blood Vessels and Blood Pressure

Name _____ *Date* _____

Lab Section _____ *Time* _____

Review Questions

1. The letters bpm stand for what phrase in cardiac measurement? _____

2. What is a sphygmomanometer and what does it measure?

_____

_____

3. To measure blood pressure, what artery would you most commonly use and why?

_____

_____

4. If you have a blood pressure of 140/80, what does the 80 represent? What phase is the heart in at this point?

_____

_____

5. What is the clinical threshold for high blood pressure in young adults? _____

6. When the first sound is heard during measurement with a blood pressure cuff, what is measured—systolic or diastolic pressure?

_____

7. Emotions have an effect on blood pressure. Predict the blood pressure of an individual who recently had a heated argument with a roommate about rent money. _____

8. Illness can also affect blood pressure. Illness tends to increase stress responses. Predict the blood pressure of an individual with a sinus headache and postnasal drip. _____

9. Nicotine and caffeine both temporarily elevate blood pressure. Explain how a long-term increase in blood pressure could have a negative effect on the cardiac output. _____

_____

10. Record your blood pressure. _____

11. How did the change in blood pressure after exercise vary from those people who were in shape versus those who were out of shape? _____

_____

12. According to the potential risk factors for hypertension, which, if any, do you have?

_____

13. What type of tissue makes up the endothelium of capillaries? What functional advantage does this tissue type provide to the capillary?_____

14. Superficial veins contain valves. Inactive people may have problems with their veins in that blood pools in the veins. Can you propose a mechanism by which blood from the veins may be returned to the heart (other than standing on your head!)?

_____

_____

Anatomy of the Respiratory System

INTRODUCTION

The respiratory system provides oxygen to the cells of the body and removes carbon dioxide. Cells use oxygen as the terminal electron acceptor, and carbon dioxide is a metabolic waste product of cellular respiration.

Atmospheric oxygen moves into the lungs and diffuses into the circulatory system. This vital exchange occurs due to the respiratory system, discussed in the Saladin text in chapter 22, "The Respiratory System." The oxygen subsequently reaches the individual cells of the body while carbon dioxide is released from the intercellular environment and travels via the blood to the lungs, where it is released by exhalation. Too much carbon dioxide in the blood increases the hydrogen ion concentration, producing acidosis of the blood. Too few hydrogen ions in the blood cause alkalosis. Acidosis or alkalosis disrupts normal metabolic processes in the blood. The respiratory system is therefore integrated with the other body systems. As you study the anatomy of the respiratory system in this exercise, be aware of the role that other systems play in respiration, such as the cardiovascular system, which is the system that transports oxygen and carbon dioxide, and the muscular system, which increases demand for oxygen during times of exertion.

OBJECTIVES

At the end of this exercise, you should be able to

1. list the organs and significant structures of the respiratory system;
2. discuss the role of the respiratory system in terms of the overall function of the body;
3. explain the physical reason for the tremendous surface area of the lungs;
4. identify the cartilages of the larynx;
5. distinguish among a bronchus, bronchiole, and respiratory bronchiole.

MATERIALS

Lung models or detailed torso model, including median section of head

Model of larynx

Microscopes

Prepared microscope slides of

    normal lung tissue

    smoker's lung

    trachea

Charts and illustrations of the respiratory system

Cats

Materials for Cat Dissection

Dissection trays

Scalpel and two or three extra blades

Protective gloves

Blunt (mall) probe

Forceps and sharp scissors

First aid kit in lab or prep area

Sharps container

Animal waste disposal container

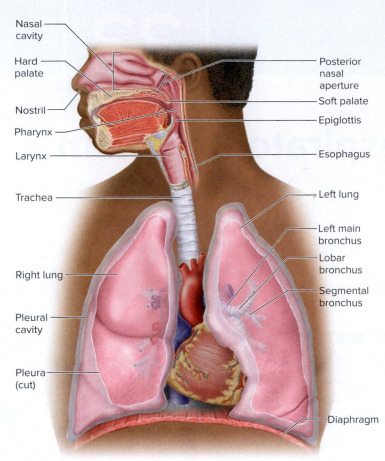

FIGURE 33.1 Overview of the Anatomy of the Respiratory System.

Labels (Figure 33.1):
Nasal cavity
Hard palate
Nostril
Pharynx
Larynx
Trachea
Right lung
Pleural cavity
Pleura (cut)
Posterior nasal aperture
Soft palate
Epiglottis
Esophagus
Left lung
Left main bronchus
Lobar bronchus
Segmental bronchus
Diaphragm

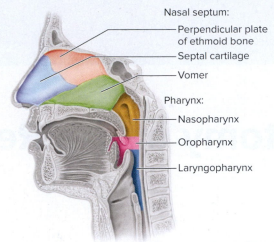

FIGURE 33.2 Structures that Make Up the Nasal Septum.

Labels (Figure 33.2):
Nasal septum:
Perpendicular plate of ethmoid bone
Septal cartilage
Vomer
Pharynx:
Nasopharynx
Oropharynx
Laryngopharynx

PROCEDURE

Overview of the Respiratory System

Ⓐ **Activity**

Look at the charts and models of the respiratory system for a general orientation and compare them to fig. 33.1. Locate the following structures:

bronchus	nasal cavity	right lung
larynx	nose	trachea
left lung	pharynx	

Nose and Nasal Cartilages

Ⓐ **Activity**

Examine a median section of a model or chart of the head and look for the **nose, nasal cartilages, external nares** (nostrils), and **nasal septum.** The nasal septum separates the nasal fossae and is composed of the perpendicular plate of the ethmoid bone, the vomer, and the septal cartilage. Examine these features in figs. 33.2 and 33.3.

The entrance of the external nares is protected by guard hairs. The region of the nose just posterior to the external nares is the **nasal vestibule,** lined with stratified squamous epithelium. Behind the vestibule is the nasal cavity lined with a **mucous membrane,** the **nasal mucosa,** that moistens and warms air entering the respiratory system. The membrane consists of **respiratory epithelium,** composed of **pseudostratified ciliated columnar epithelium with goblet cells.** This mucous membrane overlies a superficial venous plexus that warms the air. The lateral walls of the cavity have three protrusions that push into the nasal cavity. These are the **nasal conchae.** They cause the air to swirl in the nasal cavity and come into contact with the nasal mucosa. Locate the **superior, middle,** and **inferior conchae** in the nasal cavity. The nasal cavity ends where two openings, the **posterior nasal apertures,** or **choanae** (co-AH-nee), lead to the **pharynx** (FAIR-inks). These are seen in fig. 33.3.

Pharynx

The pharynx can be divided into three regions based on location. The uppermost area is the **nasopharynx** (NAZE-oh-FAIR-inks), directly posterior to the nasal cavity. The nasopharynx has two openings on the lateral walls, which are the openings of the **auditory, pharyngotympanic,** or **eustachian, tubes.** As the pharynx descends behind the oral cavity, it becomes the **oropharynx.** The oropharynx is a common passageway for food, liquid, and air. The **uvula** is a small, pendulous structure that partially separates the oral cavity from the oropharynx. The uvula is used in phonation, particularly in trilled "r's" and keeps food in the oral cavity until swallowed. The most inferior portion of the pharynx is the **laryngopharynx** (la-RING-go-FAIR-inks), located posterior to the larynx.

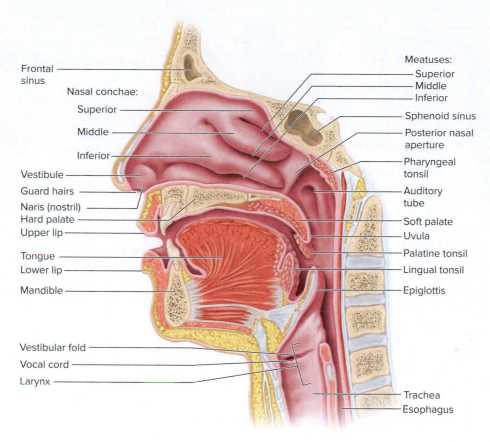

Frontal sinus

Nasal conchae:
Superior
Middle
Inferior

Vestibule
Guard hairs
Naris (nostril)
Hard palate
Upper lip

Tongue
Lower lip
Mandible

Vestibular fold
Vocal cord
Larynx

Meatuses:
Superior
Middle
Inferior

Sphenoid sinus
Posterior nasal aperture
Pharyngeal tonsil
Auditory tube
Soft palate
Uvula
Palatine tonsil
Lingual tonsil
Epiglottis

Trachea
Esophagus

FIGURE 33.3 Upper Respiratory Tract. Median section of the head with nasal septum removed.

Ⓐ Activity

Use models or charts to find the regions of the pharynx in fig. 33.2.

Larynx

The **larynx** (LAIR-inks) is commonly known as the "voice box" because it is an important organ for sound production in humans. It controls the pitch of the voice, while the shape of the oral cavity and the placement and size of the paranasal sinuses are responsible for the sonority (sound quality) of the voice. The larynx occurs at about the level of the fourth through sixth cervical vertebrae and consists of a number of cartilages. The most prominent cartilage in the larynx is the **thyroid cartilage,** a shield-shaped structure made of hyaline cartilage.

The thyroid cartilage is more prominent in males because it increases in size under the influence of testosterone. Inferior to the thyroid cartilage is the **cricoid** (CRY-coyd) **cartilage.** The cricoid cartilage is also composed of hyaline cartilage, and it is relatively narrow when seen from the anterior but increases in size at its posterior surface. Superior to the cricoid cartilage

in the posterior wall of the larynx are the paired **arytenoid** (AR-ih-TEE-noyd) **cartilages.** These cartilages attach to the posterior end of the **vocal cords** (**vocal folds).** Movement of the arytenoid cartilages pulls on the vocal cords, causing them to stretch and increasing the pitch of the voice. This occurs by the contraction of intrinsic muscles attached to the arytenoid cartilages from the posterior, while the vocal cords are held stationary by the thyroid cartilage anteriorly. Superior to vocal cords are the **vestibular folds** (fig. 33.4).

At the very posterior, superior edge of the larynx are the **corniculate** and **cuneiform** (cue-NEE-ih-form) **cartilages.** These are also made of hyaline cartilage. The most superior cartilage of the larynx is the **epiglottis,** composed of elastic cartilage and mucous membrane. During swallowing, the epiglottis protects the opening of the larynx, known as the **glottis** (fig. 33.3). In the swallowing reflex, muscles pull the epiglottis down over the glottis. This is not a perfect system, as anyone knows who has started swallowing a liquid and responded by laughing at a joke. Inhalation at the beginning of the laugh causes fluid to move into the larynx and trachea, irritating the respiratory lining. This causes another reflex called the **cough reflex,** which propels the liquid out of the respiratory system.

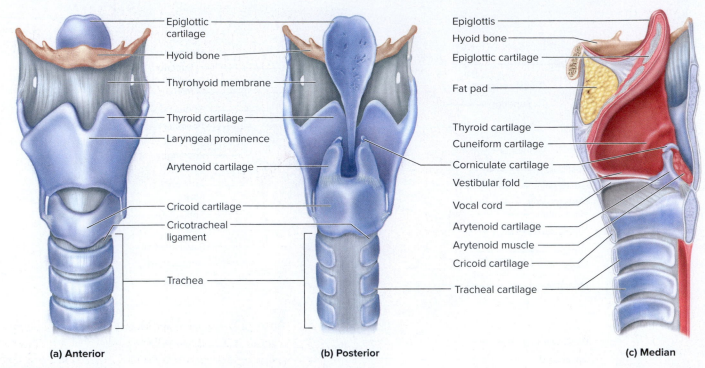

(a) Anterior   **(b) Posterior**   **(c) Median**

FIGURE 33.4 Larynx.

Ⓐ **Activity**

Locate the structures of the larynx on models or charts in the lab and in fig. 33.4.

Trachea and Bronchi

Ⓐ **Activity**

The **trachea (TRAY-kee-uh)** is commonly known as the "windpipe" because it conducts air from the larynx to the lungs. The trachea is a straight tube whose lumen is kept open by C-shaped **tracheal cartilages.** Examine these cartilages by running your fingers gently down the outside of your throat. Palpate the cartilage rings below the larynx. The tracheal cartilages are composed of hyaline cartilage. At the most inferior portion of the trachea is a center point known as the **carina** (ca-RY-na) (keel). Locate the features of the trachea in figs. 33.5 and 33.6.

Ⓐ **Activity**

The trachea is also lined with respiratory epithelium. Obtain a prepared slide of the trachea and find the tracheal cartilage, respiratory epithelium, and **posterior tracheal membrane** (with the trachealis, a muscle—absent in some slide preparations) (fig. 33.6).

The trachea splits into two tubes, which enter the lungs. These tubes are the **main bronchi** (BRONK-eye). Each lung receives air from a main bronchus. The main bronchi of the lung

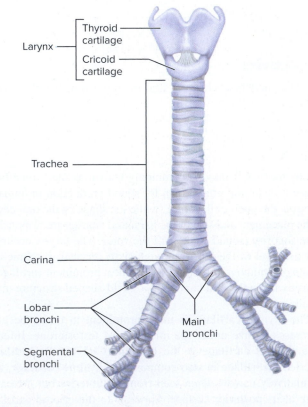

FIGURE 33.5 Larynx, Trachea, and Bronchi, Anterior View.

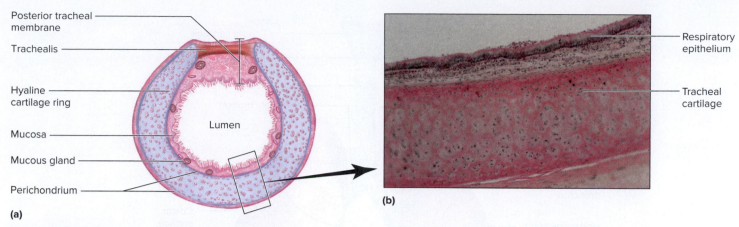

Posterior tracheal membrane

Trachealis

Hyaline cartilage ring

Mucosa

Mucous gland

Perichondrium

Lumen

(a)

Respiratory epithelium

Tracheal cartilage

(b)

FIGURE 33.6 **Trachea, Cross Section.** (a) Diagram; (b) photomicrograph (100×).
(b) ©Eric Wise

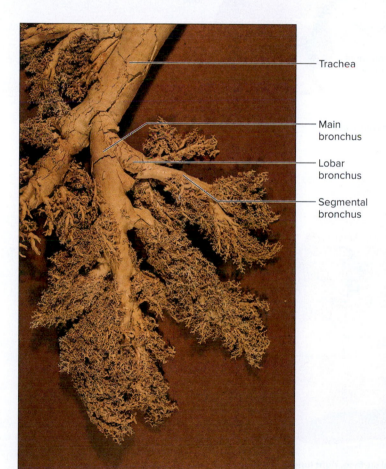

Trachea

Main bronchus

Lobar bronchus

Segmental bronchus

FIGURE 33.7 **Cast of Bronchial Tree.**
©Eric Wise

divide into the **lobar bronchi,** and these further divide to form **segmental bronchi.** These bronchi contain hyaline cartilage and are lined with respiratory epithelium. The extensive branching of the bronchi produces a structure called the **bronchial tree** (figs. 33.5 and 33.7).

Lungs

Of the two lungs in humans, the right lung has three lobes: a **superior lobe,** a **middle lobe,** and an **inferior lobe.** The left lung has two lobes: a **superior lobe** and an **inferior lobe.** The left lung also has an indentation, called the **cardiac impression,** where the apex of the heart rests.

Ⓐ Activity

Look at models or charts in the lab and identify the major features, as shown in fig. 33.8.

The lungs are surrounded by the **pleural cavities** on each side of the **mediastinum.** The **parietal pleura** is the outer membrane on the chest cavity wall, and the membrane that adheres to the surface of the lungs is the **visceral pleura.** The space between the membranes is the pleural cavity. These are shown in fig. 33.9.

Histology of the Lung

The bronchi continue to divide until they become **bronchioles,** small respiratory tubules with smooth muscle in their walls, no cartilage, and an inner lining of respiratory epithelium.

Ⓐ Activity

Obtain a prepared slide of lung and scan it first under low power and then under higher powers. The bronchioles in the lung further divide into **respiratory bronchioles,** so named due to small structures called alveoli attached to their walls. The respiratory bronchioles lead to passageways known as **alveolar ducts,** which branch into **alveoli.** Alveoli are air sacs in the lung that exchange oxygen and carbon dioxide with the blood capillaries of the lungs. From the bronchioles to the alveoli, the respiratory epithelium progressively decreases in height, eventually becoming simple squamous epithelium. Examine your slide and locate the bronchi, bronchioles, respiratory bronchioles, **alveolar ducts,** and alveoli.

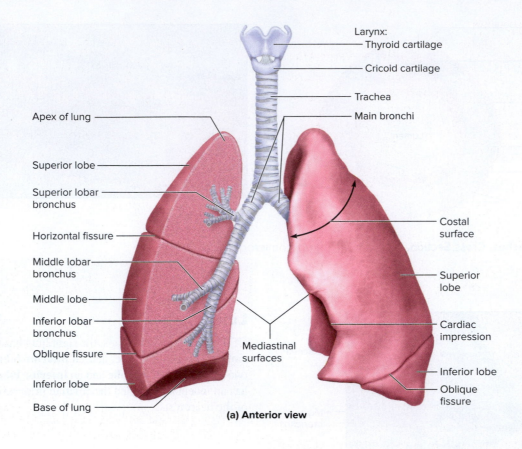

(a) **Anterior view**

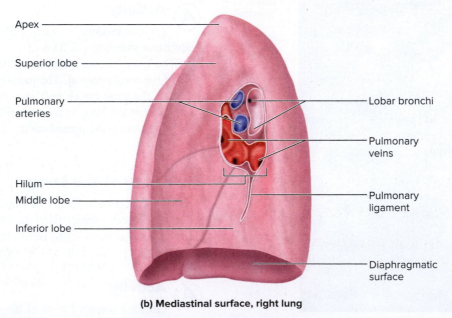

(b) **Mediastinal surface, right lung**

FIGURE 33.8 Lungs. (a) Right lung showing bronchial tree, left lung surface view; (b) medial view.

Alveoli clustered around an alveolar duct are collectively known as an **alveolar sac** (fig. 33.10).

The alveoli are lined primarily with simple squamous epithelium called **squamous (type I) alveolar cells** and are in close proximity to the vascular endothelium of the capillaries surrounding the alveoli. The division of the lung into numerous alveoli tremendously increases the surface area of the lung. This increase is vital for the rapid and extensive *diffusion* of oxygen across the respiratory membranes.

Other cell shapes that occur in the prepared lung sections are cuboidal **(type II) (septal) cells,** which decrease the surface tension of the lung by secreting pulmonary surfactant. The lung also contains **alveolar macrophages,** which phagocytize foreign material entering the lungs and produce a significant number of blood platelets.

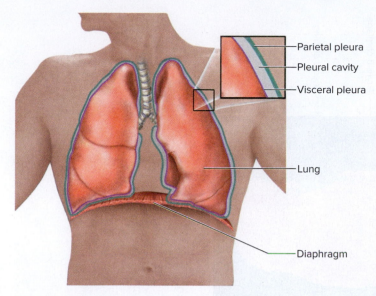

FIGURE 33.9 **Pleural Membranes and Cavity.**

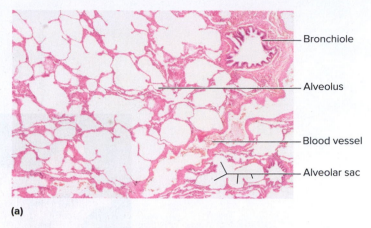

(a)

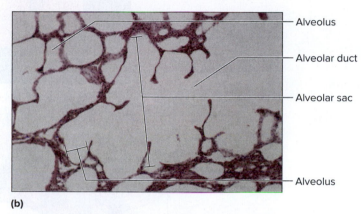

(b)

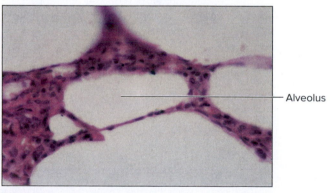

(c)

FIGURE 33.10 **Histology of the Lung.** (a) Photomicrograph of lung (40×); (b) photomicrograph of alveolar duct (100×); (c) photomicrograph of alveolus (400×).
©Eric Wise

(A) Activity

Examine a prepared slide of smoker's lung. Note the dark material in the lung tissue and the general destruction of the alveoli. Breakdown of the alveoli leads to a disease known as **emphysema.**

Cat Dissection

(A) Activity

Prepare your cat for dissection and remember to place all excess tissue in the appropriate waste container and not in a standard wastebasket or down the sink!

1. If you have not opened the chest cavity in your study of the cat, you should do so now. Removal of the skin is discussed in Exercise 11, and the procedure for opening the thoracic cavity is described in Exercise 29.
2. Locate the **larynx** of the cat above the **trachea.** Notice the broad, wedge-shaped structure in the front. This is the **thyroid cartilage.**
3. Make a median incision through the thyroid cartilage and continue carefully cutting until you have cut completely through the larynx. To examine the larynx more completely, you may wish to continue your median cut partway through the trachea.
4. Open the larynx and find the **epiglottis** of the cat. Notice how the elastic cartilage is lighter in color than that of the thyroid cartilage. This is because the epiglottis is made of elastic cartilage, while the thyroid cartilage is composed of hyaline cartilage.
5. Find the **cricoid cartilage,** the **arytenoid cartilages,** and the **vestibular folds** and **vocal cords** (fig. 33.11).
6. Examine the trachea as it passes from the larynx into the thoracic cavity. Ask your instructor for permission before you cut the trachea in cross section. If you do, you should see the **tracheal cartilages** and the **posterior tracheal membrane.** The posterior portion of the trachea is ventral to the esophagus. Notice how the trachea splits into the two **bronchi,** which then enter the lungs. The lobes of

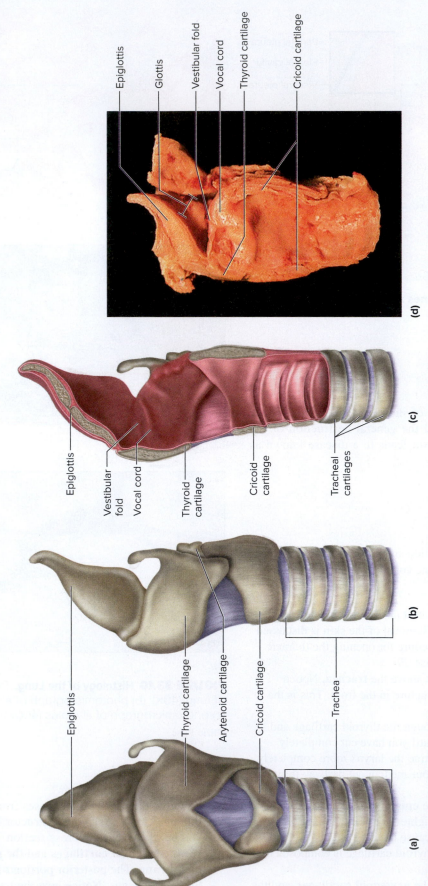

FIGURE 33.11 Larynx of the Cat. Diagram (a) anterior view; (b) left lateral view; (c) median view; (d) photograph, median view.

(d) ©Eric Wise

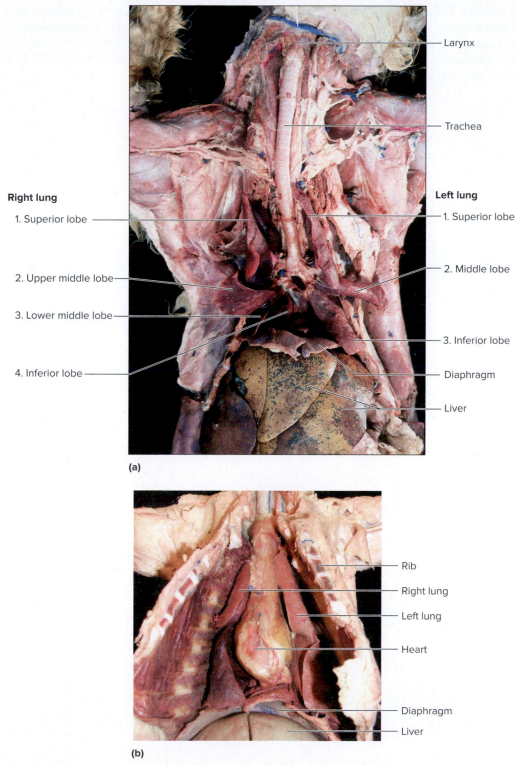

Right lung

1. Superior lobe

2. Upper middle lobe

3. Lower middle lobe

4. Inferior lobe

Larynx

Trachea

Left lung

1. Superior lobe

2. Middle lobe

3. Inferior lobe

Diaphragm

Liver

(a)

Rib

Right lung

Left lung

Heart

Diaphragm

Liver

(b)

FIGURE 33.12 Respiratory Structures of the Cat. (a) Larynx, trachea, and lungs; (b) lungs.
(a) ©Eric Wise; *(b)* Cordi Smith/McGraw Hill

the lungs are different in the cat than in the human. The right lung in the cat has four lobes, while the left lung has three lobes. How does this compare to the pattern in humans?

7. Examine the structure of the lungs in your specimen and compare it to fig. 33.12. The lungs are covered with a thin serous membrane called the visceral pleura, while the outer covering (on the deep surface of the ribs and intercostal muscles)

is called the parietal pleura. The space between these two membranes is the pleural cavity.

8. Cut into one of the lungs of the cat and examine the lung tissue. Note how the lungs appear like a very fine mesh sponge. The alveoli of the lungs are microscopic.

9. At the inferior portion of the thoracic cavity is the diaphragm. As it contracts, the pressure in the thoracic cavity decreases and air fills the lungs. Examine the diaphragm in your specimen and compare it to fig. 33.12.

REVIEW SECTION

Anatomy of the Respiratory System

Name _____ *Date* _____

Lab Section _____ *Time* _____

Review Questions

1. What is the common name for the external nares? _____

2. Some of the nasal cartilages are made of hyaline cartilage. What functional adaptation does cartilage have over bone in making up the external framework of the nose?

   _____

3. The nasal fossae are separated from each other by what structure? _____

4. In lung cancer, there are frequently tumors in the lymphoid tissue. These may reduce the flow in the pulmonary arteries. What impact would this have on the respiratory system?

   _____

   _____

5. What is the function of respiratory epithelium and the superficial blood vessels in the nasal cavity?

   _____

   _____

6. Name the openings between the nasal cavity and the pharynx.

   _____

7. What is the name of the space behind the oral cavity and above the laryngopharynx?

   _____

8. What is the name of the large cartilage of the anterior larynx? _____

9. What membrane attaches directly to the lungs? _____

10. The trachea branches into two tubes that go to the lungs. What are these tubes called?

   _____

11. Where is the bronchial tree found? _____

12. What small structure in the lung is the site of exchange of oxygen with the blood capillaries? _____

13. The surface area of the lungs in humans is about 70 square meters. How can this be so if the lungs are located in the small space of the thoracic cavity? What role do alveoli play in the nature of surface area?

_____

_____

14. Emphysema is a destruction of the alveoli of the lungs. What effect does this have on the surface area of the lungs?

_____

15. Fill in the following illustration of the human respiratory system using the terms provided.

cricoid cartilage   nasal cavity

epiglottis   posterior nasal apertures

inferior lobe   superior lobe

main bronchus   trachea

middle lobe

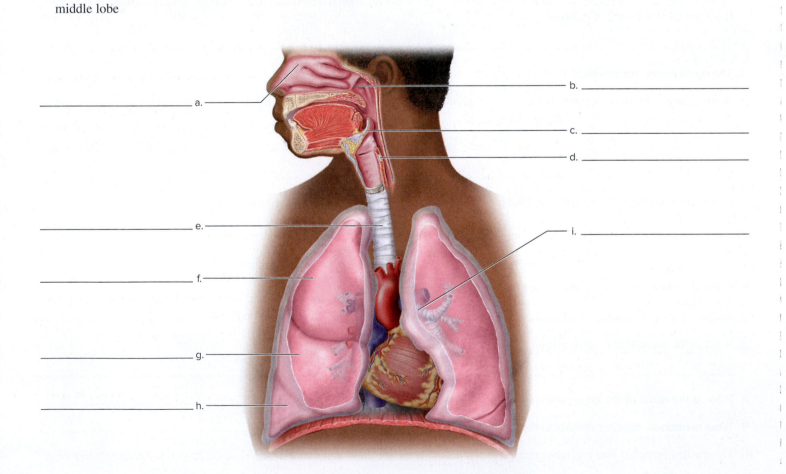

Respiratory Physiology

INTRODUCTION

The function of the respiratory system is to exchange gases between the external air and the blood. Respiration involves both **pulmonary ventilation,** movement of air into and out of the lungs, and the **exchange of gases** (oxygen and carbon dioxide) across the respiratory membrane. The amount of air that enters the lungs can be altered by the **respiratory rate** (number of breaths per minute), the change in the **tidal volume** (amount of air inhaled and exhaled with each breath), or both. Reduction in the lung volume may be reflective of poor health conditions, such as fibrosis in the lungs, which limits the ability of the lungs to fully expand.

Adequate pulmonary ventilation is critical in clinical settings. The measurement of arterial blood gases is important when patients are artificially ventilated because it allows the respiratory therapist to determine if adequate oxygen is being delivered to the patient. Reduction in pulmonary ventilation can lead to a buildup of carbon dioxide (hypercapnia) in the blood, producing acidosis. Increased pulmonary ventilation (as results from being at high altitudes, having severe anemia, or ingestion of certain drugs) can lead to a reduction in carbon dioxide levels (hypocapnia) in the blood, producing alkalosis.

Just as changes in the ventilation rate or tidal volume can alter the pH of the blood, the reverse is true. Excess carbon dioxide in the blood stimulates an increase in pulmonary ventilation, which subsequently reduces the carbon dioxide level in the blood. This negative feedback mechanism is important in maintaining homeostasis in the blood. These topics are discussed in the Saladin text in chapter 22, "The Respiratory System." In this exercise, you measure respiratory rates, lung volumes and capacities, and the effects of carbon dioxide on the acid-base balance of a solution.

OBJECTIVES

At the end of this exercise, you should be able to

1. measure the pulmonary volumes and calculate the pulmonary capacities;
2. describe the relationship between the tidal volume, vital capacity, inspiratory reserve volume, and expiratory reserve volume;
3. describe the mechanical process of breathing;
4. describe the use of the spirometer or airflow transducer available in your lab;
5. identify the tidal volume, expiratory reserve volume, and vital capacity on a spirogram;
6. determine whether a person will inhale or exhale based on the differences in air pressure between the lungs and the external air;
7. explain how resistance in the airways changes the flow of air into or out of the lungs;
8. describe how carbon dioxide in solution changes the pH of the solution;
9. demonstrate the use of the stethoscope in obtaining respiratory sounds;
10. predict the flow of air in an individual, with changes in air pressure between the lungs and the external air or with changes in the resistance in the respiratory passages.

MATERIALS

Respiration Model

Bell jar respiration model

Pulmonary Volume Setup

BIOPAC MP30 or MP150 data acquisition unit, wet spirometer, or handheld spirometers

Disposable mouthpieces to fit respirometer or spirometers

Noseclips

Watch with accuracy in seconds or timer function on smartphone

Biohazard bag

Two meter sticks taped against a wall to measure height in cm.

BIOPAC equipment

AFT6 or AFT6A 600 mL calibration syringe

AFT1—disposable bacteriological filter

AFT2 disposable mouthpiece

SS11LA airflow transducer

13-gallon plastic bags

Plastic dishpan (11.5 quart)

1-foot length of ⅜-inch or ½-inch plastic aquarium tubing

1-liter graduated cylinder

Breathing Sounds and Breathing Rate Setup

Stethoscope

Alcohol wipes

Acid-Base Setup (One Setup per Table)

Litmus solution (2 g litmus powder in 600 mL water)

Household ammonia in dropper bottles

Straws

100 mL Erlenmeyer flasks

Safety glasses

⫼PhILS Version 4.0

PROCEDURE

Mechanics of Breathing

Air moves from regions of higher pressure to regions of lower pressure. The lungs fill with air or release air due to changes in air pressure. Normal atmospheric air pressure is measured in centimeters of water (cm H_2O), with standard air pressure at sea level designated as zero cm H_2O. When there is no movement of air into or out of the lungs, the pressure in the lungs is equal to that of the surrounding air, as illustrated in fig. 34.1(1). During inspiration, the diaphragm contracts, along with the external intercostal muscles, increasing the volume in the thoracic cavity. This leads to a decrease in the pressure in the lungs, and air moves from the atmosphere into the lungs. This is illustrated in fig. 34.1(2). If the diaphragm relaxes, the abdominal pressure forces the diaphragm upward and increases the pressure in the thoracic cavity beyond that of the atmospheric pressure, and the air moves out of the lungs, as illustrated in fig. 34.1(3).

 Activity

You can use a bell jar model to demonstrate this in lab.

This model illustrates the movement of air into or out of the lungs, depending on air pressure differentials between the lungs and the external environment. The clear housing (bell jar) represents the thoracic cage; the balloons represent the lungs; and the latex sheeting represents the diaphragm. Pull *gently* on the latex diaphragm. As you do this, the volume in the pleural cavity (the space between the balloons and the jar) increases, causing a decrease in the pressure of the pleural cavity.

When the pressure in the pleural cavity decreases, air in the external environment is at a higher pressure than that in the jar and it moves into the bell jar, filling the balloon lungs. As you release the diaphragm, the volume decreases and the pressure increases in the pleural cavity bell jar, exceeding the external air pressure. Air moves

out of the lungs and through the tubing. Although the mechanics of human breathing differ somewhat from this model, the bell jar model is valuable to demonstrate the basic principle of air moving from a region of high pressure to a region of low pressure.

Measurement of Relaxed Breathing Rate

 Activity

It is important that your lab partner be distracted from thinking about breathing.

1. Therefore, your lab partner should read a page from this laboratory exercise while you count the number of breaths he or she takes for a total of 2 minutes.
2. Divide the number by 2 to calculate the average number of breaths per minute.
3. Record that number for your lab partner in the space provided and in the Chapter Summary Data section at the end of the exercise.

❓ Breaths per minute: _____ (1)

Breathing rate varies with oxygen demand and/or carbon dioxide levels in the blood. Before doing any exercise, *estimate* the breaths per minute your lab partner might take after completing 2 minutes of strenuous exercise. Record your estimation.

❓ Estimation: _____ (2)

Caution!

![CAUTION] If you have a heart condition, a family history of heart failure, or another medical condition that prevents you from doing strenuous exercise, then have your lab partner do the exercise sections of this lab.

Have your lab partner do strenuous exercise for 2 minutes (jumping jacks, running in place or stepping up and down on a bench step, etc.). As soon as your lab partner is finished, record the number of breaths per minute for the *first minute* after exercise and write this number in the space below and in the Chapter Summary Data section.

❓ Number of breaths per minute after exercise:

_____ (3)

❓ How does this compare with your estimation?

_____ (4)

❓ What kind of exercise did your lab partner perform?

_____ (5)

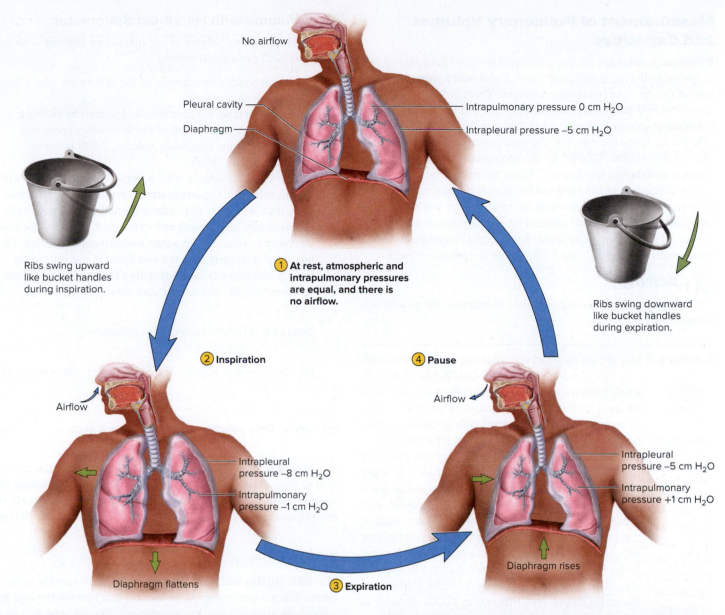

No airflow

Pleural cavity

Diaphragm

Intrapulmonary pressure 0 cm H_2O

Intrapleural pressure −5 cm H_2O

Ribs swing upward like bucket handles during inspiration.

Ribs swing downward like bucket handles during expiration.

① At rest, atmospheric and intrapulmonary pressures are equal, and there is no airflow.

② **Inspiration**

④ **Pause**

Airflow

Airflow

Intrapleural pressure −8 cm H_2O

Intrapulmonary pressure −1 cm H_2O

Intrapleural pressure −5 cm H_2O

Intrapulmonary pressure +1 cm H_2O

Diaphragm flattens

③ **Expiration**

Diaphragm rises

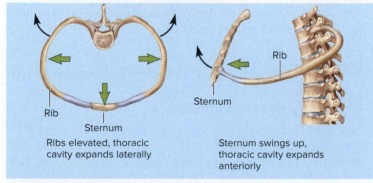

Rib

Sternum

Rib

Sternum

Ribs elevated, thoracic cavity expands laterally

Sternum swings up, thoracic cavity expands anteriorly

② In inspiration, the thoracic cavity expands laterally, vertically, and anteriorly; intrapulmonary pressure drops 1 cm H_2O below atmospheric pressure, and air flows into the lungs.

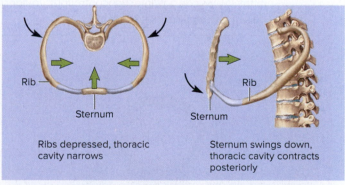

Rib

Sternum

Sternum

Rib

Ribs depressed, thoracic cavity narrows

Sternum swings down, thoracic cavity contracts posteriorly

③ In expiration, the thoracic cavity contracts in all three directions; intrapulmonary pressure rises 1 cm H_2O above atmospheric pressure, and air flows out of the lungs.

FIGURE 34.1 Mechanics of Breathing.

Measurement of Pulmonary Volumes and Capacities

Pulmonary volumes are the amount of air that flows into or out of the lungs during a particular event. **Pulmonary capacity** is the sum of two or more pulmonary volumes. There are many different ways to record pulmonary volumes. An affordable way to measure pulmonary volume is to use a handheld spirometer in which the exhaled air spins the vanes of the spirometer, estimating the volume of air exhaled. This is like the anemometer used by weather stations to measure wind speed. These are relatively inexpensive pieces of equipment and are good for measuring vital capacity, but they are not as accurate as other equipment. The wet spirometer measures the volume of air exhaled into a chamber, so it measures the actual amount of air exhaled. Some digital recorders, such as those made by BIOPAC, use a flow meter to estimate volume.

 Activity

Perform the following activities to measure the pulmonary volumes and capacities.

Caution!

You should never inhale while using any spirometer. The risk of contracting SARS-CoV-2 (the COVID-19 virus), tuberculosis, or other respiratory diseases is of concern for this lab. Any equipment used in this lab to measure respiratory ventilation (breathing) must be sterilized before **any student** uses the equipment. Your instructor will determine whether the following exercises will be done by demonstration, if a virtual lab is to be conducted, or if the exercises will be done by individual students if enough equipment and protocols are in place to perform these exercises in the safest possible manner. Under no circumstances should non-sterile equipment be used in performing these exercises. Recommendations for the sterilization of equipment can be found on the Centers for Disease Control and Prevention (CDC) website at https://www.cdc.gov/infectioncontrol/pdf /guidelines /disinfection-guidelines-H.pdf.

FIGURE 34.2 Handheld Spirometer.
©Eric Wise

Tidal Volume with Handheld Spirometer

Tidal volume is the volume of air inhaled or exhaled with each breath without extra effort applied.

1. Place a disposable mouthpiece on the spirometer tube. Close your nose with a noseclip.
2. Make sure you set the indicator dial to zero by twisting the knurled ring on the top of the spirometer (zero may be the same as the maximum volume, 7,000 in some spirometers) (fig. 34.2).
3. As you exhale, estimate what a normal breath volume will be and exhale this amount rather forcefully. Do not exhale more than what you would for a normal breath. If you exhale gently into a handheld spirometer, you may not cause the vanes to spin enough to get a significant recording. Do not put your fingers over the exit holes in the spirometer.
4. Exhale five times (without resetting the spirometer to zero between breaths) and record your results.

Total of five breaths recorded by spirometer: _____

Divide the total number by 5 and enter the average tidal volume.

? Average tidal volume: _____ (6)

You may find that the tidal volume is variable among members of your class. This is due to the difficulty of measuring tidal volumes with a standard lab apparatus and variations in body size and physical condition. The average tidal volume is about 500 mL.

Tidal Volume with a Wet Spirometer

The tidal volume can be measured if your wet spirometer is accurate. Place a sterile disposable mouthpiece into the flexible hose and exhale normally into the mouthpiece. Record your data in the following space. Throw the disposable mouthpiece into the biohazard container when you are finished.

? Tidal volume: _____ (6)

Expiratory Reserve Volume

You can determine your **expiratory reserve volume (ERV)** with the wet spirometer or the handheld spirometer. The expiratory reserve volume is the maximal amount of air you can exhale *after a normal exhalation*. This is typically around 1,000 mL.

1. Make sure to close off your nostrils with a noseclip and, after a normal exhalation, forcibly expel the remainder of your pulmonary volume through the mouthpiece into the spirometer.
2. Repeat this two more times for a total of three exhalations, and calculate the average expiratory reserve by dividing the sum of the volumes by 3. Alternatively, you could not reset the spirometer and divide the final reading by three.

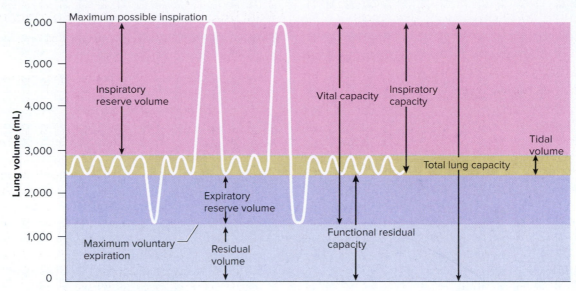

FIGURE 34.3 Lung Volumes and Capacities.

3. Record your results.

Trial 1: _____

Trial 2: _____

Trial 3: _____

❓ Average expiratory reserve volume: _____ (7)

Vital Capacity (VC)

The vital capacity (VC) is the total volume of air that can be forcefully expelled from the lungs after a maximum inhalation. Measuring vital capacity is like participating in the national championship of exhalation.

1. Measure the vital capacity with the use of a wet spirometer or handheld spirometer by first inhaling as deeply as you possibly can.
2. Close your nostrils with a noseclip, and then exhale through the mouthpiece completely until you cannot exhale anymore.
3. Force as much air from your lungs as you can.
4. Record your vital capacity.

❓ Vital capacity: _____ (8)

Lung volumes and capacities are illustrated in fig. 34.3.

Percent of Expected Vital Capacity

You can compare your vital capacity to those of other individuals of your gender, height, and age by using one of the two charts (one for females, one for males) in table 34.1. As you examine the chart for a person of your height, notice that the vital capacity decreases with age.

Calculate your percent of expected vital capacity by dividing your data by the expected data for a person of your age, gender, and height and multiplying the result by 100.

$$\frac{\text{Your vital capacity}}{\text{Expected vital capacity}} \times 100 = \frac{\text{Percent of expected}}{\text{vital capacity}}$$

❓ Your percent of expected vital capacity

_____ (9)

If you come out at 100% of expected, then you have an average vital capacity of individuals in your bracket. If your value is less than 100%, you have a smaller vital capacity than expected for your bracket. A percent vital capacity of 80% of average, or better, is in the expected range. If your value is larger than 100%, you have a larger vital capacity than average.

Alternative Spirometry Experiment

Ⓐ **Activity**

If your lab is not equipped with a commercial spirometer, you can make an inexpensive spirometer with a new 13-gallon plastic bag (one usually used for kitchen trash cans), a plastic dishpan (11.5 quart), 1 foot of clean ⅜-inch or ½-inch plastic aquarium tubing, a 1-liter graduated cylinder, and a sink!

Place the dishpan into the lab sink and fill the pan until it just begins to overflow. Standard dishpans are usually a little less than 11 liters.

Vital Capacity

1. Insert the plastic tubing into the opening of the 13-gallon plastic bag. Make sure that there is no air in the bag! Either tighten the bag around the tube with a rubber band or hold on tightly to the bag with your hand to make an airtight seal.
2. Use either plastic wrap to wrap around the plastic tubing or a disposable mouthpiece.
3. Inhale maximally, pinch off your nose, and exhale into the tube, filling the bag, until you can exhale no further.
4. Slide your hand down outside the bag and force the air into the bottom of the bag until you have the entire volume of exhaled air pushed into the bottom of the bag. Make sure you do not let any air escape as you do this.
5. While holding tightly onto the bag, slowly push the bag into the water until the volume of air you exhaled just displaces the water in the pan. Your hand should just be touching

TABLE 34.1   **Predicted Vital Capacities for Females and Males**

Females

Height in Centimeters

Age	152	154	156	158	160	162	164	166	168	170	172	174	176	178	180	182	184	186	188
16	3,070	3,110	3,150	3,190	3,230	3,270	3,310	3,350	3,390	3,430	3,470	3,510	3,550	3,590	3,630	3,670	3,715	3,755	3,800
18	3,040	3,080	3,120	3,160	3,200	3,240	3,280	3,320	3,360	3,400	3,440	3,480	3,520	3,560	3,600	3,640	3,680	3,720	3,760
20	3,010	3,050	3,090	3,130	3,170	3,210	3,250	3,290	3,330	3,370	3,410	3,450	3,490	3,525	3,565	3,605	3,645	3,695	3,720
22	2,980	3,020	3,060	3,095	3,135	3,175	3,215	3,255	3,290	3,330	3,370	3,410	3,450	3,490	3,530	3,570	3,610	3,650	3,685
24	2,950	2,985	3,025	3,065	3,100	3,140	3,180	3,220	3,260	3,300	3,335	3,375	3,415	3,455	3,490	3,530	3,570	3,610	3,650
26	2,920	2,960	3,000	3,035	3,070	3,110	3,150	3,190	3,230	3,265	3,300	3,340	3,380	3,420	3,455	3,495	3,530	3,570	3,610
28	2,890	2,930	2,965	3,000	3,040	3,070	3,115	3,155	3,190	3,230	3,270	3,305	3,345	3,380	3,420	3,460	3,495	3,535	3,570
30	2,860	2,895	2,935	2,970	3,010	3,045	3,085	3,120	3,160	3,195	3,235	3,270	3,310	3,345	3,385	3,420	3,460	3,495	3,535
32	2,825	2,865	2,900	2,940	2,975	3,015	3,050	3,090	3,125	3,160	3,200	3,235	3,275	3,310	3,350	3,385	3,425	3,460	3,495
34	2,795	2,835	2,870	2,910	2,945	2,980	3,020	3,055	3,090	3,130	3,165	3,200	3,240	3,275	3,310	3,350	3,385	3,425	3,460
36	2,765	2,805	2,840	2,875	2,910	2,950	2,985	3,020	3,060	3,095	3,130	3,165	3,205	3,240	3,275	3,310	3,350	3,385	3,420
38	2,735	2,770	2,810	2,845	2,880	2,915	2,950	2,990	3,025	3,060	3,095	3,130	3,170	3,205	3,240	3,275	3,310	3,350	3,385
40	2,705	2,740	2,775	2,810	2,850	2,885	2,920	2,955	2,990	3,025	3,060	3,095	3,135	3,170	3,205	3,240	3,275	3,310	3,345
42	2,675	2,710	2,745	2,780	2,815	2,850	2,885	2,920	2,955	2,990	3,025	3,060	3,100	3,135	3,170	3,205	3,240	3,275	3,310
44	2,645	2,680	2,715	2,750	2,785	2,820	2,855	2,890	2,925	2,960	2,995	3,030	3,060	3,095	3,130	3,165	3,200	3,235	3,270
46	2,615	2,650	2,685	2,715	2,750	2,785	2,820	2,855	2,890	2,925	2,960	2,995	3,030	3,060	3,095	3,130	3,165	3,200	3,235
48	2,585	2,620	2,650	2,685	2,715	2,750	2,785	2,820	2,855	2,890	2,925	2,960	2,995	3,030	3,060	3,095	3,130	3,160	3,195
50	2,555	2,590	2,625	2,655	2,690	2,720	2,755	2,785	2,820	2,855	2,890	2,925	2,955	2,990	3,025	3,060	3,090	3,125	3,155
52	2,525	2,555	2,590	2,625	2,655	2,690	2,720	2,755	2,790	2,820	2,855	2,890	2,925	2,955	2,990	3,020	3,055	3,090	3,125
54	2,495	2,530	2,560	2,590	2,625	2,655	2,690	2,720	2,755	2,790	2,820	2,855	2,885	2,920	2,950	2,985	3,020	3,050	3,085
56	2,460	2,495	2,525	2,560	2,590	2,625	2,655	2,690	2,720	2,755	2,790	2,820	2,855	2,885	2,920	2,950	2,980	3,015	3,045
58	2,430	2,460	2,495	2,525	2,560	2,590	2,625	2,655	2,690	2,720	2,750	2,785	2,815	2,850	2,880	2,920	2,945	2,975	3,010
60	2,400	2,430	2,460	2,495	2,525	2,560	2,590	2,625	2,655	2,685	2,720	2,750	2,780	2,810	2,845	2,875	2,915	2,940	2,970
62	2,370	2,405	2,435	2,465	2,495	2,525	2,560	2,590	2,620	2,655	2,685	2,715	2,745	2,775	2,810	2,840	2,870	2,900	2,935
64	2,340	2,370	2,400	2,430	2,465	2,495	2,525	2,555	2,585	2,620	2,650	2,680	2,710	2,740	2,770	2,805	2,835	2,865	2,895
66	2,310	2,340	2,370	2,400	2,430	2,460	2,495	2,525	2,555	2,585	2,615	2,645	2,675	2,705	2,735	2,765	2,800	2,825	2,860
68	2,280	2,310	2,340	2,370	2,400	2,430	2,460	2,490	2,520	2,550	2,580	2,610	2,640	2,670	2,700	2,730	2,760	2,795	2,820
70	2,250	2,280	2,310	2,340	2,370	2,400	2,425	2,455	2,485	2,515	2,545	2,575	2,605	2,635	2,665	2,695	2,725	2,755	2,780
72	2,220	2,250	2,280	2,310	2,335	2,365	2,395	2,425	2,455	2,480	2,510	2,540	2,570	2,600	2,630	2,660	2,685	2,715	2,745
74	2,190	2,220	2,245	2,275	2,305	2,335	2,360	2,390	2,420	2,450	2,475	2,505	2,535	2,565	2,590	2,620	2,650	2,680	2,710

Data Source: "Predicted Vital Capacities for Females and Males (tables)" by E. A. Gaensler, M.D. and G. W. Wright, M.D., AEH., Vol. 12, pp. 146–189, February 1966. From the Helen Dwight Reid Educational Foundation. Published by Heldref Publications, 1319 18th Street NW, Washington, DC 20036-1802. 1966.

TABLE 34.1

Continued

Males

Height in Centimeters

Age	152	154	156	158	160	162	164	166	168	170	172	174	176	178	180	182	184	186	188
16	3,920	3,975	4,025	4,075	4,130	4,180	4,230	4,285	4,335	4,385	4,440	4,490	4,540	4,590	4,645	4,695	4,745	4,800	4,850
18	3,890	3,940	3,995	4,045	4,095	4,145	4,200	4,250	4,300	4,350	4,405	4,455	4,505	4,555	4,610	4,660	4,710	4,760	4,815
20	3,860	3,910	3,960	4,015	4,065	4,115	4,165	4,215	4,265	4,320	4,370	4,420	4,470	4,520	4,570	4,625	4,675	4,725	4,775
22	3,830	3,880	3,930	3,980	4,030	4,080	4,135	4,185	4,235	4,285	4,335	4,385	4,435	4,485	4,535	4,585	4,635	4,685	4,735
24	3,785	3,835	3,885	3,935	3,985	4,035	4,085	4,135	4,185	4,235	4,285	4,330	4,380	4,430	4,480	4,530	4,580	4,630	4,680
26	3,755	3,805	3,855	3,905	3,955	4,000	4,050	4,100	4,150	4,200	4,250	4,300	4,350	4,395	4,445	4,495	4,545	4,595	4,645
28	3,725	3,775	3,820	3,870	3,920	3,970	4,020	4,070	4,115	4,165	4,215	4,265	4,310	4,360	4,410	4,460	4,510	4,555	4,605
30	3,695	3,740	3,790	3,840	3,890	3,935	3,985	4,035	4,080	4,130	4,180	4,230	4,275	4,325	4,375	4,425	4,470	4,520	4,570
32	3,665	3,710	3,760	3,810	3,855	3,905	3,950	4,000	4,050	4,095	4,145	4,195	4,240	4,290	4,340	4,385	4,435	4,485	4,530
34	3,620	3,665	3,715	3,760	3,810	3,855	3,905	3,950	4,000	4,045	4,095	4,140	4,190	4,225	4,285	4,330	4,380	4,425	4,475
36	3,585	3,635	3,680	3,730	3,775	3,825	3,870	3,920	3,965	4,010	4,060	4,105	4,155	4,200	4,250	4,295	4,340	4,390	4,435
38	3,555	3,605	3,650	3,695	3,745	3,790	3,840	3,885	3,930	3,980	4,025	4,070	4,120	4,165	4,210	4,260	4,305	4,350	4,400
40	3,525	3,575	3,620	3,665	3,710	3,760	3,805	3,850	3,900	3,945	3,990	4,035	4,085	4,130	4,175	4,220	4,270	4,315	4,360
42	3,495	3,540	3,590	3,635	3,680	3,725	3,770	3,820	3,865	3,910	3,955	4,000	4,050	4,095	4,140	4,185	4,230	4,280	4,325
44	3,450	3,495	3,540	3,585	3,630	3,675	3,725	3,770	3,815	3,860	3,905	3,950	3,995	4,040	4,085	4,130	4,175	4,220	4,270
46	3,420	3,465	3,510	3,555	3,600	3,645	3,690	3,735	3,780	3,825	3,870	3,915	3,960	4,005	4,050	4,095	4,140	4,185	4,230
48	3,390	3,435	3,480	3,525	3,570	3,615	3,655	3,700	3,745	3,790	3,835	3,880	3,925	3,970	4,015	4,060	4,105	4,150	4,190
50	3,345	3,390	3,430	3,475	3,520	3,565	3,610	3,650	3,695	3,740	3,785	3,830	3,870	3,915	3,960	4,005	4,050	4,090	4,135
52	3,315	3,353	3,400	3,445	3,490	3,530	3,575	3,620	3,660	3,705	3,750	3,795	3,835	3,880	3,925	3,970	4,010	4,055	4,100
54	3,285	3,325	3,370	3,415	3,455	3,500	3,540	3,585	3,630	3,670	3,715	3,760	3,800	3,845	3,890	3,930	3,975	4,020	4,060
56	3,255	3,295	3,340	3,380	3,425	3,465	3,510	3,550	3,595	3,640	3,680	3,725	3,765	3,810	3,850	3,895	3,940	3,980	4,025
58	3,210	3,250	3,290	3,335	3,375	3,420	3,460	3,500	3,545	3,585	3,630	3,670	3,715	3,755	3,800	3,840	3,880	3,925	3,965
60	3,175	3,220	3,260	3,300	3,345	3,385	3,430	3,470	3,500	3,555	3,595	3,635	3,680	3,720	3,760	3,805	3,845	3,885	3,930
62	3,150	3,190	3,230	3,270	3,310	3,350	3,390	3,440	3,480	3,520	3,560	3,600	3,640	3,680	3,730	3,770	3,810	3,850	3,890
64	3,120	3,160	3,200	3,240	3,280	3,320	3,360	3,400	3,440	3,490	3,530	3,570	3,610	3,650	3,690	3,730	3,770	3,810	3,850
66	3,070	3,110	3,150	3,190	3,230	3,270	3,310	3,350	3,390	3,430	3,470	3,510	3,550	3,600	3,640	3,680	3,720	3,760	3,800
68	3,040	3,080	3,120	3,160	3,200	3,240	3,280	3,320	3,360	3,400	3,440	3,480	3,520	3,560	3,600	3,640	3,680	3,720	3,760
70	3,010	3,050	3,090	3,130	3,170	3,210	3,250	3,290	3,330	3,370	3,410	3,450	3,480	3,520	3,560	3,600	3,640	3,680	3,720
72	2,980	3,020	3,060	3,100	3,140	3,180	3,210	3,250	3,290	3,330	3,370	3,410	3,450	3,490	3,530	3,570	3,610	3,650	3,680
74	2,930	2,970	3,010	3,050	3,090	3,130	3,170	3,200	3,240	3,280	3,320	3,360	3,400	3,440	3,470	3,510	3,550	3,590	3,630

the surface of the water. Make sure the entire volume of enclosed air in the bag is submerged (remember that the dishpan should be in the lab sink).

6. Remove the bag and carefully pour water from a graduated cylinder into the pan. You may have to fill the graduated cylinder several times. Record this total volume in the following space.

    Amount of water used to fill the dishpan: _____

7. The amount of water you used to refill the dishpan gives you an estimation of your vital capacity. Record your vital capacity.

 Vital capacity: _____ (8)

You can determine tidal volume and forced expiratory volume with the following modifications of this procedure.

Tidal Volume

1. Insert the plastic tube into a flattened bag (there should be no air in the bag) or squeeze all the air out of the bag from the first experiment.
2. *After a normal inhalation,* **exhale normally** into the tube that is firmly attached to the collapsed bag.
3. Do this for a total of 5 times, filling the bag with 5 breaths.
4. Take that volume of air in the bag and displace the water in the full dishpan.
5. Record the amount of water needed to refill the dishpan.

Amount of water needed to fill the dishpan: _____

6. Divide that number by 5 (since you took 5 breaths) and record your tidal volume below.

 Tidal volume: _____ (6)

Expiratory Reserve Volume

1. The expiratory reserve volume is measured by *exhaling normally* and then forcing the remaining air in your lungs into the plastic tube and plastic bag.
2. Displace the water in the dishpan as done previously; measure the volume to obtain the expiratory reserve volume. Record your expiratory reserve volume below.

 Expiratory reserve volume: _____ (7)

Calculation of the Inspiratory Reserve Volume (IRV)

 Activity

The vital capacity consists of the expiratory reserve volume, the tidal volume, and the inspiratory reserve volume. You can *indirectly* determine the **inspiratory reserve volume (IRV)** by subtracting the expiratory reserve volume (ERV) and the tidal volume (TV) from the vital capacity (VC). This is indicated in fig. 34.3. You cannot measure the inspiratory reserve volume directly with the use

of a handheld spirometer (it records exhalations only). An average IRV is around 3,000 mL. Calculate your inspiratory reserve.

$$IRV = VC - (ERV + TV)$$

 Your IRV: _____ (10)

A decrease in lung capacity may be due to a decrease in lung elasticity (which reduces the pulmonary compliance) caused by disorders such as tuberculosis or pulmonary fibrosis. The obstructive disorders, such as asthma and emphysema, not only produce high lung volumes but also a higher residual lung capacity.

BIOPAC Respiration Lab

 Activity

Pulmonary volumes and capacities can be measured directly with the use of a respirometer or spirometer, or they can be measured indirectly with the use of a flow meter. One such device is an airflow transducer made by BIOPAC. You should have all the necessary material for measuring pulmonary flow, including an MP30 or MP150 data acquisition unit, transducer hardware, sterile mouthpieces, clean bacteriological filters, a compatible computer, and software (BIOPAC Pro or BIOPAC Student Lessons). Every student in lab should record a spirogram. Make sure everyone has access to the equipment prior to doing any data analysis.

Caution! When you change subjects, always use a new disposable bacteriological filter.

Calibrating the Machine

1. Make sure the computer is turned on, that the MP30 or MP150 data acquisition unit is connected to the computer and turned on, and that the transducer is plugged into the data acquisition unit.
2. Select **Lesson 12: Pulmonary 1—Volumes & Capacities (L012-Lung-1)** and type your name where appropriate. If you have a folder with your name already on the computer, select **Use it** if prompted.
3. Place a new bacteriological filter into the end with the sterile calibration syringe. Insert the syringe into the side of the sterile transducer labeled "Inlet."
4. Pull the plunger of the calibration syringe all the way out by holding the syringe by the body. Do not hold the unit by the transducer.
5. Click **Calibrate,** read all the directions, and click **OK.**
6. Push the plunger in and out for a total of 5 cycles (5 in and 5 out). This should take about 30 seconds.
7. Click on **End Calibration.**
8. The peaks should be even. If the calibration looks abnormal on the screen, click **Redo.** If it looks OK, begin recording the data.

Recording Data

1. Place a new sterile mouthpiece into the transducer where the calibration syringe was.
2. Pinch off your nose with a noseclip.
3. Click on **Record** and begin breathing easily into the transducer.
4. Take three normal breaths (**tidal volume**) and then inhale maximally and exhale normally (**inspiratory reserve volume**), as shown in fig. 34.3.
5. Take a few normal breaths again and then exhale maximally (**expiratory reserve volume**).
6. Take a few normal breaths; then inhale maximally and exhale maximally (**vital capacity**).
7. Click on **Stop**.

If you are not satisfied with your recording (compare what you see with fig. 34.3), click on **Redo.** If you are happy with your recording, click on **Done** and remove the bacteriological filter from the apparatus.

You can select **Review Saved Data** from the menu.

Data Analysis

Select **CH2** (which is the volume) and select **p–p** (peak to peak), which selects the range of minimum and maximum values. Select CH2 and click on "max," then select CH2 and click on "min." Select CH2 again and click on "delta." Compare your data with fig. 34.3.

Select the I-beam and highlight one of the tidal volume measurements from the peak to the depth of a breath. Save this data as "Tidal Volume."

Select the area from the peak of a normal inhalation to the maximum air inhaled and save the data as "Inspiratory Reserve Volume."

Find the point of a normal exhalation (after a tidal volume) to the complete exhalation and save the data as "Expiratory Reserve Volume."

Find the point of the maximum inhalation to the maximum exhalation and save the data as "Vital Capacity."

Enter your data from the BIOPAC Respiration Lab in the following section.

❓ Tidal volume: _____ (6)

❓ Inspiratory reserve volume: _____ (10)

❓ Expiratory reserve volume: _____ (7)

❓ Vital capacity: _____ (8)

Your percent expected vital capacity can be calculated the same as it was using wet spirometry or handheld spirometers. Record your percent expected vital capacity in the following space.

❓ Percent expected vital capacity: _____ (9)

Residual Volume

There is still air left in the lungs after a maximal exhalation. This is known as the residual volume (RV), and it is approximately 1,000 mL. The total lung capacity (TLC) is the sum of the vital capacity and the residual volume.

Minute Ventilation

Ⓐ **Activity**

The total amount of air inspired and expired in 1 minute of normal (tidal volume) pulmonary ventilation is known as the minute ventilation. This is calculated by multiplying the number of breaths per minute by the tidal volume. Record your minute volume in the following space.

❓ Minute volume: _____ (11)

Graphing Your Data

Ⓐ **Activity**

Construct a bar graph in chart 34.1 using your data. The vital capacity should be composed of the ERV, TV, and IRV.

Flow and Resistance

The flow of air is proportional to the difference in pressure of the air and inversely proportional to the resistance in the airways (Flow = Pressure/Resistance). As the pressure difference between the outside air and the air inside the lungs increases, the flow of air increases. If the resistance in the air passageways increases, then the flow decreases.

Ⓐ **Activity**

You can demonstrate the effects of changing the resistance of the airways in a simple demonstration. Do not do this experiment if you have a respiratory condition, such as asthma, a cold, or severe sinus allergies.

1. Using a stopwatch, a watch with a second hand, or the stopwatch function on a smartphone, record how long it takes to forcibly inhale the maximum amount of air your lungs can hold. Do this by breathing through both your mouth and your nose. Record the time in seconds in the following space.

❓ Time for maximum inhalation: _____ (12)

2. Close your mouth and one nostril and try the experiment again. Breathe only through one nostril and record the time it takes to maximally inhale. Record the time in seconds in the following space.

❓ Time for inhalation through one nostril: _____

_____ (13)

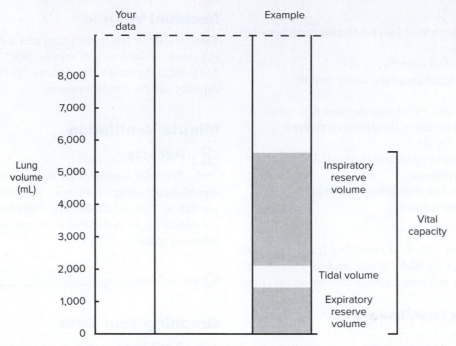

Your data   Example

Lung volume (mL)

8,000
7,000
6,000
5,000 — Inspiratory reserve volume
4,000
3,000 — Vital capacity
2,000 — Tidal volume
1,000 — Expiratory reserve volume
0

CHART 34.1 Shade in the chart with your own data from **BIOPAC** or spirometry. In the example, the IRV is 3,500 mL, the TV is 500 mL, and the ERV is 1,500 mL, for a VC of 5,500 mL.

? Closing off the respiratory passages (mouth and one nostril) increases the resistance in the respiratory system. How does this change the rate of airflow?

_____ (14)

? Asthma occurs due to a constriction of the bronchioles in the lung. If this occurs, is there a change in the pressure between the alveoli and the external air or an increase in the resistance to airflow into the lungs?

_____ (15)

Respiratory Sounds

(A) Activity

In this section, you listen to breathing sounds, which can be heard with a stethoscope.

1. Before you begin the experiment, clean the earpieces of the stethoscope with an alcohol wipe and let them dry. The stethoscope earpieces should point toward the anterior as you insert them into your ears.
2. Locate the larynx of your lab partner and place the diaphragm of the stethoscope just inferior to it.
3. Listen for the sound as your lab partner inhales and exhales. These are the tracheal and bronchial sounds.
4. Locate the triangle of auscultation (fig. 34.4), an area just medial to the inferior angle of the scapula. This is an ideal area for listening to sounds because the thoracic cage is covered by fewer muscles in this location.
5. Have your lab partner inhale and exhale deeply several times.

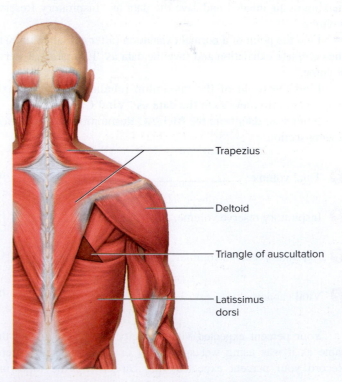

Trapezius

Deltoid

Triangle of auscultation

Latissimus dorsi

FIGURE 34.4 Triangle of Auscultation, Located in the Shaded Region.

6. Listen for a smooth flow of air into and out of the lungs. Wheezing and rattling are indicators of congestion in the lungs.
7. Record the sounds you hear in the following space and indicate the condition of your lab partner.

❓ Sounds heard: _____

_____ (16)

Acid-Base Effects of the Respiratory Gases

Carbon dioxide is a waste product of respiration. Increased carbon dioxide levels in the blood change the pH of the blood, and this can cause metabolic problems not only in the blood but elsewhere in the body such as in the brain, producing states of disorientation or confusion. In this part of the exercise, you examine the impact of elevated carbon dioxide levels in solution.

Carbon dioxide combines with water to form carbonic acid, which subsequently dissociates into bicarbonate and hydrogen ions, as represented in the following equation:

$$CO_2 \;+\; H_2O \longrightarrow H_2CO_3 \longrightarrow HCO_3^- \;+\; H^+$$

| Carbon dioxide | Water | Carbonic acid | Bicarbonate ion | Hydrogen ion |

Ⓐ **Activity**

You can see the effects of increasing carbon dioxide levels in the blood in the following experiment.

1. Add about 50 mL of litmus solution to a small (100 mL) Erlenmeyer flask. If the solution is red, add the household ammonia solution drop by drop until the color just begins to turn blue.
2. Wear safety goggles and insert a drinking straw into the flask and *gently* blow air into the flask.
3. Bubble your exhaled breath into the flask and look for a color change. This may take a few minutes. A blue color indicates an alkaline condition, and a red color indicates an acid condition.
4. Using the preceding formula, determine how the exhalation into the flask alters the acid-base conditions.

Cardiopulmonary Resuscitation

Cardiopulmonary resuscitation, or **CPR,** is typically used for people suffering from myocardial infarcts (heart attacks), drug overdoses, drowning or trauma, and obstruction of the airways, among other things. This is a technique that uses chest compression of about 100 times per minute on the body of the sternum. Trained professionals follow 30 chest compressions with two mouth-to-mouth ventilations. If untrained in CPR, then the chest compressions alone may save someone's life. When the lungs are temporarily nonfunctional, CPR may keep a person alive until

medical help arrives. This is the current CPR procedure for health professionals at the time of this writing.

It is important to have proper training in CPR before you perform it. Incorrect technique can cause injury.

Virtual Lab

𝗣𝗵𝗜𝗟𝗦

Respiratory System Pulmonary Function Tests

1. Select "Respiratory System Pulmonary Function Tests" assigned by your instructor.
2. Read "Before you begin."
3. In the Understand the Graphs and Spirometry section of this exercise, note that this graph is a little different than most graphs. First of all, these three loops, in reality, are overlapping and not sequential from left to right. Second, in the graph itself the flow on the vertical (y-axis) represents the flow of air out of the lungs (fig. 34.5). When you inhale, the flow out of the lungs is negative and the curve moves down (below the baseline, which is the horizontal or x-axis). When you exhale, the flow out of the lungs is positive and the curve moves up. For the volume, the graph might seem backward because the right-hand side of the graph represents low volume, while the left-hand side of the horizontal (x-axis) line is maximum lung capacity. The maximal lung capacity is variable due to the size and gender of the individual. The tip of the blue arrow in the Normal graph in the program represents about 2 liters, which is the residual capacity of the lungs. When you inhale, the flow "out of the lungs" is negative and the curve dips below the baseline. The lungs increase in volume, so the graph is moving in a clockwise direction. When you exhale, the curve moves sharply up as the air flows out of the lungs quickly, but as you continue to exhale, the flow slows down and the curve returns to the baseline. The flow stops at the baseline and the volume returns to about 2 liters.
4. Continue to the Laboratory Simulation.
5. In Phase 1, select the magnifying glass with the plus sign icon on the simulated computer monitor. Read the patient history.
6. Click on the Lab Data Icon below the Patient screen and enter the preliminary diagnosis.
7. Click on the magnifying glass with the minus sign to return to the patient.
8. Click on the spirometer, and then begin lung volume analysis.
9. Compare the lung volumes of the patient with normal volumes and make a diagnosis in the lab data section.
10. Go back to the patient and proceed with flow-volume loop analysis.
11. Make a determination of the final diagnosis and if there is lung disease or not.
12. Continue to Phase 2 to repeat the same process for the remaining patients.

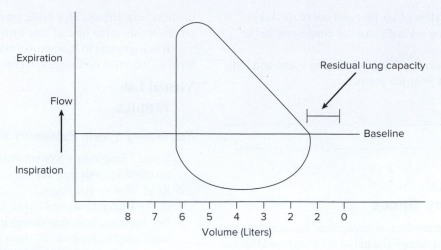

FIGURE 34.5 Pulmonary Function Test.

REVIEW SECTION

Respiratory Physiology

Name _____ *Date* _____

Lab Section _____ *Time* _____

❓ Chapter Summary Data

Use this section to record your results from questions within the exercise.

1. _____   9. _____

2. _____   10. _____

3. _____   11. _____

4. _____   12. _____

5. _____   13. _____

6. _____   14. _____

7. _____   15. _____

8. _____   16. _____

Review Questions

1. What is pulmonary ventilation? _____
_____

2. What is the pressure difference between the external air and the pleural cavity when inhalation just begins? _____
_____

3. What was your lab partner's measured breathing rate in breaths per minute? _____

4. Define *tidal volume*. _____

5. What instrument do you use to measure breathing volumes? _____

6. If you inhale maximally, what is the name of the capacity of air that you completely exhale?

_____

7. Examine the predicted vital capacity chart. What is the approximate percent decrease of vital capacity in the same individual from age 25 to age 75? _____

8. How does the decrease in vital capacity potentially influence an individual's athletic performance or aerobic condition as aging occurs?

_____

9. Calculate the IRV of an individual with a vital capacity of 4,400 mL, an expiratory reserve volume of 1,300 mL, and a tidal volume of 500 mL.

_____

10. How does excess carbon dioxide change the acid-base condition of a solution?

_____

_____

Physiology of Exercise and Pulmonary Health

INTRODUCTION

An understanding of exercise physiology is important not only for athletic training but also for wellness and general health. Significant changes occur during exercise that affect many organ systems of the body. The changes vary, depending on the type of exercise. Exercise is broadly classified as either **aerobic exercise** or **anaerobic exercise.** Aerobic exercises increase heart rate and breathing rates at moderate levels for extended periods of time. In aerobic exercise, the cardiovascular and respiratory systems of the body deliver adequate amounts of oxygen to cells as the cells increase their demand for oxygen. This occurs, for example, when muscles contract and increase their metabolic demands. Anaerobic exercises result in the consumption of available oxygen faster than it can be supplied to the muscle tissue. The tissue uses glucose anaerobically and produces lactic acid. Exercise has a profound effect on the muscular, skeletal, cardiovascular, and respiratory systems. Consider the increased metabolic demands placed on skeletal muscles during repeated contractions. Skeletal muscle is more metabolically active when contracting than when at rest; it uses additional oxygen and nutrients. Oxygen diffuses from the blood to the muscle tissue due to the concentration gradient of oxygen between these two areas. In addition to the diffusion of oxygen, the arterioles of the cardiovascular system dilate, and the precapillary sphincters open, providing greater blood flow to the muscles. This additional blood flow requires an increase in the volume of blood passing through the heart with each contraction. Heart size increases and pulse rate decreases with long-term rigorous exercise.

The respiratory system responds to this greater oxygen demand with increased volume per breath and a greater number of breaths per minute. This increases the minute volume. The lung capillaries expand as well, and a greater diffusion of oxygen occurs between the alveoli and the blood capillaries. This respiratory response is covered in the Saladin text in chapter 22, "The Respiratory System." In this exercise, you examine respiratory health, the effects of exercise on the body, and the body's responses to aerobic exercise.

OBJECTIVES

At the end of this exercise, you should be able to

1. list the major organ systems directly involved in fitness;
2. describe basic physiological differences between a person who is physically active and one who is physically inactive;
3. determine the forced expiratory volume exhaled in 1 second;
4. use a hand-held flow air transducer to measure FEV_1;
5. calculate the personal fitness index of a subject who performs the Harvard step test or Cooper's 12-minute run test;
6. record the Waist/Hip ratio of a person and determine the health risk of that individual.

MATERIALS

BIOPAC Setup

AFT6 or AFT6A 6,000 mL calibration syringe

AFT1 disposable bacteriological filter

AFT2 disposable mouthpiece

SS11L or SS11LA—airflow transducer

MP30 or MP150 data acquisition unit

Compatible computer

Noseclips

16-inch step

20-inch step

Metronome with second hand or smartphone with timer function

Treadmill

Caution!

The risk of contracting SARS-CoV-2 (the COVID-19 virus), tuberculosis, or other respiratory diseases is of concern for this lab. Any equipment used in this lab to measure respiratory ventilation (breathing) must be sterilized before **any student** uses the equipment. Your instructor will determine whether the following exercises will be done by demonstration, if a virtual lab is to be conducted, or if the exercises will be done by individual students if enough equipment and protocols are in place to perform these exercises in the safest possible manner. Under no circumstances should non-sterile equipment be used in performing these exercises. Recommendations for the sterilization of equipment can be found on the Centers for Disease Control and Prevention (CDC) website at https://www.cdc.gov/infectioncontrol/pdf/guidelines/disinfection-guidelines-H.pdf.

PROCEDURE

Pulmonary Health

Forced Expiratory Volume (FEV)

Indications of health can be roughly correlated with the forced expiratory volume in one second (FEV_1) or amount of air expelled from the lungs in 1 second. This can have clinical value when it is compared to the vital capacity and is expressed as a percentage of the person's vital capacity (VC) as percent FEV_1/VC. In healthy adults, this is approximately 75% of the VC. In this exercise, you will use the BIOPAC setup, which is similar to the one used in Exercise 34. A smaller-than-normal forced expiratory volume may be caused by asthma, emphysema, or other pulmonary conditions.

BIOPAC Lesson 13—Pulmonary Function II

(A) Activity

1. Setup

Make sure the computer is on but the MP30 or MP150 data acquisition unit is off. Plug in the sterile airflow transducer to the data acquisition unit and turn the data acquisition unit on. Open the BIOPAC Student Lab (BSL) Software and select **BIOPAC Lesson 13—Pulmonary Function II.** You will collect data for your respiratory volumes by breathing into a flow meter that measures the force of air passing through it, which it then will convert to a volume. The resulting "spirogram" is displayed on the computer monitor and saved as a file in your BIOPAC folder. The following are the steps you will take.

Click on the BIOPAC icon on the computer desktop to start BIOPAC. A menu of lessons will appear. Select **BIOPAC Lesson 13—Pulmonary Function II.** Type in your (folder) **name.** If you have a folder on this computer station, a window

should appear with the message "A folder with this name already exists. Would you like to use it or create a new folder?" Choose **Use it.**

Place the new bacterial filter onto the end of the calibration syringe. Insert the sterile calibration syringe/filter assembly into the side of the airflow transducer labeled **"Inlet."**

2. Calibration

Pull the calibration syringe plunger all the way out and hold the calibration syringe horizontal, so that the airflow transducer is upright as in fig. 35.1. It must remain vertical for the calibration and experimentation, because the wire screen is sensitive to changes in deflection due to gravity. Hold the calibration syringe by the barrel (body), not by the airflow transducer. Click on **Calibrate.** After the first stage of the calibration is recorded, open the dialog box and read all the directions. Do so and click on **Yes.** Cycle the syringe plunger in and out 5 times (10 strokes). Use a rhythm of about 1 second per stroke, with 2 seconds between strokes. Click on **End Calibration.** If the calibration looks good as seen in fig. 35.2, then detach the calibration syringe and proceed to Step 3 **(Record).** (You should be able to see 5 downward and 5 upward deflections on the screen.) If the calibration peaks are uneven or significantly vary from one another, click **Redo.**

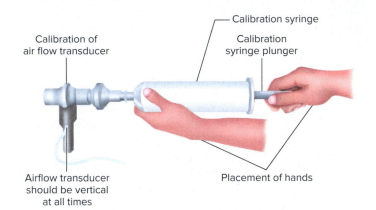

FIGURE 35.1 Calibration of Air Flow Transducer.

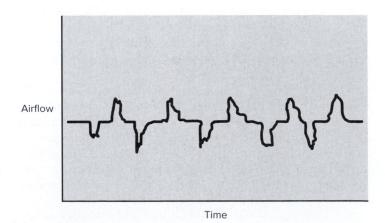

FIGURE 35.2 Calibration Should Have Level Baseline with Upper and Lower Peaks.

3. Recording Data

Insert a sterile mouthpiece (where the calibration syringe had been attached) and a new bacteriological filter on the end of the transducer. Place a clean noseclip on your nose. Make sure you hold the transducer vertically at all times. Click on **Record** and breathe as follows (be sure to start the recording before you start the breathing experiment):

1. Take three normal breaths (three inhales and three exhales).
2. Inhale as deeply as you can (take in as much air as you can). Hold it for an instant.
3. Exhale as fast and as much as you can.
4. Click on **Stop**.

If the recording is not good, click on **Redo** to repeat it. If the recording looks good, as shown in fig. 35.3, click on **Done**. Select **Record from Another Subject** or, if you have recorded everyone's data, select **Analyze Current Data File**.

4. Data Analysis for FEV

Open your folder in the **Data Files** folder (probably listed as *"your name FEV-L13"*) and then open the data file. Select **Display Preferences** from the **File** menu, choose **Grids,** and click on **Show Grids** and then **OK**. Use the **I-beam** to highlight your graph. The **p–p** (peak to peak) will represent your vital capacity, as represented in fig. 35.4.

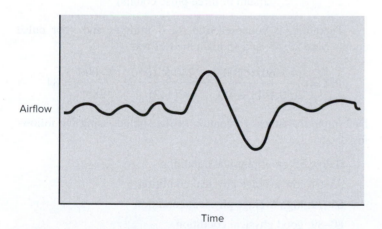

FIGURE 35.3 Spirogram Recording.

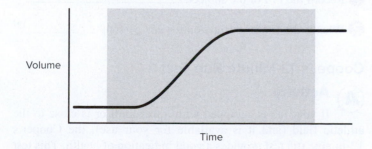

FIGURE 35.4 Peak-to-Peak Determination of Vital Capacity.

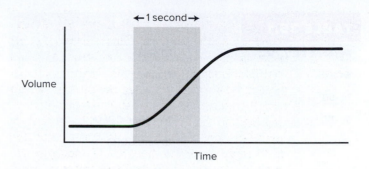

FIGURE 35.5 FEV$_1$ Highlight 1 Second of Time from Beginning of Expiration.

Use the **I-beam** to select the volume of air you exhaled for the first second. This will be the FEV$_1$, as shown in fig. 35.5. Determine the percent of the volume of air exhaled in 1 second compared to the vital capacity by the following:

$$\text{Percent} = \frac{\text{Volume exhaled in 1 second}}{\text{Forced vital capacity volume}} \times 100\%$$

Calculate your percent FEV$_1$ compared to the vital capacity and enter the value here:

? _____ (1)

Save and print the entire spirogram you record to include in your lab report.

5. Interpretation

Spirometry is an important measurement tool to assist in diagnosing asthma, chronic obstructive pulmonary disease (COPD), and cystic fibrosis.

Pulmonary obstruction occurs when the respiratory passages are narrowed. This can be due to conditions such as asthma, excess mucus, and inflammation, such as bronchitis. Most COPD in North America is caused by smoking. *In these cases, the vital capacity of the individual is normal, but the percent FEV$_1$/VC is low.* It takes the subject longer to exhale completely.

Pulmonary restriction occurs when the lungs cannot fully inspire or expire the full volume of air. In these cases, the vital capacity of the individual is reduced. This can be due to fibrosis of the lungs (cystic fibrosis or fibrosis due to asbestos or silica); scarring of the lung tissue, as when a subject has had chronic lung infections; adhesions of the lung to the chest wall due to extreme emphysema; or removal of a section of lung. It may also be due to damage to the phrenic nerve, which stimulates the diaphragm. In these cases, the vital capacity of the individual is low, but the percent FEV$_1$/VC is normal. Clinical values for FEV$_1$ compared to vital capacity are listed in table 35.1.

TABLE 35.1	
Percent FEV₁/VC	**Status**
100–75	Normal
74–60	Mild COPD
59–50	Moderate COPD
<50	Severe COPD

Exercise Physiology

The measurement of fitness in this part of the exercise is on a voluntary basis. It is best if the class can obtain data from people who regularly participate in aerobic exercises (three to six times per week) and from people who do not exercise. The Harvard step test and the Cooper's 12-minute run test are two reliable measurements of fitness.

Caution! Do not do these exercises if you are at risk for heart disease or have a family history of heart disease or another condition, such as asthma, for which exercise is harmful. Stop if you feel exhausted or faint. If you develop chest pain or pain radiating down the left arm, seek medical attention immediately.

A correlation exists between heart rate and fitness level. The resting heart rate of people involved in regular, active aerobic exercise is lower than the rate of sedentary people. In addition, the heart recovers faster in people who have a regular exercise program than in those who do not exercise. In this experiment, record the measurements of at least two volunteers from the class. The greater the number of students who participate, the better the data. Read the entire exercise before beginning.

Harvard Step Test

(A) Activity

The Harvard step test was developed during World War II at Harvard University to determine a person's physical fitness. In this exercise, students should select either the 20-inch step for people 5′8″ or taller or the 16-inch step for people under 5′8″. With one person acting as an observer, the subject should step up on the step in 1 second and down on the floor in another second, thus completing 30 complete cycles in 1 minute. The subject should keep the body upright and keep pace with the observer's count or with a metronome set at 60 beats per minute. The subject should exercise for at least 3 minutes but no more than 5 minutes. If the subject stops due to exhaustion, then the observer should note the time of exercise. After the period of exercise, the subject should sit and rest for 1 minute. The pulse should be taken from 1 minute to 1 minute 30 seconds and recorded in the following space. (See "A" in fig. 35.6.)

Pulse from 1 minute to 1 minute 30 seconds: _____

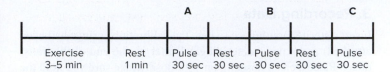

FIGURE 35.6 Harvard Step Test Periods.

The subject should rest for another 30 seconds before the pulse is taken from 2 minutes to 2 minutes 30 seconds after exercise and recorded here. (See "B" on fig. 35.6.)

Pulse from 2 minutes to 2 minutes 30 seconds: _____

The subject should remain seated for another 30 seconds before the pulse is taken from 3 minutes to 3 minutes 30 seconds and recorded here. (See "C" on fig. 35.6.)

Pulse from 3 minutes to 3 minutes 30 seconds: _____

Add the three pulse counts recorded.

? Sum of three pulse counts: _____ (2)

To determine the personal fitness index (PFI), use the following formula:

$$PFI = \frac{\text{Number of seconds of exercise}}{2(\text{sum of three pulse counts})} \times 100$$

Therefore, if you exercised for 4 minutes and your pulse counts were 50, 48, and 45, then the PFI was

$$PFI = \frac{(4 \times 60) \times 100}{2(50 + 48 + 45)} = \frac{240 \times 100}{2(143)} = \frac{24,000}{286} = 84$$

Determine the fitness evaluation of the subject using the following data:

Below 55: poor physical condition

55–64: low average physical condition

65–79: high average physical condition

80–89: good physical condition

90 and above: excellent physical condition

? Record the PFI of the subject: _____ (3)

? What is the fitness evaluation for this person? _____ (4)

Cooper's 12-Minute Run Test

(A) Activity

If your lab is equipped with a treadmill or is close to the athletic field (and it is available for your use), the Cooper's 12-minute run test provides a good indication of health. This test involves determining how far a person can run and/or walk in 12 minutes.

Caution! As with the Harvard step test, do not do these exercises if you are at risk for heart disease or have a family history of heart disease or another condition, such as asthma, for which exercise is harmful.

Procedure Warm up for 15 minutes prior to doing this test. Find a suitable running track or use a treadmill that is set to no incline and run or walk as fast and as far as you can in 12 minutes. You probably will not be able to sprint for the entire time, so pace yourself. If you cannot run anymore during the test, you should finish the test by walking or jogging or a combination of running and walking.

❓ Record the distance you traveled in 12 minutes:

_____ (5)

The results of your test and general fitness can be judged by the following table.

Under Age 40

Distance in 12 Minutes	Physical Condition
Over 2,700 m	Excellent
2,300 m to 2,700 m	Good
1,900 m to 2,299 m	Average
1,500 m to 1,899 m	Below average
Less than 1,500 m	Poor

Over Age 40

Distance in 12 Minutes	Physical Condition
Over 2,500 m	Excellent
2,100 m to 2,500 m	Good
1,700 m to 2,099 m	Average
1,500 m to 1,699 m	Below average
Less than 1,400 m	Poor

Target Heart Rate Zone for Exercise

(A) **Activity**

The target heart rate zone for exercise is 60–80% of the maximum heart rate (MHR) for healthy adults. The maximum heart rate for an individual is his or her age subtracted from 220. You can calculate a basic target heart rate zone with the following procedure:

❓ 1. To calculate your maximum heart rate, subtract your age from 220. Enter the MHR value here: _____ (6)

2. For 60% of the MHR, multiply the MHR by 0.6 and enter it in the space: _____

3. For 80% of the MHR, multiply the MHR by 0.8 and enter it in the space: _____

4. Your target heart rate zone for exercise should be between these two values.

Health Guidelines

The American Heart Association recommends that healthy adults from the ages of 18 to 65 need moderate exercise (such as brisk walking) 30 minutes a day, 5 days a week, or vigorous exercise (such as jogging or running) for 20 minutes 3 days a week. Physical inactivity is a major public health issue. Any person of normal mobility who is inactive can increase his or her fitness with a program of exercise that may start simply with walking. Gradual increase in the intensity and duration of exercise will improve the general health of the individual.

Body Mass Index

(A) **Activity**

The body mass index (BMI) is a general guide to fitness that makes a couple of assumptions. One is that the person is of average build. Fitness level, gender, muscle mass, bone structure, and ethnicity can all influence the BMI. One way to get a general idea of the BMI is to use this calculation:

$$BMI = \frac{\text{Weight in pounds}}{(\text{Height in inches})^2} \times 703$$

You can also use the metric system calculator:

$$BMI = \frac{\text{Mass (Kg)}}{\text{Height (M)}^2}$$

❓ Record your BMI here: _____ (7)

According to the National Heart, Lung, and Blood Institute, the BMI for average adults can be interpreted this way:

- Underweight = <18.5
- Normal weight = 18.5–24.9
- Overweight = 25–29.9
- Obese = BMI of 30 or greater

If a person has a high muscle mass, the BMI values may be overestimated because muscle will show a person with higher weight than predicted. In older individuals who have less muscle mass, the BMI may be underestimated.

Waist/Hip Ratio

(A) **Activity**

Another way to calculate fitness is to use the waist/hip ratio (WHR). According to the American Heart Association (AHA), people who carry more weight in their waist region (with "apple-shaped bodies") are more at risk for health problems than people with more weight in their hips (with "pear-shaped bodies"). Of course, increased weight of any kind is a health risk. You can use a flexible tape measure (US standard or metric) and do the following:

1. Measure the circumference of your hips at their widest part. Record the value.

Circumference of hips: _____

2. Measure your waist just superior to the umbilicus (belly button). Record the value.

   Circumference of waist: _____

3. Use the formula WHR = $\dfrac{\text{Circumference of waist}}{\text{Circumference of hips}}$

   Calculate the WHR and record the value.

❓ WHR: _____ (8)

4. Compare it with table 35.2 and determine your health risk.

❓ Health risk: _____ (9)

TABLE 35.2		Waist/Hip Ratio
Females	**Males**	**Health Risk Based on WHR**
0.80 or below	0.95 or below	Low risk
0.81 to 0.85	0.96 to 1.00	Moderate risk
0.85+	1.0+	High risk

Personal Goals

If your personal fitness index or body mass index indicates you are out of shape or overweight, list three things you can do to improve your fitness.

REVIEW SECTION

Physiology of Exercise and Pulmonary Health

Name _____ *Date* _____

Lab Section _____ *Time* _____

❓ Chapter Summary Data

Use this section to record your results from questions within the exercise.

1. _____ 6. _____

2. _____ 7. _____

3. _____ 8. _____

4. _____ 9. _____

5. _____

Review Questions

1. Define percent FEV_1/VC. _____

2. Record your percent FEV_1/VC or the one you measured in lab. _____

3. Does your percent FEV_1/VC value fall within normal limits? _____

4. How does the percent FEV_1/VC compare in a person with a pulmonary obstructive condition, such as asthma? Explain this difference.

   _____

   _____

5. How does the percent FEV_1/VC compare in a person with a pulmonary restrictive condition, such as asbestosis? Explain this difference.

   _____

   _____

6. If a person had a smaller body and therefore a smaller vital capacity, would the percent FEV_1/VC necessarily change?

   _____

7. What is your personal fitness index or the one measured in lab? _____

8. If the heart rate after 5 minutes of exercise was 70, 68, and 66 beats in the consecutive 30-second trials, what was the personal fitness index and what condition does that represent? _____

_____

9. Record the results of the Harvard step test or the Cooper's 12-minute run test in the spaces provided below for the members of the class who performed them. Next to the Personal Fitness Index (Harvard) or Physical Condition (Cooper's) for your own results, write an *S* if you smoke cigarettes or an *N* if you are a nonsmoker. Evaluate the results with your classmates.

PFI (Personal Fitness Index)
Physical Condition

1. _____   13. _____

2. _____   14. _____

3. _____   15. _____

4. _____   16. _____

5. _____   17. _____

6. _____   18. _____

7. _____   19. _____

8. _____   20. _____

9. _____   21. _____

10. _____   22. _____

11. _____   23. _____

12. _____   24. _____

Urinary System

INTRODUCTION

The organs of the urinary system consist of two kidneys, two ureters, a single urinary bladder, and a single urethra. The urinary system filters dissolved material from the blood, regulates electrolytes and fluid volume, concentrates and releases waste products, and reabsorbs metabolically important substances back into the circulatory system. **Filtration** occurs when one or more substances pass through a selectively permeable membrane while others do not. Filtration in the kidney involves both metabolic waste products (urea) and material beneficial to the body. Not all filtered material is desirable to have **excreted** from the body. Glucose and other materials, such as sodium and potassium ions, are **reabsorbed** from the kidney back into the circulatory system. The kidney **secretes** urea; some drugs; hydrogen and hydroxide ions; hormones, such as erythropoietin; and enzymes such as renin. Finally the kidneys excrete metabolic wastes, hydrogen ions, toxins, water, and salts. These topics are discussed in the Saladin text in chapter 23, "The Urinary System." In this exercise, you examine the gross and microscopic anatomy of the urinary system, study the major organs as represented in humans, and dissect a mammal kidney if available.

OBJECTIVES

At the end of this exercise, you should be able to

1. list the major organs of the urinary system;
2. describe the blood flow through the kidney;
3. describe the flow of filtrate through the kidney;
4. name the major parts of the nephron;
5. trace the flow of urine from the kidney to the exterior of the body;
6. distinguish between the parts of the nephron in histological sections;
7. compare male and female urinary anatomy.

MATERIALS

Models and charts of the urinary system

Models and illustrations of the kidney and nephron system

Microscopes

Microscope slides of kidney and bladder

Samples of renal calculi (if available)

Preserved specimens of sheep or other mammal kidney

Dissection Trays and Materials

Scalpels

Forceps

Blunt (mall) probes

Protective gloves

Waste container

PROCEDURE

Overview

You can begin the study of the urinary system by locating its principal organs. Look at fig. 36.1 and compare it to the material available in the lab. Find the **kidneys,** the **ureters,** the **urinary bladder,** and the **urethra.**

Kidneys

The kidneys are **retroperitoneal** (located posterior to the parietal peritoneum) and are embedded in the **perirenal fat capsule.** The capsule cushions the kidneys, which are found mostly below the protection of the rib cage. The kidneys are located adjacent to the vertebral column about at the level of T12 to L3. The right kidney is slightly more inferior than the left.

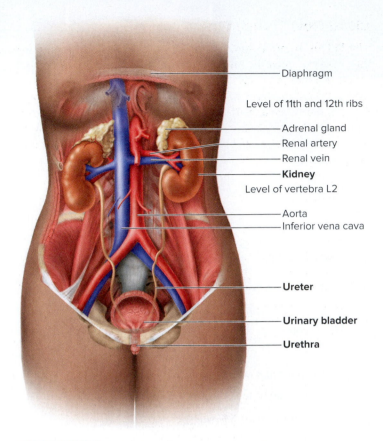

FIGURE 36.1 **Major Organs of the Urinary System**

Labels:
- Diaphragm
- Level of 11th and 12th ribs
- Adrenal gland
- Renal artery
- Renal vein
- **Kidney**
- Level of vertebra L2
- Aorta
- Inferior vena cava
- **Ureter**
- **Urinary bladder**
- **Urethra**

 Activity

1. Examine a model of the kidney and compare it to fig. 36.2.
2. Locate the outer **fibrous capsule,** a tough connective tissue layer; the outer **cortex;** and the inner **medulla** of the kidney. The kidney has a depression on the medial side where the renal artery enters the kidney and the renal vein and the ureter exit the kidney. This depression is called the **hilum.**
3. Examine a coronal section of the kidney, and find the **renal (medullary) pyramids,** separated by the **renal columns.** Each renal pyramid ends in a blunt point called the **renal papilla** (see fig. 36.2). Urine drips from many papillae toward the middle of the kidney.

   The urine drips into the **minor calyces** (singular, *calyx*), which enclose the renal papillae. Minor calyces are somewhat like funnels that collect fluid. Minor calyces lead to the **major calyces,** and these, in turn, conduct urine into the large **renal pelvis.** The renal pelvis is found in the **renal sinus.** The renal pelvis is like a glove in a coat pocket. The pocket is the renal sinus, and the membranous glove that occupies the space is the renal pelvis.
4. Locate these structures in fig. 36.2. The renal pelvis is connected to the ureter at the medial side of the kidney.

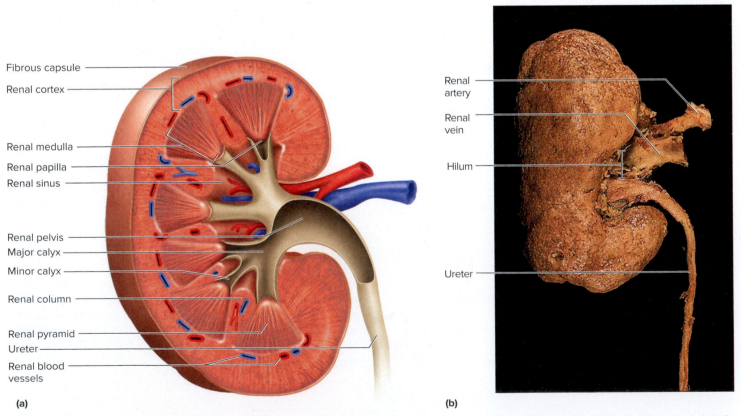

Labels (a):
- Fibrous capsule
- Renal cortex
- Renal medulla
- Renal papilla
- Renal sinus
- Renal pelvis
- Major calyx
- Minor calyx
- Renal column
- Renal pyramid
- Ureter
- Renal blood vessels

Labels (b):
- Renal artery
- Renal vein
- Hilum
- Ureter

(a)   (b)

FIGURE 36.2 **Gross Anatomy of the Kidney, Entire and Coronal Section.** (a) Diagram of coronal section of the kidney and details of renal pyramid; (b) photograph of entire kidney; (c) photograph of coronal section.
(b), (c) ©Eric Wise

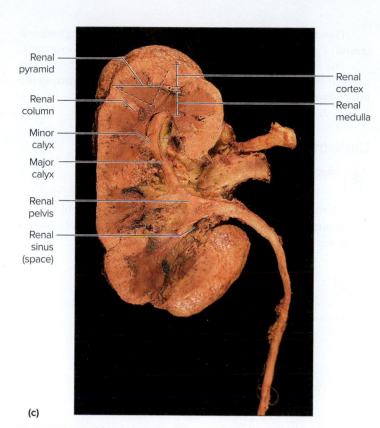

Renal pyramid
Renal column
Minor calyx
Major calyx
Renal pelvis
Renal sinus (space)
Renal cortex
Renal medulla

(c)

FIGURE 36.2 *Continued.*

Blood Flow Through the Kidney

The kidney filters material from the blood and returns important material such as water, glucose, and sodium back to the blood. It is not a perfect system because some urea is also returned to the circulatory system. The blood flow in the kidney forms a portal system (defined as a group of blood vessels in which blood flows from one capillary bed to another capillary bed with an arteriole or venule between them prior to returning to the heart).

(A) Activity

Examine models or charts in lab and look at the renal vascular system in fig. 36.3. The first vessel to enter the kidney comes from the abdominal aorta, and is the **renal artery,** which branches in the kidney to form the **segmental arteries,** located in the renal sinus. Blood flows from here to the **interlobar arteries,** which pass between the lobes of the kidney in the renal columns before bending abruptly to form the **arcuate arteries,** named for forming an arc between the cortex and the medulla. From here the arteries

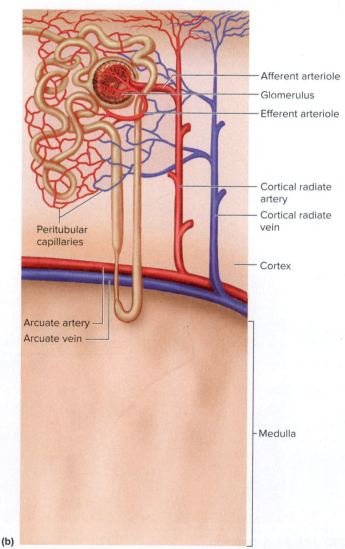

Afferent arteriole
Glomerulus
Efferent arteriole
Cortical radiate artery
Cortical radiate vein
Cortex
Peritubular capillaries
Arcuate artery
Arcuate vein
Medulla

(b)

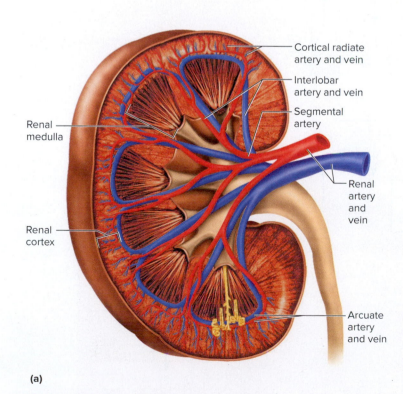

Cortical radiate artery and vein
Interlobar artery and vein
Segmental artery
Renal artery and vein
Renal medulla
Renal cortex
Arcuate artery and vein

(a)

FIGURE 36.3 Blood Flow Through the Kidney. (a) Major arteries and veins of the kidney; (b) detail of smaller vessels.

become the **cortical radiate (interlobular) arteries** that enter the renal cortex.

The cortical radiate arteries branch and form the **afferent arterioles,** which take blood to the **glomerulus,** a cluster of capillaries where filtration occurs. From there, blood travels through the **efferent arterioles** and then to the **peritubular capillaries,** where reabsorption and secretion take place. In regions of the cortex near the medulla, there are other vessels that branch from the efferent arteriole. These are the **vasa recta.** They represent only a small number of capillaries in the kidney, but they are important in producing a concentrated urine by reabsorption and secretion. The vasa recta are found in association with special nephrons called **juxtamedullary nephrons.**

Thus there are three capillary beds in the kidney, the glomeruli (plural of *glomerulus*), the peritubular capillaries, and the vasa recta.

The return flow to the heart occurs as blood returns via the **cortical radiate (interlobular) veins,** to the **arcuate veins,** to the **interlobar veins,** and to the **renal vein,** which leads to the inferior vena cava.

Ultrastructure of the Kidney

Ⓐ **Activity**

1. Before you examine the sections of kidney under the microscope, first become familiar with the structure of the **nephron.** Examine models and charts in the laboratory and compare them to fig. 36.4.

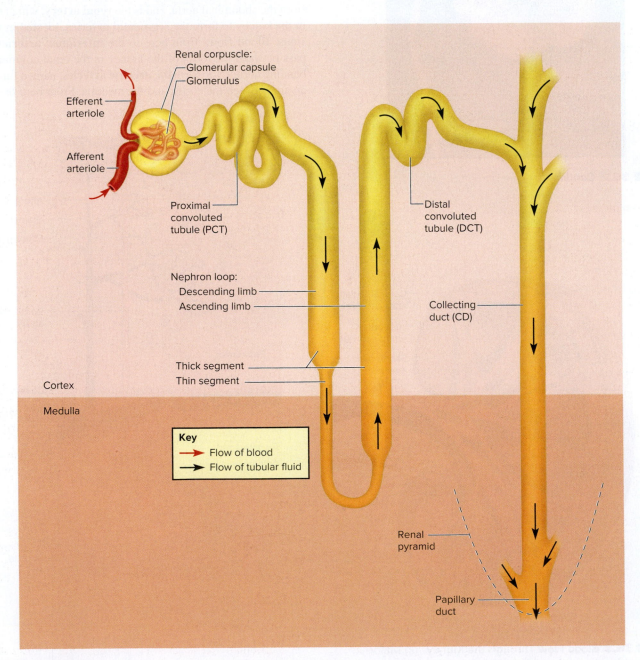

FIGURE 36.4 Nephron.

2. Locate the **glomerular capsule (Bowman capsule),** the **proximal convoluted tubule,** the **nephron loop (loop of Henle),** and the **distal convoluted tubule.** The glomerulus and the glomerular capsule are known together as the **renal corpuscle.** The nephron consists of the renal corpuscle, proximal convoluted tubule, nephron loop, and distal convoluted tubule. Blood travels to the nephron via the afferent arteriole. When the blood reaches the glomerulus, the plasma is filtered by blood pressure, forcing fluid across the capillary membranes. Blood pressure is the driving force for renal filtration, and this pressure is opposed by hydrostatic pressure in the glomerular capsule space outside of the glomerular capillaries and the osmotic pressure that occurs due to the proteins left in the blood plasma. This filtered fluid, present in the nephron, is called **filtrate.**

As the filtrate flows through the nephron, water, glucose, and many electrolytes are returned to the blood. The urea is concentrated as it passes through the nephron and the **collecting duct,** a tube that receives the fluid from the nephrons. The amount of reabsorption in the kidney varies with the material that flows through it. In normal conditions, glucose is completely reabsorbed into the blood. About 99% of the water in the filtrate is reabsorbed. Only 50% of the urea is excreted into the urine. The remainder of the urea returns to the blood and is filtered again. Urea is concentrated in the collecting duct, and some of it diffuses into the medulla, increasing the osmolarity in the medulla. This increases the flow of water out of the nephron, thus concentrating the urine.

There are two types of nephrons in the kidney. The majority of the nephrons are found in the cortex of the kidney and are thus called **cortical nephrons.** Those found in the medulla are called **juxtamedullary nephrons** and are far fewer in number, some extending to near the tip of the renal pyramids.

In summary, urine is produced from filtered blood. Some of the liquid portion of blood flows from the glomerulus across the glomerular capsule. Materials valuable to the body, such as glucose and other solutes, are reabsorbed by the nephron and return to the circulatory system by the peritubular capillaries. The main metabolic by-product, urea, is removed from the kidney and passes as urine from the collecting ducts to the minor calyces. The volume of urine and some of the constituents found in urine are controlled by hormones such as aldosterone and antidiuretic hormone (ADH). Aldosterone increases the reabsorption of sodium and reduces water volume. Antidiuretic hormone causes the distal convoluted tubules to reabsorb water, also decreasing the urine output.

Microscopic Examination of the Kidney

Ⓐ **Activity**

1. Examine a kidney slide under low power. You should see the cortex of the kidney, which has a number of round **glomeruli** composed of capillary tufts. The medullary region of the kidney slide has open, parallel spaces called **collecting ducts.** Compare the slide to fig. 36.5a.

2. Examine the slide under higher magnification and locate the glomerulus and the **glomerular capsule.** The capsule is composed of simple squamous epithelium and specialized cells called podocytes.

3. Examine the outer edge of the capsule around the glomerulus. If you move the slide around in the cortex, you should find the proximal convoluted tubules with the **brush border** or microvilli on the inner edge of the tubule. The inner surface of the tubule appears fuzzy. The microvilli increase the surface area of the proximal convoluted tubule.

The distal convoluted tubules do not have brush borders; therefore, the inner surface of a tubule does not appear fuzzy. The cells of the distal convoluted tubules generally have darker nuclei and relatively clear cytoplasm when compared to the cells of the proximal convoluted tubules, as seen in fig. 36.5b.

4. Examine the medulla of the kidney under high magnification and locate the thin-walled nephron loop and the larger-diameter collecting ducts, as seen in fig. 36.5c.

Dissection of the Sheep Kidney

Ⓐ **Activity**

1. Rinse a sheep kidney in a bucket of clean water and place it on a dissection tray along with dissection equipment. Examine the outer **capsule** of the kidney. You may see some tubes coming from a dent in the kidney. The dent is the **hilum,** and the tubes are the **renal artery, renal vein,** and **ureter.** The renal artery is smaller in diameter and has a thicker wall than the renal vein. The ureter has an expanded portion near the hilum.

2. Make an incision in the sheep kidney a little off center in the coronal plane (fig. 36.6). This section allows you to better see the interior structures of the kidney.

3. Locate the **renal cortex,** the outer layer of the kidney. You should also locate the **renal medulla,** which has triangular regions known as the **renal pyramids.** At the tip of each pyramid is a **papilla.** Urine from the papillae drips into the **minor calyx.** Many minor calyces lead to a **major calyx.** The **major calyces** take fluid to the **renal pelvis.**

4. Lift the renal pelvis somewhat to pull it away from the **renal sinus.** The sinus is the space in the kidney, which may be filled with adipose tissue. You should also examine the exit of the renal pelvis as it becomes the **ureter.**

5. When you are finished with the dissection, place the material in the proper waste container provided by your instructor.

Ureters

The ureters are long, thin tubes that conduct urine from the kidneys to the urinary bladder. The ureters have **urothelium (transitional epithelium)** as an inner lining and smooth muscle in their wall. Urine is expressed by peristalsis from the kidney to the urinary bladder.

Ⓐ **Activity**

Examine the models in the lab and compare them to figs. 36.1 and 36.6.

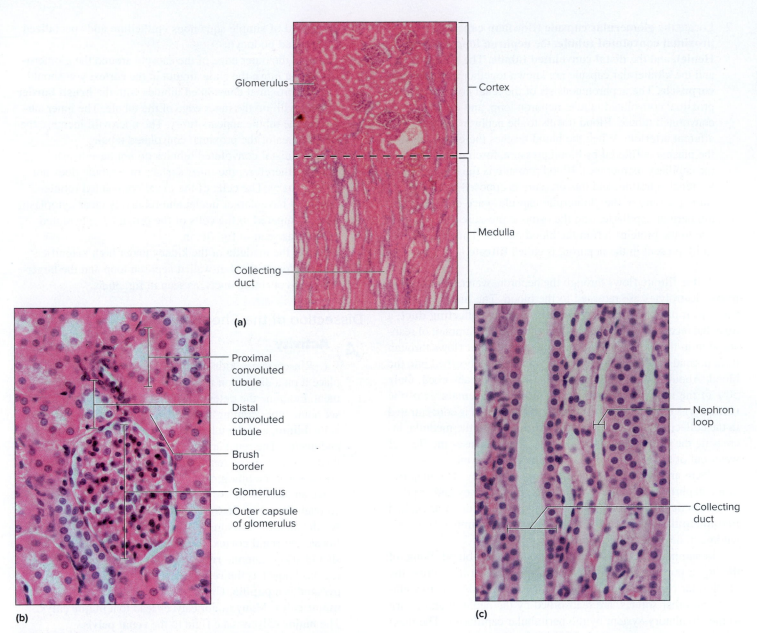

FIGURE 36.5 **Photomicrographs of a Kidney.** (a) Overview (40✕); (b) cortex (400✕); (c) medulla (400✕).
©Eric Wise

Urinary Bladder

The **urinary bladder** is located anterior to the parietal peritoneum and is thus described as being **anteperitoneal.** Locate the urinary bladder in a torso model in lab. It is found just posterior to the pubic symphysis. On the inferior portion of the urinary bladder is a triangular region known as the **trigone.** The trigone is defined by the superior entrances of the ureters and the inferior exit of the urethra. Compare models in lab to fig. 36.7.

Histology of the Bladder

Urothelium (transitional epithelium) lines the inner surface of the bladder, while layers of smooth muscle known as the **detrusor** are found in the wall of the bladder. Urothelium has a special role in the urinary bladder. This epithelium can withstand a significant amount of stretching (distention) when the bladder fills with urine.

 Activity

1. Examine a slide of urothelium under the microscope and compare it to fig. 36.8. Urothelium is also discussed and illustrated in Exercise 5.
2. Look at the inner surface of the section for the epithelial layer. The cells are shaped somewhat like teardrops. Urothelium can be distinguished from stratified squamous epithelium in that

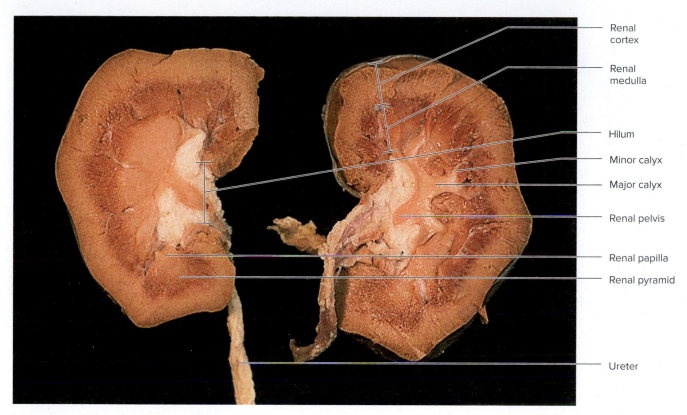

Renal cortex

Renal medulla

Hilum

Minor calyx

Major calyx

Renal pelvis

Renal papilla

Renal pyramid

Ureter

FIGURE 36.6 Dissection of a Sheep Kidney.
©Eric Wise

the cells of urothelium from an empty bladder do not flatten at the surface of the tissue. Below the urothelium is an underlying layer known as the **lamina propria.**

3. Examine the smooth muscle layers of the urinary bladder that make up the detrusor.

Urethra

The terminal organ of the urinary system is the **urethra.** The urethra is approximately 3 to 4 cm long in females. It passes from the urinary bladder to the **external urethral orifice** located anterior to the vagina and posterior to the clitoris, as seen in figs. 36.7b and 36.9b. In fig. 36.9 you can also see the position of the urinary bladder in relationship to the pubic symphysis. The urethra is about 20 cm long in males (figs. 36.7a and 36.9a). It begins at the urinary bladder and passes through the prostate gland as the **prostatic urethra.** It continues and passes through the body wall as the **membranous urethra,** then exits through the penis to the external urethral orifice at the tip of the glans penis. This terminal portion of the urethra is known as the **penile,** or **spongy, urethra.** Urinary bladder infections are more common in females than in males because of the difference in length in the urethra between males and females.

Cat Dissection

(A) **Activity**

1. Open the abdominal region of the cat, if you have not done so already. The kidneys are on the dorsal body wall of the cat and are located dorsal to the parietal peritoneum. Gently move the digestive structures out of the way but be careful to leave them intact for future studies.

2. Examine the kidneys and the structures that lead to and from the hilum of the kidney.

3. Find the renal veins that take blood from the kidney. The veins are larger in diameter than the renal arteries, and they are attached to the posterior vena cava of the cat, which runs along the ventral, right side of the vertebral column.

4. Find the renal arteries that take blood from the aorta to the kidneys. The aorta lies to the left of the posterior vena cava.

5. Locate the ureters as they run posteriorly from the kidney to the urinary bladder.

6. Locate the urinary bladder. Do not dissect the urethra at this time. You can locate the urethra during the dissection of the reproductive structures in Exercises 40 and 41. You should compare your dissection to fig. 36.10.

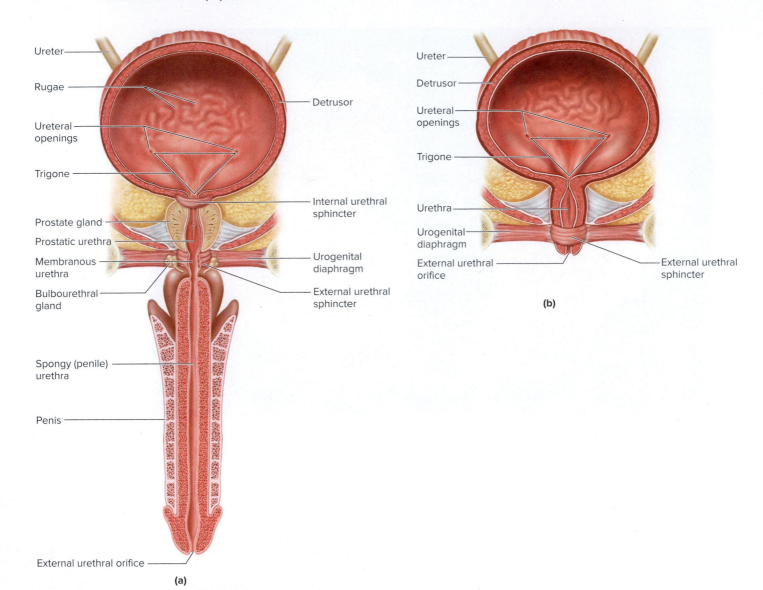

Ureter

Rugae

Ureteral openings

Trigone

Prostate gland

Prostatic urethra

Membranous urethra

Bulbourethral gland

Spongy (penile) urethra

Penis

External urethral orifice

Detrusor

Internal urethral sphincter

Urogenital diaphragm

External urethral sphincter

(a)

Ureter

Detrusor

Ureteral openings

Trigone

Urethra

Urogenital diaphragm

External urethral orifice

External urethral sphincter

(b)

FIGURE 36.7 Bladder and Urethra, Coronal Sections. (a) Male; (b) female.

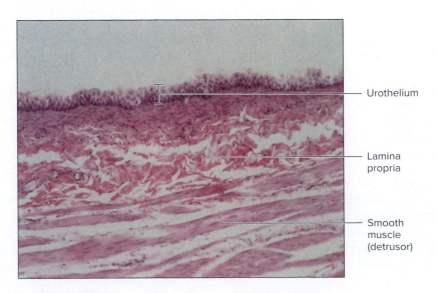

Urothelium

Lamina propria

Smooth muscle (detrusor)

FIGURE 36.8 Histology of the Bladder (100×).

©Eric Wise

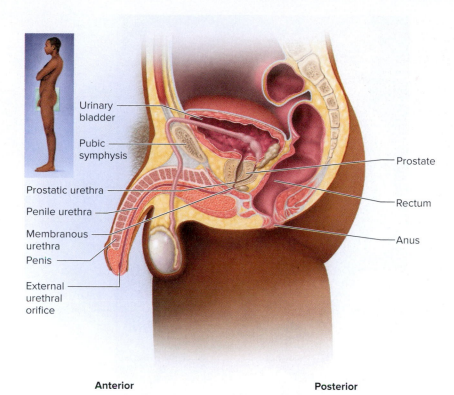

Urinary
bladder

Pubic
symphysis

Prostatic urethra

Penile urethra

Membranous
urethra

Penis

External
urethral
orifice

Prostate

Rectum

Anus

Anterior

Posterior

(a)

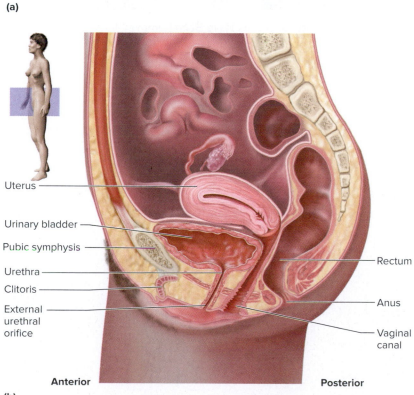

Uterus

Urinary bladder

Pubic symphysis

Urethra

Clitoris

External
urethral
orifice

Rectum

Anus

Vaginal
canal

Anterior

Posterior

(b)

FIGURE 36.9 Median Section of the Male and Female Pelves. (a) Male pelvis; (b) female pelvis.

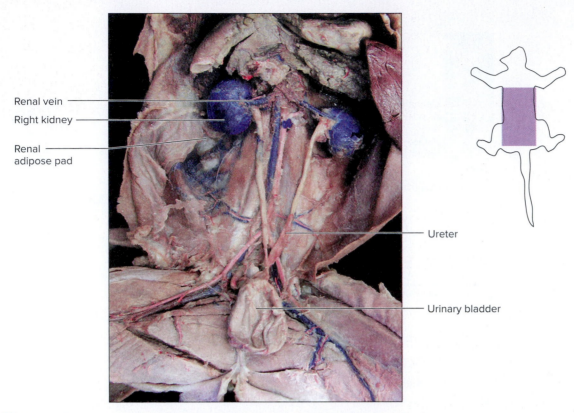

Renal vein

Right kidney

Renal
adipose pad

Ureter

Urinary bladder

FIGURE 36.10 Urinary System in the Cat. Digestive Organs Have Been Removed for Clarity.
Cordi Smith/McGraw Hill

REVIEW SECTION

Urinary System

Name _____   *Date* _____

Lab Section _____   *Time* _____

Review Questions

Matching

Match the descriptions in the left column with the terms in the right column.

1. outermost part of the kidney   a.   renal vein

2. storage organ of the urinary system   b.   major calyx

3. takes blood from the kidney   c.   minor calyx

4. separates the renal cortex from the medulla   d.   urinary bladder

5. receives urine from the renal papilla   e.   fibrous capsule

6. leads directly to the renal pelvis   f.   arcuate arteries

7. The _____ is found between the kidney and the urinary bladder.

8. The _____ is the terminal part of the nephron.

9. Filtration occurs at the _____ part of the nephron.

10. What is the outer region of the kidney called that contains glomeruli? _____

11. Describe the kidneys with regard to their position in relation to the parietal peritoneum.

_____

12. What takes urine from the bladder to the exterior of the body? _____

13. What blood vessel takes blood to the kidney? _____

14. On the inferior bladder is a triangular region. What is it called? _____

15. Distal convoluted tubules take urinary filtrate directly into what structures? _____

16. What is a renal papilla? _____

17. Name the fatty material that covers a kidney. _____

18. Blood in the glomerulus travels next to what arteriole? _____

19. Which urinary structure shows the greatest anatomical difference between the sexes: ureters, urinary bladder, or urethra?

_____

20. What histological feature distinguishes a proximal convoluted tubule from a distal convoluted tubule?

_____

21. Name the parts of the nephron from proximal to distal.

_____

_____

22. What cell type lines the inside of the bladder? _____

23. Trace the flow of blood from the renal artery through the kidney to the renal vein.

_____

_____

_____

24. What anatomical feature is responsible for the higher level of urinary tract infections in females?

_____

25. Trace the path of filtrate and urine from the glomerulus to the external urethral opening.

_____

_____

_____

26. Fill in the following illustration using the terms provided.

fibrous capsule   renal artery   renal pyramid

major calyx   renal cortex   renal vein

minor calyx   renal pelvis   ureter

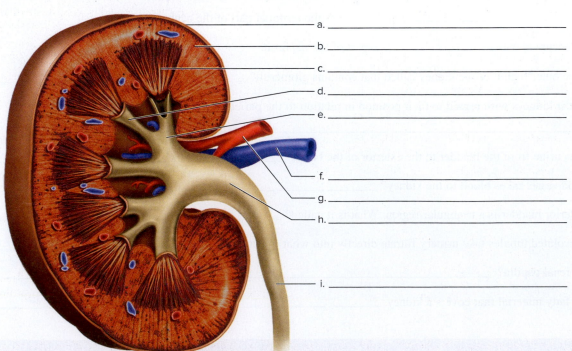

a. _____

b. _____

c. _____

d. _____

e. _____

f. _____

g. _____

h. _____

i. _____

Urinalysis

INTRODUCTION

Urinalysis is important in clinical assessment of an individual's physical condition. Traces of blood in the urine can indicate the presence of kidney stones or renal damage. Urinalysis can also test for elevated levels of glucose, diabetes mellitus; high bilirubin levels, liver problems; and high bacterial counts, bladder or kidney infections. The analysis of urine provides a broad diagnostic indicator of general health. You can read more about the urinary system in the Saladin text in chapter 23, "The Urinary System."

Normal values of urine vary based on water intake, medications used, and diet. A person who drinks 3 quarts of water per day will have a different urine composition than one who drinks 3 cups of water per day. Not only will the volume of water in the urine vary, but the concentration of other urinary solutes will vary as well. Some cations, such as potassium and sodium, are also diet-dependent. The properties of urine are listed in table 23.2 in the Saladin text. In this exercise you will analyze a sample of urine for materials dissolved in the urine or suspended in it. The suspended material may include cells.

OBJECTIVES

At the end of this exercise, you should be able to

1. use a Multistix or Chemstrip test and determine if the values obtained are within the standard range;
2. list the sediments commonly found in urine;
3. discuss the importance of urinalysis as a general diagnostic tool;
4. distinguish among casts, crystals, and microbes in a urine sample;
5. prepare a stained sediment slide and identify major components of the sediment.

MATERIALS

Sterile urine collection containers

Permanent marker or wax pencil

Microscope slides

Coverslips

Urine sediment stain (Sedistain, Volusol, etc.)

Chemstrip or Multistix 10SG urine test strips

Tapered centrifuge tubes

Test tube racks

Pasteur pipettes and bulbs

Centrifuge

Protective gloves

Protective eyewear

Biohazard bag

Microscopes

Urine specimen (yours or synthetic/sterilized urine)

10% bleach solution or other disinfecting solution

 Caution! *Urine is potentially contaminated with pathogens.* **Wear protective barrier gloves and protective eyewear during the entire exercise.** Place all disposable material that comes into contact with urine in the biohazard bag. Work only with your own urine and avoid any contact between urine and an open wound or cut. If you spill your sample, notify your instructor and wipe up the spill and swab the countertop with a 10% bleach or other disinfecting solution. When you are finished with the exercise, place reusable glassware in a 10% bleach or other disinfecting solution. Read all procedures before beginning this exercise.

PROCEDURE

You can use either your own urine for urinalysis or simulated urine depending on which your instructor directs you to use. If you use simulated urine, you can test for color, pH, specific gravity, glucose, and protein. Some tests are provided in kit form from biological supply houses. If you use your own urine, follow the directions given next. If you are using simulated urine, you can proceed to the Multistix/Chemstrip procedure.

(A) Activity

Using a marker or wax pencil, write your name on a sterile collection container. Proceed to the restroom and obtain a urine sample for testing. If you collect your urine at home, do so in the morning and store the sample in the refrigerator. The sample should be a **midstream** collection because the first volume of urine will contain abnormally elevated levels of microorganisms as they occur in or around the urethral opening. Take the specimen back to the lab and note the color of the sample.

Urine color:

_____ Pale

_____ Light yellow

_____ Deep yellow

_____ Orange yellow

_____ Bright yellow

The color of urine should normally be a pale yellow. This is due to the pigment **urochrome,** a metabolic product of hemoglobin breakdown. High levels of B vitamins may artificially color the urine bright yellow, and a low fluid intake may cause the urine to be a deep yellow.

Typically, freshly voided urine is clear or slightly cloudy. The factors that affect urine turbidity are increased numbers of red blood cells, white blood cells, epithelial cells, bacteria, mucus, lipids, or crystals.

Crystals make the urine cloudy, usually when the urine cools and the crystals precipitate. Examine your sample and determine into which category your urine falls.

Urine turbidity:

_____ Clear

_____ Slightly cloudy

_____ Cloudy

_____ Opaque

If your urine is cloudy or opaque, you should be able to see evidence of that in the microscopic examination portion of the lab.

Urine should have a faint but characteristic odor. Consumption of certain foods, such as asparagus, may produce sulfur compounds in urine, which produce stronger odors.

Smell the urine sample and record the odor.

Odor: _____

Urine Characteristics

Gently swirl the urine before testing and pour a 10 mL sample into a clean test tube. Follow your instructor's directions for determining

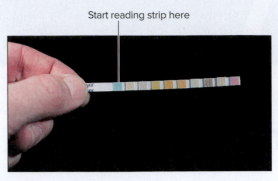

FIGURE 37.1 Reading a Urine Test Strip. Read the strip near the end where you hold onto the strip with your fingers and progress to the other end of the strip.
©Eric Wise

standard values in urine. You can test for specific compounds in urine individually, or you can use a Chemstrip or Multistix to run a battery of tests in a few minutes. The Multistix procedure is outlined next. Before you proceed with the test, you should have the urine sample, the test strip, a pencil or pen, and a container handy (so you can read the values). The test covers 10 specific exams in 2 minutes, so you have to be organized to record the results. Consider any result read after 2 minutes invalid. Multistix test strips are plastic with sections of paper with test reagents embedded into the fibers. They react with urine components if present. Examine fig. 37.1 and note the way the stick is read. Read the test strip first from the region near where you hold the strip and continue away from the area to the end of the strip.

Multistix/Chemstrip Procedure

(A) Activity

If you are using a Multistix or Chemstrip, be sure to pay attention to the time limits provided on the container for reading the results. Wear protective gloves as you dip the strip into the urine sample, and do not leave the test strip in the urine for longer than 1 second. Follow directions precisely on the test container, and record the results from your test strip in the spaces provided. Use the container the test strips came in as a reader to determine your results. Evaluations for the test results follow each item tested. Some drugs and vitamins can give false positive results in urinalysis tests. Any test that indicates renal disease should be further validated by professionals prior to any course of action. After you have finished your results, complete table 37.1 in the review section.

Glucose: _____

Although glucose is present in blood plasma, there is normally no glucose in urine. **Glycosuria** is the condition of having glucose in the urine. Fasting levels of glucose should be minimal (less than 40 mg/dL) or none. Trace amounts of glucose can be found after meals high in carbohydrates, especially sugar, and significant levels can be found in individuals with diabetes mellitus.

Bilirubin: _____

The bilirubin test should be negative, since bilirubin is present in urine in very small amounts. Bilirubin is a metabolic waste product from the destruction of erythrocytes. In the condition of **bilirubinuria,** bile pigments and bilirubin are present in urine. This can be due to erythrocyte destruction (hemolytic anemia), blockage of the bile duct, or liver damage, such as hepatitis or cirrhosis.

Ketones: _____

Normally, there should be a trace (5 mg/mL) or no ketones (one type of ketone is acetone) in the urine. Ketone bodies in urine **(ketonuria)** represent the general body condition known as **ketosis.** This is due to mobilization and use of fat stores in people during periods of starvation, in individuals with diabetes mellitus, or in people with an abnormally high-fat diet.

Specific gravity: _____

The specific gravity of pure water is 1.000. The normal values for urine specific gravity are variable between clinical labs, but they generally range from a dilute 1.001 to 1.028 in concentrated urine. Urine that has a high specific gravity contains more dissolved solutes. A high urine specific gravity may be due to dehydration or diabetes mellitus (where there is more sugar in the urine) among other things. A low urine specific gravity may be due to diabetes insipidus (where the body is either secreting too little ADH or the kidneys are not responding to ADH), increased water intake, or renal failure.

Hematuria: _____

Normally there should be no blood present in urine. The presence of erythrocytes or hemoglobin in urine **(hematuria)** may be due to some forms of anemia, erythrocyte destruction after incompatible blood transfusions, renal disease or infection, kidney stones, or urinary contamination during the menses (bleeding period) of a female's menstrual cycle. If the indicator strip shows a spotted pattern, then the erythrocytes are intact (nonhemolyzed). If the red blood cells have ruptured, blood shows up as uniform color changes on the indicator strip.

pH: _____

The pH of urine can vary from around 4.5 to 8.2 (mean 6.0). This represents almost a 10,000-fold difference in hydrogen ion concentration. The dramatic fluctuation in pH occurs because kidneys regulate blood pH by removing ions from the blood. Urine is normally slightly acidic. In a diet high in protein, the urine is more acidic, while a diet high in vegetable material produces a urine that is more alkaline.

Protein: _____

Normally, there should be no protein in urine. The most common blood protein is albumin, and the presence of protein in urine is known as **proteinuria (albuminuria).** This may be due to damages caused by diabetes mellitus, renal damage (kidney disease), extreme physical activity, or hypertension.

Urobilinogen: _____

Normal ranges of urobilinogen are 0.2 to 1 mg/dL. Increases in the secretion of urobilinogen indicate significant **hemolysis** of erythrocytes to the point that the liver cannot process the bilirubin. The bilirubin increases in the plasma, and the formation of urobilinogen in the intestines increases. The urobilinogen diffuses into the blood, where it is filtered by the kidneys.

Nitrites: _____

The normal blood value of nitrites should be zero. A positive test for nitrites indicates the presence of large numbers of bacteria in the urinary tract.

Leukocytes: _____

The normal value for leukocytes is zero or trace amounts. Elevated levels of leukocytes in urine is known as **pyuria.** This indicates a possible urinary tract infection, typically bladder or kidney infection.

Sediment Study

(A) Activity

1. Obtain a centrifuge tube from the supply area and write your name on the tube for identification.
2. Pour 10 mL of urine into the tube and place it in the centrifuge opposite another tube containing an equal volume.
3. The centrifuge must be balanced before it begins to rotate. An unbalanced centrifuge can cause damage to the rotor and is extremely dangerous. If there is an uneven number of tubes, make sure to fill a tube with water to the same level as the unpaired tube and place it opposite that tube in the centrifuge. This balances the centrifuge while it is operating.
4. Spin the tubes for about 4 minutes at a slow speed (1,500 rpm [revolutions per minute]) and let the rotor of the centrifuge slow down over time. If you brake the spinning, you may resuspend the sediments.
5. Carefully remove your tube and place it in a test tube rack on your desk.
6. Examine the tube. You should have an upper fluid layer and a small amount of urine sediment at the bottom of the tube.
7. Place one drop of urine sediment stain (Sedistain or other urine stain) on a clean microscope slide.
8. Withdraw a small sample from the bottom of the centrifuge tube with a Pasteur pipette and place a drop of the sediment

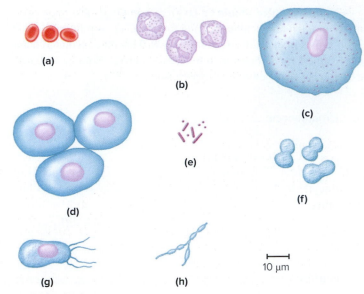

FIGURE 37.2 Urine Sediments—Cells. (a) Red blood cells; (b) white blood cells; (c) squamous epithelial cells; (d) uroepithelial cells; (e) bacteria; (f) yeast; (g) *Trichomonas;* (h) *Candida.*

on the slide. **Do not put the urine on the slide before you put on the stain because you may contaminate the stain if the stain bottle comes in contact with the sediment.**

9. Place a coverslip on the sample and examine the slide under the microscope. Examine the slide under low power first and look at the free edge of the coverslip. Red-orange, elongated crystals may appear here later as the stain evaporates. These are stain crystals. Do not confuse them with sediment crystals.

10. Now examine the slide under high power and compare your sample to the common urine sediments in figs. 37.2, 37.3, and 37.4. Fig. 37.3 is not to scale.

11. Record the material you find in urine sediment in number 17 in the review section at the end of this laboratory exercise. Some of the common items found in urine are as follows:

Organisms (fig. 37.2)

Yeasts (other than *Candida*)—Many yeasts are a normal constituent.

Trichomonas vaginalis—common protozoan infection

Candida albicans—causes vaginal yeast infections

Cells (fig. 37.2)

Epithelial cells

Cells of urethra (squamous cells)—normal constituent

Cells of bladder, ureter (urothelial cells)—normal constituent

Bacterial cells in urine indicate a condition known as bacteriuria.

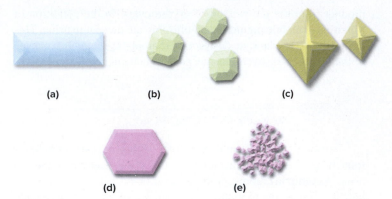

FIGURE 37.3 Urine Sediments—Crystals. Sizes are variable: (a) struvite; (b) calcium carbonate; (c) calcium oxalate; (d) cystine; (e) ureate.

Erythrocytes (RBCs)—menstrual blood in females or blood from the passage of kidney stones

Leukocytes (WBCs)—indicative of urinary tract (bladder/kidney) infections

Crystals (fig. 37.3) (usually indicative of reduced water intake)

Struvite (magnesium ammonium phosphate)

Calcium carbonate

Calcium oxalate—small, green crystals

Cystine—oxidation product of an amino acid

Ureates

Casts and Artifacts (fig. 37.4)

Casts are conglomerations of cells and other materials from the kidneys molded into elongated shapes by the renal tubules

Hyaline—from sclerosis of arteries or glomerulus

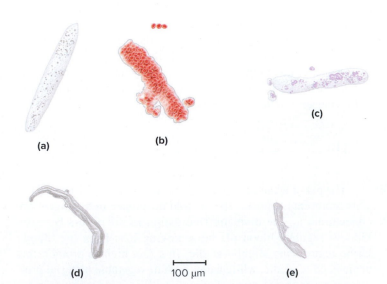

FIGURE 37.4 Urine Sediments—Casts and Fibers. (a) Hyaline cast; (b) erythrocyte cast; (c) leukocyte cast; (d) vegetable fiber; (e) mucous thread.

Leukocytes (indicate pyelonephritis, inflammation of the kidney)

Erythrocytes (indicate glomerulonephritis, inflammation of glomeruli)

Artifacts

Fibers—clothing fibers, mucous threads

Skin oil, powders

Renal Calculi

Kidney stones, or **renal calculi,** are due to mineral buildup in the kidneys, amino acid deposition, or amino acid oxidative products, such as tyrosine or cystine crystals. Kidney stones are commonly composed of calcium oxalate or calcium phosphate and occasionally uric acid or magnesium ammonium phosphate (struvite).

Calculi may occur from not drinking enough fluids, excessive mineral uptake, or metabolic disorders, such as an inability or a decreased ability to metabolize some amino acids. The calculi are masses of crystals that fuse into a larger form and can block the ureter, causing a retention of fluid in the kidney. Calculi can be exceptionally painful.

 Activity

Examine renal calculi in the lab, if available.

 Clean Up Place urine-contaminated material to be disposed of in the biohazard container and all the urine-contaminated material to be reused in the 10% bleach or other disinfecting container. Swab the countertops with 10% bleach or other disinfecting solution, which destroys infectious agents.

Notes

REVIEW SECTION

Urinalysis

Name _____ Date _____

Lab Section _____ Time _____

Review Questions

1. What metabolic by-product from hemoglobin colors the urine yellow? _____

2. How can adequate water intake be judged by the color of urine?

   _____

3. Why is it important to ask a woman if she is menstruating when doing a urinalysis? _____

4. Which urine sediment material is probably due to reduced water intake? _____

   a. vegetable fibers   b. crystals   c. epithelial cells   d. bacterial cells

5. There are about 200 grams of protein in the blood plasma and normally none in urine. What mechanism keeps protein out of the urine, and what structure would be potentially damaged (review Laboratory Exercise 36 if necessary) if protein were found in significant amounts in urine?

   _____

   _____

6. The kidneys are efficient at balancing the blood pH. It is critical that blood acidity remain constant. If excess hydrogen ions are present in the blood and increase blood acidity, the kidneys secrete the hydrogen ions. What effect would this have on the pH of urine?

   _____

7. If you process 180 liters of water through the kidney each day yet produce only 1.8 liters of urine, what is the percent reabsorption of water?

   _____

8. Assume that a person did not collect a midstream sample of urine but collected a sample from the beginning of urination. What additional materials might be in greater numbers in this sample?

   _____

9. How much water do you normally drink each day? Does this correspond to the specific gravity of your urine?

   _____

10. What is the name of the condition of having measurable amounts of glucose in urine? _____

11. What must be present in urine to have the condition of ketonuria? _____

12. Elevated levels of white blood cells produce what condition in urine? _____

13. What cells would be found in urine that originally came from the urethra? _____

14. What cells would be found in urine that came from the bladder? _____

15. If you had large numbers of calcium crystals in the urine, what could this tell you about the amount of water you drink?

_____

16. List your results from urinalysis here. Indicate if they are within the normal range or, if not, what this may indicate.

TABLE 37.1

Characteristic	Normal Value or Range	Your Results	Indication
Appearance	Almost clear to deep amber		
Odor	Faint odor		
Glucose	None to trace		
Bilirubin	None		
Ketones	None to trace (5 mg/mL)		
Specific gravity	1.001–1.028		
Free hemoglobin	None		
pH	4.5–8.2		
Albumin	None to trace		
Urobilinogen	Trace (0.2–1 mg/dL)		
Nitrites	None		
Leukocytes	None to trace		

17. List the sediments that you found in your urine sample.

_____

_____

Anatomy of the Digestive System

INTRODUCTION

The digestive system can be divided into two major parts, the **digestive tract (alimentary canal)** and the **accessory organs.** The digestive tract is a long tube that runs from the mouth to the anus and comes into contact with food or the breakdown products of digestion. Some of the organs of the digestive tract are the esophagus, stomach, intestines, rectum, and anus. The accessory organs are important in that they secrete many important substances necessary for digestion, yet these organs do not come into direct contact with food. Examples of accessory organs are the salivary glands, liver, gallbladder, and pancreas.

The functions of the digestive system are many and include ingestion of food, physical breakdown of food, chemical breakdown of food, food storage, water absorption, absorption of digested material, and elimination of indigestible material. These topics are covered in the Saladin text in chapter 25, "The Digestive System."

In this exercise, you examine the anatomy of the digestive system in both human and cat, correlating the structures of the digestive organs with their functions.

OBJECTIVES

At the end of this exercise, you should be able to

1. list, in sequence, the major organs of the digestive tract;
2. describe the basic function of the accessory digestive organs;
3. note the specific anatomical features of each major digestive organ;
4. describe the layers of the wall of the gastrointestinal tract;
5. describe the major functions of the stomach and small and large intestines;
6. distinguish among different regions of the digestive tract by their histology.

MATERIALS

Models, charts, or illustrations of the digestive system

Mirror

Materials for Cat Dissection

  Cats

  Dissection trays

  Scalpels

  Protective gloves

  Waste container

Skull, human teeth, or cast of teeth

Cadaver (if available)

Microscopes

Microscope slides

  Esophagus

  Stomach

  Small intestine

  Large intestine

  Liver

PROCEDURE

Overview of the Digestive Organs

(A) Activity

Begin this exercise by examining a torso model or charts in the lab and compare them to fig. 38.1. Locate the major digestive organs and place a check mark in the appropriate space.

_____ Mouth

_____ Teeth

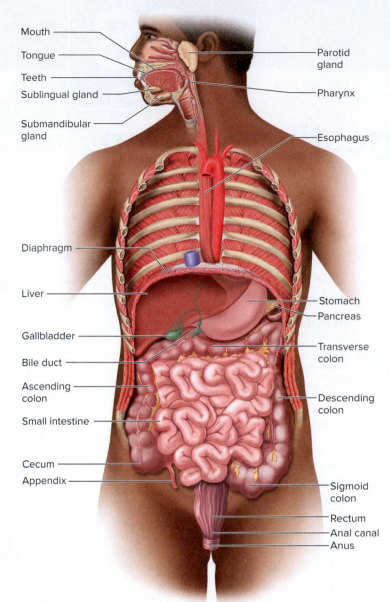

Mouth
Tongue
Teeth
Sublingual gland
Submandibular gland
Parotid gland
Pharynx
Esophagus
Diaphragm
Liver
Gallbladder
Bile duct
Ascending colon
Small intestine
Cecum
Appendix
Stomach
Pancreas
Transverse colon
Descending colon
Sigmoid colon
Rectum
Anal canal
Anus

FIGURE 38.1 Overview of the Digestive System.

_____ Pharynx
_____ Esophagus
_____ Stomach
_____ Small intestine
_____ Large intestine (colon)
_____ Ascending colon
_____ Transverse colon
_____ Descending colon
_____ Sigmoid colon
_____ Rectum
_____ Anal canal
_____ Anus

_____ Liver
_____ Appendix
_____ Pancreas
_____ Gallbladder
_____ Salivary glands (sublingual and submandibular)

Digestive Tract

(A) Activity

Begin your study of the digestive tract with the mouth. Examine a median section of the head, as represented in fig. 38.2, and locate the major anatomical features.

Mouth

At the beginning of the digestive tract is the **mouth (oral** or **buccal cavity).** The opening of the mouth is surrounded by **lips,** or **labia.** The **labial frenulum** is a membranous structure that keeps the lip adhered to the gums, or **gingivae.** The mouth is a space bordered in front by the lips, behind by the oropharynx, and on the sides by the inner wall of the cheeks. The hard and soft palates form the roof of the mouth, and the floor of the chin is the inferior border. The **hard palate** is composed of the palatine bones and the palatine processes of the maxillae. The **soft palate** is composed of connective tissue, a mucous membrane, and skeletal muscle that is important for swallowing. At the posterior portion of the mouth is the **uvula,** a small, grapelike structure suspended from the posterior edge of the soft palate. The uvula keeps food or liquid in the mouth until swallowed. The mouth is lined with **nonkeratinized** and **lightly keratinized stratified squamous epithelium,** which protects the underlying tissue from abrasion. The **tongue** is made of epithelial and connective tissue, glands, the lingual tonsils, and

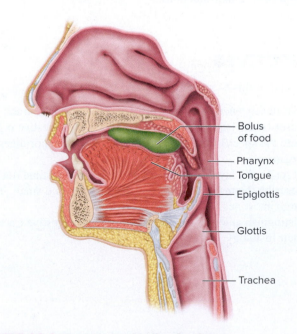

Bolus of food
Pharynx
Tongue
Epiglottis
Glottis
Trachea

FIGURE 38.2 Oral Cavity, Median Section.

skeletal muscle. The tongue is important in speech, taste, moving food toward the teeth for chewing, and swallowing. The tongue propels food to the **oropharynx,** the space behind the oral cavity. The tongue is held down to the floor of the mouth by a thin mucous membrane called the **lingual frenulum.** Use a mirror to examine the **papillae,** or raised areas, on the tongue. Three common ones are **fungiform, filiform,** and **vallate** (circumvallate) papillae. Papillae increase the frictional surface of the tongue. **Taste buds** occur on many parts of the tongue, including the sides of the vallate papillae. The sense of taste is covered in Laboratory Exercise 20.

A primary digestive function of the mouth is the physical breakdown of food. This process is driven by powerful muscles called the **muscles of mastication.** The **masseter** and the **temporalis muscles** are involved in the closing of the jaws, and the **pterygoid muscles** are important in the sideways grinding action of the molar and premolar teeth.

(A) Activity

Teeth   Examine models of teeth, dental casts, or real teeth on display in the lab and compare them to fig. 38.3. A tooth consists of a **crown, neck,** and **root.** The crown is the exposed part of the tooth; the neck is a constricted portion of the tooth that normally occurs at the surface of the gingivae; and the root is embedded in the jaw. Examine a model or an illustration of a longitudinal section of a tooth and find the outer **enamel,** an extremely hard material. Inside this layer is the **dentin,** which is made of bonelike

material. The innermost portion of the tooth consists of the **pulp cavity,** which leads to the **root canal,** a passageway for nerves and blood vessels into the tooth. The nerves and blood vessels enter the tooth through the **apical foramen** at the tip of the root of the tooth. The teeth occur in depressions in the mandible or maxilla called **alveoli** and are anchored into the bone by the periodontal ligament.

There are four different types of teeth in the adult mouth:

- **Incisors** are flat, bladelike front teeth that nip food. There are eight incisors in the adult mouth.
- **Canines,** or **cuspids,** are the pointed teeth just lateral to the incisors that shear food. There are four canines in the adult, and they can be identified as the teeth with just one cusp, or point.
- **Premolars,** or **bicuspids,** are lateral to the canines; they grind food. There are typically eight premolars in adults, and they can be identified by their two cusps.
- **Molars** are found closest to the oropharynx. These grind food, and there are 12 molar teeth in the adult mouth (including the wisdom teeth). Molars typically have three to five cusps.

Examine a human skull or dental cast and identify the characteristics of the four different types of teeth (fig. 38.4).

Humans have two sets of teeth: the **deciduous,** or "milk," **teeth** appear first, and these are replaced by the **permanent teeth.** There are 20 deciduous teeth. In children there are no deciduous premolar teeth, and there are only 8 deciduous molar teeth. In adults there are 8 premolar teeth and 12 molar teeth. Compare fig. 38.4, the adult pattern, to fig. 38.5, the deciduous teeth.

Frequently, the pattern of tooth structure is represented by a **dental formula,** which describes the teeth by quadrants. The dental formula for the deciduous teeth is illustrated here.

I = incisor, C = canine,
P = premolar, and M = molar.

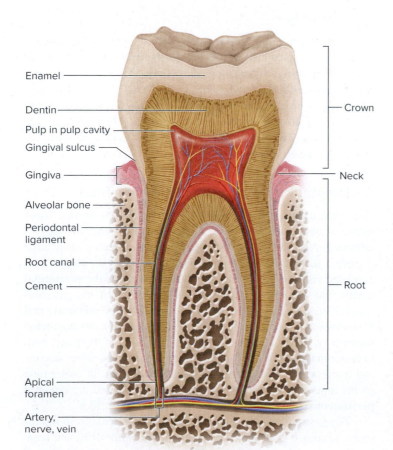

FIGURE 38.3 Tooth, Longitudinal Section.

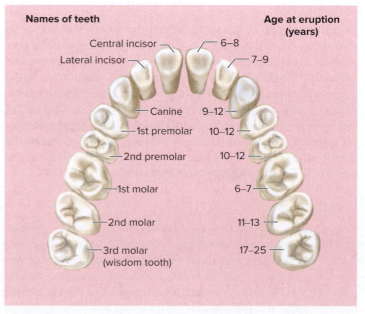

FIGURE 38.4 Upper Teeth of an Adult.

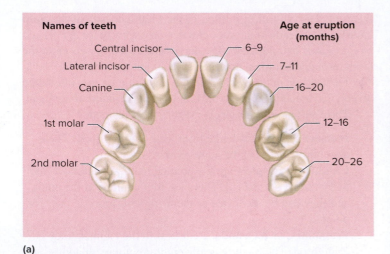

(a)

FIGURE 38.5 Deciduous Teeth. (a) Upper teeth;
(b) replacement of teeth.
(b) ©Eric Wise

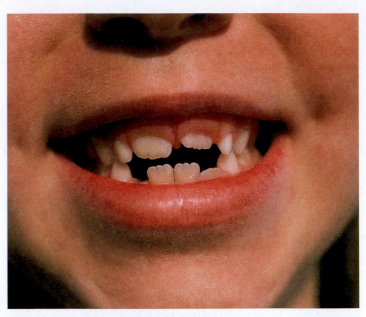

(b)

Deciduous teeth (20 total)

dental formula	I	C	P	M	One side (quadrant)
	2	1	0	2	Top
	2	1	0	2	Bottom

The upper numbers refer to the number of teeth in the maxilla on one side of the head. The lower numbers refer to the number of teeth in the mandible on one side of the head. The adult dental formula is as follows:

Adult (permanent) teeth (32 total)

I	C	P	M
2	1	2	3
2	1	2	3

Oropharynx

The **oropharynx** is the space behind the mouth that serves as a common passageway for air, food, and liquid. Above the oropharynx is the **nasopharynx,** which leads to the nasal cavity, and below the oropharynx is the **laryngopharynx,** which leads to the larynx and the esophagus. The oropharynx is lined with nonkeratinized stratified squamous epithelium. Muscles around the wall of the oropharynx and laryngopharynx are the **pharyngeal constrictor muscles** and are involved in swallowing. Food is moved by the tongue to the region of the pharynx, where it is propelled into the esophagus. Locate the pharynx and the esophagus in fig. 38.1.

Esophagus

The esophagus conducts food and liquid from the pharynx, through the diaphragm, and into the stomach. Normally, the esophagus is a closed tube that begins at about the level of the sixth cervical vertebra. As a lump of food, or **bolus,** enters the esophagus, skeletal muscle begins to move it toward the stomach. The middle portion of the esophagus is composed of both skeletal and smooth muscle, while the lower portion of the esophagus is made of smooth muscle.

In the lower region of the esophagus, the smooth muscle contracts, moving the bolus by a process known as **peristalsis.** The esophagus has an inner epithelial lining of **stratified squamous epithelium** and an outer connective tissue layer called the **adventitia.**

(A) Activity

Identify the four layers of the esophagus—mucosa, submucosa, muscularis, and adventitia—under the microscope. The space inside the esophagus where food passes through is called the **lumen,** which continues through the gastrointestinal tract. The lower portion of the esophagus has a **lower esophageal sphincter,** which prevents stomach acids from moving into the esophagus. **Heartburn** occurs if the stomach contents pass through the lower esophageal sphincter and irritate the esophageal lining.

Abdominal Portions of the Digestive Tract

The inner structure of the body has been referred to as a "tube within a tube." The body wall forms the outer tube; the gastrointestinal tract, including the stomach, small intestine, and large intestine, forms the inner tube. Specialized serous membranes cover the various organs and line the inner wall of the **peritoneal cavity.** The membrane lining the outer surface of the gastrointestinal tract is called the **visceral peritoneum (serosa),** and it continues as a double-folded membrane called the **posterior mesentery,** which attaches the tract to the back of the body wall. In between the linings of the mesentery are arteries, veins, nerves, and lymphatics. The mesentery is continuous with the membrane on the inner side of the body wall, where it is called the **parietal peritoneum.** Locate these three membranes in fig. 38.6.

The gastrointestinal tract has a series of layers seen in microscopic sections. In general, the stomach, small intestine, and large intestine have the same layers from the lumen to the peritoneal cavity. The innermost layer is the **mucosa,** which consists of a

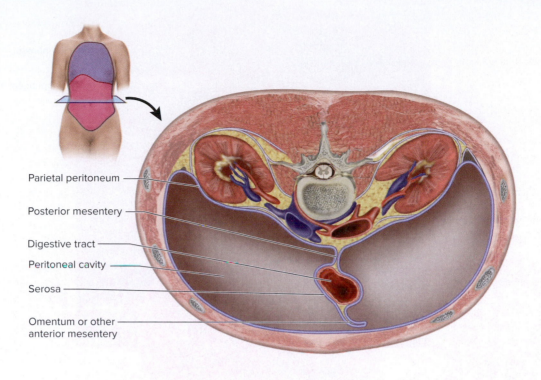

FIGURE 38.6 Membranes of the Gastrointestinal Tract (Idealized Drawing).

mucous membrane closest to the lumen, a connective tissue layer called the **lamina propria,** and an outer, muscular layer called the **muscularis mucosae.**

The next layer is called the **submucosa,** mostly made of connective tissue and containing numerous blood vessels. The next layer is the **muscularis externa,** typically made of two or three layers of smooth muscle. The muscularis externa propels material through the gastrointestinal tract and mixes ingested material with digestive juices. The outermost layer is the **serosa,** or **visceral peritoneum,** and this layer is closest to the peritoneal cavity.

(A) **Activity**

Examine a microscope slide of the gastrointestinal tract (small intestine) and locate these three membranes in fig. 38.7.

Stomach

The **stomach** is located on the left side of the body and receives its contents from the esophagus. The food that enters the stomach is stored and mixed with the enzyme pepsin and hydrochloric acid to form a soupy material called **chyme.** The stomach can have a pH as low as 1 or 2. Chyme remains in the stomach as the acids denature proteins and pepsin reduces proteins to shorter fragments. The acid of the stomach also has an antibacterial action, as most microbes do not grow well in conditions of low pH.

(A) **Activity**

Examine a model of the stomach or charts in the lab and compare them to fig. 38.8. Locate the upper portion of the

stomach called the **cardia,** or **cardiac part.** The region of the stomach that extends superiorly as a domed section called the **fundic region,** or **fundus.** The main part of the stomach is called the **body** and the terminal portion of the stomach, closest to the small intestine, is called the **pyloric part.** The pyloric part has an expanded area called the **antrum** and a narrowed region called the **pyloric canal.** The pylorus leads to the duodenum, and this opening is controlled by the **pyloric sphincter.** The left side of the stomach is arched and forms the **greater curvature,** while the right side of the stomach is a smaller arch, forming the **lesser curvature.** The inner surface of the stomach has a series of folds called **rugae,** which allow for expansion of the stomach. The contents are held in the stomach by two sphincters. The proximal lower esophageal sphincter prevents stomach contents from moving into the esophagus. The distal pyloric sphincter prevents the premature release of stomach contents into the small intestine. Locate the pyloric sphincter at the terminal portion of the stomach.

(A) **Activity**

Stomach Histology   Examine a prepared slide of stomach. Identify the four primary layers: **mucosa, submucosa, muscularis externa,** and **serosa.** Notice how the mucosa in the prepared slide has numerous indentations. These depressions are **gastric pits.** These occur in the inner lining of the stomach. The mucous membrane contains **simple columnar epithelium. Surface mucous cells** occur in the membrane and secrete **mucus,** which protects the stomach lining from erosion by stomach acid and the **proteolytic** (protein-digesting) enzyme **pepsin.** Other specialized cells that you might find in the mucosa are **chief cells,** which secrete

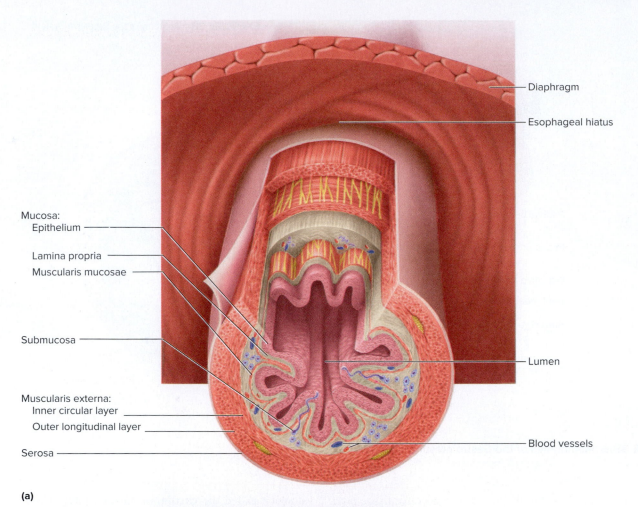

Mucosa:
Epithelium
Lamina propria
Muscularis mucosae

Submucosa

Muscularis externa:
Inner circular layer
Outer longitudinal layer

Serosa

Diaphragm

Esophageal hiatus

Lumen

Blood vessels

(a)

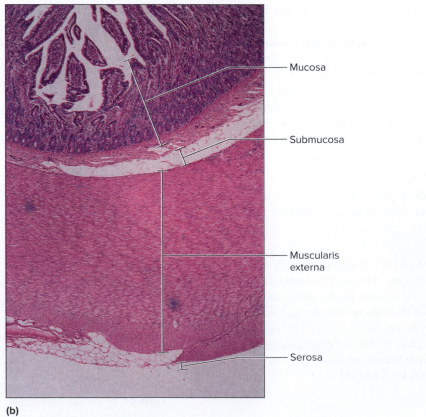

Mucosa

Submucosa

Muscularis
externa

Serosa

(b)

FIGURE 38.7 **Gastrointestinal Tract, Cross Section.** (a) Diagram; (b) photomicrograph (100×).
(b) ©Eric Wise

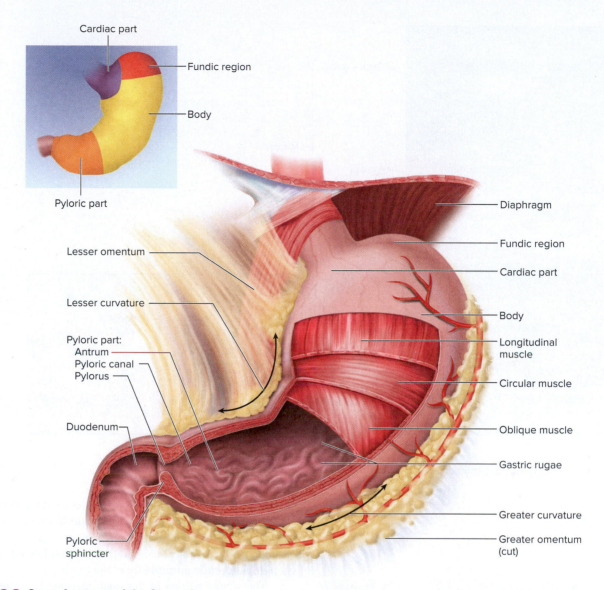

FIGURE 38.8 Gross Anatomy of the Stomach.

gastric lipase and **pepsinogen** (the inactive state of pepsin). Chief cells contain blue-staining granules in some prepared slides. Other cells are **parietal cells,** which secrete HCl. They typically contain orange-staining granules. When pepsinogen comes into contact with HCl, it is activated as pepsin.

Deeper to the epithelium, locate the **lamina propria,** usually lighter in color because it is made of areolar tissue. The **muscularis mucosae** is even farther away from the lumen and moves the mucous membrane. The **submucosa** is the next layer and is typically lighter in color in prepared slides.

The next layer of the stomach is the **muscularis externa.** The muscularis externa consists of an **inner oblique layer,** a **middle circular layer,** and an **outer longitudinal layer.** The muscularis externa mixes chyme in the stomach and moves it from the stomach through the pyloric sphincter and into the small intestine.

The outermost layer is the **serosa,** and it is composed of a thin layer of connective tissue and **simple squamous epithelium.** Locate these structures and compare them to fig. 38.9.

Small Intestine

The **small intestine,** named because it is small in diameter, is approximately 4–5 m long. It is typically 3 to 4 cm in diameter when empty. Movement through the small intestine occurs by peristalsis, smooth muscle contraction. The primary function of the small intestine is nutrient absorption.

Ⓐ **Activity**

Locate the three major regions of the small intestine on a model or chart and compare them to fig. 38.10. The first part of the small intestine is the **duodenum,** a C-shaped structure attached to the pyloric part of the stomach. The duodenum is approximately 25 cm long. It receives fluid from both the **pancreas** and the **gallbladder.** The gallbladder releases **bile,** which emulsifies lipids, in the duodenum. The lipids break into smaller droplets, which increase the surface area for digestion. The bile is transported to the duodenum by the bile duct. The pancreas secretes many digestive

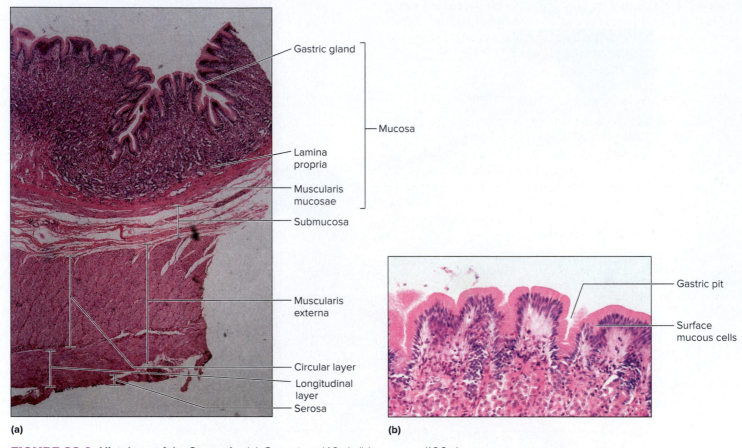

FIGURE 38.9 Histology of the Stomach. (a) Overview (40×); (b) mucosa (100×).
©Eric Wise

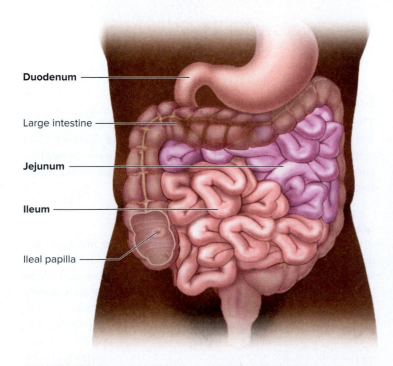

FIGURE 38.10 Gross Anatomy of the Small Intestine.

enzymes (proteases, amylases, and lipases) and bicarbonate, which neutralizes the stomach acids. These materials are secreted into the duodenum by the **pancreatic duct.** The pancreatic duct can lead to the duodenum or it can feed into the common bile duct to form the **hepatopancreatic ampulla** (ampulla of Vater).

The second portion of the small intestine is the **jejunum,** approximately 1.0 to 1.7 m in length. The junction between the jejunum and the duodenum is known as the **duodenojejunal flexure.** The terminal portion of the small intestine is the **ileum,** which is 1.6 to 2.7 m in length. A closure between the small intestine and large intestine is called the **ileal papilla (ileocecal valve),** which reduces the chance of material in the large intestine from reentering the small intestine. Locate the small intestine and associated structures in fig. 38.10.

Histology of the Small Intestine   **Villi** distinguish the small intestine from both the stomach and the large intestine. Villi are fingerlike projections that increase the surface area of the mucosa. Each villus contains **blood vessels,** which transport sugars and amino acids from the intestine to the liver. In addition to this, the villi contain **lacteals,** which transport lipids as chylomicrons via lymphatic vessels to the blood stream. When seen with the naked eye, villi give the lining of the small intestine a velvety appearance. The inner lining of the small intestine contains **simple columnar epithelium** with **goblet cells.** You can distinguish the three sections of the small intestine by noting that the duodenum has **duodenal (Brunner's) glands** in the submucosa. The jejunum and the ileum lack these glands. The ileum is distinguished by the presence of clustered **lymphoid nodules** (Peyer patches) in the submucosa. These lymphoid nodules contain lymphocytes, which are activated and protect the body from the bacterial flora in the lumen of the small intestine. Compare the prepared slides of the small intestine to fig. 38.11.

Large Intestine

The **large intestine** is so named because it is large in diameter. The large intestine is approximately 6.5 cm in diameter and 1.5 m in length. The large intestine absorbs water, some vitamins, and solutes and forms feces. The mucosa of the large intestine contains **simple columnar epithelium** with a large number of **goblet cells.** There are no villi present, yet the wall of the large intestine is heavily populated with lymphocytes, many of them aggregated into nodules.

(A) Activity

Look at a model or chart of the large intestine and note the major regions. Compare them to fig. 38.12.

- **Cecum:** first part of the large intestine. The cecum is a pouch that joins with the small intestine at the level of the ileal papilla.
- **Ascending colon:** found on the right side of the body. It becomes the transverse colon at the right colic (hepatic) flexure.
- **Transverse colon:** traverses the body from right to left. It leads to the descending colon at the left colic (splenic) flexure.
- **Descending colon:** passes inferiorly on the left side of the body and joins with the sigmoid colon.
- **Sigmoid colon:** an S-shaped segment of the large intestine in the pelvic region.
- **Rectum:** a straight section of colon in the pelvic cavity. The rectum has superficial veins in its wall called **hemorrhoidal veins.** These may enlarge and cause the uncomfortable condition known as **hemorrhoids.**

The large intestine has some unique structures. The longitudinal layer of the muscularis externa of the large intestine thickens in three places, forming the **taenia coli.** These muscles contract and form pouches in the intestinal tract called **haustra** (singular, *haustrum*). Another unique feature of the outer wall of the large intestine are fat lobules called **omental appendices.** Locate these structures in fig. 38.12.

Fecal material passes through the large intestine by peristalsis and is stored in the rectum and sigmoid colon. **Defecation** occurs as **mass peristalsis** causes a bowel movement.

(A) Activity

Histology of the Large Intestine Examine a prepared slide of the large intestine and compare it to fig. 38.13. The large intestine is distinguished from the small intestine by the absence of villi and from the stomach by the presence of large numbers of goblet cells. Examine your slide for these characteristics.

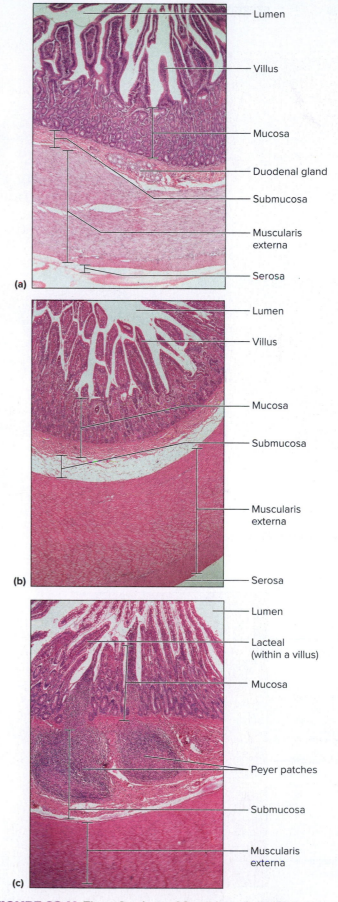

FIGURE 38.11 **Three Sections of Small Intestine (40✕).**
(a) Duodenum; (b) jejunum; (c) ileum.
©Eric Wise

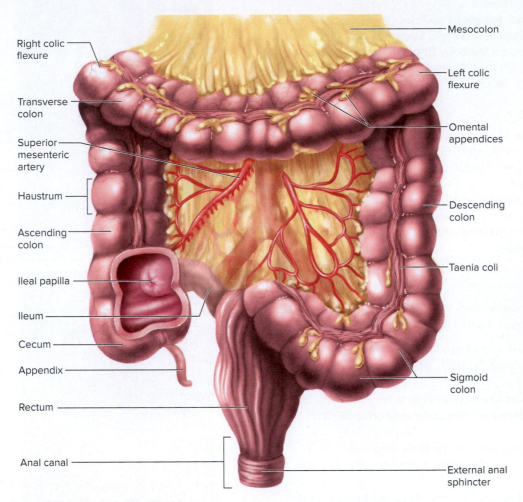

Right colic flexure

Transverse colon

Superior mesenteric artery

Haustrum

Ascending colon

Ileal papilla

Ileum

Cecum

Appendix

Rectum

Anal canal

Mesocolon

Left colic flexure

Omental appendices

Descending colon

Taenia coli

Sigmoid colon

External anal sphincter

FIGURE 38.12 Gross Anatomy of the Large Intestine.

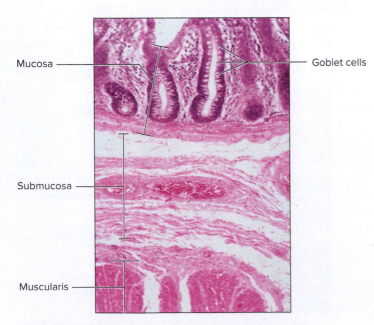

Mucosa

Submucosa

Muscularis

Goblet cells

FIGURE 38.13 Histology of the Large Intestine (100×).

©Eric Wise

Anal Canal

The **anal canal** is not part of the large intestine but is a short tube that leads to an external opening, the anus. The anal canal is lined with stratified squamous epithelium, which resists abrasion during defecation. Locate the anal canal in fig. 38.12.

Accessory Organs

Salivary Glands

The **salivary glands,** located in the head, secrete **saliva** into the mouth. Saliva is a watery secretion that contains **mucus** (a protein lubricant), **salivary amylase,** a starch-digesting enzyme, and lingual lipase, a lipid-digesting enzyme. The average adult secretes about 1.5 liters of saliva per day.

A Activity

There are three pairs of salivary glands. The first of these, the **parotid glands,** are located just anterior to the ears. Each gland secretes saliva through a **parotid duct,** a tube that traverses the buccal (cheek) region and enters the mouth

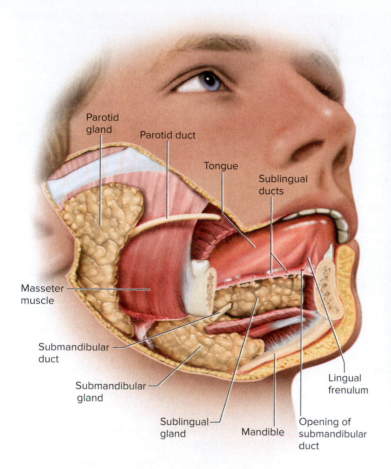

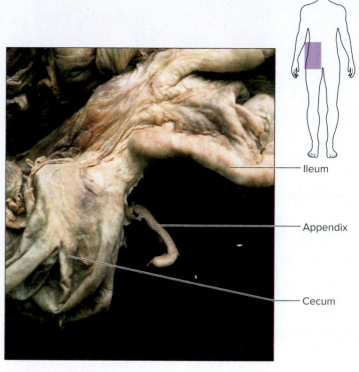

FIGURE 38.15 **Appendix.**
©Eric Wise

FIGURE 38.14 Salivary Glands.

just posterior to the upper second molar. The second pair, the **submandibular glands,** are located just medial to the mandible on each side of the face. The submandibular glands secrete saliva into the mouth by a single duct on each side of the mouth inferior to the tongue. The third pair, the **sublingual glands,** are located inferior to the tongue, and open into the mouth by several ducts. Locate these salivary glands on a model of the head and in fig. 38.14.

Appendix

(A) Activity

The **appendix** is about the size of your little finger and is located near the junction of the small and large intestines (at the region of the ileal papilla). Locate the appendix on a torso model or chart in the lab and compare it to fig. 38.15.

Omenta

The **lesser omentum** is an extension of the peritoneum that forms a double fold of tissue between the stomach and the liver. The **greater omentum** is a section of peritoneum between the stomach

and transverse colon and drapes over the intestines as a fatty apron. Locate the lesser and greater omenta in fig. 38.16.

Liver

The **liver** is a complex organ with numerous functions; one of these functions is digestive (the secretion of bile), but most are not. The liver processes digestive products from the vessels returning blood from the intestines and has a role in either moving material into the bloodstream or storing it in the liver tissue. The liver produces blood plasma proteins; detoxifies harmful material that has been produced by, or introduced into, the body; and produces bile.

(A) Activity

Examine a model or chart of the liver and note its relatively large size. The liver is located on the right side of the body and is divided into four lobes, the **right, left, quadrate,** and **caudate lobes.** Only two lobes of the liver can be seen from the anterior side, and these are the large right lobe and the smaller left lobe. These two lobes are separated by a slip of mesentery called the falciform ligament, which suspends the liver from the diaphragm. In an inferior view of the liver, all four lobes can be seen. The quadrate lobe is a small, rectangular lobe adjacent to the gallbladder, and the caudate lobe is at the posterior edge of the liver near the inferior vena cava. Locate the **gallbladder,** on the inferior aspect of the liver, and compare the gallbladder and liver structures to fig. 38.17.

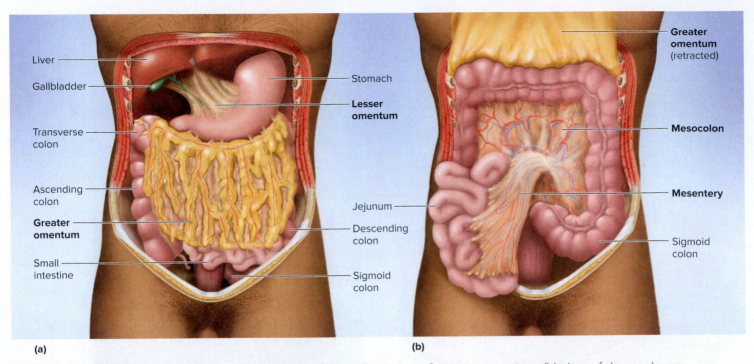

FIGURE 38.16 Omenta, Mesentery, and Mesocolon. (a) Superficial view of greater omentum; (b) view of deeper layers.

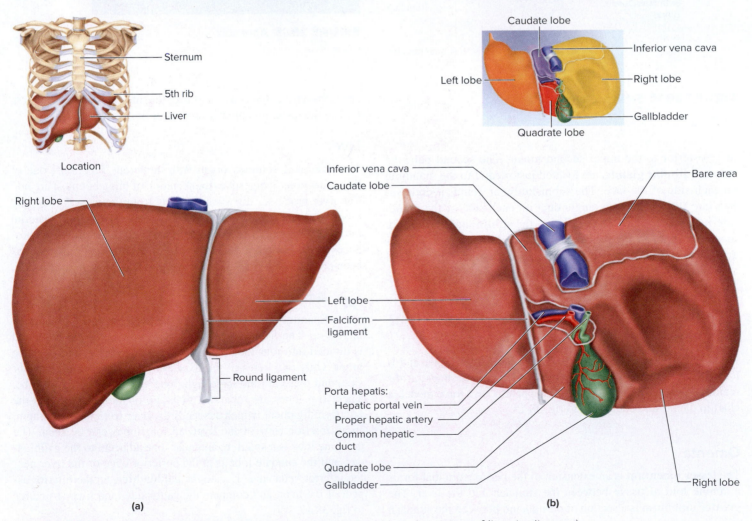

FIGURE 38.17 Liver. (a) Anterior view; (b) Infero-posterior view (dorsal is at top of liver in diagram).

(A) **Activity**

Liver Histology   Examine a prepared slide of liver tissue. Note the hexagonal structures in the specimen. These are known as **hepatic lobules.** Each lobule has a blood vessel in the middle called the **central vein.** Vessels that carry blood to the central vein are the **hepatic sinusoids,** and these are lined with a double row of cells called **hepatocytes.** Hepatocytes carry out the various functions of the liver as just described. The tissue of the liver is extremely vascular and functions as a sponge. Fresh, oxygenated blood from the hepatic artery and deoxygenated blood from the hepatic portal vein mix in the liver. **Stellate (hepatic) macrophages (Kupffer cells)** occur in hepatic sinusoids and function as phagocytic cells. Locate the hepatic lobules, central vein, hepatic sinusoids, bile ductule, and hepatocytes in a prepared slide, using fig. 38.18 to aid you.

Gallbladder

The **gallbladder** is located just inferior to the liver. The liver is the site of bile production, and the bile flows from the liver through **left** and **right hepatic ducts** to enter the **common hepatic duct.** Once in the common hepatic duct, the bile flows into the **cystic** duct and then is stored in the gallbladder, where it is concentrated. As the stomach begins to empty its contents into the duodenum, the gallbladder constricts, and bile flows from the gallbladder back into the cystic duct and into the **bile duct,** which empties into the duodenum.

(A) **Activity**

Locate these structures on a model and compare them to fig. 38.19.

Pancreas

The **pancreas** is located inferior to the stomach on the left side of the body. It has both endocrine and exocrine functions. The hormonal function of the pancreas is covered in Laboratory Exercise 23. The exocrine function of the pancreas is studied in Laboratory Exercise 39. The pancreas consists of a **tail,** found near the spleen; an elongated **body;** and a rounded **head,** located near the duodenum. Enzymes and buffers pass from the tissue of the pancreas into the **pancreatic duct** and then into the duodenum. Locate the pancreatic structures, as represented in fig. 38.19.

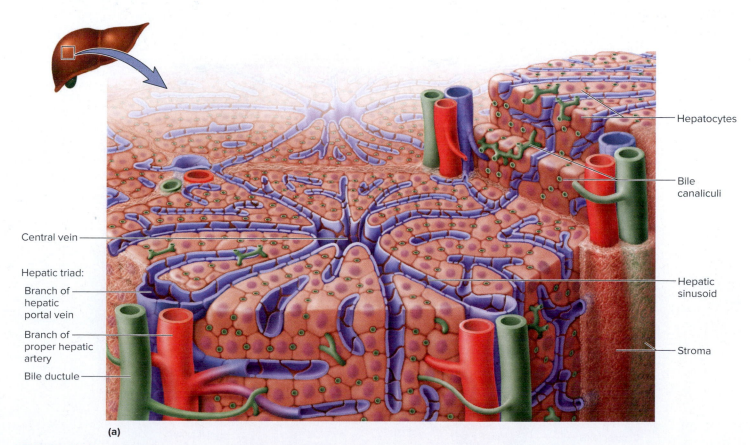

(a)

FIGURE 38.18 Histology of the Liver, Liver Lobule. (a) Diagram; (b) photomicrograph (100×).

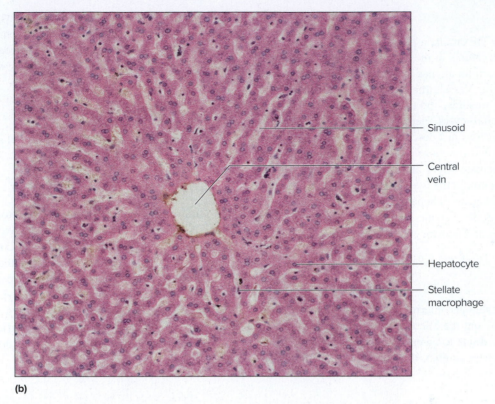

(b)

FIGURE 38.18 *Continued.*

(b) ©Eric Wise

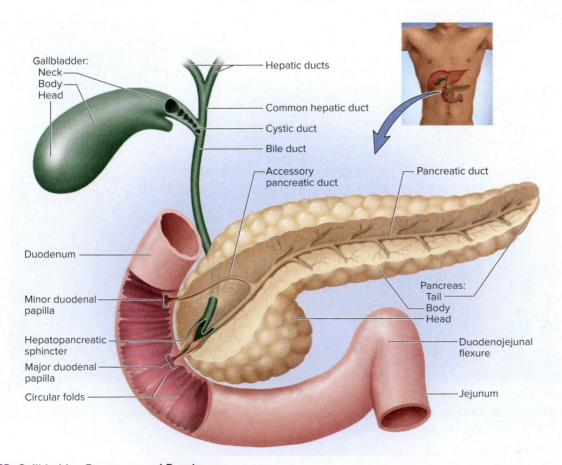

FIGURE 38.19 **Gallbladder, Pancreas, and Duodenum.**

Cat Dissection

Ⓐ Activity

Prepare for the dissection by obtaining a cat and dissection equipment. Remember to place all excess tissue in the appropriate waste container and not in a standard wastebasket or down the sink.

Begin your study of the digestive system of the cat by locating the large **salivary glands** around the face. You must first remove the skin from in front of the ear. If you have dissected the face for the musculature, you may have already removed the **parotid gland,** a spongy, cream-colored pad anterior to the ear. You can also locate the **submandibular gland** as a pad of tissue slightly anterior to the angle of the mandible and lateral to the digastric muscle. The **sublingual gland** is just anterior to the submandibular gland and is an elongated gland that parallels the mandible. Use fig. 38.20 to help you locate these glands.

If your instructor instructs you to do so, start your dissection of the head by cutting through the skin with your scalpel from the forehead to the occipital region in the median plane. Cut through any overlying muscle in the cat. You should then use a small saw and gently cut through the cranial region of the skull, being careful to stay in the median plane. Once you make an initial cut through the dorsal side of the head, you should cut the mental symphysis of the cat, thus separating the mandible.

You can then use a long knife (or scalpel) to cut through the softer regions of the skull, brain, and tongue. Use a scalpel to cut through the floor of the oral cavity to the hyoid bone. It is best to use scissors or a small bone cutter to cut through the hyoid

bone. This should allow you to open the head and examine the structures seen in a median section, as seen in fig. 38.21. Locate the hard palate, tongue, oral cavity, and oropharynx. Note how the larynx in the cat is closer to the tongue than in humans.

Examine the **tongue** of the cat and note the numerous papillae on the dorsal surface. These are **filiform papillae.** They are used to groom the fur of cats. You can lift up the tongue and examine the **lingual frenulum** on the ventral surface. Notice how the teeth of the cat are adapted for seizing prey with the elongated canines and how the molar teeth are adapted for shearing and not grinding as they are in humans.

Look for the muscular tube of the esophagus by carefully lifting the trachea ventrally away from the neck. You should be able to insert your blunt dissection probe into the oral cavity and gently wiggle it into the esophagus. The tongue and esophagus are represented in fig. 38.21.

Abdominal Organs

To get a better view of the abdominal organs it is best to cut the **diaphragm** away from the anterior body wall. Carefully cut the lateral edges of the diaphragm from the ribs. You should see a fatty drape of material covering the intestines. This is the **greater omentum.** Just caudal to the diaphragm, note the dark brown, multilobed **liver.** In the middle of the liver is the green **gallbladder.** To the left of the liver (in reference to the cat) is the J-shaped **stomach.** Lift the liver and locate the **lesser omentum,** a fold of tissue that connects the stomach to the liver. Locate these structures in fig. 38.22.

Make an incision into the stomach and locate the folds known as rugae. Place a blunt probe inside the stomach and move it anteriorly. Notice how the lower esophageal sphincter makes it difficult to pass through the diaphragm. If you do move the probe into the esophagus, you should be able to see it as you look anterior to the diaphragm.

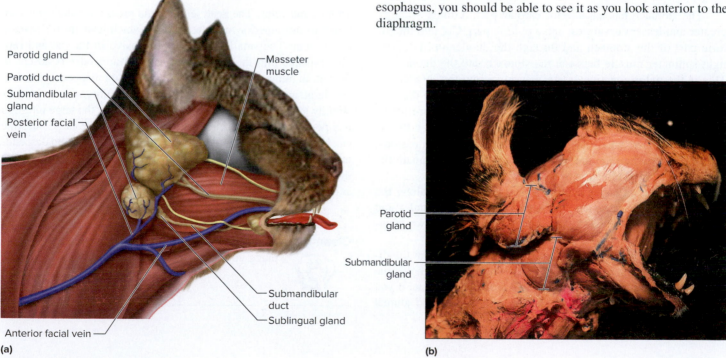

(a)

(b)

FIGURE 38.20 **Salivary Glands of the Cat.** (a) Diagram; (b) photograph.
(b) ©Eric Wise

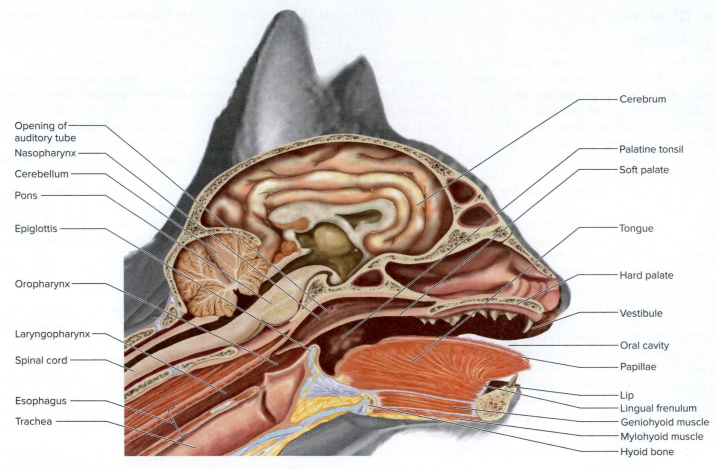

Opening of auditory tube
Nasopharynx
Cerebellum
Pons
Epiglottis
Oropharynx
Laryngopharynx
Spinal cord
Esophagus
Trachea

Cerebrum
Palatine tonsil
Soft palate
Tongue
Hard palate
Vestibule
Oral cavity
Papillae
Lip
Lingual frenulum
Geniohyoid muscle
Mylohyoid muscle
Hyoid bone

FIGURE 38.21 **Median Section of the Head and Neck of a Cat.**

The stomach has an anterior cardiac part, a domed fundus, greater and lesser curvatures, and a pyloric part. Cut into the pyloric part of the stomach and through the duodenum. Locate a tight sphincter muscle between the stomach and the duodenum. This is the pyloric sphincter. As you move into the duodenum, you may have to scrape some of the chyme away from the wall of the small intestine to be able to see the fuzzy texture of the intestinal wall. This texture is due to the presence of villi. As in the human, the small intestine is composed of three regions, the duodenum, the jejunum, and a terminal ileum. Compare the stomach and small intestine to fig. 38.22.

Elevate the greater omentum to see the pancreas, along the caudal side of the stomach. The pancreas appears granular and brown. The tail of the pancreas is near the spleen, a brown, elongated organ on the left side of the body.

The small intestine is an elongated, coiled tube about the diameter of a wooden pencil. Note the **mesentery,** which holds the small intestine to the posterior body wall near the vertebrae. If you fan out the mesentery you will see the colic, ileal, and jejunal arteries and veins. The ileal, jejunal, and proximal colic veins take blood to the superior mesenteric vein, which leads to the hepatic portal vein. The small intestine is extensive in the cat and rapidly expands into the large intestine. The appendix is absent in the cat. In humans the appendix is found at the ileocecal junction. The large intestine in the cat is a fairly short tube, with a diameter slightly larger than your thumb. The first part of the large intestine is a pouch called the cecum. The remainder of the large intestine can be further divided into the ascending, transverse, and descending colon and the rectum. Compare these to fig. 38.22. Examine the **parietal peritoneum** along the inner surface of the body wall and the visceral peritoneum that envelops the intestines.

Clean Up   When you are done with your dissection, carefully place your cat back in the plastic bag.

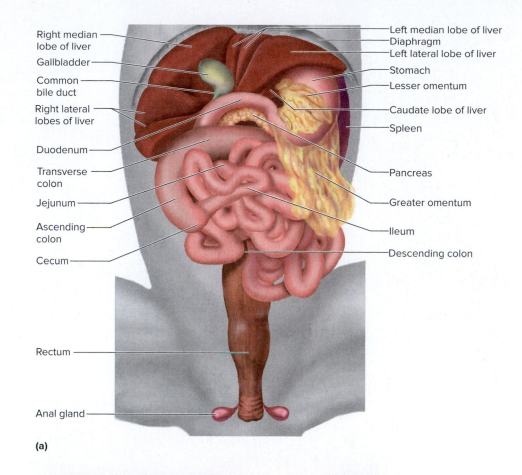

Right median lobe of liver
Gallbladder
Common bile duct
Right lateral lobes of liver
Duodenum
Transverse colon
Jejunum
Ascending colon
Cecum

Left median lobe of liver
Diaphragm
Left lateral lobe of liver
Stomach
Lesser omentum
Caudate lobe of liver
Spleen
Pancreas
Greater omentum
Ileum
Descending colon

Rectum

Anal gland

(a)

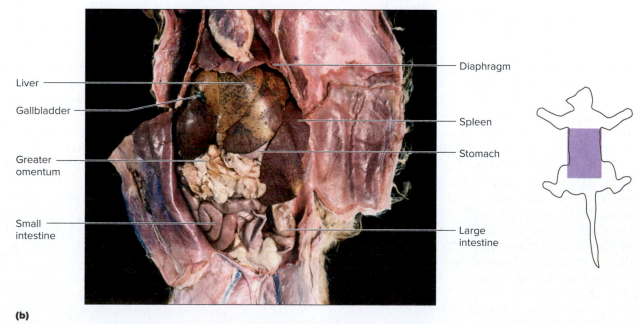

Liver
Gallbladder
Greater omentum
Small intestine

Diaphragm
Spleen
Stomach
Large intestine

(b)

FIGURE 38.22 **Abdominal Organs of the Cat.** (a) Diagram; (b) photograph.
(b) ©Eric Wise

Notes

REVIEW SECTION

Anatomy of the Digestive System

Name _____ *Date* _____

Lab Section _____ *Time* _____

Review Questions

Matching

Match each term or description on the left with a term on the right.

Set One

_____ 1. pancreas

_____ 2. descending colon

_____ 3. part of the stomach closest to the small intestine

_____ 4. middle portion of the small intestine

_____ 5. distal portion of the small intestine

a. jejunum

b. pyloric part

c. ileum

d. digestive tract

e. accessory organ

f. salivary gland

g. anal canal

Set Two

_____ 6. outer surface of the stomach

_____ 7. layer adjacent to the lumen of intestine

_____ 8. cell type in the muscularis

_____ 9. location of the villi

a. submucosa

b. serosa

c. mucosa

d. smooth muscle

e. peritoneal cavity

f. skeletal muscle

10. What is semidigested food in the stomach called? _____

11. Where do you find lacteals in the digestive tract? _____

12. What membrane holds the tongue to the floor of the mouth? _____

13. What part of the tooth is found above the neck? _____

14. What is the layer of a tooth superficial to the dentin? _____

15. What are the adult teeth directly posterior to the canine teeth called? _____

16. The segments or pouches of the large intestine have what particular name? _____

17. What are the names of the salivary glands adjacent to the ear? _____

18. Where is the lesser omentum found? _____

19. Where does the cystic duct take bile for storage? _____

20. Stomach acidity is in the range of pH 1 to 2. How might this inhibit the growth of ingested bacteria?

_____

21. Trace the flow of bile from the liver to the duodenum, listing all the structures that come into contact with the bile on its journey.

_____

_____

_____

22. How does the large intestine differ from the small intestine in length?

_____

23. How does the large intestine differ from the small intestine in diameter?

_____

24. Name two functions of the pancreas.

_____

_____

25. How would the fats in a high-fat diet make their way from the digestive system to the circulatory system? What would be their pathway, and why would this be significant to a person's health?

26. Label the following illustration using the terms provided.

appendix
ascending colon
duodenum
esophagus

liver
mouth
parotid gland
rectum

sigmoid colon
small intestine
stomach
tongue

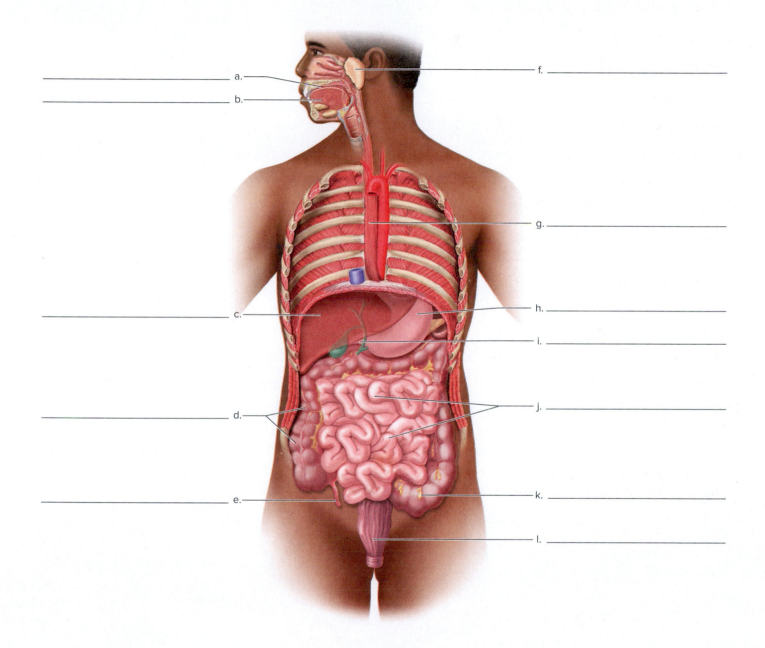

a. _____

b. _____

c. _____

d. _____

e. _____

f. _____

g. _____

h. _____

i. _____

j. _____

k. _____

l. _____

Notes

Digestive Physiology

INTRODUCTION

An understanding of the digestive process is fundamental to the study of human physiology. We obtain almost all the raw materials the body uses for growth, development, life-sustaining energy, and other metabolic functions from digestion. Digestive physiology involves the physical and chemical breakdown of material, absorption of nutrients, coordination between sections of the digestive tract and accessory organs (liver secretions and processing of absorbed foods), and elimination of fecal material. In this exercise, you look at the physical (mechanical) and chemical digestion of food. Digestive physiology is covered in the Saladin text in chapter 25, "The Digestive System," and chapter 26, "Nutrition and Metabolism."

In this experiment, you analyze several enzymes important in digestion and their effectiveness in digesting four materials commonly found in plant or animal tissue: starch, lipid, cellulose, and protein. The goal of these experiments is to determine if various enzymes in the human digestive tract can break down the four substances into smaller products. If the enzymes are effective, then starch should be broken down into sugars, lipids broken down into monoglycerides and fatty acids, cellulose into sugars, and proteins into amino acids.

Furthermore, you examine the mechanical breakdown of food that precedes most of the chemical processes. By increasing the surface area of solid food, you will determine if the rate of digestion increases.

OBJECTIVES

At the end of this exercise, you should be able to

1. discuss the importance of hydrolysis in the digestive process;
2. describe how an enzyme is important in chemical digestion;
3. list possible human digestive enzymes, as determined by experiment;
4. demonstrate the effectiveness of enzymes in food digestion;
5. describe the mechanisms involved in the mechanical digestion of food;
6. state one of the limitations of human digestive enzymes.

MATERIALS

Microscopes

General Supplies

Test tubes (15 mL each)

Test tube holders

Test tube racks

Test tube brushes and soap

Warm water bath, set at 35°C with test tube racks and thermometer

Hot water bath or hot plates and 400 mL beakers for hot bath (100°C) and thermometer

Permanent markers

Pipettes of 5 mL each

Pipette pumps

7 100 mL beakers to pour from stock bottle for distribution

1 250 mL bottle of tap water

Small spatulas

100 mL 1% pancreatin solution, fresh

300 mL 1% alpha amylase solution, fresh

100 mL 0.1% alpha amylase solution, fresh

Dropper bottles of ammonia

Dropper bottles of vinegar

Porcelain spot plates

300 mL Benedict's reagent (6 50 mL bottles)

Parafilm® squares (10 per table)

Biohazard container or container with 10% bleach or other disinfecting solution

Starch Digestion

Dropper bottles of iodine solution

250 mL of 0.5% potato starch solution

Microscope slides (1 per table)

Coverslips (1 per table)

Eyedroppers

Sugar Test

100 mL of 1% maltose solution

10 mL graduated cylinders

Cellulose Digestion

Cellulose—2 cotton balls cut into small pieces

Small dropper bottles of water

Lipid Digestion

200 mL litmus cream

Dropper bottles of "acid solution" (lemon juice or vinegar)

Dropper bottles of household ammonia

Protective gloves

Protein Digestion

25 mL of 1% BAPNA solution; BAPNA (benzoyl-DL-arginine-p-nitroanilide) is available from Sigma-Aldrich B4875

Surface Area and Digestion

2 large potatoes per class

2 breadboards

2 #4 cork borers

2 small rods (applicator sticks) to push through the borers

15 cm rulers

Razor blades

Mortars and pestles

PROCEDURE

Read all the experiments before beginning! Your instructor may want you to do these experiments in student pairs or as members of a larger group. If you do these experiments as part of a larger group, make sure you *witness* and *record* the results of your group. **Do not discard any material until everyone in your group has seen the result** or until your instructor directs you to do so. While you are waiting for one set of materials to incubate, you can begin the next set of experiments.

Label all test tubes with your *group name* and the *test tube number.* This is important because removal of the wrong tube will give you erroneous results, and the removal of a tube belonging to another group will produce the wrong result not only for your group but for the other group as well.

Tests in this exercise are divided into **reagent tests** and **experimental tests.** The reagent tests allow you to see the results of an experiment when using known samples. For example, if you are testing for a sugar, you want to use a reagent test with a known sample of sugar so you know what a positive test looks like. The experimental tests determine if a digestive reaction has taken place.

The Nature of Enzymes

The function of a digestive enzyme is to break down **macromolecules** in the stomach and small intestine into smaller molecules in a process known as **hydrolysis.** Enzymes allow chemical changes to occur at body temperatures by lowering the activation energy of chemical reactions. These reactions occur faster because of the presence of enzymes.

The material acted on by enzymes is called the **substrate,** and the result of the reaction produces the **product.** This is shown in fig. 39.1.

Starch Digestion

Starch digestion occurs by the breakdown of a large molecule of starch into smaller molecules of sugar by enzymes called **amylases.** Amylases catabolize starches as represented by fig. 39.2. One way to analyze the effectiveness of amylases is to see if the enzyme

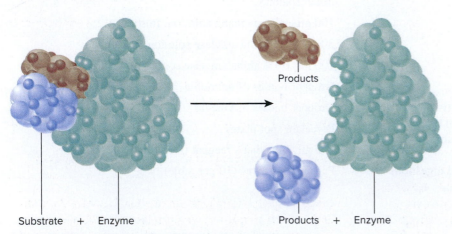

FIGURE 39.1 Catabolic Reactions with Enzymes.

Substrate + Enzyme   Products + Enzyme

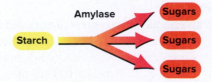

FIGURE 39.2 Starch Digestion.

removes starch from solution. The decrease in or absence of starch from solution after the reaction indicates that amylases are present.

Experiment 1: Reagent Test—Iodine

 Activity

You will use a *reagent test* to determine the presence of starch. This test uses **iodine.** If starch is present, iodine produces a *blue-black color,* a **positive result.** If starch is absent, then the solution remains *yellow,* and this is a **negative result.**

1. Label a test tube with your group name and the number 1.
2. Pour a small amount of 0.5% starch solution from the stock bottle into a small beaker. *Never return excess solution to the stock bottle!*
3. Take a pipette pump and pipette and place 1 to 2 mL of starch solution from the beaker in the test tube labeled 1.

Caution!   Never pipette material by mouth—use a pipette pump or bulb.

CAUTION

---

4. Add 4 or 5 drops of iodine solution to the starch solution in test tube 1. The solution should turn blue or black if starch is present. Save this test result for future comparison and record the outcome of your test.

Color of reaction (blue/black or yellow): _____

Test result (positive/negative for starch): _____

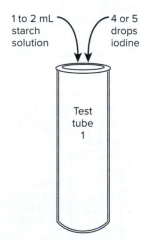

Experiment 2: Determination of Starch Digestion by Amylase

 Activity

1. Using a pipette, remove 5 mL of 0.5% starch solution from the beaker that received the stock solution, and place it in a small test tube that you have labeled with your group name and the number 2a. Label another test tube 2b and one more 2c. Add 5mL of 0.5% starch solution to each tube. In tube 2b add 1 mL of vinegar, and in tube 2c add 1 mL of ammonia.

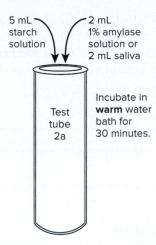

2. Into a small beaker pour a small amount of 1% amylase from the stock bottle. With a new pipette, withdraw 6 mL of amylase solution from the beaker and add 2 mL each to the starch solutions in test tubes 2a, 2b, and 2c. Discard any remaining solution in your pipette in the sink. Do not return it to the beaker. You can substitute 2 mL of saliva for the amylase if directed by your instructor. Do not use your saliva if you recently had sugar in your mouth. Remember to treat bodily fluids with standard safety precautions. Mix the solutions in the test tubes by covering the test tubes with Parafilm®, place your thumb over the Parafilm® and invert each tube a few times.
3. Incubate the mixture in test tube 2a, 2b, and 2c in a *warm* water bath (35°C) for 30 minutes.
4. After 30 minutes, take a few drops of the solution from each test tube and place them in a separate depression in a porcelain spot plate.
5. Test the solutions in the spot plate with a couple of drops of iodine solution and record your results.
6. Compare the results of your experiment with test tube 1.

Color of the solutions with iodine for:

2a _____

2b _____

2c _____

Test results (positive/negative) for:

2a _____

2b _____

2c _____

How does a change in the pH affect the enzyme activity?

_____

7. From the spot plate holding the solution from tube 2a, withdraw a small sample with an eyedropper and place a drop on a microscope slide. Add a drop from test tube 1 next to it on a slide and add a coverslip to each.

8. Examine the slide under low power and look for starch grains. Compare the sample stained with iodine from test tube 2 with that from the *undigested* starch solution (from test tube 1). Is there a difference? Record your observations in the following space.

Observations: _____

_____

Do **not** discard the contents of tube 2a at this time.

Sugar Production

The preceding experiment tested for the *decrease* in starch as a way to determine if a hydrolysis reaction had taken place. Another way to look at the effectiveness of amylase is to determine the *presence* of sugar after the exposure of starch to the enzyme. In this experiment, instead of looking for the absence of the **substrate** (starch) to determine enzyme effectiveness, you can test for the formation of the **product** (sugar). If amylase is effective, then the presence of sugar in solution, after exposure to the enzyme, indicates the conversion of starch to sugar.

Experiment 3: Reagent Test—Benedict's Test

Ⓐ **Activity**

The first test in this section involves the detection of sugar in a solution. This is done with the Benedict's test.

Caution!   Benedict's reagent is poisonous. Use caution when mixing the solutions in the test tube.

△ CAUTION

1. Label a test tube with the number 3 and your group name and then pipette 1 to 2 mL of 1% maltose solution into the test tube. Maltose is a reducing sugar.
2. Using a pipette and a pipette pump (*never* by mouth), add 1 mL of Benedict's reagent to the maltose solution in the test tube. Hold the test tube by the top and gently flick the bottom of the tube with the pad of your index finger to mix the contents.
3. Place this mixture in a *hot* (100°C) water bath for about 10 minutes. If the mixture turns green, yellow, orange, or brick red, then a reducing sugar is present.

The sequence of colors indicates the presence of sugar in increasing amounts:

Blue = 0

Green = trace

Yellow = moderate

Orange, red = large amounts

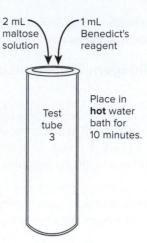

If there is no color change (for example, if the solution stays blue) after 10 minutes, then remove the tube from the bath and note the absence of sugar (a negative result). Record your results.

Color of test solution: _____

Test result (positive/negative for sugar): _____

Experiment 4: Determination of Sugar (Maltose) Production by Amylase

Ⓐ **Activity**

1. Place 4 mL from test tube 2a (the starch and amylase reaction) and put it in a test tube labeled with your group name and the number 4.
2. To the solution in test tube 4, add 4 mL of Benedict's reagent.
3. Place the tube in a hot water bath (100°C) for 10 minutes. Record your results.

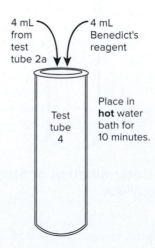

Color of solution with Benedict's reagent: _____

Presence of sugar: _____

Effectiveness of digestion: _____

Testing for Cellulose Digestion

Cellulose is composed of repeating **glucose** units. Starch is also composed of repeating glucose units. If amylase converts cellulose to sugar, then the presence of sugar in solution indicates that the enzyme amylase is effective in breaking down cellulose into glucose. This is the same process as experiment 4, except that the substrate is changed from starch to cellulose. To test for the effectiveness of amylase in digesting cellulose, incubate a cellulose mixture with amylase or saliva in a warm water bath. A positive reaction indicates the presence of sugar after incubation.

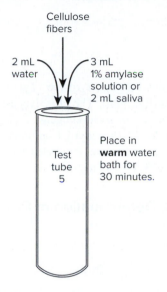

Cellulose fibers

2 mL water

3 mL 1% amylase solution or 2 mL saliva

Test tube 5

Place in **warm** water bath for 30 minutes.

Experiment 5: Determination of Sugar Production from Cellulose

(A) **Activity**

1. Place a small pinch of cellulose fibers (the size of a green pea) in a test tube and add 2 mL of water and either 3 mL of 1% amylase solution or 2 mL of saliva. You should have a total volume of 4–5 mL. Make sure the majority of cellulose fibers are not stuck to the sides of the test tube. You can use a small spatula to push the fibers into the liquid, if necessary. The test tube should be labeled with your group name and the number 5.
2. Incubate for 30 minutes in a *warm* water bath.
3. After 30 minutes, remove the solution from the bath and add 4 mL of Benedict's reagent.
4. Place the tube in a *hot* water bath (100°C) for 10 minutes. Record the presence or absence of sugar. A positive reaction to the Benedict's test indicates the presence of sugar after incubation.

Color of Benedict's test: _____

Presence of sugar: _____

Effectiveness of digestion: _____

Fat Digestion

In the digestive process, fats and oils (**triglycerides**) are broken down into **monoglycerides** and **free fatty acids.** The fatty acids make the solution more **acidic,** thus lowering the **pH.** See fig. 39.3.

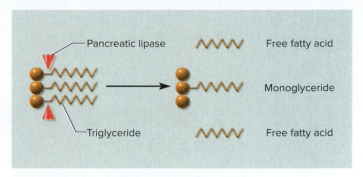

Pancreatic lipase

Triglyceride

Free fatty acid

Monoglyceride

Free fatty acid

FIGURE 39.3 Breakdown of a Triglyceride into Two Free Fatty Acids and a Monoglyceride.

The decrease in pH can be used as a measure of digestion. The greater the digestion, the greater the acidity of the solution.

Experiment 6: Reagent Test—Litmus Test

(A) **Activity**

1. To determine the effectiveness of the litmus reagent, pour 3 mL of **litmus cream** into a test tube labeled with your group name and the number 6. Litmus cream consists of dairy cream (containing fat) and litmus powder (a pH indicator).
2. Add the acid solution (lemon juice or vinegar) drop by drop while flicking the bottom of the test tube until the color changes from *blue to pink*. Do not add too much acid but just enough to change the color. A pink color indicates an acidic condition.
3. Now add household ammonia a drop at a time until you reverse the color. This indicates that the solution is alkaline.

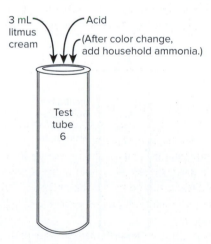

3 mL litmus cream

Acid (After color change, add household ammonia.)

Test tube 6

Experiment 7: Determination of Lipid Digestion by Pancreatin

 Activity

1. Pipette 3 mL of litmus cream into a small test tube labeled with your group name and the number 7.
2. Add 1 mL of pancreatin solution (pancreatin contains many digestive enzymes, including enzymes that digest lipids, called lipases) to the litmus cream and stir well with a spatula.

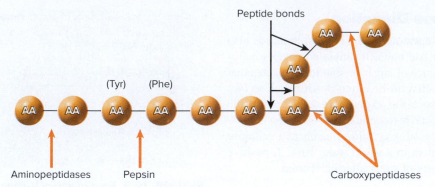

FIGURE 39.4 Enzymatic Catabolism of Proteins. "AA" represents amino acids. Carboxypeptidase removes amino acids on the carboxyl end and aminopeptidase removes amino acids on the amino end.

3. Incubate for 60 minutes in the *warm* (35°C) water bath.
4. After 60 minutes, look for a color change. If the cream turns pink, digestion has occurred. If the color remains blue, no digestion has occurred. As cream is digested, a couple of short chain fatty acids, such as butyric acid and caproic acid, are liberated. Butyric acid has the smell of rancid butter, and caproic acid is named for its similarity to the smell of goats. Record your results.

Color of solution after incubation (pink/blue): _____

Condition of the litmus cream (acidic/alkaline): _____

Extent of digestion: _____

Change in smell from normal cream: _____

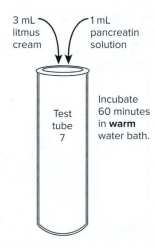

Protein Digestion

Proteins are made of **amino acids.** These amino acids are linked by **peptide bonds** to form **polypeptide chains.** If the chains are long enough, they form **proteins.** Proteins are split into amino acids by a number of digestive enzymes called proteases. Aminopeptidases cleave amino acids from the amine end of polypeptide chains. Carboxypeptidases cleave amino acids from the carboxyl end. Pepsin

cleaves polypeptides between tyrosine (Tyr) and phenylalanine (Phe) amino acids (fig. 39.4). The effectiveness of proteases can be studied with the use of a chromogenic (color-producing) substance known as BAPNA (benzoyl-DL-arginine-p-nitroanilide). If proteases are present and active, they release a yellow aniline dye from the larger BAPNA molecule.

Experiment 8: Determination of Protein Digestion by Pancreatin

Ⓐ **Activity**
  1. Label a test tube with your group name and the number 8.
  2. Pipette 1 mL of BAPNA solution into the tube. To this add 1 mL of pancreatin solution and stir by flicking the bottom of the test tube.
  3. Place the tube in the *warm* water bath (35°C) for 15 minutes.
  4. After 15 minutes, remove the tube from the bath and examine the test tube to see if it has turned yellow. Yellow indicates protein-digesting abilities of the enzyme. Record your results.

Color of the tube (yellow/clear): _____

Digestion capability (yes/no): _____

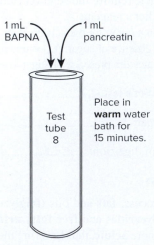

Physical Factors in Digestion

Experiment 9: Surface Area and Digestion

Ⓐ **Activity**

As food is broken down physically, the surface area increases relative to the volume of the food. This experiment examines the effectiveness of an increase in surface area for digestion.

1. Label two test tubes, each having your group name and the number 9. Write "E" for "entire" on one tube and "M" for "mashed" on the other.
2. Carefully plunge a number 4 cork borer into the center of a raw potato. *Make sure your hand is not on the receiving end of the cork borer! Use a cuttingboard.*
3. Remove the cork borer from the potato and, using a small rod, push the potato cylinder from the borer.
4. Using a ruler and a razor blade, cut the potato cylinder into two pieces, each 1 cm in length. *It is important that the two pieces are the same length!*
5. Thoroughly rinse both pieces, and place one piece of potato in the test tube labeled "E."
6. Chop the other piece into several little pieces and mash these with a mortar and pestle until they are well pulverized.
7. Carefully place *all* the mashed potato in the other test tube with the "M" label.
8. Pipette exactly 4 mL of 0.1% amylase solution* (or 0.5 mL of saliva and 4 mL water) into each test tube and incubate them for 20 minutes in the *warm* (35°C) water bath.
9. After incubation, place exactly 2 mL of Benedict's reagent into each tube and place these two tubes in the hot water bath for 10 minutes.

The amount of digestion can be estimated by the color of the test. As described in experiment 4, blue indicates no sugar, with green, yellow, orange, and brick red indicating increasing amounts

of sugar. In which tube did the greatest amount of digestion take place? In which tube did the least amount of digestion take place? Record your results.

Color of tube (entire potato): _____

Color of tube (mashed potato): _____

Relative digestion of intact piece: _____

Relative digestion of mashed piece: _____

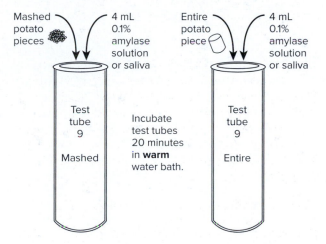

When you clean up, place all the test tubes, pipettes, or other material in either a biohazard container or a 10% bleach solution.

*You can use a solution of 0.1% alpha amylase or dilute the 1% stock solution by adding 1 mL of 1% amylase solution to 9 mL of water.

Notes

REVIEW SECTION

Digestive Physiology

Name _____ Date _____

Lab Section _____ Time _____

Review Questions

Matching

Set One

Match the enzyme in the left column with the product produced in the right column.

_____ 1. amylase

_____ 2. protease

_____ 3. lipase

a. amino acid

b. monoglycerides and free fatty acids

c. sugar

Set Two

Match the test result in the left column with the conclusion drawn in the right column. An answer may be used more than once.

_____ 4. Blue/black starch test with iodine

_____ 5. Red sugar test with Benedict's solution

_____ 6. Pink lipid test with litmus cream

_____ 7. Yellow protein test with BAPNA

_____ 8. Yellow starch test with iodine

a. starch digested

b. starch not digested

c. fat digested

d. fat not digested

e. protein digested

f. protein not digested

9. Substrates are converted into what substances by enzymes? _____

10. What is hydrolysis? _____

_____

11. What are the bonds that hold amino acids together? _____

12. Explain why a negative iodine test for starch would indicate a positive result for the enzymatic degradation of starch by amylase.

_____

_____

13. What would a positive iodine test indicate in the preceding reaction?

_____

14. Amylase is now frequently prepared with lactose or other sugars as extenders. Determine which experiments would give
erroneous results if contaminated amylase were used.

_____

_____

15. Record all your negative results. Determine if these negative results indicate digestion or no digestion.

_____

_____

_____

16. Explain why cellulose could or could not be digested by amylase. Could cellulose be considered a dietary nutrient? Is there any
value to having cellulose in the diet?

_____

_____

17. What effect does chewing your food have on digestion? What experiment did you perform to simulate the effectiveness
of chewing for digestion?

_____

_____

18. How does aminopeptidase differ from carboxypeptidase in terms of protein digestion?

_____

_____

Male Reproductive System

INTRODUCTION

The male reproductive system produces **male gametes, sperm,** or **spermatozoa;** transports the gametes to the female reproductive tract; and secretes the male reproductive hormone, **testosterone.** The **gonads,** or gamete-producing structures of the male reproductive system, are the testes (singular, *testis*). The structure and function of the male reproductive system are covered in the Saladin text in chapter 27, "The Male Reproductive System." In this exercise, you examine the gross anatomy of the male reproductive system, the histology of the system, and the male reproductive system of the cat.

OBJECTIVES

At the end of this exercise, you should be able to

1. describe the structures and functions of the male reproductive system;
2. describe the formation of sperm cells in the testis;
3. list the pathway that sperm cells follow from production to expulsion;
4. describe the anatomy of the spermatic cord;
5. list the four components of semen and where they are produced;
6. name the three cylinders of erectile tissue in the penis;
7. compare and contrast the anatomy of the cat reproductive system with that of the human.

MATERIALS

Charts, models, and illustrations of the male reproductive system

Microscopes

Prepared slides of a cross section of testis and sperm smear

Male cat

Materials for Cat Dissection

Dissection trays

Scalpel and two or three extra blades

Protective gloves

Blunt (mall) probe

Forceps and sharp scissors

First aid kit in lab or prep area

Sharps container

Animal waste disposal container

PROCEDURE

Overview of the Gross Anatomy of the Male Reproductive System

A **Activity**

Examine the models and charts of the male reproductive system available in the lab and locate the following structures in fig. 40.1:

testis

epididymis

scrotal sac (scrotum)

ductus (vas) deferens

seminal vesicle

prostate gland

bulbourethral gland

penis

urethra

Testes

A **Activity**

Examine charts and models of the testes and compare them to figs. 40.1 and 40.2. The testes are outside of the body cavity. The **testes** are paired organs wrapped in a tough connective tissue sheath called the **tunica albuginea** (fig. 40.2). They are

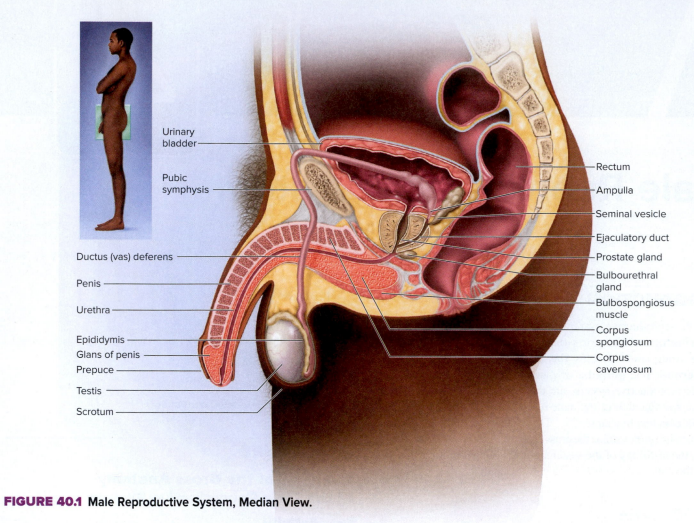

Urinary
bladder

Pubic
symphysis

Ductus (vas) deferens

Penis

Urethra

Epididymis

Glans of penis

Prepuce

Testis

Scrotum

Rectum

Ampulla

Seminal vesicle

Ejaculatory duct

Prostate gland

Bulbourethral
gland

Bulbospongiosus
muscle

Corpus
spongiosum

Corpus
cavernosum

FIGURE 40.1 Male Reproductive System, Median View.

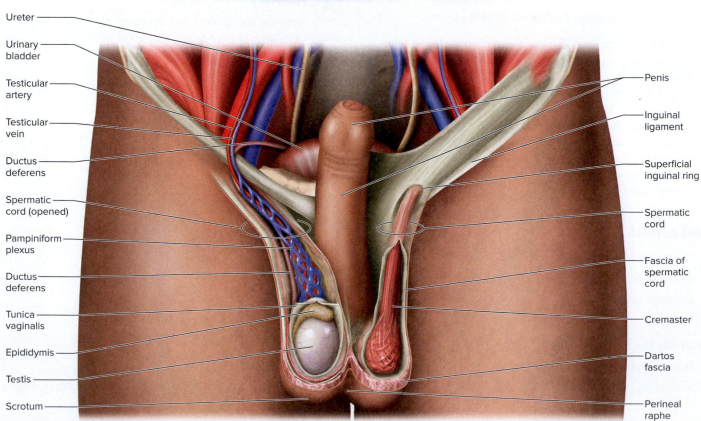

Ureter

Urinary
bladder

Testicular
artery

Testicular
vein

Ductus
deferens

Spermatic
cord (opened)

Pampiniform
plexus

Ductus
deferens

Tunica
vaginalis

Epididymis

Testis

Scrotum

Penis

Inguinal
ligament

Superficial
inguinal ring

Spermatic
cord

Fascia of
spermatic
cord

Cremaster

Dartos
fascia

Perineal
raphe

FIGURE 40.2 Scrotum, Testis, Spermatic Cord, and Penis, Anterior View.

surrounded by the **scrotal sac (scrotum),** which keeps the testes on the exterior of the body cavity, where the temperature is somewhat cooler. The testes are the site of **spermatozoa** (sperm) production, and this process must occur at about 35°C. The scrotal sac contains smooth muscle called the **dartos muscle.** These muscles contract when the testes are cold, tightening the scrotal sac and bringing the testes closer to the body. When the environment around the testes is warm, the dartos muscle relaxes and the testes descend from the body, thus becoming cooler.

The **cremaster** is located in the spermatic cord and consists of skeletal muscle. When it contracts, it brings the testes closer to the body, which increases the temperature of the testes. When it relaxes, the testes descend.

Histology of the Testis

(A) Activity

1. Examine a prepared slide of the testis under low power. Numerous tubules are seen in cross section. These are the **seminiferous tubules.** The sperm are produced in seminiferous tubules in the testis (fig. 40.3).
2. Find the triangular clusters of cells in between the tubules. These are the **interstitial (Leydig) cells.** They produce the male sex hormone, **testosterone.**
3. Examine the seminiferous tubules under high magnification. You should be able to see the outer row of cells called the **spermatogonia.** These cells reproduce by mitosis to produce **primary spermatocytes.** The primary spermatocytes undergo meiosis, or reduction division, to eventually produce

spermatozoa. The primary spermatocytes divide to form **secondary spermatocytes,** found closer to the lumen. The secondary spermatocytes become **spermatids.** Spermatids lose their remaining cytoplasm and mature into **spermatozoa.**

4. Compare this process to figs. 40.3 and 40.4 and locate the spermatogonia, primary and secondary spermatocytes, spermatids, and spermatozoa. You may see **sustentacular (Sertoli) cells,** which nourish, support, and move the sperm cells during their development.

Draw what you see in the following space. You should include the interstitial cells, seminiferous tubules, spermatogonia, spermatocytes, spermatids, and spermatozoa. You may need to look at several sections to see all these items. Your drawing may be a composite of many tubules.

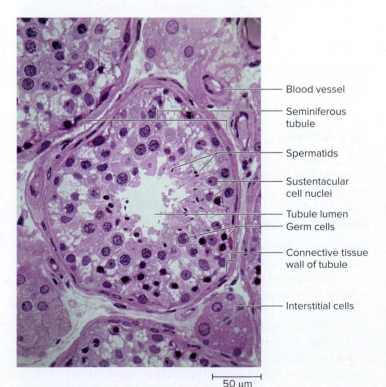

FIGURE 40.3 Histology of Testis (400×).
Ed Reschke/Getty Images

Labels on figure:
- Blood vessel
- Seminiferous tubule
- Spermatids
- Sustentacular cell nuclei
- Tubule lumen
- Germ cells
- Connective tissue wall of tubule
- Interstitial cells
- 50 μm

Sperm

The structure of an individual sperm cell consists of a **head, middle piece,** and **tail.** The head contains the genetic information (DNA), as well as a cap known as the **acrosome.** The acrosome contains digestive enzymes that digest the exterior covering of the female gamete. The middle piece of the sperm contains mitochondria that provide ATP to the tail of the sperm. The tail is a flagellum that propels the sperm forward.

(A) Activity

Examine a prepared slide of sperm and compare it to fig. 40.5. Your instructor may want you to look at this slide with the oil immersion lens. After you focus the slide on high power, swing the high-power lens away from the slide, add a drop of immersion oil to the slide, and use ONLY the fine-focus knob to examine the slide. Make sure to wipe the slide clean of immersion oil when you are done.

Epididymis

Sperm from each testis travel from seminiferous tubules through the **efferent ductules** to the **epididymis,** where they are stored and mature. Each epididymis has a blunt, rounded **head;** an elongated **body;** and a tapering **tail** that leads to the **ductus deferens** (or **vas deferens**).

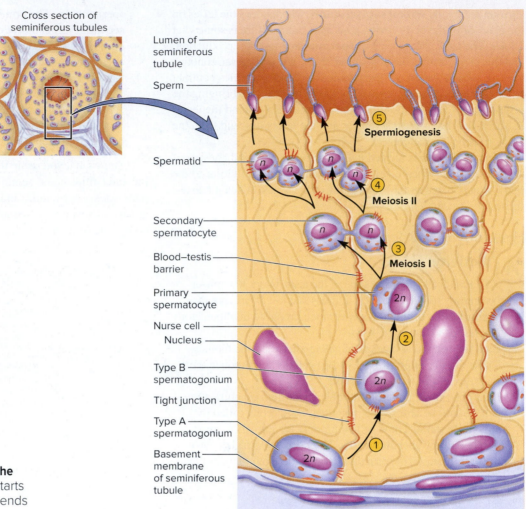

Cross section of
seminiferous tubules

Lumen of
seminiferous
tubule

Sperm

⑤
Spermiogenesis

Spermatid

④
Meiosis II

Secondary
spermatocyte

Blood–testis
barrier

③
Meiosis I

Primary
spermatocyte

Nurse cell

Nucleus

②

Type B
spermatogonium

Tight junction

Type A
spermatogonium

①

Basement
membrane
of seminiferous
tubule

**FIGURE 40.4 Spermatogenesis in the
Seminiferous Tubule.** The process starts
at number 1 (spermatogonium) and ends
with number 5 (sperm).

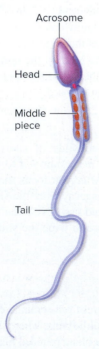

Acrosome

Head

Middle
piece

Tail

FIGURE 40.5 Sperm Cell.

Sperm maturation occurs in the epididymis. If sperm are removed from
the testis proper, they are not capable of fertilizing the female oocyte
(egg). Sperm move slowly through the coiled tubule of the epididymis.

Ⓐ **Activity**

Examine a model or chart of the longitudinal section of a
testis and epididymis and locate the structures by comparing them
to fig. 40.6.

Spermatic Cord

Sperm travel from the epididymis into the ductus deferens. The
ductus deferens is enclosed in the **spermatic cord,** a complex
cable consisting of the ductus deferens, the **testicular artery** and
vein, and the testicular **nerves** and the cremaster. The spermatic
cord is longer on the left side than on the right; therefore, the left
testis is lower than the right. Locate the structures of the spermatic
cord in figs. 40.2 and 40.6.

As the spermatic cord reaches the body wall, the ductus deferens travels around the posterior surface of the urinary bladder.
You can trace the course of the ductus deferens until it reaches the
inferior portion of the bladder. The ductus deferens enlarges somewhat here to form the **ampulla** of the ductus deferens. Each ductus

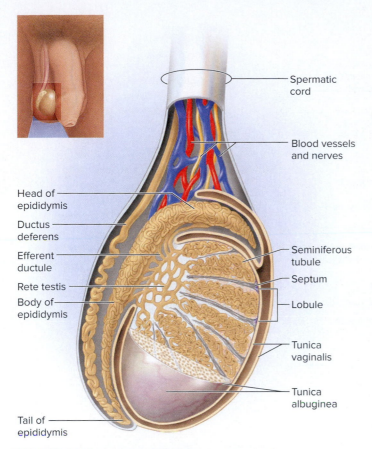

FIGURE 40.6 Testis and Epididymis, Lateral View.

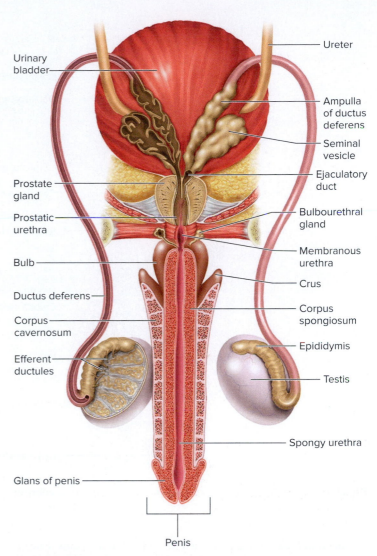

FIGURE 40.7 Urinary Bladder with Seminal Vesicles, Posterior View.

deferens joins with a **seminal vesicle,** a gland that adds fluid to the sperm. The union of the ductus deferens and the seminal vesicle produces the **ejaculatory duct.** Locate the seminal vesicle and the ejaculatory duct in fig. 40.7. The seminal vesicle fluid adds about 65% to 70% to the final volume of semen.

From this location the ejaculatory duct leads to the inferior portion of the bladder and joins with the urethra, which passes through the **prostate gland.** The prostate gland is located just inferior to the urinary bladder. The prostate gland adds a buffering fluid to the secretions of the testes and seminal vesicles. The prostate fluid makes up around 25% to 30% of the final semen volume. The seminal fluid passes through the prostatic urethra to the membranous urethra. Here the paired **bulbourethral (Cowper's) glands** are found, adding a lubricant to the seminal fluid. **Seminal fluid (semen)** consists of secretions from the seminal vesicles, prostate gland, and bulbourethral glands plus spermatozoa from the testes. Sperm and spermatic duct fluid make up about 2% to 5% of the total volume of semen. The semen passes out of the body cavity through the spongy urethra of the penis. Locate the portions of the urethra and the accessory glands in fig. 40.7.

External Genitalia

Penis

The **penis** consists of the **root, bulb,** an elongated **shaft,** and a distally expanded **glans penis.** The glans is covered with the **prepuce,** or **foreskin,** removed in some males by a procedure called a

circumcision. At the inferior portion of the glans is a ridge of tissue, the **frenulum,** which is richly supplied with sensory receptors. The glans penis is an expanded region that, along with the shaft of the penis, stimulates the genitalia of the female. The erect penis is, on average, about 16 cm in length. The anatomy of the penis can be seen in figs. 40.2 and 40.8.

The penis contains three cylinders of **erectile tissue.** The **corpus spongiosum** is the cylinder of erectile tissue that contains the **spongy urethra.** The two **corpora cavernosa** are located dorsal to the corpus spongiosum.

Ⓐ Activity

Examine a model or chart of a cross section of penis and compare it to fig. 40.8. The proximal parts of the cylinders of erectile tissue are anchored to the body. Locate the **crus,** an expansion of the corpora cavernosa. The bulb of the penis is an extension of the corpus spongiosum. The crus and the bulb form the root of the penis, as illustrated in fig. 40.7. The corpus spongiosum expands distally to form the glans penis. Note the **dorsal arteries**

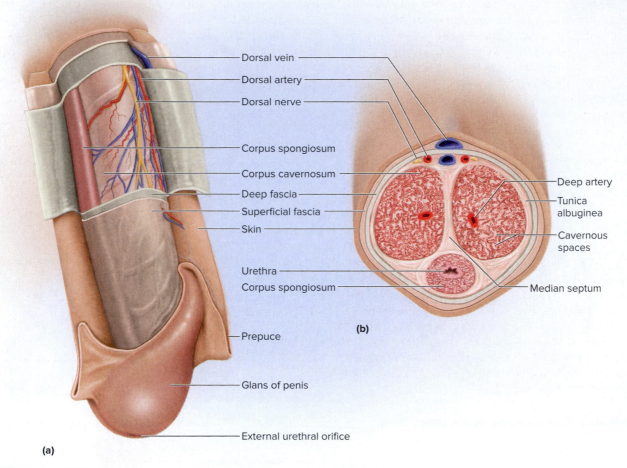

FIGURE 40.8 Penis. (a) Three-quarter view; (b) cross section.

and **deep arteries** of the penis that take blood to the penis. Locate the **dorsal vein** and the **deep dorsal vein** of the penis. When the arteries of the penis dilate, the erectile tissues engorge with blood and the penis becomes erect. The erection subsides as the arteries constrict, decreasing blood flow into the penis. Examine a model of the penis and find the features listed in fig. 40.8.

Perineum

The floor of the pelvis as seen from the outside is referred to as the **perineum.** It is a diamond-shaped structure defined by the pubic symphysis at the anterior point, the lateral points being the ischial tuberosities and the posterior point being the coccyx. The perineum can be divided into a posterior **anal triangle** and an anterior **urogenital triangle.** The anal triangle surrounds the anus, and the urogenital triangle encloses the penis and scrotum (fig. 40.9).

Cat Dissection

Ⓐ Activity

Prepare for the cat dissection by obtaining a cat and dissection equipment. Check to see whether your cat is male or female. Team up with a lab partner or group that has a cat of a different gender than your specimen, so you can learn both male

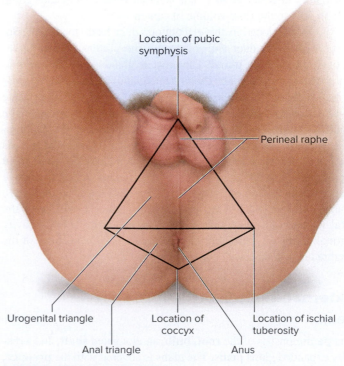

FIGURE 40.9 Male Perineum.

and female reproductive systems. The penis may be retracted in your specimen, so look for the opening of the penile urethra and the scrotal sac. Once you are sure you have a male cat, locate the **scrotal sac** and paired **testes.** Make an incision on the lateral side of the scrotal sac and locate the testis. If your cat was neutered, you will not be able to locate the testis. If the testes are present, cut through the connective tissue (the **tunica vaginalis**) of the scrotal sac and observe both the testis and the **epididymis.** The testis is covered by a tough connective tissue membrane called the **tunica albuginea.** Use fig. 40.10 as a guide. Spermatozoa move from the testis into the epididymis, where they mature.

Once you have located the epididymis, proceed in an anterior direction and trace the thin **ductus deferens** from the epididymis into the spermatic cord. Outside the body cavity the ductus deferens is contained within the spermatic cord. It is much smaller in diameter than the spermatic cord. The spermatic cord traverses the body wall on the exterior and enters the body of the cat at the inguinal ring (an opening through the inguinal ligament). You may want to gently insert a blunt probe into the inguinal ring so you can locate the ductus deferens as it passes into the body cavity. Locate the ductus deferens as it enters the body cavity and notice how it arches around the ureter on the dorsal side of the urinary bladder. You may have cut the ductus deferens in an earlier exercise, so if you cannot find it on one side, look for it on the other side.

You may want to look for the accessory organs of the male reproductive system, but this takes some significant dissection. *Check with your instructor before cutting through the pelvis of your cat.* If your instructor directs you to do so, begin by cutting through the musculature of the cat at the level of the pubic symphysis. Make a median incision through the groin, and carefully cut through the cartilage of the pubic symphysis. You should now be able to open the pelvic cavity and locate the single **prostate gland** and the paired **bulbourethral glands** (fig. 40.10). Much of the anatomy of the cat is similar to that of the human except that there are no seminal vesicles in the cat. Trace the ductus deferens from the posterior surface of the bladder through the prostate gland and the penis. You can make either a longitudinal section through the penis to trace the urethra in the erectile tissue or a cross section of the penis to see the three cylinders of erectile tissue—the corpus spongiosum and the two corpora cavernosa.

Clean Up   Remember to place all excess tissue in the appropriate waste container and not in a standard wastebasket or down the sink!

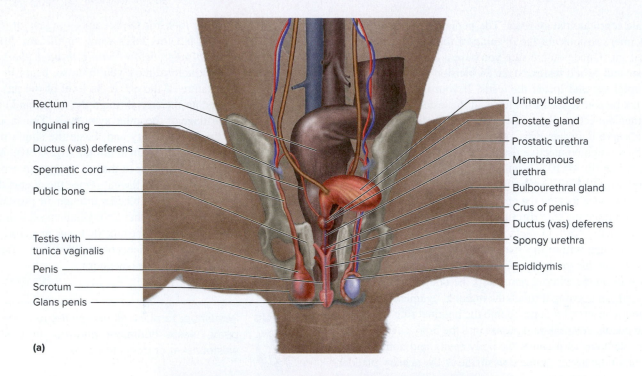

(a)

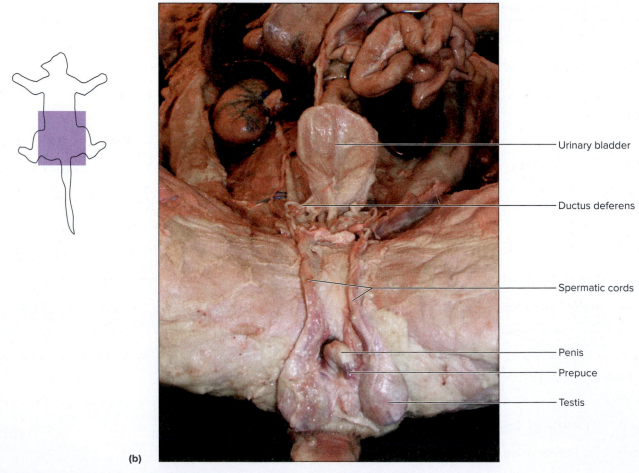

(b)

FIGURE 40.10 Male Reproductive Organs of the Cat. (a) Diagram; (b) photograph.
(b) Cordi Smith/McGraw Hill

REVIEW SECTION

Male Reproductive System

Name _____ *Date* _____

Lab Section _____ *Time* _____

Review Questions

1. What are the gonads in the male reproductive system? _____

2. The testes have both endocrine and exocrine functions. Describe the endocrine and exocrine products that come from the testes.

   _____

   _____

3. Proper sperm production must occur at what temperature? _____

4. Male sterility can result from excessively high temperatures around the testes. What adjustment is made to counteract the effects of high temperature?

   _____

5. Name the layer in the scrotal sac that consists of a smooth muscle layer. _____

6. What structure of the testis produces sperm? _____

7. What cells initiate sperm production? _____

8. Where do sperm move after being in the epididymis? _____

9. Where is the cremaster found? _____

10. List all the structures involved in producing semen. _____

    _____

11. A vasectomy is the cutting and tying of the two ductus deferens at the level of the spermatic cords. Review the percent of sperm that composes semen and determine what effect a vasectomy has on semen volume.

    _____

12. Which one of the seminal fluid glands is not a paired gland? _____

13. Name the three sections of the urethra and where they occur.

    _____

    _____

    _____

14. Where is the glans penis located? _____

15. What is the cylinder of erectile tissue ventral to the corpora cavernosa? _____

16. What male reproductive gland is missing in the cat but present in the human? _____

17. Match the structure in the left column with the function in the right column.

Structure

_____ 1. Testis

_____ 2. Corpus cavernosum

_____ 3. Epididymis

_____ 4. Prostate gland

_____ 5. Bulbourethral gland

Function

a. secretes lubricant

b. secretes buffering solution

c. produces sperm

d. erectile tissue in the penis

e. place for maturation of sperm

18. Label the following illustration using the terms provided.

bulb of penis	ductus (vas) deferens	prepuce	seminal vesicle
bulbourethral gland	epididymis	prostate gland	testis
corpus cavernosum	glans penis	scrotum	urinary bladder

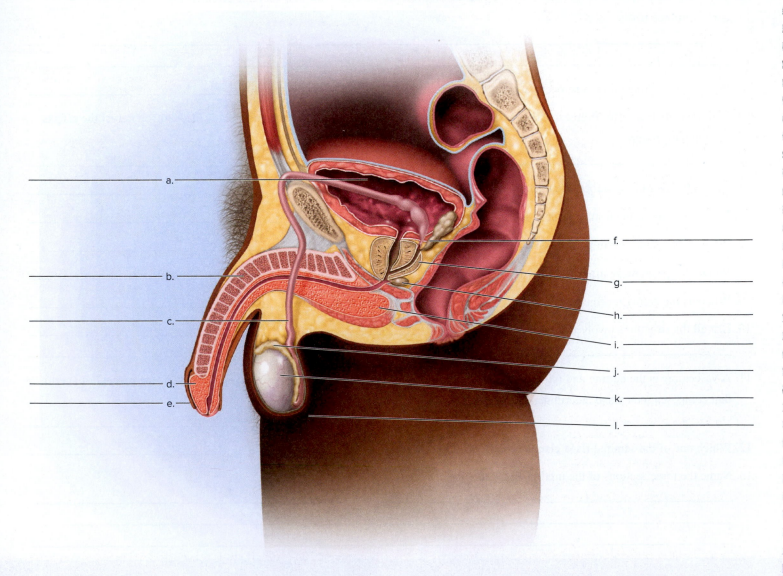

Female Reproductive System and Early Development

INTRODUCTION

The female reproductive system is functionally more complex than the male reproductive system. In the male, the reproductive system produces gametes and delivers them to the female reproductive system. The female reproductive system not only produces gametes and receives the gametes from the male but also provides space and maternal nutrients for the developing **conceptus.** Finally, the female reproductive system delivers the child into the outer environment. The female reproductive system is covered in the Saladin text in chapter 28, "The Female Reproductive System." The ovaries are the gonads or gamete-producing organs of the female reproductive system. They produce oocytes and the female sex hormones, estrogen (estradiol) and progesterone. In the first part of this exercise, you learn about the structure and function of the female reproductive system.

One of the key characteristics of life is that organisms develop, reproduce, and age. Reproduction concerns the union of sperm and oocyte and subsequent development. Sperm and oocytes are haploid, because they have half the genetic material of a somatic cell, which is diploid. Once the sperm and oocyte unite, a complete set of genetic information is produced in the cell, called a **zygote,** which is a combination of genetic material from both parents. It is amazing that this one cell develops into a structurally and functionally complex organism such as a human. All the fine anatomical details of the human body and the ability to convert food, water, and air to different molecules are nothing less than astounding.

This development must occur in a sequence organized in both time and space. In the first stage of development, increasing the number of cells is the preeminent event. After this, the cells begin to specialize and form tissues. Tissues form organs and finally organ systems.

Tissues and organs are formed in a coordinated spatial arrangement and follow a timeline of development. If the development of the body organs is out of sequence or the organs are imperfect in their formation, the results can be debilitating or even fatal. The study of the development of a human from a zygote to the fetal stage is known as **embryology. Fetal development** occurs from there to birth. The increase in size and maturation of the organs is characteristic of fetal development. Development occurs after birth as well. There are changes in height, weight, organ maturation, and sexual development. The body continues to change with age, with different organs showing different effects of aging.

These changes are covered in the Saladin text in chapter 29, "Human Development and Aging."

OBJECTIVES

At the end of this exercise, you should be able to

1. describe the structures and functions of the female reproductive system;
2. trace the pathway of a gamete from the ovary to the usual site of implantation;
3. list the structures of the vulva;
4. name the layers of the uterus;
5. compare and contrast the anatomy of the cat reproductive system with that of the human;
6. identify and describe the stages of development from fertilization to blastocyst;
7. list each of the germ layers and what develops from them.

MATERIALS

Charts, models, and illustrations of the female reproductive system

Microscopes

Prepared slides of ovary and uterus

Prepared slides of sea urchin development

Charts and models of human development

Female cat

Materials for Cat Dissection

Dissection trays

Scalpel and two or three extra blades

Protective gloves

Blunt (mall) probe

Forceps and sharp scissors

First aid kit in lab or prep area

Sharps container

Animal waste disposal container

PROCEDURE

Overview of the Gross Anatomy of the Female Reproductive System

A Activity

Examine a model or chart of the female reproductive system and locate the following major reproductive organs there and in fig. 41.1. Place a check mark next to the following structures when you locate them on materials in lab.

_____ ovary

_____ uterine (fallopian) tube

_____ uterus

_____ vaginal canal

_____ clitoris

_____ labia minora (singular, *labium minus*)

_____ labia majora (singular, *labium majus*)

Ovaries

Each oblong **ovary** is approximately 3 to 4 cm in length. The ovaries produce **oocytes,** shed from the outer surface of the ovary during **ovulation.** From here the oocytes move into the **uterine (fallopian) tube.** The oocytes do not travel in a sealed tube from the ovary to the uterine tube but are shed into the pelvic cavity. They commonly (but not exclusively) move from the outer surface of the ovary into the uterine tube.

Histology of the Ovary

A Activity

Examine a prepared slide of the ovary under the microscope on low power. Locate the **medulla,** which is a vascular, fibrous tissue in the middle of the ovary. Look for circular structures in the ovary. These are the **ovarian follicles.** Locate the **primordial follicles** in your slide and compare them to the follicles in fig. 41.2. The primordial follicles contain **primary oocytes.** More mature follicles contain **secondary oocytes.**

You should also locate the **primary, secondary,** and **tertiary follicles.** Some of the follicles may contain **oocytes.** A primary follicle has a single layer of follicular cells surrounding an oocyte, secondary follicles have multiple layers of follicular cells surrounding oocytes. Tertiary follicles contain significant amounts of fluid. The largest follicle in the ovary is the **mature ovarian**

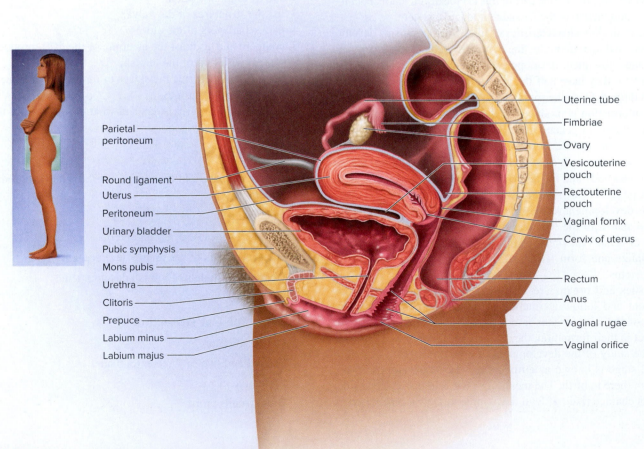

Parietal peritoneum

Round ligament

Uterus

Peritoneum

Urinary bladder

Pubic symphysis

Mons pubis

Urethra

Clitoris

Prepuce

Labium minus

Labium majus

Uterine tube

Fimbriae

Ovary

Vesicouterine pouch

Rectouterine pouch

Vaginal fornix

Cervix of uterus

Rectum

Anus

Vaginal rugae

Vaginal orifice

FIGURE 41.1 Female Reproductive System, Midsagittal View.

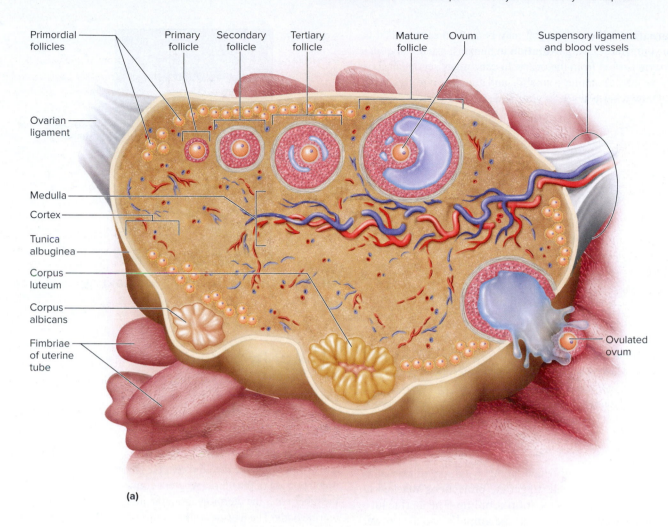

Primordial follicles

Primary follicle

Secondary follicle

Tertiary follicle

Mature follicle

Ovum

Suspensory ligament and blood vessels

Ovarian ligament

Medulla

Cortex

Tunica albuginea

Corpus luteum

Corpus albicans

Fimbriae of uterine tube

Ovulated ovum

(a)

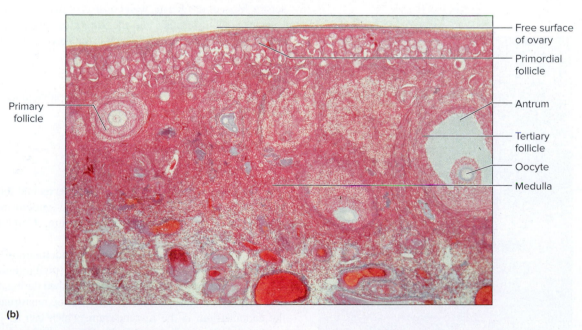

Free surface of ovary

Primordial follicle

Antrum

Tertiary follicle

Oocyte

Medulla

Primary follicle

(b)

FIGURE 41.2 **Histology of the Ovary.** (a) Diagram of human ovary; (b) photomicrograph of cat ovary (40✕).
(b) ©Eric Wise

follicle (graafian follicle), and you may be able to see one if it is present in your slide. During **ovulation** in humans, usually one secondary oocyte is shed from the ovary. In cats, many secondary oocytes may be shed. Examine your slide and compare it to figs. 41.2 and 41.3. Draw what you see in your slide in the following space.

FIGURE 41.4 Post-Ovulatory Ovary (40×).
©Eric Wise

After ovulation, the remains of a mature ovarian follicle become a **corpus luteum,** which primarily secretes **progesterone.** If pregnancy does not occur, the corpus luteum decreases in size and becomes the **corpus albicans.** Examine a prepared slide of the ovary with a corpus luteum or corpus albicans and compare it to figs. 41.2 and 41.4.

Uterine Tubes

The uterine tube has a small fringe on the distal region known as the **fimbriae.** These are small, fingerlike projections attached to an expanded region known as the **infundibulum.** The uterine tube also has an enlarged region known as the **ampulla** and a narrower portion toward the uterus.

(A) Activity
Examine a model or chart of the female reproductive system and compare it to fig. 41.5.

Uterus

The **uterus** is a pear-shaped organ with a domed **fundus;** a **body;** and an inferior end called the **cervix.** The cervix is a cylindrical, terminal portion of the uterus where it connects to the vagina. The uterine tubes enter the uterus at about the junction of the fundus with the uterine body. The uterine wall is composed of three layers. The outer surface of the uterus is called the **perimetrium.** This is the uterine serosa. Most of the uterine wall consists of the **myometrium,** a thick layer of smooth muscle, and the innermost (deepest) layer of the uterus is the **endometrium,** which is the mucosa of the uterus.

(A) Activity
Examine the models or charts in the lab and locate the structures of the uterus in fig. 41.5.

Histology of the Uterus

(A) Activity
Examine a prepared slide of the **uterus** and locate its three layers. Locate the outer perimetrium, the smooth muscle of the myometrium, and the inner endometrium. Compare these to figs. 41.5 and 41.6.

Now examine the two layers of the endometrium of the uterus under higher magnification. These contain spiral arterioles and uterine glands. The **functional layer** is the one shed during menstruation. The **basal layer** is deeper and not shed during menstruation. Deep to the endometrium is the myometrium, which can be distinguished by the smooth muscle found in it. The myometrium also contains straight arteries and arterioles.

Pap smears are scrapings of the cervix to check for the presence of abnormal cells. Normal epithelial cells of the cervix have a

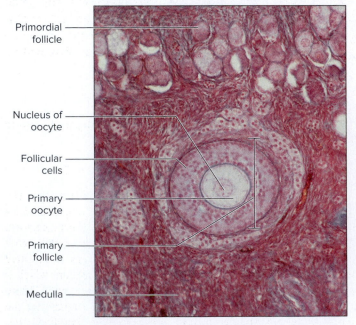

FIGURE 41.3 Primary Oocyte in Follicle (100×).
©Eric Wise

Primordial follicle

Nucleus of oocyte

Follicular cells

Primary oocyte

Primary follicle

Medulla

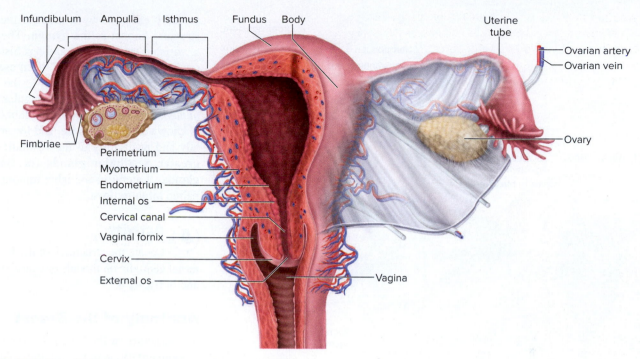

FIGURE 41.5 Female Reproductive System, Posterior View.

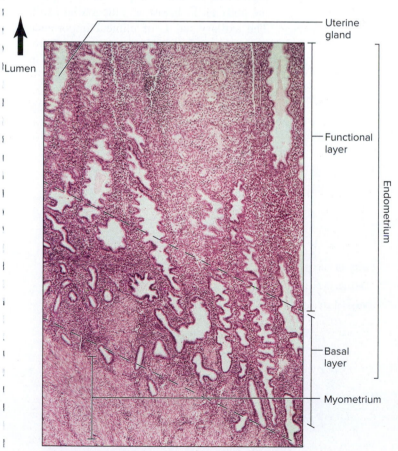

FIGURE 41.6 Histology of the Uterus, Secretory Phase (40×).
©Eric Wise

relatively small nuclear to cytoplasmic ratio. When the nucleus occupies a larger portion of the cell, there is concern for cervical cancer. Regular pap smears are important because early detection significantly reduces the threat of cervical cancer.

Ligaments

The uterus and ovaries are suspended in the pelvic cavity by a number of connective tissue sheaths called ligaments. The **broad ligament** anchors the uterus to the lateral pelvic wall. The **round ligament** attaches the uterus to the anterior body wall at about the region of the inguinal canal. The **ovarian ligament** directly attaches the ovary to the uterus, and the **suspensory ligament** attaches the ovaries to the lumbar region.

Ⓐ Activity

Locate the ligaments in models or charts in the lab and in figure 41.7.

Vagina

The vagina consists of the **vaginal canal** and the **vaginal orifice.** The uterus joins with the vaginal canal at the cervix. The vaginal canal is a tough, muscular tube with a recessed region around the cervix known as the **vaginal fornix.** The vagina is about 8–10 cm long, although it can stretch considerably during intercourse and delivery. It is located between the urethra, on the anterior side, and the rectum, which is posterior. The outer layer of the vaginal wall is the adventitia. There is a middle muscularis layer, and the layer of the vagina near the lumen is the mucosa. The mucosa is composed of stratified squamous epithelium in adult women. The vaginal canal is poorly supplied with nerves, and the wall of the vagina has cross ridges called

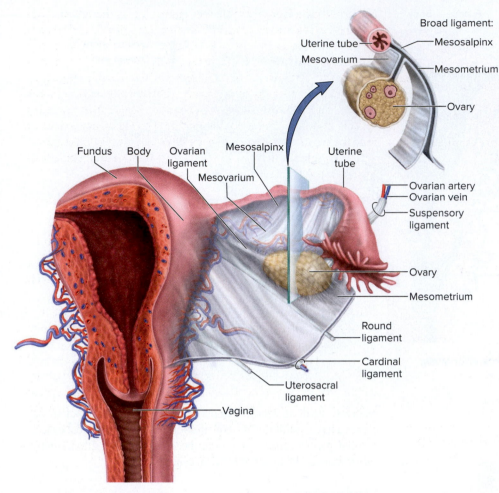

FIGURE 41.7 Ligaments of the Female Reproductive System.

vaginal rugae. The vaginal canal and orifice can expand greatly during the delivery of a child. Locate the vaginal canal, vaginal orifice, fornix, and vaginal rugae in figs. 41.1 and 41.5.

External Genitalia

As in the male, the floor of the pelvis as seen from the outside is referred to as the **perineum.** It can be divided into a posterior **anal triangle,** a region enclosing the anus, and an anterior **urogenital triangle** consisting of the reproductive and urinary structures (fig. 41.8).

The external female reproductive structures are called the external genitalia and are collectively referred to as the **vulva.** The **mons pubis** is the anterior-most structure of the vulva and is an adipose pad that overlies the pubic symphysis. Posterior to the mons pubis is the **clitoris,** a cylinder of erectile tissue embedded in the body wall that terminates anterior to the urethral orifice as the **glans clitoris.** The clitoris has the same embryonic origin as the penis in males, and like the penis it is richly supplied with nerve endings. The body of the clitoris is a curved structure illustrated in fig. 41.1, and the glans clitoris is the terminal portion. The glans clitoris is enclosed by the **prepuce,** an extension of the **labia minora.** Posterior to the clitoris is the **external urethral orifice** and posterior to this is the **vaginal orifice.** The vaginal orifice is

partially enclosed by a mucous membrane structure known as the **hymen.** The hymen is variable anatomically and has historically (and sometimes incorrectly) been used as an indicator of virginity. Lateral to the vaginal orifice are the labia minora (singular, *labium minus*). The space between the labia minora is known as the **vestibule,** and located laterally and posteriorly to the vestibule are the **greater vestibular glands** (or **Bartholin glands**). Lateral to the labia minora are the paired **labia majora.**

(A) Activity

Locate the structures of the female external genitalia on models or charts in the lab and in fig. 41.8.

Anatomy of the Breast

The structure of the breast derives from the integumentary system, yet the role of the female breast in reproduction is important as a source of nourishment for the offspring. The major structures of the external breast are the pigmented **areola,** the protruding **nipple,** the **body** of the breast, and the **axillary tail.** The axillary tail is of clinical importance in that breast tumors frequently occur there. Examine the surface features of the breast in fig. 41.9 and locate the structures listed.

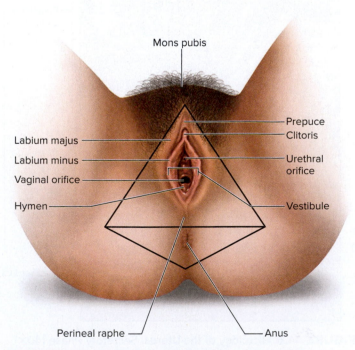

FIGURE 41.8 Female Perineum and External Genitalia.

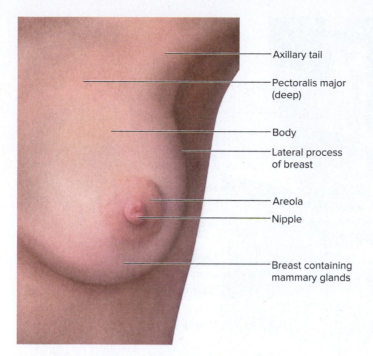

FIGURE 41.9 Surface Features of the Female Breast.

Axillary tail

Pectoralis major (deep)

Body

Lateral process of breast

Areola

Nipple

Breast containing mammary glands

Ⓐ **Activity**

Compare models or charts in the lab and locate the internal structures of the breast. The breast is anchored to the pectoralis major muscle with **suspensory ligaments.** Much of the breast is

composed of **adipose tissue,** and embedded in the adipose tissue are the **mammary glands.** The mammary glands are responsible for the production of milk in lactating females. The glands are clustered in **lobes,** and there are about 15 to 20 lobes in each breast. The mammary glands increase in size during pregnancy and lead to **lactiferous ducts,** which subsequently lead to **lactiferous sinuses (ampullae)** that exit via the nipple. Humans have several ampullae leading to each nipple. The mammary glands in females begin to undergo changes prior to puberty and become functional glands after delivery of a child. Note the features illustrated in fig. 41.10.

Cat Dissection

Ⓐ **Activity**

Prepare for the cat dissection by obtaining a dissection tray, scalpel, scissors or bone cutter, string, forceps, and plastic bag with label.

You probably already removed the multiple mammary glands of the female cat during the removal of the skin. If this is the case, then examine the external genitalia for the **urogenital orifice.** *Check with your instructor before cutting through the pelvis of your cat.* If your instructor directs you to do so, then begin by cutting through the musculature of the cat at the level of the pubic symphysis. Continue to cut in the midsagittal plane through the cartilage of the pubic symphysis and then cranially through the abdominal muscles, as described in the dissection of the male cat. This should expose the reproductive organs (fig. 41.11).

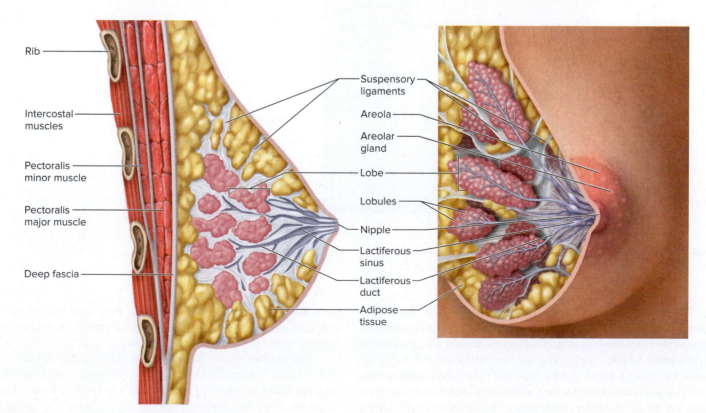

Rib

Intercostal muscles

Pectoralis minor muscle

Pectoralis major muscle

Deep fascia

Suspensory ligaments

Areola

Areolar gland

Lobe

Lobules

Nipple

Lactiferous sinus

Lactiferous duct

Adipose tissue

FIGURE 41.10 Interior of the Female Breast.

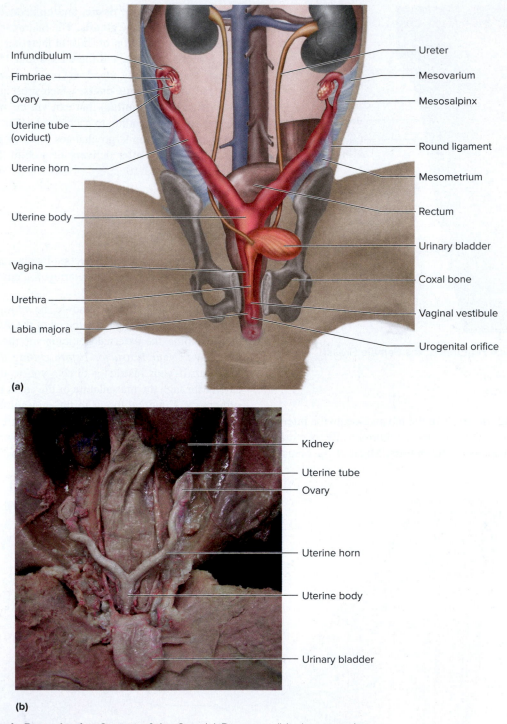

(a)

Kidney

Uterine tube

Ovary

Uterine horn

Uterine body

Urinary bladder

(b)

FIGURE 41.11 **Female Reproductive Organs of the Cat.** (a) Diagram; (b) photograph.
(b) Cordi Smith/McGraw Hill

Unlike the human female, the cat has a **horned (bipartite) uterus.** The uterus of a cat has the appearance of a Y, with the upper two branches being the **uterine horns** and the stem of the Y the **body** of the uterus. The uterine horns are not fallopian tubes but extensions of the uterus itself. They divide the uterus into two sections. The uterine (fallopian) tubes in cats are just adjacent to the fimbriae. Humans normally have single births from a pregnancy, while the expanded uterus in cats facilitates multiple births. If your cat is pregnant, the uterus will be greatly enlarged, and you may find many fetuses inside. If your cat was spayed, then the ovaries were removed.

The **ovaries** in cats are caudal to the kidneys and are relatively small organs. Examine the paired ovaries and the short **uterine tubes (oviducts)** in the cat. If you have difficulty locating them, trace the uterus toward the uterine horns and locate the ovaries. The uterine tubes in the human female are proportionally longer than those in the cat. The opening of the uterine tube near the ovary is called the **ostium,** and it receives the **oocytes** during **ovulation.**

At the termination of the uterus is the **cervix,** which leads to the **vaginal canal.** The vagina in cats is different than in humans in that the **urethral opening** is internally enclosed in the vaginal canal. This region where the vagina and the urethral opening occurs is called the **vaginal vestibule.** Thus, the opening to the external environment is a common urinary and reproductive outlet called the **urogenital orifice.** Locate these structures in fig. 41.11.

Clean Up Remember to place all excess tissue in the appropriate waste container and not in a standard wastebasket or down the sink!

Stages of Development

Overview

Sperm and egg unite in a process known as **fertilization** that initiates a remarkable phenomenon of growth and differentiation from the single-celled zygote to the adult human. These early cells undergo mitotic divisions, producing cells with the same amount of DNA as the parent cell. Fertilization usually occurs in the uterine tube, and the zygote divides into two cells called blastomeres. These divide to produce more blastomeres. The first division occurs in a vertical plane producing two cells, and the second division occurs in the horizontal plane, producing a total of four cells (fig. 41.12). The cells continue to divide, forming a solid cluster of cells called a **morula** (derived from Latin in reference to the cells resembling a mulberry). From fertilization, it takes about 72 hours to become a morula, which has not yet implanted in the endometrium of the uterus. Note that the size of the four-cell stage, eight-cell stage, and morula is not significantly different from the size of the zygote. This is due to the simple division of cells without increasing the volume of the cytoplasm in each cell.

Eventually, the cluster of cells forms a hollow center, and this developmental stage is called a **blastocyst.** At about six days after fertilization, the blastocyst implants in the uterus. There are more cells clustered at one end of the blastocyst, and this creates a difference in the organization of the cells. It establishes the polarity, or axis, of the body, which has implications for development. The covering of cells on the outside of the blastocyst is called the **trophoblast,** and the cluster of cells on the inside is known as the **inner cell mass.**

In the next stage in development, the blastocyst becomes a **gastrula;** this occurs when a pocket of cells forms inside the

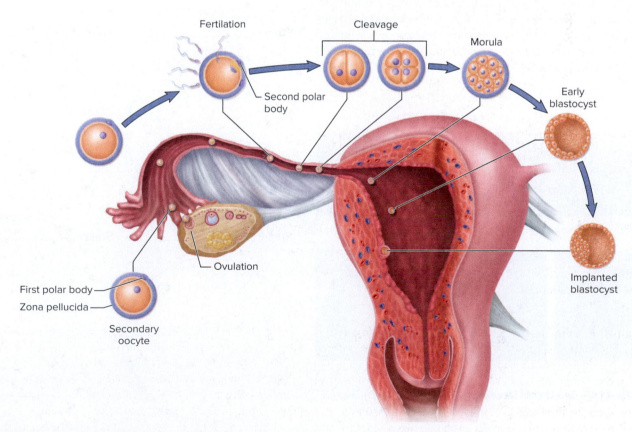

FIGURE 41.12 Early Stages of Human Development.

hollow ball of cells (figure 41.13). This is similar to pushing your fist into a large balloon. This invagination produces cells that develop into an embryonic tissue known as the **endoderm.** On the outer surface of the gastrula, the cells form the **ecto-derm,** and some cells develop between these two layers to form the **mesoderm.** These tissues will be discussed later in the lab exercise. In humans the opening to the pocket becomes the site of the anus, and a second opening develops at the opposite end and becomes the mouth.

Because of the polarity in the body pattern, cells in certain parts of the developing gastrula shut off some genes while turning on other genes. These genetic triggers lead to the development of organs in specific regions of the body. The first part of develop-ment involves increasing the number of cells. The second part is specialization of the cells to form tissues. Later the tissues assem-ble into organs and organ systems.

Nourishment of the conceptus first takes place by glycogen storage in the endometrium. Later the placenta takes over the nourishment of the conceptus, providing oxygen and removing carbon dioxide as well. The placenta also produces hormones that maintain pregnancy.

Microscopic Examination of Development

(A) Activity

Much of the early work on embryology was done on sea urchins as an animal model because of the similarities in their development to that of humans. In both sea urchins and humans, cleavage is the fundamental first step in development. Exam-ine prepared slides in microscopes or pictures of sea urchin embryo development (fig. 41.13) and find the different stages of development. In the following circles, draw and label these stages.

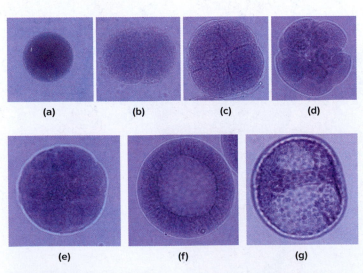

(a)   (b)   (c)   (d)

(e)   (f)   (g)

FIGURE 41.13 Sea Urchin Development. (a) Zygote. (b) Two cells. (c) Four cells. (d) Eight cells. (e) Morula. (f) Blastocyst. (g) Gastrula.

©Eric Wise

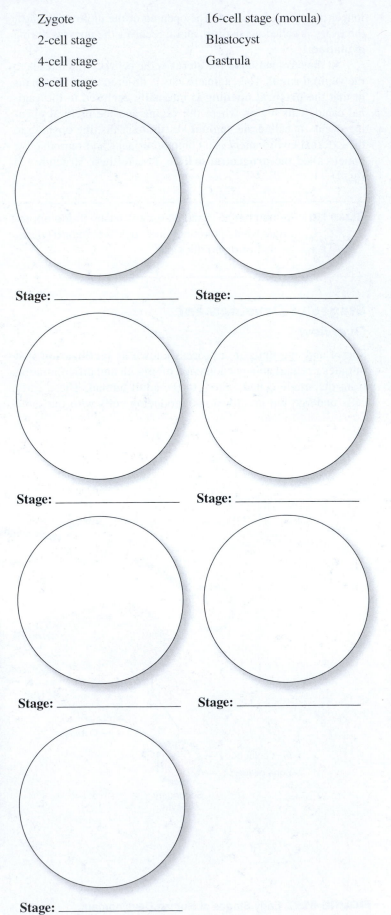

Zygote   16-cell stage (morula)

2-cell stage   Blastocyst

4-cell stage   Gastrula

8-cell stage

Stage: _____   Stage: _____

Stage: _____   Stage: _____

Stage: _____   Stage: _____

Stage: _____

Embryonic Tissues

The early stage of development continues to progress, and the formation of three embryonic tissues occurs. These are the **ectoderm,** the **mesoderm,** and the **endoderm.** The ectoderm gives rise to the outer layer of skin and the nervous tissue, the mesoderm gives rise to bones and muscles, and the endoderm gives rise to many internal organs, such as digestive and respiratory organs. The early development of these layers is illustrated in fig. 41.14.

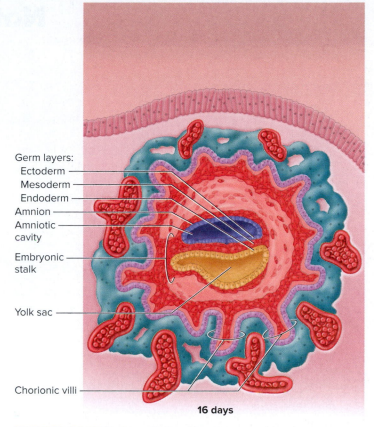

FIGURE 41.14 Embryonic Development.

Notes

REVIEW SECTION

Female Reproductive System and Early Development

Name _____   *Date* _____

Lab Section _____   *Time* _____

Review Questions

1. What are the gonads in the female reproductive system? _____

2. What is the inner layer of the uterus called? _____

3. The ovaries attach to the uterus by what structure? _____

4. Where is the fornix in the female reproductive system?

   _____

5. Which is more anterior, the urethral opening or the clitoris? _____

6. What is the material in the center of the ovary called? _____

7. What is the name for the expulsion of the oocyte from the ovary? _____

8. What is the layer of the endometrium closest to the myometrium called? _____

9. What is the name of the part of the breast that is near the axilla? _____

10. What are the milk-producing glands of the breast called? _____

11. A zygote is formed from the fusion of what two cells? _____

12. Embryonic tissue consists of three layers. What are these called?

    _____

13. What is a morula? _____

14. Trace the pathway of milk from the mammary glands to expulsion.

    _____

    _____

15. Ectopic pregnancies are those that occur outside of the endometrial layer of the uterus. Provide an explanation for how pregnancies may occur in the uterine tube (thus a tubal pregnancy) or in the abdominopelvic cavity.

_____

_____

16. How is the uterus in the cat different from the uterus in humans in terms of structure and function?

_____

_____

_____

17. Label the following illustration using the terms provided.

cervix	labium minus	vaginal canal	vaginal orifice
clitoris	urinary bladder	vaginal fornix	vaginal rugae
fundus			

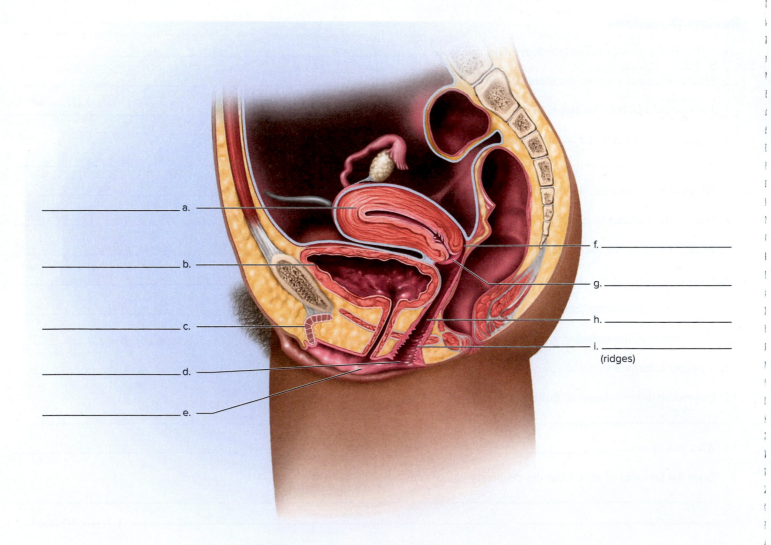

18. What is embryology?

_____

19. Define blastomeres.

_____

20. How can you distinguish between the blastocyst and the gastrula stage?

_____

21. Why is it critical to divide the chromosome number in half prior to reproducing?

_____

22. What is the stage called when the cell mass forms a hollow cavity?

_____

23. Examine your text or a reputable online source and determine what embryonic layers form the following structures:

   a. brain _____

   b. urinary bladder _____

   c. femur _____

Notes

APPENDIX A

MEASUREMENT CONVERSION

The **metric system** is universally used in science to measure certain **values** or **quantities.** These values are **length, volume, mass** (weight), **time,** and **temperature.** The metric system is based in units of 10, and conversion to higher or lower values is relatively easy when compared to using the U.S. customary system. In the United States, length is typically measured in inches, feet, yards, or miles. The **base unit** for length in the metric system is the meter. As shown in table A.1, 1 meter is equivalent to 1.09 yards, or 39.4 inches. When the measurement of great lengths makes the use of meters impractical, then the kilometer (1,000 meters) is used. When the measurement of extremely small lengths is needed, then the centimeter (1/100 of a meter), millimeter, micrometer, nanometer, and angstrom are used. The common multiples or fractions for the values are outlined in table A.1.

TABLE A.1	Conversions of Standard and Metric Systems				
Value	**Base Unit**	**×1,000**	**1/100**	**1/1,000**	**1/1,000,000**
Length	Meter (m)	Kilometer (km)	Centimeter (cm)	Millimeter	Micrometer
	1 m = 39.4 in.	1 km = ⅝ mi	2.54 cm = 1 in.		(1 nanometer = 10^{-9} m)
	1 m = 1.09 yd				(1 angstrom = 10^{-10} m)
Mass	Gram (g)	Kilogram (kg)	Centigram	Milligram	Microgram
	454 g = 1 lb	1 kg = 2.2 lb			
	1 g = .036 oz				
Volume	Liter (L)	Kiloliter	Centiliter	Milliliter (mL)	Microliter
	1 liter = 1.06 qt			1 mL = 1 cc	
	1 gal = 3.78 liters				
Time	Second (sec)	Kilosecond	Centisecond	Millisecond	Microsecond
Temperature	Degrees/Celsius				
	0°C = freezing				
	100°C = boiling				
	°F = ⅘° C + 32				

PREPARATION OF MATERIALS AND SOLUTIONS

The following solutions are designed for a lab of 24 students. It is best to estimate how many milliliters of stock solutions are required for all the students in a lab and then double that amount. The preparations are listed alphabetically, and the solutions are the approximate volume for the needs of a class of 24. The number of the lab exercise follows the solution description for cross-referencing.

Acetylcholine Chloride Solution (0.1%)

Add 0.1 g of acetylcholine chloride to 100 mL frog Ringer's solution. Pour into a small, labeled dropper bottle. (Laboratory Exercise 28)

Acid Solution

Pour 200 mL lemon juice or vinegar into dropper bottles labeled "Acid Solution." (Laboratory Exercise 39)

Agar Plates

Add 20 g agar to enough water to make 1 liter of solution. Boil and stir the agar until it all dissolves. Pour into Petri dishes for three dishes per table. (Laboratory Exercise 4)

Alpha Amylase Solution (0.1%)

Add 0.1 g of alpha amylase to a graduated cylinder and fill to 100 mL with water or make a 0.1 serial dilution by taking 10 mL of the following solution and adding 90 mL of water. Pour into a small, labeled bottle or beaker. (Laboratory Exercise 39)

Alpha Amylase Solution (1%)

Add 3 g of alpha amylase to a graduated cylinder and add water to the 300 mL mark. Pour into a small, labeled bottle or beaker. (Laboratory Exercise 39)

Ammonia

Small bottle of ammonia. (Laboratory Exercises 20, 34, and 39)

BAPNA Solution (1%)

N-alpha-benzoyl-DL-arginine-p-nitroanilide hydrochloride. Add 0.5 g of BAPNA powder in 50 mL of water. Pour into two bottles of 25 mL each. BAPNA is an expensive material, but you do not need much of it for the lab. Available from Sigma-Aldrich B4875. (Laboratory Exercise 39)

Benedict's Reagent

A copper sulfate solution that turns color if reducing sugars are present and remains blue in the absence of reducing sugars. Add 35 g sodium citrate and 20 g sodium carbonate (Na_2CO_3) to 160 mL water. Filter through paper into a glass beaker. Dissolve 3.5 g copper sulfate ($CuSO_4$) in 40 mL water. Pour the copper sulfate solution into the 160 mL, stirring constantly. Add to six 50 mL bottles. Label "Benedict's Reagent." (Laboratory Exercise 39)

Bleach Solution (10%)

Mix 100 mL household bleach (sodium hypochlorite) with 900 mL tap water. (Laboratory Exercises 24, 25, and 37)

Caffeine Solution, Saturated

Add small amounts of caffeine to 50 mL water until no more will dissolve. Decant the solution into small dropper bottles. (Laboratory Exercise 28)

Calcium Chloride Solution (2%)

Weigh 5 g calcium chloride and place in a graduated cylinder. Add frog Ringer's solution to make 250 mL. Pour in dropper bottles. (Laboratory Exercise 28)

Cat Wetting Solution

Numerous formulations are available for keeping preserved specimens moist. Some commercial preparations are available that reduce the exposure of students to formalin or phenol. You may not need any wetting solution if the cats are kept in a securely closed plastic bag. You can make a wetting solution by putting 75 mL formalin, 100 mL glycerol, and 825 mL distilled water in a 1-liter squeeze bottle. Another mixture consists of equal parts Lysol and water. (Laboratory Exercises 11–14, 29–30, 40, and 41)

Cellulose

Cut several (3–4) g pure cotton (cotton wool, cotton balls) into fine pieces (0.5 cm or less). Label "Cellulose." (Laboratory Exercise 39)

Epinephrine Solution (0.1%)

Add 0.1 g of adrenalin chloride to 100 mL frog Ringer's solution. Label and pour the solution into small dropper bottles. (Laboratory Exercise 28)

Essential Oils Preparation

Fill several small screw-top vials with peppermint, almond, wintergreen, and camphor oils (available from local drug stores). Label "Peppermint," "Almond," "Wintergreen," and "Camphor," respectively, and keep vials in separate wide-mouthed jars to prevent cross-contamination of scent. (Laboratory Exercise 20)

Fill four small vials halfway to the top with cotton and color them red with food coloring. Label the vials "Wild Cherry" and add benzaldehyde solution until the cotton is moist. Fill four small vials halfway to the top with cotton. Add benzaldehyde solution until the cotton is moist. Label "Almond." (Laboratory Exercise 20)

Filtration Solution (1% Starch, Charcoal, and Copper Sulfate Solution)

Place 5 g starch, 5 g powdered charcoal, and 5 g copper sulfate ($CUSO_4$) in a 1-liter beaker. Add enough water to make 500 mL. Stir well and pour into a 500 mL bottle. Label "Filtration Solution." (Laboratory Exercise 4)

Frog Ringer's Solution

Weigh and place the following materials in a 1-liter graduated cylinder:

6.5 g NaCl (sodium chloride)

0.2 g $NaHCO_3$ (sodium bicarbonate)

0.1 g CaCl$_2$ (calcium chloride)

0.1 g KCl (potassium chloride)

To these add enough water to make 1,000 mL. This solution should be prepared fresh and used within a few weeks. (Laboratory Exercises 18 and 28) (In Laboratory Exercise 28, three solutions are needed—one at room temperature, one at 37°C, and one in an ice bath.)

Hydrochloric Acid Solution (0.1%)

Add 1 mL concentrated HCl to 1 liter of water. Remember "AAA"—Always Add Acid to water. (Laboratory Exercise 18)

India Ink

Dropper bottles of India ink. (Laboratory Exercise 4)

Iodine Solution

Prepare by adding 10 g I$_2$ (iodine) and 20 g KI (potassium iodide) to 1 liter of distilled water. Store in small, dark dropper bottles. Label "Iodine Solution." (Laboratory Exercises 4 and 39)

Litmus Cream

Use approximately 250 mL heavy cream. To this add powdered litmus until the cream is a light blue. Pour into two separate bottles and label "Litmus Cream." (Laboratory Exercise 39)

Litmus Solution

Weigh 2 g litmus powder and dissolve in 600 mL water. Titrate HCl into the solution until it begins to turn red; then add NaOH solution by drops until it just turns back to blue. Pour into two bottles. (Laboratory Exercise 34)

Maltose Solution (1%)

Add 1 g maltose in enough water to make 100 mL. Stir until dissolved and pour into two clean bottles. Label "1% Maltose Solution." (Laboratory Exercise 39)

Methylene Blue (1%)

Add 5 g methylene blue powder to 500 mL distilled water. Pour into dropper bottles. (Laboratory Exercise 3)

Methylene Blue Solution (0.01 M)

Add 3.2 g methylene blue (MW 320) to distilled water to make 1 liter of solution. Place in dropper bottles. (Laboratory Exercise 4)

Molasses or Concentrated Sucrose Solution (20%)

Use undiluted molasses or a 20% sugar solution. To make the sugar solution, add 100 g table sugar (sucrose) to water to make 500 mL of solution. Make sure the sucrose is completely dissolved. (Laboratory Exercise 4)

Nitric Acid (1 N) (for Decalcifying Bones)

Add 64 mL concentrated nitric acid (70%) slowly to water to make 1 liter of solution. Remember "AAA"—Always Add Acid to water. (Laboratory Exercise 7)

Pancreatin Solution (1%)

Place 1 g pancreatin powder in a graduated cylinder and add water to make 100 mL. Stir well. Adjust the pH with 0.05 M sodium bicarbonate until neutral (pH 7). Pour into different stock bottles and label "1% Pancreatin Solution." Preparation note: Use fresh pancreatin. Pancreatin may be stored frozen (not in a frost-free freezer that regularly cycles between freezing and defrosting). Commercially prepared pancreatin has an optimum pH. If the pH is too low, the reaction will be slowed or stopped. (Laboratory Exercise 39)

Perfume, Dilute Solution

Add 10 mL inexpensive perfume to 50 mL denatured ethyl alcohol. (Laboratory Exercise 20)

Phosphate Buffer Solution

Add 3.3 g potassium phosphate (monobasic) and 1.3 g sodium phosphate (dibasic) to 500 mL water. Place in squeeze bottles. (Laboratory Exercise 24)

Potassium Permanganate Solution (0.01 M)

Add 1.58 g potassium permanganate (MW 158) crystals to water to make 1 liter of solution. Label and pour into dark brown dropper bottles. (Laboratory Exercise 4)

Procaine Hydrochloride

Place 1 g procaine hydrochloride solution in 1 mL water. Add to this 30 mL pure ethanol. Place in small screw-capped bottle. (Laboratory Exercise 18)

Quinine Solution

(see Tonic Water)

Salt Solution (3%)

Add 15 g NaCl crystals (food-grade table salt) to water to make 500 mL solution. Fill clean, food-grade dropper bottles labeled "Salty." (Laboratory Exercise 20)

Sodium Chloride Solution (0.9%) (Physiological Saline)

Put 9.0 g NaCl in 1,000 mL water and pour into small dropper bottles. (Laboratory Exercise 4)

Sodium Chloride Solution (5%)

Add 25 g NaCl crystals to water to make 500 mL of solution. Label the solution and place in small dropper bottles. (Laboratory Exercise 4)

Sodium Chloride Solution (5%)

Add 2.5 g NaCl crystals to water to make 50 mL solution. Label the solution and place in small beaker. (Laboratory Exercise 18)

Starch Solution (0.5%)

A potato starch solution is made by first boiling 500 mL water. Remove the water from the hot plate and add 2.5 g potato starch powder. Stir and cool the mixture. Do not boil the starch and water mixture because this will lead to some hydrolysis of starch to sugar. Test for the presence of sugar by using Benedict's reagent. There should be no sugar present. Place into 250 mL bottles and label "0.5% Starch Solution." (Laboratory Exercise 39)

Starch Solution (1%)

Boil 1 liter of distilled water. Remove the water from the heat and add 10 g cornstarch (or 10 g potato starch). Filter the mixture through cheesecloth into bottles. (Laboratory Exercise 4)

Sugar Solution (3%)

Using a clean container, for food use, dissolve 15 g table sugar in enough water to make 500 mL of solution. Fill clean, food-grade dropper bottles labeled "Sweet." (Laboratory Exercise 20)

Sugar Solutions

Four table sugar solutions of 2 liters each. (Laboratory Exercise 4)

0%—2 liters of water

5%—dissolve 100 g sugar in enough water to make 2 liters of solution

15%—dissolve 450 g sugar in enough water to make 3 liters of solution; place 2 liters in one bottle and 1 liter in a bottle labeled "15% sucrose solution"

30%—dissolve 600 g sugar in enough water to make 2 liters of solution

Tonic Water

Select a clean, food-grade dropper bottle that holds 100 mL and label "Bitter." Fill with commercial tonic water. (Laboratory Exercise 20)

Umami Solution

Add 15 g monosodium glutamate (MSG) to 500 mL water in a clean, food-grade container. Fill clean, food-grade dropper bottles labeled "Umami." (Laboratory Exercise 20)

Vinegar Solution

Use household vinegar or make a 5% food-grade acetic acid solution by adding 5 mL concentrated acetic acid to about 50 mL water and then adding additional water to make 100 mL. Fill clean, food-grade dropper bottles labeled "Sour." (Laboratory Exercise 20)

Wright's Stain

Wright's stain is available as a commercially prepared solution from a number of biological supply houses. Place in labeled dropper bottles. (Laboratory Exercise 24)

APPENDIX C

LAB REPORTS

Part of working in science involves writing lab reports. You should write lab reports in a certain style and follow basic guidelines that are generally accepted in the field. The general format for the lab reports falls into four categories: **introduction, materials and methods, results,** and **conclusion** sections. Each of these parts is important in the write-up, and each of these four sections must be included in the lab report. Your report should have a title, your name, your instructor's name, the course name and semester, and your lab section.

The purpose of a scientific report is to explain an investigation. Your instructor is your primary audience for your report in this class, so you should communicate that you understand the basic information. Part of a grade in a lab report is dependent on doing the experiment correctly and demonstrating that you are aware of the outcome of the experiment and its relationship to theory or applications you have learned in lecture.

If your experiment did not come out as you thought it might, you still can do well in your lab report. Results from your experiment may have come out satisfactorily and be fine even if you think that the data should have been different from what you obtained.

Introduction

The introduction section consists of a description of the problem and the subject of your study. In professional journals, the introduction often includes a history of past experimentation or current knowledge in the field, but you will probably not have this in your lab report. You should pose the question or hypothesis that your experiment is trying to resolve in the introduction. You may be conducting an experiment to determine whether an enzyme functions on a substrate, whether a particular effect occurs when you perform an action.

Materials and Methods

The materials and methods section consists of a clear description of what equipment, animals, chemicals, and so on were used and the experimental procedure followed. You must write this section in clear and precise terms. From this section you should expect that a person could repeat your experiment and produce the same results.

In one way this is the "recipe" for your experiment. As in baking a cake, it is not good enough simply to list the materials. You must include quantity, what sequence the materials were added, how long things were stirred, how long the cake baked, and so on. You must provide specific details to your procedure. Make sure you state how much material was added to a sample. For example, you should write: "We added 5 mL." This is a known quantity that people can duplicate, whereas "We added a little" is vague. You must make sure you include all steps in your procedure. If you leave something out in your description, a person following your directions might get different results.

Results

The results section is where the outcome from your experiment is listed in a clear and defined way. Your data must be clearly presented. This may consist of the tabulation of data that you acquired from your experiment in the form of a line graph, bar graph, or table. Remember that any graph should have a complete description. If you study the effect of exercise on heart rate, then exercise is the independent variable and is listed on the horizontal axis, while heart rate is the dependent variable and is listed on the vertical axis, as illustrated in the following graph.

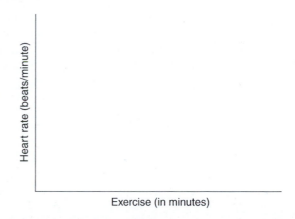

The data for your lab result section may be in raw form (direct measurements from your experiment), or you may want to organize the data by calculating the mean (average) data, range (high and low), and so on, as determined from your experiment. There should be no interpretation of the results at this point. You should not try to explain your results but simply list what the outcome was. Save the interpretation of the data for the conclusion.

Conclusion

In the conclusion, you analyze and interpret your data and determine whether the results of your experiment proved your initial question (make sure to answer your original question). This is the most important part of your lab report in terms of finding out if you understood the lab. Did you prove your hypothesis, did you disprove your hypothesis, or are the results inconclusive? Where should you go from here?

Sometimes experiments go awry, and the results do not support the original hypothesis. This can be due to many things, such as experimental error, a big factor in lab experiments, especially at the undergraduate level. Someone may not have followed the experimental procedure correctly and added too much, too little, or, in some cases, none of the materials that should have been part of the procedure. Sometimes it is a matter of timing; the process did not go on as long as it should have or went on too long. In some cases, the data are taken down incorrectly or read wrong. A "3" in the data might be mistaken for an "8," or the data might be entered in the wrong location. There are hundreds of possibilities as to why error plays a part in experiments, and your instructor is probably familiar with many of them.

Other factors are faulty equipment, supplies, and setup. Some problems are caused by the variation in living organisms. Humans and other animals have idiosyncratic physiological responses. Not all people and not all experimental animals respond in the same way. For example, almost all product information sheets that come with pharmaceutical drugs list adverse reactions seen in some people. If everyone responded the same way, there would be no need for warnings on drug products.

If you obtain variances from the results you expect to get, write this in your report. Sometimes experimental procedures are included that produce results other than what you expect to see. You should be honest in your recording of data and resist the temptation to "fudge" the data so you can get a better result and, therefore, a better grade on your report.

Sometimes people in industry change their data to produce more desirable results. If this is discovered by independent investigations, the corporation that allowed this to happen usually pays significant fines. Students who change data and are discovered are also subject to penalties far more severe than what they would have received due to the "bad" data they think they might have gotten. In addition to being honest in the reporting of data, you also want to make sure your words are your own. Even though you may have performed an experiment word for word from this lab manual or another source, make sure you do not plagiarize the material. You should paraphrase material if you want to have the same meaning but not the same words as someone else.

Your lab report should follow the format described here unless your instructor decides to make significant changes in the write-up. Lab reports tend to take a long time to write, but the analysis in the lab report is where you work toward understanding the experiment.

COMMON PREFIXES, SUFFIXES, AND ROOT WORDS IN ANATOMY AND PHYSIOLOGY

Anatomy and physiology have words that may seem long and confusing. Much of the vocabulary has a Latin or Greek origin. The **root word** is the core portion of the term. In the word *histology,* the *histo-* part is the root word, which means "tissue." **Prefixes** are parts that come before the root word. These are indicated in the following list with a hypen after the term. In the term *cytoskeleton,* the prefix is written *cyto-* to show that it occurs first. **Suffixes** are parts that come after the root word and they are indicated with a hyphen before the term. In the term *fungicide,* the suffix is *-cide,* indicating that it occurs after the root word. Words that have hyphens before and after can be either prefixes or suffixes.

Term	Meaning	Example
a-	Without	Acellular (without cells)
ab-	To take away	Abduct (to take away)
acoust-	Hearing	Acoustic meatus (ear canal)
acro-	Tip, peak	Acromion (tip of shoulder)
ad-	Next to	Adduct (to bring a limb next to the body)
adipo-	Fat	Adipose (fat tissue)
-al	Relating to	Costal (relating to the ribs [costa])
alb-	White	Tunica albuginea (white layer)
-algia	Pain	Neuralgia (nerve pain)
an-	Without	Anaphylaxis (without protection—a systemic reaction)
ana-	Up	Anatomy (to cut up)
andro-	Male	Androgens (male sex hormones)
angio-	Vessel	Angiogram (radiograph of the blood vessels)
ante-	In front of	Antebrachial (forearm)
anti-	Against	Antiviral (against viruses)
arthro-	Joint	Arthrology (study of joints)
-ase	An enzyme	Lipase (lipid-digesting enzyme)
-aur-	Ear	Auricle (part of the ear)
auto-	Self	Autoimmune (against one's own immune system)
basi-	Base or bottom	Basicranial (base of the skull)
bi-	Two	Bipolar neuron (neuron with two poles)
bio-	Life	Biology (study of life)
brady-	Long, slow	Bradycardia (slow heart rate)
bucco-	Cheek	Buccinator (muscle of the cheek)
carcin-	Cancer	Carcinoma (cancerous tumor)
cardio-	Heart	Cardiology (study of the heart)
cata-	Downward	Catabolism (to break down)
-cephal-	Head	Hydrocephaly (water on the brain)
-cele	Space	Blastocele (space in a blastocyst)
celi-	Abdomen	Celiac artery (artery in the abdomen)
cerebro-	Brain	Cerebrum (large structure of the brain)
chondros-	Cartilage	Chondrocyte (cartilage cell)
-cide	To kill	Fungicide (something that kills fungus)
circum-	Around	Circumcise (to cut around, such as the foreskin of the penis)
-clast-	To break	Osteoclast (cell that dissolves bone)
co-	Together	Coenzyme (molecule that functions with an enzyme)
colpo-	Vagina	Colposcope (scope used to see the vagina)
com-	Join together	Gray commissure (part that joins two halves of the spinal cord)
con-	Join together	Conduct (to go with)
contra-	Opposite	Contralateral (on the opposite side)

corp-	Body	Corpus callosum (callous [tough] body in the brain)
cort-	Bark	Renal cortex (outer part of the kidney)
crypto-	Hidden	Cryptorchidism (hidden testes)
cyano-	Blue	Cyanosis (low oxygen level, causing blue color)
cyst-	Bladder	Cystitis (inflammation of the bladder)
-cyte- (cyto-)	Cell	Leukocyte (white blood cell)
de-	Without	Deoxyribonucleic acid (RNA without oxygen)
demi-	Half	Demifacet (half of one face)
derma-	Skin	Dermatology (study of the skin)
di-	Two	Diploe (spongy bone between two hard sections of bone)
dia-	Through, across	Diapedesis (move through cell spaces)
dis-	Apart	Distend (to move apart from resting condition)
-duct-	Lead	Conduct (to lead to)
dys-	Bad, painful	Dysentery (pain in the intestines)
e-	Away from	Evaporate (to take vapor away from water)
ec-	Out	Eccrine (to take out, as in a sweat gland takes out liquid from the body)
ecto-	Outside	Ectoderm (outer germ layer)
-ectomy	Surgical removal	Appendectomy (to remove the appendix)
-edem-	Swell	Edema (swelling of tissue)
-emia	Blood	Anemia (not enough blood)
en-	Inside	Encephalitis (inflammation in the brain)
endo-	Within	Endocytosis (within the cell)
-entero-	Intestine	Gastroenterology (study of the stomach and intestines)
epi-	On, above	Epidermis (on the skin)
erythro-	Red	Erythrocyte (red blood cell)
eu-	True, good	Eukaryotic (cells with a true nucleus)
ex-	Out, away from	Expiration (to breathe out)
exo-	Outside	Exocytosis (moving outside the cell)
extra-	Outside	Extracellular (outside of the cell)
-facet	Face	Costal facet (flat surface where a rib attaches)
-fere	Carry	Efferent (to carry away)
-ferous	Carry	Calciferous (producing calcium)
-form	To look like	Fungiform (looks like a fungus or mushroom)
gastro-	Stomach	Gastritis (inflammation of the stomach)
-genesis	To make	Spermatogenesis (formation of sperm)
-glossus	Tongue	Genioglossus (muscle of the tongue)
glyc-	Sugar	Hypoglycemic (low in blood sugar)
-gram	A recording	Electocardiogram (recording of electrical activity of the heart—ECG)
-graph	Recording instrument	Electrocardiograph (instrument measuring ECG)
gyno-	Female	Gynecology (study of the female reproductive system)
hemo-	Blood	Hemolytic (that which destroys blood)
hemi-	Half	Hemisphere (half of a sphere)
hepato-	Liver	Hepatocyte (liver cell)
hex-	Six	Hexagonal (six-sided)
hist-	Tissue	Histology (study of tissues)
hydr-	Wet, water	Hydrate (to give water to)
hyper-	More, above	Hyperactive (excessive activity)
hypo-	Less, below	Hypodermis (under the skin)
hyster-	Uterus	Hysterectomy (surgical removal of the uterus)
-id	State of being	Putrid (state of being rotten)
in-	Into	Infected (to have disease move into the body)
infra-	Below	Infraspinous (below a spine)
inter-	Between	Intercellular (between cells)
intra-	Within	Intravenous (within the veins)
ipsi-	Same	Ipsilateral (on the same side)
-ism	State or condition	Hypothyroidism (having low thyroid hormone levels)
iso-	Equal	Isometric (same length)
-itis	Inflammation	Hepatitis (inflammation of the liver)
juxta-	Near	Juxtaglomerular (near the glomerulus)

kerato-	Hornlike	Keratinocyte (toughened skin cell)
-kin-	Move	Kinase (enzymes that cause movement)
lact-	Milk	Lactose (milk sugar)
-lacrima-	Tear	Lacrimal glands (tear gland)
leuko-	White	Leukocyte (white blood cell)
liga-	Bind	Ligand (chemical that binds to a receptor)
lipo-	Fat	Liposuction (surgical removal of fat)
litho-	Stone	Lithotripsy (breaking up of kidney stones)
-logos	Study	Endocrinology (study of the endocrine system)
-lysis	To break up, destroy	Hemolysis (destruction of red blood cells)
macro-	Big	Macrophage (large, phagocytic cell)
mal-	Bad	Malnourished (with bad nutrition)
-malacia	Softening	Osteomalacia (softening of bone)
mamma-	Breast	Mammogram (X-ray of the breast)
mast-	Breast	Mastitis (inflammation of the breast)
mega-	Big	Megacolon (enlarged colon)
melano-	Black	Melanin (black pigment)
meso-	Middle	Mesothelium (endothelium arising from the middle embryonic layer)
-metrium	Uterus	Endometrium (tissue in the uterus)
meta-	Change	Metastasis (change from original)
micro-	Small	Microsurgery (surgery on a small thing)
mono-	One, single	Mononucleosis (disease with many cells, having one nucleus each)
multi-	Many, much	Multipolar neuron (neuron with many poles)
myo-	Muscle	Myometrium (muscle of the uterus)
necro-	Death	Necrotic (dead material)
neo-	New	Neonate (newborn)
nephro-	Kidney	Nephritis (inflammation of the kidney)
neuro-	Nerve	Neurology (study of the nervous system)
oculo-	Eye	Oculomotor nerve (nerve that moves the eye)
odonto-	Tooth	Odontoid process (toothlike process)
-oid	Looks like	Sigmoid (S-shaped)
-ole	Small	Bronchiole (smaller than a bronchus)
oligo-	Few, medium	Oligosaccharide (carbohydrate of a few sugars)
-oma	Tumor	Lymphoma (tumor of the lymphatic system)
oo-	Egg	Oocyte (egg cell)
-opia-	To see	Hyperopia (farsighted)
ophthalmo-	Eye	Ophthalmologist (person who studies the eye)
-orchid-	Testis	Orchidectomy (removal of the testis)
-ory	Belonging to	Respiratory (belonging to the lungs, etc.)
-osis	A condition	Kyphosis (condition of having a bent spine)
osteo-	Bone	Osteology (study of bones)
oto-	Ear	Otoliths (bones of the ear)
-ous	Producing material	Mucous membrane (membrane that produces mucus)
pan-	All	Pandemic (worldwide)
para-	Near, next to	Paravertebral (near the vertebrae)
-pathy	Illness	Neuropathy (nerve disease)
ped-	Child	Pediatrician (doctor who treats children)
-penia	Lacking	Leukopenia (low white blood cell number)
penta-	Five	Pentagonal (with five sides)
per-	Through	Perfuse (to pass liquid through)
peri-	Around	Perichondrium (around the cartilage)
-phago-	Eat	Phagocytosis (process in which cells "eat" material)
-pharyn-	Throat	Glossopharyngeal (of the tongue and throat)
-phas	Speech	Aphasia (lack of speech)
-phil-	To love	Hydrophilic (water-loving)
phleb-	Vein	Phlebitis (inflammation of a vein)
-phobia	Fear	Hydrophobic (does not mix well with water)
physio-	Nature	Physiology (study of the nature of the body)
-plasm-	Cell, tissue	Plasma membrane (cell membrane)

-plegia	Paralysis	Quadriplegia (paralysis of the four limbs)
pneumo-	Air	Pneumonia (inflammation of the lungs)
pod-	Foot	Podiatrist (doctor who deals with conditions of the feet)
-poie-	Make	Erythropoiesis (make red blood cells)
poly-	Many	Polycythemia (too many red blood cells)
post-	Behind, after	Posterior (back side of something)
pre-	Before	Precentral gyrus (bump before the central sulcus)
pseudo-	False	Pseudostratified epithelium (epithelium with false layers)
psycho-	Mind, soul	Psychology (study of the mind)
py-	Pus	Pyuria (pus in urine)
quad-	Four	Corpora quadrigemina (brain structure with four parts)
re-	Return, back	Relaxation (return to being loose)
recto-	Straight	Rectum (straight tube of the colon)
ren-	Kidney	Renal vein (vein of the kidney)
retro-	Backward, behind	Retroperitoneal (in back of the peritoneum)
rhino-	Nose	Rhinovirus (cold in the nose)
-rrhagia	Outpouring	Hemorrhage (to bleed profusely)
-rrhea	Flow, discharge	Diarrhea (watery bowel movement)
sarco-	Flesh	Sarcoplasm (cytoplasm of muscle cells)
scler-	Hard	Sclera (hard, white covering of the eye)
-scopy	See, look	Microscopy (to see with a microscope)
semi-	Half	Semicircular ducts (ducts resembling half a circle)
sigma-	The letter *S*	Sigmoid colon (part of the large intestine that takes the shape of an *S*)
soma-	Body	Somatic nerves (nerves going to the body)
sperm-	Male sex cells	Spermatogenesis (formation of male sex cells)
sphygm-	Pulse	Sphygmomanometer (instrument that measures blood pressure)
-stasis	Stop, maintain	Homeostasis (steady state condition)
steno-	Narrow	Mitral stenosis (narrowing of the mitral valve)
-stoma	Mouth	Tracheostomy (to make a mouth [opening] in the trachea)
sub-	Beneath, under	Subdermal (under the skin)
super-	Above, more	Superficial (on top of)
supra-	Above	Supraspinous (above a spine)
sym-	Together	Sympathetic (with emotion)
syn-	Together	Synapse (where two nerves join together)
tachy-	Fast, quick	Tachycardia (fast heart rate)
tetra-	Four	Tetracycline (antibiotic with four hydrocarbon rings)
therm-	Heat	Thermogenesis (to produce heat)
-tomy	Cut	Anatomy (to cut up)
tox-	Poison	Toxicology (study of toxins)
trans-	Across	Transmembrane proteins (proteins that occur across membranes)
tri-	Three	Triceps brachii (three-headed muscle of the arm)
-troph-	Feed	Trophic hormone (one that "feeds" other hormones to produce hormones)
-tropic	To influence	Thyrotropic hormones (hormone that influences the thyroid gland)
uni-	One	Unipolar neuron (neuron with one pole)
-uria	Urine	Ketonuria (ketones in the urine)
vaso-	Vessel	Vasodilator (substance that opens blood vessels)
viscer-	Gut	Visceral peritoneum (lining of the viscera)

INDEX

Note: Page references followed by the letters *f* and *t* indicate figures and tables, respectively.

F